Minerals and Rocks   16

Editor in Chief
P. J. Wyllie, Chicago, IL

Editors
A. El Goresy, Heidelberg
W. von Engelhardt, Tübingen · T. Hahn, Aachen

James B. Gill

Orogenic Andesites and Plate Tectonics

With 109 Figures

Springer-Verlag
Berlin Heidelberg New York 1981

Professor JAMES B. GILL
University of California, Santa Cruz
Earth Sciences Board and Oakes College
Santa Cruz, CA 95064, USA

Volumes 1 to 9 in this series appeared under the title
Minerals, Rocks and Inorganic Materials

ISBN 3-540-10666-9 Springer-Verlag Berlin Heidelberg NewYork
ISBN 0-387-10666-9 Springer-Verlag NewYork Heidelberg Berlin

Library of Congress Cataloging in Publication Data. Gill, James, 1944–. Orogenic
andesites and plate tectonics. (Minerals and rocks; 16) Bibliography: p. Includes index.
1. Andesite. 2. Plate tectonics. I. Title. II. Series. QE462.A5G54. 552'.2. 81-2608.
AACR2.

Typesetting, printing and binding: Brühlsche Universitätsdruckerei, Gießen.
2132/3130-543210

81 010695

Preface

Students of a phenomenon as common but complex as andesite genesis often are overwhelmed by, or overlook, the volume and diversity of relevant information. Thus there is need for periodic overview even in the absence of a dramatic breakthrough which "solves the andesite problem" and even though new ideas and data keep the issues in a state of flux. Thus I have summarized the subject through mid-1980 from my perspective to help clarify the long-standing problem and to identify profitable areas for future research.

Overviews are more easily justified than achieved and there are fundamental differences of opinion concerning how to go about them. It is professionally dangerous and therefore uncommon for single authors, especially those under 35 such as I, to summarize a broad, active field of science in book-length thoroughness. Review articles in journals, multi-authored books, or symposia proceedings appear instead. The single-authored approach is intimidating in scale and can result in loss of thoroughness or authority on individual topics. The alternatives lack scope or integration or both.

At the editor's request, this book summarizes the entire field of andesite genesis from a single perspective so as to weave together threads of argument which often are lost or ignored in reviews of more modest scope. Furthermore, because the review article mode seems to weight summaries of speculative models more than of relevant data, I have emphasized the latter. Thus Chapters 1 through 7 are the core of this book. At best they provide critical syntheses of diverse data, written with the consistant objective of constraining theories of andesite genesis. At worst they annotate the over 1100 references cited. The book is written for graduate students and other professional earth scientists in a variety of specialities.

In one sense I began this book in 1966 to 1968 when, as a new graduate student attending AGU meetings in Washington, I heard the annunciation of plate tectonics and learned that potash contents of andesites increase across island arcs. By temperament and age I prefer new explanations to conventional ones and therefore was kindly disposed to believe that orogenic andesites were linked closely to subduction by being primary melts either of underthrust ocean crust or of the overlying upper mantle. I began writing this manuscript in 1976, combining field and laboratory study of particular suites (mostly in California, Fiji, and Indonesia) with a review of the literature about andesites in general. Despite the predisposition admitted above, I was unconvinced by any genetic model, had little ego at stake in defending a particular position, and assumed the issue of andesite genesis was not then resolvable. However, four years later I have finished more convinced than anticipated that conventional explanations better account for more data than do the more elegantly simple theories spawned by plate tectonic theory.

I dedicate this book to my family: my wife and closest friend, Catharine; my daughters Emily and Eleanor whose births accompanied the book's preparation; and my parents, Ray and Edith, whose fiftieth year together coincides with the book's publication. I thank many for reviews, ideas, and data, but especially faculty and students at UCSC, The University of Auckland, and IPG, Paris. Suzanne Harris, Maureen Leimbach, and Dotty Hollinger typed the manuscript; Judy Fierstein and Annette Whelan drafted the figures; and Felix Chayes gave access to this data bank.

Santa Cruz, August 1980   James B. Gill

Contents

Chapter 4 Andesite Magmas, Ejecta, Eruptions, and
           Volcanoes

Chapter 5 Bulk Chemical Composition of Orogenic
           Andesites

Chapter 6 Mineralogy and Mineral Stabilities

Chapter 7 Spatial and Temporal Variations in the Composition
           of Orogenic Andesites

Chapter 8 The Role of Subducted Ocean Crust in the
           Genesis of Orogenic Andesites

Chapter 9 The Role of the Mantle Wedge

Chapter 10 The Role of the Crust

Chapter 11 The Role of Basalt Differentiation

Chapter 12 Conclusions

List of Symbols and Acronyms

a	activity
Ab	albite
ab	normative albite
An	anorthite
an	normative anorthite
AND	Andean or active continental margin plate boundary
ARC	island arc plate boundary
ATL	Atlantic of passive continental margin plate boundary
Bt	biotite
c	normative corundum
Cpx	clinopyroxene
Di	diopside
di	normative diopside
D_i	distribution coefficient for element i between crystal and liquid (see p. 198)
$\overline{D}_i$	bulk distribution coefficient $(\overline{D}_i = D_i^\alpha X^\alpha + D_i^\beta X^\beta + \ldots)$
En	enstatite
Eu*	fictive Eu content obtained by extrapolating between Sm and Gd
FeO*, $Fe_2O_3^*$	total Fe as FeO or Fe_2O_3
f_ℓ	wt. fraction liquid
f_s	wt. fraction solid
FMQ	fayalite + magnetite + quartz
Fo	forsterite

f_{O_2}	oxygen fugacity
Gar	garnet
Hb	hornblende
HFS	high field strength (charge/radius)
hy	normative hypersthene
Ilm	ilmenite
IRS	incompatible elements + radiogenic isotopes + silica
kb	kilobar
LIL	large ion lithophile
MORB	mid-ocean ridge basalt
Mt	magnetite
η	viscosity
N	number of analyses
NNO	nickel + nickel oxide
Opx	orthopyroxene
Or	orthoclase
P_{H_2O}	water pressure
P_{TOTAL}	load pressure
Pl	plagioclase
POAM	plagioclase + orthopyroxene or olivine + augite + magnetite
qz	normative quartz
ρ	density
REE	rare earth elements
v	vapor
V_P	velocity of P waves
V_S	velocity of S waves
X	mol fraction

Chapter 1 What is "Typical Calcalkaline Andesite"?

1.1 Introduction

Currently, active volcanoes on Earth erupt andesite more than any other rock type. These andesite sources are frequently stunning in beauty yet are sometimes lethal in effect. Although their eruptions cause more damage to property and environment than to persons, some people, including volcanologists, died violently due to eruptions of andesite while this book was being written in academic safety.

Andesites are as yet unknown elsewhere in our solar system, with possible exceptions on Mars (Malin 1977). Even on Earth, basalt is much more voluminous and more analyzed than andesite (Chayes 1975, Fig. 87). However, andesite is cited as a main rock type at 442 (or 61%) of the 721 volcanoes included in Katsui's (1971) *List of World Active Volcanoes* for which rock types are known, whereas basalt is cited at only 334 volcanoes. Also, andesites constitute about 25% of the volcanic rocks for which chemical analyses are stored in various electronic data banks (e.g., Chayes and Le Maitre 1972). Yet, as frequently happens with common-place things, the origin of andesite is not agreed upon and its significance is obscure.

Andesites have sustained interest for several reasons besides their ubiquity. First, andesites have a distinctive tectonic setting. They are associated primarily with convergent plate boundaries and occur elsewhere only in small amounts. Of the 561 active volcanoes in the above *List* which occur within 500 km of a sub-duction zone and for which rock types are known, 439 (or 78%) include andesite. In contrast, only three active volcanoes not near a convergent plate boundary do so (Tarso Voon in Chad, Kilimanjaro in Tanzania, and Hekla in Iceland). Second, andesitic volcanism is studied because eruption prediction has substantial social value but is risky, in part because andesitic volcanism is poorly understood. Public disagreement between scientists over a 1976 prediction and resulting evacuation in the Antilles caused considerable professional discussion of accountability and also contributed to a thoughtfully conceived film by Werner Herzog ("La Soufrière"). Third, andesites have bulk compositions similar to estimates of the composition of terrestrial crust. This, coupled with the tectonic setting of andesites, suggests that they have been important agents of crustal development or at least maintenance. Where andesites are absent, as on the Earth's moon, crust has a very different composition and mode of origin. Fourth, andesitic magmatism seems intimately related to the formation of many ore deposits. Andesites, whether as magmas or as previously crystallized volcanic rocks, seem the source of these metals.

Several issues in the preceding paragraph may alarm critical readers. For example, how similar are rocks called andesites in those 442 volcanoes cited? Is citation in that *List* a useful way of estimating relative abundance? What constitutes "association" with a convergent plate boundary (e.g., does Colorado, USA, qualify?). These and other problems cause difficulty in adequately posing, much less answering, the question: what relationship is there between plate tectonics and andesite genesis?

1.2 Definition of Orogenic Andesite

To most geologists, andesites are gray porphyritic volcanic rocks containing phenocrysts of plagioclase, but not quartz or sanidine or feldspathoid. The name andesite was coined in 1835 in Berlin, an example of the intellectual precociousness and global interests of newly federated Germany. Volcanic rocks from the Andes Mountains of Bolivia and Chile and from Kamchatka had been returned to Germany by the expeditions of Alexander von Humboldt and others, and differed from European volcanic rocks. Leopold von Buch, who had been a student of Werner in Freiberg and was, at age 61, the foremost Prussian geologist of that time, proposed the name andesite for these exotic hornblende-bearing rocks in order to distinguish them from the "trachytes" of Europe (von Buch 1836). ("Trachytes" then encompassed what are now called dacites and rhyolites as well as andesites and trachytes.) However, the criteria for classification were unclear from the start. There was confusion whether the andesites contained oligoclase, albite, or sanidine, in addition to the less abundant hornblende (Johannsen 1937), and andesites originally were distinguished from trachytes by the presence of albite instead of feldspar in andesite! Subsequent recognition that the characteristic mineral of andesites was a member of the plagioclase feldspar solid solution series, and inclusion of pyroxene-bearing as well as hornblende-bearing rocks as andesites, allowed the predominance of alkali feldspar over plagioclase to be used as a criterion for trachyte but led to ambiguity in distinguishing andesite from basalt and dacite.

Andesite and dacite have been distinguished for over a century by the presence of quartz phenocrysts in dacite, and three still-used criteria arose before 1930 for making the distinction between andesite and basalt. In each, andesite denotes a volcanic rock lacking quartz or feldspathoid phenocrysts and in which at least 2/3 of the feldspar phenocrysts are plagioclase. Rosenbusch restricted andesite to rocks in which the average plagioclase phenocryst composition was < An50. He was followed in this by Johannsen, Holmes, and many North American petrographers. Sometimes the distinction is based on normative versus modal plagioclase compositions. In contrast, Zirkel restricted andesite to rocks with pyroxene, amphibole, or biotite phenocrysts but without olivine; plagioclase and pyroxene compositions were unspecified. This practice was followed by Iddings and many European petrographers. Shand ignored both plagioclase composition and olivine content, and identified andesites as rocks in which < 30% of the minerals are mafic (i.e., color index < 30). This principle was adopted by the IUGS Subcommission on the Systematics of Igneous Rocks, although the basalt/andesite boundary was placed at a color index of 35 by volume or 40 by weight, e.g., when using CIPW norms (Streckeisen 1979).

Andesites defined by these variable and sometimes mutually inconsistent criteria predictably have widely varying bulk chemical compositions. From amid the diversity, several authors have independently selected compositions of "average andesite" (Table 1.1). Clearly, rocks which are andesites by various petrographic criteria represent ranges of compositions whose estimated averages are quite similar.

Table 1.1. Estimates of average andesite composition (wt.% except col. 5 which is in mol%)

	1	2	3	4	5
SiO_2	60.3	58.2	57.9	57.6	61.4
TiO_2	0.78	0.82	0.87	0.77	0.64
Al_2O_3	17.5	17.2	17.0	17.3	10.9
Fe_2O_3	3.4	3.1	3.3	3.1	1.2
FeO	3.2	4.0	4.0	4.3	3.8
MnO	0.18	0.15	0.14	0.15	0.13
MgO	2.8	3.2	3.3	3.6	5.7
CaO	5.9	6.8	6.8	7.2	8.2
Na_2O	3.6	3.3	3.5	3.2	3.3
K_2O	2.1	1.7	1.6	1.5	1.0
P_2O_5	0.26	0.23	0.21	0.21	0.10
H_2O	–	1.3	1.2	1.0	3.6
N	87	2177	2600	2500	–

1. Daly (1933)
2. Chayes (1975)
3. Le Maitre (1976a)
4,5. This work. Average of the orogenic andesites shown in Fig. 1.1 and 5.6. Col. 5 is col. 4 expressed as mol%. Its molecular weight on the basis of 8 oxygens is 291, determined from 1/n Al after Burnham (1979)

The limited correlation between mineralogy and chemical composition is a common problem in igneous petrology and there is no agreement whether modal mineralogy, normative mineralogy, bulk chemical composition, historical precedent, oxide activities, or democratic plebiscite should form the basis for classification. Gradational variations between rock types abound and rarely is there close correspondence between rocks defined by one of these criteria and rocks defined by another. As a result, nomenclature is the Pandora's Box of igneous petrology. While it is hazardous to define any isolated portion of the continuum of volcanic rocks, it is necessary to be explicit about this book's subject. Because this study is

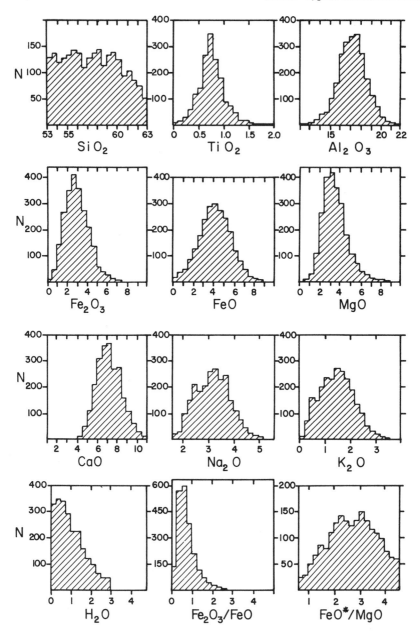

Fig. 1.1. Histograms of the wt.% oxide compositions of the 2500 orogenic andesites, as defined in text, which are in electronic data file RKOC76 of F. Chayes. Analyses with more than 3 wt.% H_2O^+ were excluded

of relationships between plate tectonics and andesite *magma,* and because much of the evidence is chemical, andesites are defined here in terms of rock chemical composition. Specifically, the definition is primarily in terms of silica contents because these are the best single discriminants for distinguishing previously defined andesites from basalts or dacites (Le Maitre 1976b), they increase regularly relative to various differentiation indices in most andesites, and they have more immediacy to most people than do combinational parameters.

Andesites are defined here simply as all hypersthene-normative volcanic rocks with SiO_2 from 53 to 63 wt.% calculated on an anhydrous basis. (For consistency, basalt will denote rocks with $< 53\%$ SiO_2; dacite, 63% to 70% SiO_2; and rhyolite, $> 70\%$ SiO_2). Andesites are further subdivided into basic (or basaltic, or mela-, or low-silica) andesites with 53% to 57% SiO_2, and acid (or high-silica) andesites with 57% to 63% SiO_2.

As discussed in Section 7.1, K_2O content is the most significant variable in major element composition between andesites for tectonics. Furthermore (perhaps causally), per mol, K_2O seems to affect polymerization of andesite liquid more than does any common oxide other than SiO_2 itself, resulting in variable physical properties of magma, mineral solubilities, oxidation states, and eutectic liquid compositions (e.g., Kushiro 1975; Lauer and Morris 1977). Water, whose concentration in andesite magma crudely correlates with K_2O (Sect. 5.3.1), has a similar effect. Thus, there is reason to adopt, as I have done in Fig. 1.2, the combination of K_2O and SiO_2 used by several authors after Taylor (1969) as criteria for further subdivision of andesite nomenclature.

Unfortunately, the boundaries shown in Fig. 1.2 are quite arbitrary so that little compels anyone to adopt them. The 53% and 63% SiO_2 values lie approximately one standard deviation from the mean of rocks called basalt and dacite, respectively (Middlemost 1972; Chayes 1975; Le Maitre 1976a), and sometimes mark the onset of precipitation of orthopyroxene and quartz phenocrysts, respectively, but they have limited statistical significance as discriminants (Le Maitre 1976b). Likewise, although most andesites in which REE patterns are flat or of positive slope (i.e., La/Yb < 1.8) are low-K according to Fig. 1.2 (some high-Mg andesites are exceptions), the converse is not true and the K_2O-SiO_2 boundaries shown do not represent discriminant functions separating populations which are rigorously distinctive in mineralogy, trace elements, or anything else. For example, the boundaries between low, medium and high-K andesites correspond poorly with the boundaries between K-poor, average, and K-rich andesite proposed by Irvine and Baragar (1971, Fig. 8). Variations in bulk composition of andesites are continuous and there appear to be no thermal maxima or minima within this portion of chemical composition space which separate liquids of consistently different origin. Thus, the boundaries are used only to specify and compare portions of the continuum.

While use of the noun andesite to denote all hypersthene-normative volcanic rocks with 53% to 63% SiO_2 simplifies matters considerably, it excludes some

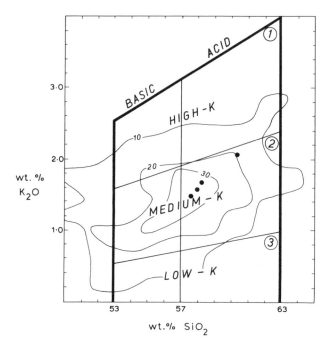

Fig. 1.2. Nomenclature of orogenic andesites using K_2O and SiO_2. Equations of lines *1, 2* and *3* are, respectively: $K_2O = 0.145 (SiO_2) - 5.135$; $K_2O = 0.0818 (SiO_2) - 2.754$; $K_2O = 0.0454 (SiO_2) - 1.864$. *Contours* enclose the number of nominal andesites, by original authors' definitions, whose analyses are stored in data file RKOC76 and which lie within each rectangle of a grid having dimensions of 0.2% SiO_2, 0.1% K_2O. *Dots* indicate average andesite compositions from Table 1.1

rocks called andesites by petrographers and includes some not called andesite in initial descriptions: e.g., basalts, dacites, latites, icelandites, shoshonites, mugearites, trachytes, etc. Thus "andesite" and "intermediate volcanic rock" become almost synonymous. Although such a definition loses some semantic resolution [Mac-Donald (1960) proposed it be abandoned for just that reason], it has the virtue of simplicity and, by appropriate use of adjectives, gains clarity. For example, most rocks called andesites by petrographers form a subset of hypersthene-normative volcanic rocks with 53% to 63% SiO_2 which can be distinguished chemically from other subsets by stipulating that members of it have $K_2O < (0.145 \times SiO_2 - 5.135)$ and $TiO_2 < 1.75\%$, thereby eliminating most rocks likely to be called shoshonites, latites, trachy-something, or icelandites. This subset I shall call "oro-genic andesite" and is the subject of this book. Other subsets of andesite can be identified and subdivided profitably (e.g., Stewart and Thornton 1975; or Wilkinson and Binns 1977).

Characteristics of the subset here called orogenic andesite were investigated in 1976 using data base RKOC76 of the electronic data bank at the Geophysical Laboratory, Washington. Of the 16,123 stored analyses, 2500 are orogenic andesites. Their average composition is given in Table 1.1, column 4; the univariant distribution of oxide concentrations is shown in Fig. 1.1. This population forms the basis for discussion of major element characteristics of orogenic andesites in Sections 1.2, 5.1, and 5.2.

About 26% of this subset consists of rocks originally called something else — principally basalt or dacite. Similarly, about 29% of rocks originally called andesite were excluded by the criteria cited. Most of these nominal andesites (645) lay outside the 53% to 63% SiO_2 range (see Fig. 1.2); only 19 analyses within that range were ne-normative, whereas the additional K_2O and TiO_2 criteria removed 78 and 39 analyses, respectively. For example, although no chemical analysis is available of the original andesites described by von Buch, recent work in the Chilean Andes indicates, ironically, that most volcanic rocks containing oligoclase and hornblende phenocrysts are dacites, not andesites, by my criteria!

The K_2O-SiO_2 bivariant frequency distribution of nominal andesites is superimposed on Fig. 1.2 to emphasize the continuum noted above. The number and mean of analyses of orogenic andesites lying within the six andesite categories defined in Fig. 1.2 are given in Table 5.1.

1.3 Magma Series Containing Orogenic Andesites

Magma series consist of magmas which are genetically related to each other by some differentiation process(es) or by being separate partial melts of a common source under similar conditions. In belief that these processes are relatively uniform in space and time, series names have been assigned to characteristics which are thought to recur amongst consanguinous magmas; examples include the tholeiitic, calcalkaline [1], alkalic, and nephelinitic series. However, neither the criteria for assigning suites of rocks to a pre-defined series nor the efficacy of such assignment have been agreed upon (e.g., see Johnson et al. 1978a). Andesites, unfortunately, provide an excellent example of the problem.

One function of names is to separate fundamentally different objects, e.g., magmas originating in significantly different ways. However, one conclusion of this review is that although different andesites follow different evolutionary paths, the paths differ by gradation and cross. Consequently, the end product reflects an integrated history which often cannot be classified in a manner certain to reflect differences in source or path. Therefore, series names like rock names, at least for orogenic andesites, seem best used to specify and compare portions of continua.

---

1   "Calcalkaline" is a bastardized form of the adjective calcalkalic, defined by Peacock (1931). Calcalkaline is used here because it is a more euphonious adjective for andesite

Nevertheless, criteria for subdivision must be specified and, as far as possible, justified. For example, although orogenic andesites usually are thought to belong only or at least primarily to the calcalkaline series, there are persistent claims that they also occur as prominent members of both tholeiitic and alkalic series. Thus we must re-examine and choose criteria for distinguishing between andesite-containing series.

The tholeiitic and alkalic series were defined with respect to basalts and are distinguished from each other most frequently by their groundmass mineralogy (Tilley 1950), normative mineralogy (Yoder and Tilley 1962), or alkali contents (MacDonald and Katsura 1964). The calcalkaline and alkalic series were distinguished by Peacock (1931) on the basis of alkali-lime indices [the wt.% SiO_2 for which $CaO = (Na_2O + K_2O)$]: suites with indices < 51 are alkalic, 51 to 56 alkalicalcic, 56 to 61 calcalkalic, and > 61 calcic. Thus, both tholeiitic and calcalkaline series were defined as being less alkali-rich than the alkalic series; i.e., subalkalic. [Note that use of a total alkalies versus SiO_2 diagram (e.g., Kuno 1959) or of Sugimura's (1968) θ index (θ signifying tholeiite) distinguishes the tholeiitic from alkalic series, but not from calcalkaline ones. Kuno, for example, did not equate his calcalkaline and high-alumina series.]

These two subalkalic series, calcalkaline and tholeiitic, were distinguished from each other by Wager and Deer (1939) on the basis of iron-enrichment trends. Pronounced Fe-enrichment during differentiation (e.g., in the Skaergaard intrusion) typifies the tholeiitic series; absence of Fe-enrichment typifies the calcalkaline series. Commonly this difference is shown in AFM diagrams (Fig. 1.3) and was quantified in terms of this diagram by Irvine and Baragar (1971, Fig. 2 and Appendix III). The same contrast also has been shown by plotting $FeO + Fe_2O_3$ vs. MgO (e.g., Jakeš and Gill 1970) or $(FeO + Fe_2O_3)/(FeO + Fe_2O_3 + MgO)$ vs. SiO_2 (e.g., Osborn 1959). Miyashiro (1974, Fig. 1) used a modification of the latter diagram in his proposal of a quantitative distinction between tholeiitic and calcalkaline series in terms of relative Fe-enrichment (Fig. 1.4).

There are two related but separate issues here. Kuno, Osborn, and others implicitly regarded the tholeiitic series as having a high rate of increase of Fe/Mg ratios relative to SiO_2. In contrast, Miyashiro, Irvine, and Baragar explicitly defined the tholeiitic series as having simply a high Fe/Mg ratio relative to SiO_2 or $(Na_2O + K_2O)$ contents. Such tholeiitic series could, therefore, lack iron-enrichment *within* suites, instead having high but relatively constant FeO^*/MgO[2] ratios over a wide SiO_2 range; this is common in alkalic rocks such as syenites, for example.

Thus, at least four separate criteria have been used to separate the tholeiitic, calcalkaline, and alkalic series: alkali contents, iron-enrichment, groundmass mineralogy, and normative mineralogy. Moreover, the iron-enrichment criterion can mean either high values or high rate of change of FeO^*/MgO ratios. Orogenic

---

2   FeO^* is total Fe as FeO

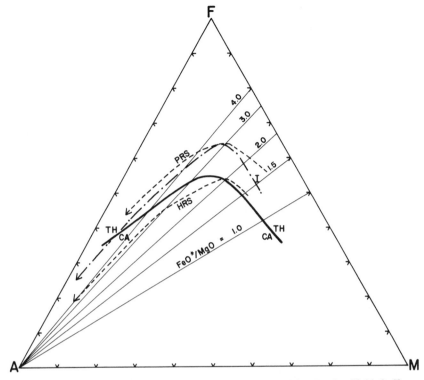

Fig. 1.3. AFM diagram of andesites. *A,* $Na_2O + K_2O$; *F,* $FeO + 0.9$ x Fe_2O_3; *M,* MgO. *Heavy solid line* separates tholeiitic *(TH)* from calcalkaline *(CA)* suites using criteria of Irvine and Barager (1971). Only 20% of the orogenic andesites shown in Fig. 1.1 are tholeiitic by this definition. *Light solid lines* show constant FeO*/MgO ratios to facilitate comparison with Fig. 1.4. *Dashed lines* connect Kuno's (1968a) average compositions of pigeonitic *(PRS)* and hypersthenic rock series *(HRS)* in Japan. The *dot-dash line* connects compositions from Thingmuli, Iceland *(I)*, which include anorogenic andesites (Carmichael 1964)

andesites show gradational variability in each of these factors so that all three series are represented and overlap. Not surprisingly, different attempts to apply these series names to orogenic andesites conflict (see Miyashiro 1975, for a clear review). Each attempt uses the series name to describe suites of rocks including andesites, dacites, and rhyolites, as well as basalts.

Kuno (1950a, 1959, 1968a) argued that there were three fundamentally different magma series in island arcs: tholeiitic, high-alumina, and alkalic. To him, the calcalkaline series was characterized by groundmass hypersthene and low FeO*/MgO ratios, and could develop from *any* of the other series. Kuno's hypersthenic rock series (volcanic rocks with groundmass hypersthene) is, therefore, synonymous with the calcalkaline series; his pigeonitic rock series (with only

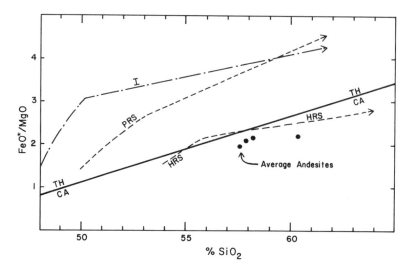

Fig. 1.4. Definition of tholeiitic *(TH)* vs. calcalkaline *(CA)* andesites, following Miyashiro (1974). Tholeiitic andesites lie *above the heavy solid line* which has the equation: FeO*/MgO = 0.1562 x SiO$_2$ – 6.685. *Dashed lines* show rock suites as in Fig. 1.3. *Dots* indicate average andesite compositions from Table 1.1

monoclinic pyroxenes in the groundmass) is synonymous with the tholeiitic ± high-alumina series. While he stressed differences in total alkali content between the tholeiitic, high-alumina, and alkalic series, no such difference between the tholeiitic and calcalkaline series is necessary (cf. Kuno 1968a, Figs. 47 and 48 or Tables 3 and 4; also see Miyashiro 1975). That is, Kuno separated iron-enrichment and alkali contents as two independent issues and used only the former to distinguish the tholeiitic from calcalkaline series (Fig. 1.3). Kuno's scheme has not been adopted widely outside Japan because of disagreement between mineralogic and chemical definitions of his series (Miyashiro 1974, Fig. 9), difficulties in identifying groundmass pyroxene, and unclear genetic significance.

Jakeš and Gill (1970) united the iron-enrichment and alkali content issues in their definition of the island arc tholeiitic series. They argued that differences in total alkali content primarily were differences in potash content, that variations in potash contents were paralleled by variations in other LIL-element [3] concentrations, and that suites with low LIL-element contents also were characterized by high FeO*/MgO ratios and vice versa (Jakeš and Gill 1970, Figs. 1 and 2). While most of the above usually is true, the vice versa conjunction frequently fails (Sect. 5.2.3) and should be abandoned. Nevertheless, it is useful to use the phrase

---

3   LIL-element means large-iron lithophile element, e.g., K, Rb, La, or Ba, having a bulk distribution coefficient <0.5

"island arc tholeiitic series" to denote those suites which are low-K according to Fig. 1.2, are tholeiitic according to Fig. 1.4, and which experience rapid iron-enrichment, i.e., have a kink in FeO^*/MgO vs. SiO_2 diagrams. Such suites are less common than claimed by Jakeš and Gill (1970), seem restricted to the volcanic front of plate boundaries characterized by convergence rates > 7 cm/year (Sect. 7.2.3), and have a variety of trace element signatures (Jakeš and Gill 1970; Masuda and Aoki 1978).

Because there seems to be no *necessary* correlation between Fe/Mg and K/Si ratios in andesites, I have adopted Miyashiro's (1974) proposal for distinguishing tholeiitic from calcalkaline behavior even though this use of "tholeiitic" need not denote iron-enrichment within a suite. I do so because Miyashiro's criterion is the most practical way to characterize individual suites using a limited number of analyses, or to characterize a single analysis. Although the criteria are somewhat arbitrary, the issue is not moot because rock series distinguished as tholeiitic or calcalkaline in this way can differ enough geochemically that they are unrelated genetically even when occurring at the same volcano (e.g., Masuda and Aoki, 1979).

In summary, I believe the three chemical differences between various orogenic andesite suites which are most important in reflecting their petrogenesis are: their LIL-element concentrations as reflected by K_2O contents; their initial FeO^*/MgO ratios; and their rate of change of FeO^*/MgO ratios versus SiO_2 contents. The first difference can be reflected in nomenclature by distinguishing low-, medium-, or high-K suites. The other two distinctions are blurred in the calcalkaline versus tholeiitic definition adopted here but I accept that definition as the best available compromise.

Thus, Fig. 1.2 provides criteria for naming an orogenic andesite on the basis of potash and silica contents (i.e., low-, medium-, or high-K; basic or acid); Fig. 1.4 provides criteria for assigning it to either the tholeiitic or calcalkaline series. For reference, at 57% SiO_2 medium-K andesites have $K_2O = 0.75\%$ to 1.9%; calcalkaline andesites have $FeO^*/MgO < 2.25$. As examples the four "average andesites" in Table 1.1 are plotted in Figs. 1.2 and 1.4. All are medium-K acid andesites of the calcalkaline series. Of the 2500 orogenic andesites whose analyses are shown in Fig. 1.1, about 1000 are tholeiitic according to Fig. 1.4. Only 500 are tholeiitic using the criterion of Irvine and Baragar (1971) shown in Fig. 1.3, the difference being in the classification of medium and high-K acid andesites.

Ambiguities still arise when summarizing the ejecta of given volcanoes or regions. For example, the orogenic andesites of Merapi volcano in Java, Indonesia (Neumann van Padang 1951) change from tholeiitic to calcalkaline with time or increasing SiO_2, or both, according to the criteria of Miyashiro (1974) and Fig. 1.4, are calcalkaline in the sense that they display little Fe-enrichment during differentiation and contain groundmass orthopyroxene, and are alkalic in terms of their total alkali content. In this book I have arbitrarily chosen to use the FeO^*/MgO ratio and K_2O content at 57.5% SiO_2 for summary classification, so that Merapi rocks are decribed as high-K tholeiitic andesites (see Appendix).

1.4 Overview

The focus of this book is the origin of orogenic andesites, as defined above. To constrain petrogenetic theory, I have summarized selected geologic, physical, chemical, and mineralogic characteristics of orogenic andesites in Chapters 2 through 6, and have discussed the spatial and temporal variations of these characteristics in Chapter 7. Evaluation of genetic hypotheses is largely deferred to Chapters 8 through 11 where the characteristics are organized pro and con and tested against several hypotheses. Chapter 12 presents a synthesis, and proposals for future research.

By way of preview, orogenic andesites can, in principle, originate from partial or total fusion of continental crust, from contamination of basalt by such crust or by rhyolites derived therefrom, from partial fusion of hydrous peridotite, from partial fusion of subducted ocean crust, or from differentiation of basalt by crystal fractionation, vapor fractionation, or magma mixing. Although it is difficult to exclude unambiguously any of these possibilities, my conclusion will be that differentiation of basalt by crystal fractionation of anhydrous minerals at low pressure is the most frequent and most fundamental process of orogenic andesite genesis, supplemented to an unknown extent by crystal fractionation of hornblende, selective crustal interactions, magma mixing, and vapor fractionation. Criteria are proposed by which to identify andesites having different ancestries.

Chapter 2   The Plate Tectonic Connection

The Earth's surface has four plate tectonic environments: divergent, convergent, and transform fault plate boundaries, and intra-plate locations. Volcanism is concentrated at plate boundaries or in linear belts or local magma floods, usually of basalt, in intra-plate locations. Andesites occur in all four environments. However, orogenic andesites are associated primarily with convergent plate boundaries, and this association is the significance of the "andesite line" frequently drawn around the Pacific Ocean. This line is a boundary seaward of which no orogenic andesites occur and is attributed to Marshall (1912) and Born (1933). Originally the andesite line was thought coincident with the boundary of the Pacific basin, but it is, instead, the western and northern boundary of the Pacific plate and the eastern boundary of the Juan de Fuca, Cocos, and Nazca plates. Sugisaki (1972) extended its meaning to include all plate boundaries converging at > 2.5 cm/yr.

The rule-of-thumb association between active orogenic andesite volcanism and convergent plate boundaries is examined in Section 2.1. I have grouped active andesite-producing volcanoes into 28 volcanic arcs whose distribution and relationship to plate boundaries are shown in Figs. 2.1 and 2.2. Geophysical and tectonic data for these arcs are given in Table 2.1; geochemical data for the arcs appear in Table 7.1 and the Appendix.

However, rules-of-thumb often are understood, as opposed to learned, by studying exceptions, some of which are discussed in Sections 2.2 to 2.8. Because these discussions rely on ambiguous and contested geologic interpretations, and because resolution of these ambiguities is far beyond this book's scope, these sections provide only heuristic constraints on genetic theories.

Table 2.1. Geophysical data for volcanic arcs which are numbered as in the Appendix. Numbers in parentheses identify references

	Arc	D (km) [a]	h (km) [b]	Dip (°) [b]	Rate (cm/yr) [c]	Crustal thickness (km) [d]	Subduction duration (m.y.) [e]
1.	New Zealand	200–300	100–200	50 (1)	5.6	36 (2)	22
2.	Kermadec	170	140	65 (3)	7.3	18 (2)	–
3.	Tonga	175	140	53 (3)	9.4	12 (2)	50
4.	Vanuatu	125	200	70 (4)	10.6–9.5 (5)	28 (58)	~10
5.	Solomon	125	200	90 (6)	10 (7)	15–20 (8)	25
6a.	New Britain	160–280	90–400	70 (6, 9)	9.2 (7)	20–40 (10)	50
b.	W. Bismarck	50 (11)	150	80 (9)	0–6 (12)	–	<10
7a.	New Guinea	90–260 (13)	125 (14)	–	–	30–35 (15)	<10
b.	Papua	100 (13)	–	–	–	8–21 (16)	<10
8a.	Sumatra	300	125	60 (17)	6.7–7.6	–	150–175
b.	Java	250–400	120–360	65 (17)	7.6–7.9	25–30 (18)	150–175
c.	Banda	200–400	100	40 (19)	–	<15 (20)	–
9.	Halmahera	80 (21)	150–200	45 (22)	–	–	~5–15
10.	Sulawesi	60 (21)	100–150	55–65 (22)	–	~20	~30
11.	SE Philippines	130	125	45 (22)	–	–	~5
12.	Luzon-Taiwan	125	100	45 (22, 59)	4.1–5.2	35 (60)	~25–50
13.	Ryuku-West Japan	150–240	100–200	45 (23)	2.7–1.8	25 (24)	~25
14.	Mariana-Izu	175–200	140	45–90 (23)	9.6–6.7	15–18 (25)	~50
15.	East Japan	250–400	100–200	30 (26)	9.9–9.6	27–36 (24)	~125
16.	Kuriles	175–200	90–170	45 (27)	9.2–9.9	15–30 (28)	75–100
17.	Kamchatka	200–400	90–270	50 (29)	8.9–9.2	25–45 (30)	125–150
18a.	Aleutians	175–240	110	55 (31)	8.7–7.0	18–25 (32)	~50
b.	Alaska	225–500	100	60 (33)	7.0–6.4	36 (34)	175+
19.	Cascades	200	–	– (35)	2–3 (36)	25–35 (37)	50
20.	Mexico	160–425	100–150	55 (38)	5.7–8.5	30 (39)	20
21.	Central America	175	105–210	60 (40)	6.9–9.2	42 (2)	~100
22.	Columbia-Ecuador	250–350	140–200	35 (41)	8.9–8.4	66 (2)	–
23.	Peru-Chile	250–500	80–250	35 (42)	10.3–10.8; 2.3 (43)	40–70 (44)	175–200

25. South Sandwich	190	120	50 (48)	9 (49)	—	8
26. Eolian	150	200–250	46 (56)	0.9	20 (51)	—
27. Aegean	120–190	125	35 (50)	3.7 (52)	25–30 (53)	5–13
28a. Turkey-NW Iran	—	<40 (54)	—	2.7 (52)	50 (55)	~25
b. SE Iran	400–600	100	45 (57)	5.1	—	—

a   Distance (D) of volcanic front from plate boundary. Where a range is given, the first figure is for the volcanic front, the second is for the boundary between the volcanic arc and backarc; the difference is the width of the volcanic arc which is unspecified if <50 km. Data mostly from sources of Fig. 2.2

b   Height (h) of volcanoes above earthquake foci of the dipping seismic zone, and dip angle of that zone at depths beneath the volcanic arc. Range in h reflects width of the volcanic arc

c   Convergence rates calculated using rotation poles of Chase (1978) unless otherwise stated. Ranges indicate the variation in convergence rate along that portion of the plate boundary coinciding with volcanoes, in a north to south or west to east direction

d   Crustal thickness beneath or near the volcanic arc. Range indicates known variations in thickness within the arc

e   Duration (in 10^6 year units) of current subduction geometry, i.e., of same direction of underthrusting beneath present volcanic arc. Mostly taken from Dickinson (1973)

References: 1. Adams and Ware (1977); 2. see Fig. 3.3; 3. Isacks and Barazangi (1977); 4. Pascal et al. (1978); 5. For PAC/IND only; 6. Denham (1969); 7. Johnson and Molnar (1972); 8. Furumoto et al. (1970); 9. Johnson (1976b); 10. Wiebenga (1973), Finlayson et al. (1972); see Fig. 3.3; 11. Distance to surface expression of north-dipping seismic zone north of New Guinea; 12. Krause (1973), rate increasing west to east; 13. Distances to plate boundaries of Krause (1973); see Fig. 2.2.4; 14. No well-defined seismic zone exists; depths are maxima of earthquakes beneath New Guinea volcanoes from Johnson et al. (1971); 15. St. John (1970); 16. Finlayson et al. (1976); 17. Hamilton (1979); 18. Ben Avraham and Emery (1973); Untung and Sato (1978); see Fig. 3.3; 19. Cardwell and Isacks (1978); 20. Curray et al. (1977); 21. Distances to surface trace of thrust faults on perimeter of Molucca Sea (Silver and Moore 1978); 22. Acharya and Aggarwal (1980); Cardwell et al. (1980); 23. Katsumata and Sykes (1969); Utsu (1971); Carr et al. (1973); 24. Sugimura and Uyeda (1973); also see Fig. 3.3; 25. Murauchi et al. (1968); Hotta (1970); see Fig. 3.3; 26. Yoshii (1979). Note that Hasegawa et al. (1978) place the volcanic front only 80 km above the seismic zone; 27. Sykes (1966); Engdahl et al. (1977); 28. Gorshkov (1970); 29. Fedotov and Tokarev (1973); 30. Gnibidenko et al. (1972); 31. Engdahl (1977); 32. Marlow et al. (1973); Grow (1973); 33. von Huene et al. (1979); Davies and House (1979); 34. Kubota and Berg (1967); 35. No well-defined plate boundary or seismic zone exists although earthquakes to 80 km depths occur (ref. 37); 36. Silver (1971); 37. Crosson (1976); Hill (1979); 38. Molnar and Sykes (1969); 39. Fix (1975); 40. Carr et al. (1979); 41. Stauder (1975); Barazangi and Isacks (1976); 42. Barazangi and Isacks (1976); only segments beneath active volcanoes are considered; 43. Higher rates are for 11° to 47° S; lower rate applies south of the triple junction at 47° S; 44. James (1971); thickness decreases southward; 45. Volcanic arc is oblique to gravity minimum, taken as the plate boundary; 46. Sykes and Ewing (1965); 47. Boynton et al. (1979); 48. Brett (1977); 49. Based on SCOTIA/SANDWICH relative movement of 8 cm/yr; 50. Papazachos (1973); 51. Barberi et al. (1973); 52. McKenzie (1972); 53. Along volcanic arc; Makris (1978); 54. Niazi et al. (1978); 55. Akashe (1972); 56. Bottari and Federico (1979); 57. Jacob and Quittmeyer (1979); 58. Ibrahim et al. (1980); 59. Seno and Kurita (1979); 60. Taiwan: Wu (1979)

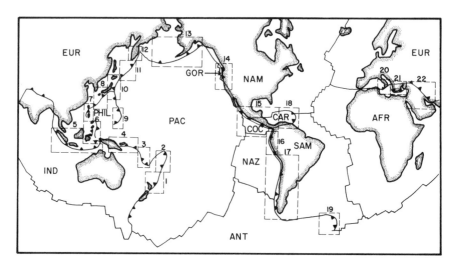

Fig. 2.1. Plates of earth's lithosphere. The location of the 28 volcanic arcs summarized in Tables 2.1 and 7.1 are shown in 22 numbered *dashed-line boxes* which are enlarged in Fig. 2.2 and the Appendix. Plate names are abbreviated as follows: *IND* Indian; *PHIL* Philippine; *EUR* Eurasian; *PAC* Pacific; *GOR* Gorda; *COC* Cocos; *NAZ* Nazca; *NAM* North America; *CAR* Caribbean; *SAM* South American; *ANT* Antarctic; *AFR* Africa

---

Figs. 2.2.1 – 2.2.22 (see pp. 17–24)

Volcanic arcs of the world. (See Fig. 2.1 for key.) Bathymetry from US Navy N.O. 66 *Chart of the World* (1961) except as noted in individual figure captions. Volcano locations given as (●) are from the *List of World Active Volcanoes* (Katsui 1971) which should be consulted for more precise volcano locations except where superceded by references cited in individual figure captions. Compare these figures with those of the Appendix (pp. 329–336)

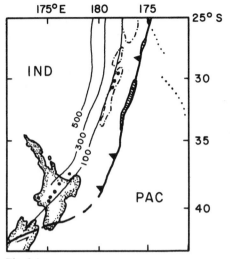

Fig. 2.2.1

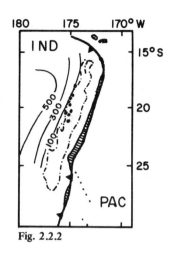

Fig. 2.2.2

Fig. 2.2.1. New Zealand and
Kermadec. Plate boundary
from Walcott (1978); seismicity contours from Isacks
and Molnar (1971). *Shaded
area* encloses >4000 fathom
depths. *Dashed line* is 1000
fathom isobath. *Dotted line*
is Louisville Ridge

Fig. 2.2.2. Tonga. Seismicity,
bathymetry, and *dotted line*
as in Fig. *2.2.1*. For more
precise volcano locations see
Ewart et al. (1973)

Fig. 2.2.3. Vanuatu. *Shaded* ►
area encloses > 5000 meter
depths. *Dashed line* is 2000
meter isobath. Seismicity
contours from DuBois (1971).
For more precise volcano
locations see Colley and Warden (1974)

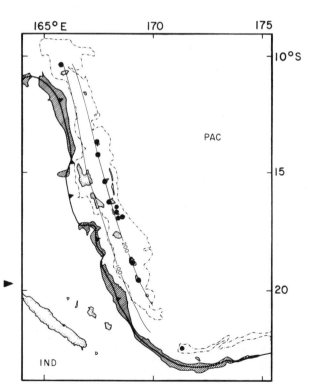

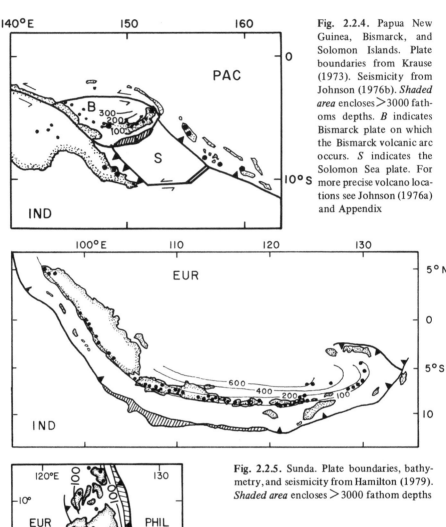

Fig. 2.2.4. Papua New Guinea, Bismarck, and Solomon Islands. Plate boundaries from Krause (1973). Seismicity from Johnson (1976b). *Shaded area* encloses >3000 fathoms depths. *B* indicates Bismarck plate on which the Bismarck volcanic arc occurs. *S* indicates the Solomon Sea plate. For more precise volcano locations see Johnson (1976a) and Appendix

Fig. 2.2.5. Sunda. Plate boundaries, bathymetry, and seismicity from Hamilton (1979). *Shaded area* encloses >3000 fathom depths

Fig. 2.2.6. Halmahera, Sulawesi, and SE Philippines. Plate boundaries, bathymetry, and seismicity from Hamilton (1979) and Cardwell et al. (1980). *Shaded area* encloses >4000 fathom depths. Thrust directions are from Silver and Moore (1978) for lithospheric plates

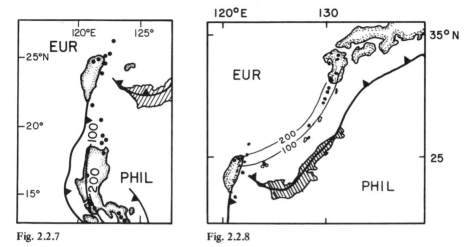

Fig. 2.2.7   Fig. 2.2.8

Fig. 2.2.7. Luzon-Taiwan. Plate boundaries and seismicity from Karig (1973), Acharya and Aggarwal (1980), and Cardwell et al. (1980). *Shaded area* encloses > 3000 fathoms

Fig. 2.2.8. Ryukus and West Japan. *Shaded area* encloses > 3000 fathom depths. Seismicity contours from Isacks and Molnar (1971) and Utsu (1971)

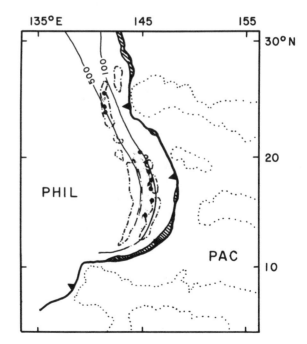

Fig. 2.2.9. Marianas. *Dashed lines* enclose depths < 2000 fathoms, forming the Mariana and West Mariana Ridges. *Shaded areas* enclose depths > 3000 fathoms. *Dotted lines* are seamount chains. Seismicity contours from Isacks and Molnar (1971). For more precise bathymetry and volcano locations see Karig (1971) and Appendix

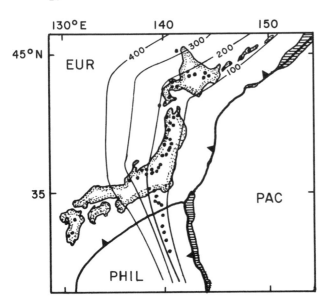

Fig. 2.2.10. East Japan. *Shaded areas* enclose > 3000 fathom depths. Plate boundaries and seismicity contours from Sugimura and Uyeda (1973) and Utsu (1971)

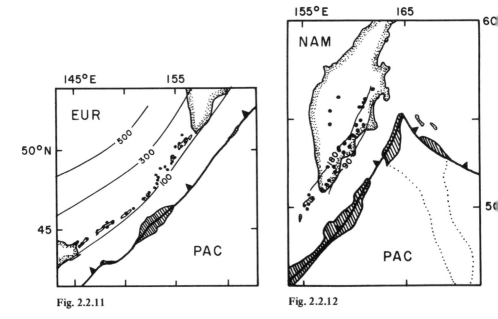

Fig. 2.2.11   **Fig. 2.2.12**

Fig. 2.2.11. Kuriles. *Shaded areas* enclose depths > 4000 fathoms. Seismicity contours from Isacks and Molnar (1971)

Fig. 2.2.12. Kamchatka. *Shaded areas* enclose depths > 3000 fathoms. *Dotted lines* show the Emperor Seamount Chain. Seismicity contours from Fedotov and Tokarev (1973)

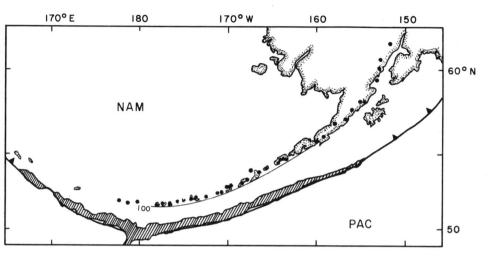

Fig. 2.2.13. Aleutians-Alaska. *Shaded area* encloses > 3000 fathom depths. Seismicity contour from Isacks and Molnar (1971). For more precise volcano locations see Coats (1962)

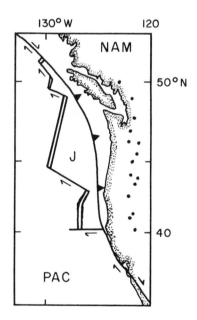

Fig. 2.2.14. Cascades. *J* is Juan de Fuca plate; Plate boundaries from Silver (1971). There is no dipping seismic zone

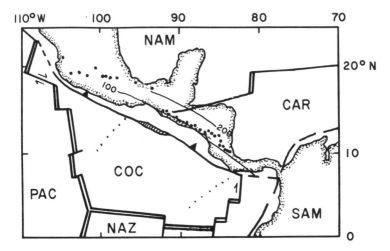

Fig. 2.2.15. Mexico and Central America. *Shaded area* encloses depths > 3000 fathom depths. Plate boundaries from Herron (1972). *Dotted lines* are seamount chains. Seismicity contours from Isacks and Molnar (1971). See Carr et al. (1979) for detailed seismicity contours

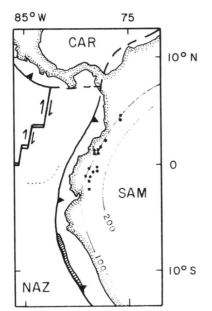

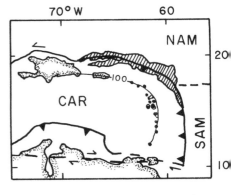

Fig. 2.2.18. Antilles. *Shaded area* encloses depths > 3000 fathoms. Plate boundaries from Silver et al. (1975). Seismicity contour from Isacks and Molnar (1971). For more precise volcano locations see Brown et al. (1977)

Fig. 2.2.16. Columbia-Ecuador. *Shaded area* encloses depths > 3000 fathoms. Plate boundaries after Herron (1972). *Dotted line* is the Carnegie Ridge. Seismicity contours from Isacks and Molnar (1971)

Fig. 2.2.17. Peru-Chile. *Shaded areas* enclose depths >3000 fathoms. Plate boundaries from Forsyth (1975). *Dashed lines* are the Nazca Ridge (north) and Chile Rise (south). Seismicity contours are from Barazangi and Isacks (1976) and are controversial

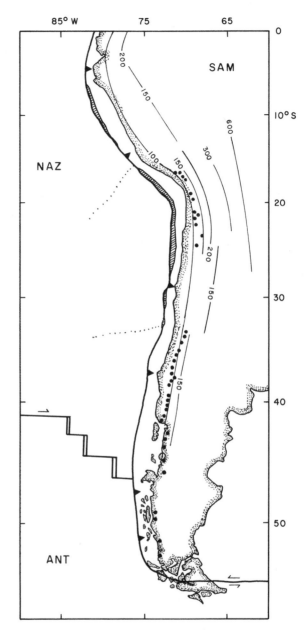

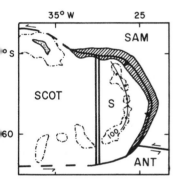

Fig. 2.2.19. South Sandwich. *Shaded area* encloses depths > 3000 fathoms. *Dashed lines* enclose depths <1000 fathoms. Plate boundaries from Forsyth (1975). Seismicity contour from Isacks and Molnar (1971). *S* indicates South Sandwich plate; *SCOT* indicates Scotia plate. For more precise volcano locations see Baker (1978)

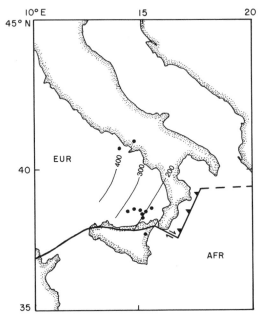

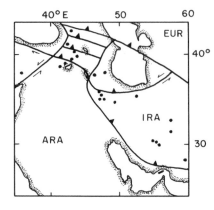

Fig. 2.2.22. Turkey, Caucasus, Iran. Plate boundaries from McKenzie (1972). Iranian volcano locations from Gansser (1971)

Fig. 2.2.20. Eolian. Plate boundaries from Barberi et al. (1973). Seismicity contours from Isacks and Molnar (1971)

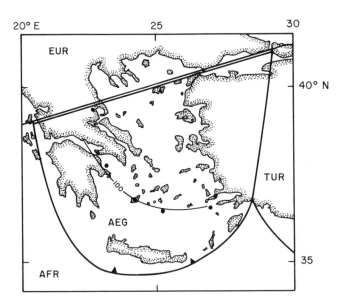

Fig. 2.2.21. Aegean. Plate boundaries from McKenzie (1972). Seismicity contour from Isacks and Molnar (1971)

2.1 Spatial Distribution of Active Orogenic Andesite Volcanoes

To sharpen discussion of convergent plate boundaries, the overthrust plate is divided in Fig. 2.3 into thirds: the forearc, volcanic arc, and backarc regions. The forearc region corresponds to the "arc trench gap". The volcanic arc is defined here as the belt in which polygenetic stratovolcanoes occur, and the backarc region as that area to the rear of the stratovolcanoes. The boundary between forearc and volcanic arc regions is the volcanic front which Sugimura (1960) defined as the trenchward limit of stratovolcanoes and which usually is an abrupt line (see Fig. 2.2). Exceptions, here called forearc volcanoes, occur uncommonly (see Sect. 2.6). In contrast, the boundary between volcanic arc and backarc regions often is ambiguous. For example, in areas such as the Tonga arc, andesite strato-

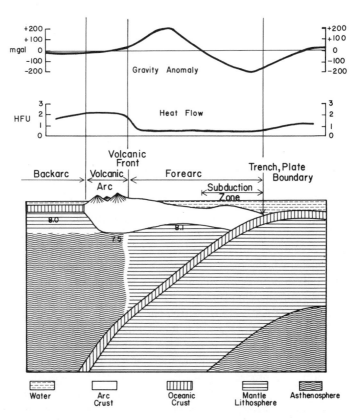

Fig. 2.3. Schematic cross section and nomenclature for a convergent plate boundary. See text and Figs. 3.1, 3.5, and 3.6 for details. *Numbers* in uppermost mantle are typical P wave velocities

volcanoes dominate the volcanic arc, while submarine basalt flows occur within an actively spreading basin in the backarc region; thus, each region is dominated by a different erupted magma type. However, in areas such as the Kermadec, Mariana, Scotia, and Vanuatu arcs, basalts dominate both regions although andesites are known only in the volcanic arc. Fields of monogenetic cones at the rear of the volcanic arc in New Zealand, Japan, Alaska, Central America, and Argentina, for example. are dominated by alkali basalts with $> 1.75\%$ TiO_2 which frequently contain lherzolite nodules and are associated with trachytic intermediate rocks. Backarc basalts, whether tholeiitic or alkalic, differ in isotopic and trace element ratios from basalts or andesites within volcanic arcs, indicating differences in source composition (Sect. 5.7). Mantle beneath these three regions also may have different physical characteristics (Sect. 3.3), and stress orientations within the regions may differ (Sect. 4.6).

Three aspects of the spatial distribution of orogenic andesite volcanoes indicate their close association with subduction: the proximity and parallelism between volcanic arcs and plate boundaries; the termination of volcanic arcs where subduction terminates; and the relationship between volcanic fronts and arc segmentation. Volcanic fronts are strikingly abrupt and linear (see Fig. 2.2) and Sugimura (1960) noted that the population of volcanoes is most dense at the front and decreases with distance away from it opposite to the trench. He also showed (Sugimura 1968, Fig. 10) that the erupted volume of volcanic rocks decreases exponentially away from the front. The volcanic arcs shown in Fig. 2.2 are 25 to 250 km wide; volcanic fronts are 166 ± 60 km ($\bar{x} \pm 1$ σ) from their associated plate boundary to which, except in Mexico, they are remarkably parallel. Thus volcanism "associated with convergent plate boundaries" means volcanism within a volcanic arc 60 to 500 km from the frontal thrust fault marking the surface expression of an active convergent plate boundary. *All* active orogenic andesite volcanoes occur within 500 km of a boundary where two plates now converge.

Secondly, it is instructive to note where volcanic fronts stop. Convergent plate boundaries terminate either by becoming transform faults between the same two plates which converge elsewhere, or by intersecting a third plate at a triple junction. In both cases, volcanic fronts stop where subduction ceases. The boundary between adjacent plates can change from being convergent (a subduction zone) to being conservative (a transform fault) rapidly over a short distance or gradually over long distances (Fig. 2.4). Rapid changes require major tear or hinge faults within the plate which in one area is subducted but in an adjacent area is not (ABC area, Fig. 2.4). Such situations presently occur north of Tonga, south of Vanuatu, and probably in several parts of eastern Indonesia and the Philippines. An alternative kind of rapid change from convergent to conservative plate boundary is referred to here as an orthogonal plate boundary and is shown in the EFGH area of Fig. 2.4; examples include both eastern and western ends of the Aleutian-Alaskan arc and Taiwan. In both types of rapid plate boundary changes, andesite volcanoes usually are restricted to locations on the overthrust plate (exceptions

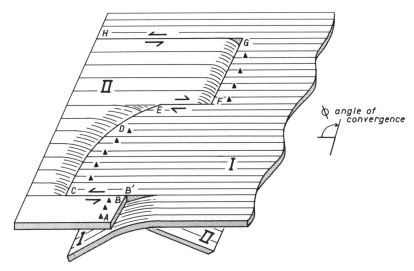

Fig. 2.4. Schematic distribution of volcanoes *(solid triangles)* along convergent plate boundaries of varying geometry. See text for discussion

are noted in Sect. 2.6). Volcanism plausibly associated with the lateral edge of subducting lithosphere is predominantly basaltic, often alkalic, and can occur either on the plate which is elsewhere underthrust (B′ in Fig. 2.4; e.g., on the Pacific plate at Samoa; Hawkins and Natland 1975) or on the overthrust plate (G in Fig. 2.4; e.g., on the Caribbean plate at Grenada: Arculus 1976). Orogenic andesites accompany alkali basalts only in the latter case.

Gradual changes from convergent to conservative plate boundaries (CDEF area, Fig. 2.4) occur as these boundaries become parallel to small circles about the rotation pole. There seems to be a minimum convergence angle necessary for orogenic andesite volcanism (e.g., point D in Fig. 2.4). Examples and minimum angles of convergence associated with orogenic andesite volcanism are listed in Table 2.2. No consistent figure occurs although convergence at $> 25°$ generally seems necessary though not sufficient for active orogenic andesite volcanoes.

Reasons for such a minimum convergence anlge are unclear. It may simply indicate absence of underthrust lithosphere due to hinge faulting or recently initiated subduction. However, if lithosphere is continuous, then a gradual decrease in the down-dip rate of underthrusting will accompany a decreasing angle of convergence. Lithosphere underthrust at a low angle of convergence will, at a given depth, move just as fast relative to surrounding mantle as lithosphere underthrust orthoganally to an arc, but most of the movement will be parallel to the strike of the arc rather than down-dip. It will take longer for ocean crust to reach a given depth but frictional heating per unit time is unaffected. Thus, if underthrust lithosphere is continuous over the gradual transition from a convergent to conservative

Table 2.2. Minimum convergence angles associated with orogenic andesite volcanism

Arc	Pole used [a]	Minimum angle [b]
1. Sumatra (NW)	IND/EUA	46
2. Mariana (S)	PHIL/PAC	55
3. Aleutians (W)	NAM/PAC	25
4. Ecuador (N)	NAZ/SAM	60
5. Antilles (N)	CAR/NAM	25
6. New Zealand (S)	IND/PAC	40

a   From Chase (1978)
b   Angle between strike of plate boundary and convergence vector at the last vol-
    cano listed as active by Katsui (1971) which has produced orogenic andesite.
    Other rocks, usually alkali basalts, occur in regions of lower convergence angles
    in Sumatra, Aleutian, and Antilles examples

plate boundary (e.g., from C to F in Fig. 2.4), then the apparent minimum convergence angle necessary for orogenic andesite volcanism may indicate either that
lithosphere has not yet reached the depth necessary to spawn andesitic volcanism
in these arcs (e.g., western Aleutians: Spence 1977), or that the longer residence
time of lithosphere in the mantle prior to reaching that depth prevents magma
genesis, e.g., by reducing per unit time the mass of volatiles lost due to compres-
sion-rate-dependent dehydration.

Convergent plate boundaries also terminate in triple junctions, and all but one
of the possible triple junctions containing a convergent boundary (see McKenzie
and Morgan 1969) occur today. In all cases where only one arm is a convergent
boundary, orogenic andesite volcanism ceases abruptly at the triple junction. As
with hinge faulting or orthogonal plate boundaries, where underthrusting is absent
so also are andesite volcanoes, and alkali basalt volcanism sometimes is associated
with the lateral edge of subducting lithosphere (e.g., north of the Cascades:
Souther 1977).

Thirdly, dependence of orogenic andesite volcanism on the geometry of subduction also can be judged within segmented arcs. The underthrust plate is not
always continuous and sometimes is broken into discrete segments which descend
into the mantle at varying dips and strikes, possibly due to non-conservation of
surface area in plates which are underthrust at dips unequal to twice the radius of
their surface trace. The overthrust plate also can be segmented into blocks 100 to
1000 km long, identifiable by the lateral extent of focal areas of great earthquakes
(Mogi 1969) or offsets of topographic features or both (e.g., Spence 1977). The
presence or position of volcanoes is influenced by these segmentations.

The clearest modern example is the South American plate boundary where
the Nazca plate apparently breaks into four segments beneath Peru and Chile with
active volcanoes lying only above the two segments being subducted at $25°$ to $35°$,
but being absent above segments being subducted at $10°$ (Barazangi and Isacks

1976, 1979a; Isacks and Barazangi 1977). Even this example is disputed by Snoke et al. (1977, 1979) and Hasegawa and Sacks (1979) who argue that subduction to 100 km occurs at about 30° throughout the region. Both groups agree that asthenosphere sandwiched between the plates at depths > 90 km always accompanies, and therefore is prerequisite to, active volcanism.

Less clear examples involve changes in dip of the underthrust plate of only a few degrees. Such arguments depend heavily on accurate focal depth determinations and on contouring. For example, Stoiber and Carr (1973) and Carr et al. (1979) argue that both underthrust and overthrust plates along the Central American margin are broken into six segments 50 to 200 km long, with the volcanic front being an en echelon composite which reflects the changing geometry. However, the hypocentral data set is small and the volcanic front is convincingly offset only at the Nicaragua/Costa Rica border. Thus, although very close correlation between the site of andesite volcanoes and locally variable depths to subducted lithosphere has been claimed, neither the focal depths nor boundaries of volcanic arc segments are yet known accurately enough for the argument to be compelling.

In conclusion, the general parallelism between volcanic fronts and trenches, the termination of volcanic fronts where plate convergence terminates or becomes too oblique, and the possible changes in volcanic fronts at boundaries of segments of underthrust lithosphere, all indicate a close relationship between currently active andesitic volcanism at convergent plate boundaries and the depth of underthrust lithosphere. This relationship is underscored by the ubiquitous occurrence of volcanic fronts above earthquake foci 100 to 200 km deep (Fig. 2.2).

2.2 Initiation of Subduction

The spatial association between orogenic andesite volcanism and underthrusting is clear. However, when in the evolving history of plate convergence does such volcanism begin? One area whose history elucidates this question is southwestern Japan, where subduction of a segment of the Philippine Sea plate beneath southern Honshu and Shikoku along the Nankai Trough has begun within the last 3 m.y. (Karig 1975). During this new phase of underthrusting the Philippine Sea plate has reached depths of about 70 km beneath the Kii Peninsula and Shikoku and extends 150 to 200 km inland at a dip of ~ 10° (Shiono 1974). While some garnetiferous andesites and other igneous rocks were emplaced in the overlying area 10 to 17 m.y. ago (Nakada and Takahashi 1979), Pliocene to Quaternary volcanism has been meagre and dominated by 3 m.y. old alkali basalts along the Sea of Japan (Fig. 2.5) which contain 3% TiO_2 and other geochemical characteristics of backarc basalts (Table 5.4), normative nepheline, and peridotite nodules. These alkali basalts may be due to edge effects of an adjacent segment of the Philippine Sea plate which has been thrusting under Kyushu throughout the Neogene. Some medium to high-K calcalkaline andesites are interbedded with the alkali basalts

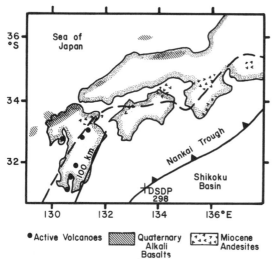

Fig. 2.5. Tectonism and volcanism of southwest Japan. *Dashed line* shows location of the leading edge of the underthrust plate (Shiono 1974). The *solid line* shows its 100 km contour (Utsu 1971). The Palau-Kyushu Ridge intersects the Nankai Trough just below the diagram base. Chemical data for the active volcanoes are in the Appendix. Distributions of Quaternary alkali basalts, discussed in Section 2.2., and of Miocene garnetiferous andesites, discussed in Sections 2.2, 2.6, and 6.6, are from the 1971 *Geological Map of Japan*. DSDP site 298 provided evidence of recently initiated subduction

but constitute < 10% of the volcanics (Aoki and Oji 1966). Active orogenic andesite volcanoes occur only in Kyushu beneath which the underthrust Philippine Sea plate has reached > 100 km depths, although andesitic volcanism in southwest Honshu ceased only within the last 4000 years.

Table 2.1 shows that andesite volcanoes usually overlie earthquake foci > 90 km deep. The situation in southwestern Japan implies that such volcanism does not occur *until* subducted lithosphere exceeds at least 70 km depths. Some examples of oblique subduction and the shallow-dipping plate segments beneath South America mentioned in the previous section also may illustrate this point.

2.3 Cessation of Subduction

Volcanic fronts terminate where underthrusting terminates; andesite volcanism begins only after underthrusting reaches a certain depth. What happens to andesite volcanims when subduction ends? Convergence at a plate boundary can end due to at least three phenomena: (1) changes of plate motion; (2) formation and migration of a triple junction; and (3) collision of nonsubductable crust. In all cases, an area previously underthrust by suboceanic lithosphere will cease to be underthrust by it. A fourth phenomenon having the same result occurs when an island arc is rifted by creation of an inter-arc basin and the previous volcanic arc becomes a remnant arc and migrates away from the subduction site.

Intermittent cessations of subduction due to intermittent changes in plate motions (e.g., Marlow et al. 1973) may be related to the several million year-long maxima and minima of arc volcanism discussed in Section 4.7, but this is neither well-documented nor understood. One late Tertiary example of terminal cessation of subduction due to plate motion changes is the South Shetland island arc, Antarctica. Spreading in the Drake passage and subduction at the South Shetland arc apparently has ceased within the last 5 m.y. (Forsyth 1975).

Late Tertiary, predominantly andesitic volcanism in the South Shetland arc (Tarney et al. 1977) has been succeeded by high-alumina or alkali basalts and subordinate calcalkaline to slightly tholeiitic orogenic andesites at active volcanoes behind the previous volcanic front and within an extensional backarc basin (Bransfield Strait) (Weaver et al. 1979). In this arc, recent cessation of subduction has resulted mostly in a change of venue for, and diminution of, andesitic volcanism which has not yet ceased.

The most studied examples of subduction cessation due to formation and migration of a triple junction are in western North America where there was a continuous convergent plate boundary prior to 29 m.y. BP when the East Pacific Rise reached southern California (Atwater 1970; Atwater and Molnar 1973). Thereafter, subduction ceased along an increasingly long segment of coast as the North America-Pacific plate boundary became a transform fault between oppositely migrating triple junctions. While the Tertiary volcanism of North America is anomalously widespread and correspondingly ambiguous tectonically (Sect. 2.7), cessation of subduction seems correlated with the cessation of predominantly orogenic andesite volcanism and with the initiation of fundamentally basaltic activity.

Christiansen and Lipman (1972) and Snyder et al. (1976) summarized numerous data concerning late Cenozoic volcanic rocks in the western USA, categorizing each occurrence as consisting predominantly of calcalkaline andesite or of basalt with or without rhyolite. In most areas, the character of volcanism changed sometime during the last 30 m.y. from being predominantly andesitic to fundamentally basaltic. These authors argued that the transition occurred in most areas when the triple junction migrated northward past the projection of that area onto the plate boundary. If so, orogenic andesite ceased to be the most abundant magma type when an area no longer had underthrust lithosphere beneath it. Andesitic volcanism on the North American plate along the Cascade trend continues for about 5 m.y. after that portion of the plate moves beyond the southern edge of the subducted Juan de Fuca plate (Snyder et al. 1976, Fig. 7). East of the Cascades the transition from andesitic to basaltic volcanism cannot be attributed to absence of subducted lithosphere but, instead, to a change from compressional to extensional tectonics (Eaton 1979).

A similar but less thoroughly studied example of the effect of triple junction migration is in British Columbia where results appear similar to those outlined above (Souther 1970, 1977).

A third circumstance of subduction cessation beneath a given area is caused by arc splitting, as proposed by Karig (1971). In his model, island arcs as well as continents are split by upwelling asthenosphere which creates new oceanic crust in extensional basins between halves of previously united sialic crust. Gill (1976a) confirmed that this process separated Fiji from Tonga about 5 m.y. BP and since has been creating the intervening Lau Basin. On the now-remnant arc (Fiji), orogenic andesite predominated before rifting commenced but was replaced initially by shoshonitic volcanism and more recently by tholeiite and alkali basaltic volcanism (Gill 1976b). The change in volcanism occurred within 1 to 2 m.y. after the time estimated for initiation of arc rifting. Although the circumstances of the North American and Fijian examples differ, in both cases when an area adjacent to a convergent plate boundary ceased to have lithosphere thrust beneath it, orogenic andesite also ceased to be the most abundant magma type within a few million years. Some orogenic andesite continued to be erupted both in Fiji and North America (Sect. 2.8) but became subordinate to basalt.

2.4 Collisions

Continents and island arcs are relatively nonsubductable because of their low density. Seamount chains and anomalous oceanic crust such as the Ontong-Java Plateau in the western Pacific also may be nonsubductable. Thus, the appearance of such objects at a convergent plate boundary as part of the underthrust plate causes collision and terminates subduction. These collisions can be described tersely by referring to the tectonic environments being juxtaposed, using the following abbreviations: ARC – island arc; AND – active continental margin such as the Andes; ATL – passive continental margins such as those of the Atlantic; and SMT – seamount chain. For example, an ARC-ATL collision is that which results from subduction beneath an island arc of a plate containing a continent, whereas an ARC-AND collision is that resulting from subduction and ultimate disappearance of a continent-free plate beneath an island arc on the one hand and an active continental margin on the other. Examples are discussed below. Orogenic volcanism which precedes collision is part of the normal mode discussed in Section 2.1 and Chapter 3. However, andesitic volcanism sometimes continues during and after collison and is thereby divorced from a trench, dipping seismic zone, and the geophysical characteristics associated therewith.

An ARC-ARC collision has commenced in northeastern Indonesia where seismic zones dip in opposite directions beneath the facing Sulawesi-Sangihe and Halmahera arcs which are separated at most by the 200 km-wide Molucca Sea (Silver and Moore 1978; Hamilton 1979). Deformation and uplift are occurring within this intervening sea and orogenic andesite volcanoes are active on the southern part of both flanking arcs (Fig. 2.2.6; Jezek et al. 1979). Potash contents in some of the volcanic rocks seem high relative to depth to the dipping seismic zone, but information is insufficient to determine whether initial stages of ARC-ARC collision yield anomalous orogenic andesites.

ARC-ATL collisions are uncommon but especially informative because they provide opportunities for contamination by continental crust during subduction but not during magma ascent. The youngest is occurring in the Banda arc of eastern Indonesia where the Australian continental shelf has reached the convergent plate boundary (von der Borch 1979). Collision is most advanced in the vicinity of Timor, north of which volcanism ceased about 3 m.y. ago and arc reversal is now developing. The now-inactive volcanoes erupted orogenic andesites characterized by lowK/Rb ($\sim$ 250) and high $^{87}Sr/^{86}Sr$ ($\sim$ 0.708) ratios but not by uniformly high LIL-element concentrations (Whitford et al. 1977). Active volcanism continues east of Timor where orogenic andesites are enriched in LIL-elements and have Sr, Nd, and O isotopic compositions which suggest incorporation of old Australian sialic detritus (Whitford and Jezek 1979).

ARC-ATL collisions also are impending in the Mediterranean Eolian and Aegean arcs which also have high LIL-element concentrations (Appendix). $^{87}Sr/^{86}Sr$ ratios are 0.706 to 0.709 in the Aegean arc, and $\sim$ 0.705 in the Eolian (Sect. 5.5.1).

A more evolved ARC-ATL collision lies along the Luzon-Taiwan arc where collision commenced in the late Miocene and was completed in the Pleistocene in Taiwan, and continues today farther south (Chai 1972; Karig 1973; Bowin et al. 1978; Wu 1979). The 17 to 22 m.y. old low-K, slightly tholeiitic, basic and acid andesites (Chimei Volcanics) of the Eastern Coast Ranges, Taiwan, and the 13 to 6 m.y. hornblende-andesites of Lü Tao and Hungt'ou Hsü, southeast of Taiwan, apparently are relicts of the pre-collision volcanic arc which then lay above east-dipping China Basin lithosphere (Wang 1970; Chai 1972; Ho 1975).

The extent and significance of volcanism which continued as collision progressed are unclear. Active submarine volcanism occurs east of Taiwan (Kuno 1962) and southward along the Lütao-Babuyan Ridge on the Philippine Sea plate, but is unstudied petrologically. The Tatun volcanic province of Pleistocene age, in and north of northern Taiwan, is on the Eurasian plate but offset behind the strike of the Ryuku volcanic front (Fig. 2.2.8), is underlain by a complex pattern of earthquakes $\leqslant$ 130 km deep possibly related to the Ryuku arc (Katsumata and Sykes 1969; Seno and Kurita 1979), and is at the western termination of the Okinawa Trough which Fitch (1972) and others interpret as an actively opening inter-arc basin. Thus, these volcanoes lie on the previously passive Asian continental margin and may now be underthrust by Philippine Sea lithosphere, but occupy a tectonically ambiguous position. The Tatun rocks are mostly pyroxene and horn-blende-bearing, medium-K, calcalkaline andesites (Appendix).

A somewhat older ARC-ATL collision in Papua New Guinea commenced in the Early Miocene (Jaques and Robinson 1977). Prior to collision, andesites were erupted on the northern, overthrust plate, now the Adelbert and Finisterre Ranges and Huon Peninsula, above a north-dipping slab. They are 34 to 22 m.y. old high-K basalts and basic andesites, and shoshonites (Jaques 1976). Middle Miocene calcalkaline intrusive and subordinate extrusive intermediate rocks accompanied collision and were emplaced south of the suture zone on the underthrusting plate (Page 1976). Post-collision volcanism on the overthrust plate (now the South

Bismarck plate) continues in the active and linear western Bismarck arc (Fig. 2.2.4). This volcanic arc is not associated with a trench or gently inclined band of earthquake foci, but Johnson (1976a) interprets a vertically dipping seismic zone which extends to 225 km depths between the Papua New Guinea mainland and the western Bismarck arc as the remnant leading edge of once-north-dipping Indian plate lithosphere. Magmas erupted in the western Bismarck arc are pyroxene-bearing, medium-K, tholeiitic basalts and basic andesites (Johnson 1976a; Appendix).

Quaternary volcanoes have erupted on the previously Atlantic-type continental margin in two areas: the New Guinea Highlands, and from eastern Papua to the D'Entrecasteaux Islands. The tectonic context of both areas is speculative. Most volcanoes in the New Guinea Highlands are $\leqslant 1$ m.y. old, unconformably overlie large Plio-Pleistocene northeast-dipping thrust sheets (Jenkins 1974), and erupted high-K basalts and basic andesites, and shoshonites (Mackenzie 1976). Scattered intermediate depth ($\leqslant 200$ km) earthquake foci underlie portions of the eastern Highlands, presumably indicating the presence of dismembered lithosphere underthrust somewhere within the last 10 m.y. and which may dip vaguely southwest (Denham 1969; Jakeš and White 1969; Johnson et al. 1971). Several interpretations of this earthquake pattern are possible (e.g., Johnson and Molnar 1972, Fig. 6), each requiring some post-Miocene subduction beneath Papua New Guinea, either from the north or east. Note, however, that andesites constitute $\leqslant 30\%$ of most Highlands volcanoes and that recent workers have attributed the volcanism to post-collision regional uplift instead of subduction (Mackenzie 1976; Johnson et al. 1978b).

Active volcanoes in eastern Papua erupt high-K, calcalkaline, basic and acid andesites, and shoshonites (Taylor 1958; Jakeš and Smith 1970); those in the D'Entrecasteaux Islands are similar but also include alkali basalts and peralkaline rhyolites (Smith 1976). These volcanoes lie near the southwestern boundary between the Solomon Sea and Indian plates (Fig. 2.2.4) where there is convergence between the Solomon Sea and Indian plates but a west-dipping seismic zone only north of the volcanoes. Moreover, the plate boundary apparently changes toward transform movement in the vicinity of the D'Entrecasteaux Islands and divergence in the Woodlark Basin (Krause 1973). Thus, although either site of Quaternary orogenic andesite volcanism in Papua New Guinea could now be complexly underthrust by lithosphere as the result of an ARC-ATL collision, both are likely examples of andesitic volcanism not underlain by subducted lithosphere.

Thus, both in Taiwan and Papua New Guinea, eruption of typical orogenic andesite on the overthrust plate persisted after a collision. This continuation has persisted for over 10 m.y. in Papua New Guinea. Post-collision volcanism on the previously Atlantic-type continental margin has occurred in Papua New Guinea and in northern Taiwan, arguably without the benefit of subducted lithosphere. In Papua New Guinea this post-collision volcanism has been very potassic. Arc volcanism preceding or accompanying collision also has been very potassic or radiogenic or both in the Papua New Guinea, Banda, Eolian, and Aegean arcs, but not in Taiwan.

No ARC-AND or AND-AND collisions are occurring now but Schweickert and Cowan (1975) inferred an ARC-AND collision during the late Jurassic in eastern California, and Hamilton (1970) inferred an AND-AND collision during the Permian or Triassic in the Urals. Neither example provides petrologic insight into andesite genesis.

Bird and Dewey (1970) inferred an AND-ATL collision during the late Paleozoic in the Appalachians which was accompanied more by felsic and sometimes peraluminous volcanics and granites with initial $^{87}Sr/^{86}Sr$ ratios ~ 0.71 than by orogenic andesites (e.g., Lyons and Faul 1968). Similarly, the Tertiary AND-ATL collision in the Himalayas produced granites with high initial $^{87}Sr/^{86}Sr$ ratios (0.74; Hamet and Allègre 1976) but no andesite.

Various combinations of ATL-AND-ARC collisions have occurred in the European alps during the Tertiary (e.g., Dewey et al. 1973), but orogenic andesites are uncommon. Notable exceptions occur in the Carpathian Mountains from eastern Czechoslovakia to Romania where significant volumes of Miocene to Pleistocene volcanic rocks were erupted, apparently at an Andean-type continental margin (Bleahu et al. 1973). The intermediate rocks are medium to high-K, calc-alkaline, basic and acid orogenic andesites (Konečny 1971; Boccaletti et al. 1973); some have initial $^{87}Sr/^{86}Sr$ ratios ~ 0.707 (Gill, unpubl. data). In eastern Czechoslovakia this andesitic volcanism ceased about 13 m.y. BP and was succeeded in late Pliocene to Pleistocene time by basaltic, mostly alkalic, volcanism (Mihalikova and Šimova 1965; Konečny et al. 1969). In Romania, the andesitic volcanism began in the late Miocene, continued to the Pleistocene, and apparently is related to the adjacent and now vertically dipping zone of 60 to 160 km deep earthquake foci interpreted as indicating foundering, recently underthrust lithosphere (Roman 1970; Boccaletti et al. 1973). However, that even these examples of Alpine andesite neeed not have been underlain by subducted lithosphere is indicated by the arguable absence of a slab beneath the Thebes region of Greece, site of recent andesitic volcanism (Le Pichon and Angelier 1979). Pleistocene alkali basalts also occur in Romania. In these examples, replacement of orogenic andesite by alkali basalt volcanism apparently is related to cessation of underthrusting of lithosphere due to an AND-ATL collision.

A continuing AND-ATL collision is associated with active orogenic andesite volcanism in the Turkey-Iran region. Collision began there in the late Cretaceous to early Tertiary (Dewey et al. 1973). Lower Miocene to Holocene volcanism in eastern Turkey, the Caucasus, and northwestern Iran is associated with large north-dipping thrusts (McKenzie 1972; Hall 1976). Some of the volcanoes lie on or south of the proposed plate boundary (Fig. 2.2.22), but most are on the overthrust plate. The volcanic rocks are medium to high-K acid andesites, dacites, and rhyolites with $^{87}Sr/^{86}Sr$ ratios ~ 0.705 and in which potash contents increase south to north (Lambert et al. 1974; Innocenti et al. 1976; Adamia et al. 1977). Alkali basalt volcanism began ~ 6 m.y. BP and dominates many of the modern edifices such as Nemrut.

In central Iran, Quaternary volcanoes have erupted predominantly high-K acid andesites, dacites, and rhyolites (Forster et al. 1972, and Appendix) north of an area of Eocene to Pliocene north-dipping thrust faults (Haynes and McQuillan 1974). Currently there are no subcrustal earthquakes and, therefore, no indications of subducted lithosphere beneath these andesitic volcanoes (Niazi et al. 1978; Jackson and Fitch 1979). In contrast, collision has not yet occurred in the Makran region of southeast Iran where andesites are less potassic and dacites less predominant than in active volcanoes elsewhere in Iran.

Thus, orogenic andesite volcanism precedes and may or may not accompany AND-ATL collisions. When accompanying or post-dating collision, the volcanism is associated with thrust faulting but not necessarily with subcrustal earthquake foci, and usually occurs near the collision zone but on the originally overthrust plate. Whether asthenosphere separates underthrust from overthrust lithosphere in these collision zones is unknown. (In South America, andesites are absent where lithospheres abut; Section 2.1). Medium to high-K acid andesites are the most mafic magmas erupted in abundance during AND-ATL collisions; dacites, rhyolites, and granites seem more abundant than andesite. High $^{87}Sr/^{86}Sr$ ratios are common. Basalt, often alkali basalt, often succeeds the intermediate to silicic phase.

The last collision events to be discussed are SMT-ARC or SMT-AND interactions. Where some seamount chains intersect active convergent plate boundaries, as at the junction of the Louisville Ridge with the Tonga-Kermadec Trench, large magnitude and intermediate depth earthquakes are infrequent or historically absent and andesite volcanoes are missing (Kelleher and McCann 1976; Vogt et al. 1976). Some of these intersections, such as between the Nazca Ridge and Peru Trench, correspond to boundaries of segments in the adjacent volcanic arc. These features are attributed to diminution or absence of subduction at these intersections due to buoyancy of the seamount chains. If so, absence of volcanism again is associated with phenomena arresting subduction and, consequently, seamounts are accreted rather than subducted.

2.5 Reversal of Subduction Polarity

The direction of plate underthrusting is its subduction polarity. Polarity reversals can occur, usually in association with collisions. While no examples are documented unambiguously, several have been inferred. As noted in the preceding section, polarity reversal may be underway in the Philippine-Taiwan arc as the result of the ARC-ATL collision in Taiwan, and may be occurring beneath Papua New Guinea and north of Timor as the result of collision there. A change from west- to east-dipping subduction during or since the Miocene has been proposed for Vanuatu (Karig and Mammerickx 1972; Gill and Gorton 1973), the Solomon arc (Kroenke 1972), and the Halmahera arc (Katili 1975). However, data presently are insufficient to know the fate of orogenic andesite volcanism during subduction polarity reversals.

2.6 Forearc and Transform Fault Volcanism

Forearcs were defined in Section 2.1 as being nonvolcanic, but several examples of anomalously near-trench volcanism exist. While none of the examples is well understood tectonically or petrogenetically, they include orogenic andesites erupted in environments beneath which underthrust lithosphere apparently was very shallow or missing altogether.

One possible cause of forearc volcanism is collision between a spreading center and a subduction zone, causing subduction of material which is younger, hotter, less altered, and less rigid than usual. Whenever an ocean basin closes, as between Europe and North America in the Paleozoic, at least one spreading center must be consumed. Similarily, the younging of magnetic anomalies toward a convergent plate boundary, as in the Great Magnetic Bight of the northeastern Pacific, also indicates that such consumption has occurred. Consequences of these events probably vary considerably depending on relative plate motions and whether trench and ridge intersect orthogonally, obliquely, or in parallel.

Some geometries of ridge-trench interactions cause andesitic volcanism in forearcs. A modern example is the New Georgia volcanic group, Solomon Islands, which lies on the arc side of a ridge-trench-trench triple junction and is approximately 100 km closer to the adjacent plate boundary than is the local volcanic front (Fig. 2.2.4). The Benioff-Wadati Zone north of the triple junction plunges steeply to 500 km but shoals to $\leqslant$ 60 km beneath the New Georgia volcanoes (Denham 1969). The triple junction is stable and migrating southeasterly (Luyendyk et al. 1973). New Georgia magmas are picrites, olivine tholeiites, and clinopyroxene $\pm$ hornblende-bearing, high-K, tholeiitic basic andesites, all with $^{87}Sr/^{86}Sr = 0.7037 \pm 0.0003$; rocks with 49% to 54% SiO_2 predominate (Stanton and Bell 1969; Gill and Compston 1973). The rocks therefore have similar $^{87}Sr/^{86}Sr$ ratios to, but are more mafic and have higher $K_{57.5}$ values than, magmas erupted at the volcanic front (Appendix; Page and Johnson 1974).

An older, aerially more extensive, and better studied example occurs in southern California where 15 to 25 m.y. old basalts to pyroxene andesites and dacites occur in the northwestern Los Angeles Basin and on the continental borderland (Hawkins et al. 1971; Hawkins and Divis 1975; Crowe et al. 1976; Higgins 1976). These rocks were erupted within 0 to 100 km of the inferred contemporaneous plate boundary and at least some volcanism apparently occurred on the transform side of the ridge-trench-transform triple junction which developed when the East Pacific Rise reached the Pacific-North American plate boundary (Dickinson and Snyder 1979a; Pilger and Henyey 1979). Consequently, the region either was not underthrust by lithosphere during andesite genesis or such lithosphere was present only at very shallow depths. The rocks are predominantly andesitic, are mostly calcalkaline according to Fig. 1.4, range from low to medium-K, have $^{87}Sr/^{86}Sr$ ratios of 0.703 to 0.704, and locally contain pigeonite phenocrysts. The basalts and some basic andesites have $> 1.5\%$ TiO_2.

However, there is no consensus that ridge-trench interactions necessarily yield forearc volcanism. Indeed two of the best developed discussions of ridge subduction ignore the forearc altogether and instead propose diminution (De Long and Fox 1977) or enhancement (Bird and Dewey 1970; Uyeda and Miyashiro 1974) of andesitic volcanism in the volcanic arc. Either possbility has implications for andesite petrogenesis, but both are too highly speculative to warrant exploration of those implications now.

Also, some examples of forearc volcanism are inexplicable by ridge subduction. Notable instances include the medium-K calcalkaline hornblende-bearing acid andesites and dacites of early Miocene age drilled at DSDP site 439 near the Japan Trench (Fujioka and Nasu 1978), and the high-Mg calcalkaline andesites ("boninites") and interbedded tholeiitic basic andesites of Oligocene age found near and in the Mariana Trench and described in Section 5.2.3. Both suites apparently were erupted in mid-arc, away from triple junctions.

Also ambiguous are the forearc volcanic and plutonic terranes of the Shumagin-Kodiak shelf, Alaska (Hill and Morris 1977), and the Inland Sea area, southwest Japan (see Fig. 2.5). These terranes are dominated by peraluminous felsic rocks with high $^{87}Sr/^{86}Sr$ and $^{18}O/^{16}O$ ratios which arguably reflect derivation from or assimilation of surrounding trench sediments. Although both terranes have been attributed to ridge subduction (Marshak and Karig 1977), neither attribution is generally accepted and both remain enigmatic.

A second set of anomalous orogenic andesites includes those along "leaky" transform faults. Transforms can leak whenever their traces diverge from small circles around rotation poles. Transitional basalts dominate oceanic occurrences but voluminous orogenic andesites erupt from leaks above mantle recently underthrust by subducted lithosphere. Examples include the locally abundant orogenic andesites which have erupted near transform fault zones in southeastern Alaska, central California, Fiji, eastern Indonesia, and Burma during the last 10 m.y. The active Papuan volcanoes discussed in Section 2.4 also may qualify. None is conclusively unrelated to subduction but each place is unlikely to have had subducted lithosphere > 70 km beneath it during the andesitic volcanism. Few petrologic data are published for any of these cases; two are cited below.

The Quaternary Wrangell and Edgecumbe volcanic areas of southeastern Alaska lie within the Pacific-North American transform plate boundary zone, and at least Edgecumbe has flank vents whose spatial distribution seems controlled by stress orientation along the transform (Nakamura et al. 1977). No intermediate depth earthquakes occur beneath these recently active volcanoes nor is there a seaward trench or accretionary prism, in contrast to the adjacent Alaska Peninsula. Orogenic andesite is the dominant rock type of the Wrangell lavas (MacKevett 1971) and at the adjacent Mt. Drum (Miller et al. 1980), although it is subordinate to basalt at Mt. Edgecumbe (Brew et al. 1969). In both cases the andesites are calcalkaline and contain olivine ± pigeonite.

In central California, a linear belt of Upper Miocene to Quaternary volcanic rocks dominated by orogenic andesite and dacite is aligned subparallel to and 30 to 80 km east of the San Andreas transform fault system between the now-inactive Quien Sabe and Clear Lake volcanoes. The rocks become younger northward along this belt, and were erupted approximately when the Mendocino transform-transform-trench triple junction was located at the adjacent plate boundary. Intermediate rocks mostly are medium- to high-K calcalkaline acid andesites and dacites with $^{87}Sr/^{86}Sr$ to $\sim$ 0.705 (e.g., Hearn et al. 1975). The volcanism has been attributed to instablitiy near the migratory triple junction, resulting in local extension at the plate boundary (Dickinson and Snyder 1979a).

Whatever the reason(s) for forearc and transform fault volcanism, three conclusions relevant to petrogenesis seem clear. First, lithosphere at depths > 70 km beneath a volcano is not always necessary for orogenic andesite to predominate as a magma type. Second, crustal influence in andesite genesis is indicated by the restriction of felsic forearc volcanism to accretionary prisms or continental margins. Third, recent modification of the upper mantle by subduction seems necessary for voluminous orogenic andesite volcanism to occur even along leaky transforms not underlain by subducted lithosphere at the time of volcanism.

2.7 Anomalously Wide Volcanic Arcs

The farthest from a convergent plate boundary that currently active orogenic andesite volcanoes occur is 500 km (Fig. 2.2). Quaternary intracontinental orogenic andesite volcanism has occurred in Romania, Turkey, the Caucasus, and western Iran, but is associated in each case with an on-going AND-ATL collision. Two older instances apparently unrelated to collision where orogenic andesite was the predominant rock type up to 1500 km from a convergent plate boundary are the western USA and eastern USSR. This anomalous arc width has been attributed to unusually shallow ($\leq 10°$) subduction (e.g., Coney and Reynolds 1977), although the interpretation is weakened by the absence of volcanism in central Peru beneath which similarly shallow subduction is now occurring (Sect. 2.1). Nevertheless, Lipman et al. (1972), Snyder et al. (1976), Coney and Reynolds (1977) and Keith (1978) all have compiled numerous data for the western USA which, together with marine magnetic anomaly patterns, show that subduction and predominantly andesitic volcanism as far east as Colorado were at least contemporaneous.

$K_{57.5}$ values increase west to east but have a discontinuity at about 110° W longitude, resulting in repetition of the $K_{57.5}$ pattern (Lipman et al. 1972). Thus, something other than shallow subduction may have occurred which widened the volcanic arc. Possible explanations of the discontinuity include imbricate underthrusting (Lipman et al. 1972; Snyder et al. 1976), secular change in the angle of subduction (Coney and Reynolds 1977; Keith 1978) or the influence of variable crustal thickness or crustal age (Elston 1976).

2.8 Andesites Clearly Not at Convergent Plate Boundaries

The spatial and temporal dependence of orogenic andesite volcanism on subduction which was emphasized in Sections 2.1 to 2.3 suggests that subducted lithosphere is necessary for generation of orogenic andesite magmas. The collision histories sketched in Sections 2.4 and 2.6 weaken that argument, but andesitic volcanism in such circumstances at least retains spatial or temporal proximity to subduction events. However, some andesites clearly occur elsewhere than at present or former convergent plate boundaries. Whether these rocks are orogenic andesites depends on definitions, as discussed in Chapter 1. As shown below, orogenic andesites erupt infrequently in other environments where they are associated with and subordinate to various basalt types. Nevertheless, my claim that "orogenic andesites are associated primarily with convergent plate boundaries" depends not only on the spatial and temporal latitude allotted to the verb "associate", but also on one's definition of orogenic andesite and one's estimate of its relative volumetric significance.

Andesites clearly not at convergent plate boundaries occur in two general magma series: as icelandites, ferrobasalts, or granophyres associated with silica-saturated to oversaturated tholeiitic and high-alumina basalts at mid-ocean ridges, ocean islands, and continental rifts; and as trachytes, hawaiites, mugearites, trachybasalts, and trachyandesites associated with alkaline or transitional basalts[4] at continental or oceanic rifts, and oceanic islands. The bulk composition of these andesites is discussed in Section 5.6; analyses are given in Table 5.5.

Mid-ocean ridge andesites are very uncommon; e.g., none of the 590 fresh glasses from the world's mid-ocean ridge system analyzed by Melson et al. (1976) is andesite, although one is a rhyolite. However, some mid-ocean andesites have been reported, especially from the southeastern Pacific floor where Bonatti and Arrhenius (1970) suggest that andesite and silicic volcanic rocks are abundant. (Subsequent oceanographic work has not confirmed this abundance). Andesites associated with tholeiitic basalt volcanoes erupted above hot spots (e.g., from Iceland) are similar in major element chemical composition to mid-ocean ridge andesites, although they differ from them in trace element and isotopic composition just as ocean island tholeiites differ from mid-ocean ridge basalts. Both groups of andesites commonly are called icelandites and differ from tholeiitic orogenic andesites in TiO_2 contents and some other respects (Sect. 5.6).

Minor basic or acid andesite differentiates commonly are associated with continental tholeiitic basalt provinces, and basic andesites locally can be abundant (e.g., lower Yakima or Grande Ronde basalt, Columbia River Plateau). While most have $> 1.75\% \ TiO_2$, some are orogenic andesites in composition (Table 5.5).

---

4   Transitional basalts are those whose analyses plot near the tholeiite-alkali basalt boundary in the total alkali-silica diagram of MacDonald and Katsura (1964) or near the Fo-Ab-Di plate of silica-saturation in the Fo-Ne-Qz-Di tetrahedron of Yoder and Tilley (1962) or both

Andesites similar to medium- and high-K orogenic andesites sometimes accompany mildly alkaline or transitional basalt in oceanic or continental rifts or in areas previously adjacent to convergent boundaries. Locally they too are abundant (e.g., the Plio-Pleistocene flood trachytes of the Gregory Rift Valley, Kenya; or in the Pliocene shield volcanoes of the Taos plateau, Rio Grande Rift USA). Compositional differences from orogenic andesites at convergent plate boundaries again are in total alkalies, TiO_2, or subtleties of trace element ratios (see Sect. 5.6).

Several conclusions can be drawn from the above. First, andesite predominates over basalt only near convergent plate boundaries. Apparent exceptions to this rule seem to be functions of definition, or spatial or temporal scale. That is, when a broad enough scale is studied, basalt predominates elsewhere that at convergent boundaries. The most notable exceptions are the "tholeiitic andesites" which constitute large proportions of continental flood "basalt" plateaux and which consistently thwart petrological classifications, including mine (see Wilkinson and Binns 1975). Second, andesites not at convergent plate boundaries often fail to qualify as orogenic andesites, according to the definition adopted in Chapter 1, because they have higher TiO_2 contents. This frequent but inconsistent failure confirms that depletion in Ti-group elements is characteristic of, though not unique to, magmas at convergent plate boundaries (Chayes 1964a; Sect. 5.7). Third, rocks similar to orogenic andesites in many respects except TiO_2 are found in many tectonic settings but usually are subordinate to basalts from which they seem to have been derived by magmatic differentiation processes.

Finally, orogenic andesites occur in some regions both while subduction occurs and after it terminates. Examples are described by Fodor (1975), Gill (1976a), and Hausel and Nash (1977). In each case andesites are similar in composition and mineralogy during both periods but differ in relative abundance, being predominant during subduction but subordinate to basalt thereafter.

Thus, the relationship between orogenic andesites and subduction is primarily one of relative volume and secondarily one of distinctive chemical composition. Mere occurrence of andesite in the geologic record does not indicate that a subduction zone was once lurking nearby. However, there is *no* example in which basalt is volumetrically subordinate to orogenic andesite on a regional scale in a situation unambiguously far removed in time or space from a convergent plate boundary.

2.9 Conclusions

The spatial and temporal distribution of orogenic andesite suggests strongly that, either during or before the time of volcanism, intersection of subducted lithosphere with overlying asthenosphere at depths greater than 70 km beneath the site of volcanism and with a high down-dip component of motion usually is necessary, though insufficient, for orogenic andesite to predominate as a magma type.

Andesitic predominance can continue 1 to 5 m.y. after such intersection ceases beneath the site of volcanism. Reasons for the qualifying caveats stem from discussions in Section 2.1 of segmentation of the South American arcs which indicates need for subducted lithosphere to intersect asthenosphere, discussion in Section 2.2 of Shikoku which indicates the need for > 70 km depths to be reached, discussion in Section 2.1 of oblique subduction which bears on both of the above or indicates the need for a high down-dip component of motion or both, and discussion of volcanic cessations in Sections 2.3 and 2.4 which constrains the timing.

However, the occurrence of andesites, including orogenic andesites, in compromising situations clearly unrelated to subduction (Sect. 2.8) indicates that the conditions stipulated above affect the relative proportion more than the major element composition of magma produced.

Exceptions to this creed are few and restricted to situations whose relationship to subduction is ambiguous such as southeastern Alaska, the New Guinea Highlands, the Thebes area of Greece, Papua, or eastern Turkey in the Quaternary (Sects. 2.4 and 2.6), central and southern California in the Miocene to Quaternary (Sect. 2.6), and anomalously wide volcanic arcs (Sect. 2.7). However, these exceptions are sufficient to preclude confidence that mere predominance of orogenic andesite in the geologic record demonstrates the existence of underlying subducted lithosphere within 5 m.y. of the volcanism. Leaky transforms and waning collisions seem the chief rival tectonic environments implicated by voluminous orogenic andesites.

The close spatial and temporal association between predominance of orogenic andesites and the presence of subducted lithosphere 90 to 200 km beneath the volcano indicates either that subducted lithosphere contributes something to andesite or that convergence itself, e.g., compressional stress within the crust of the overthrust plate, facilitates differentiation to orogenic andesite, or both. The geochemical effects of impending ARC-ATL collisions on andesite compositions (Sect. 2.4) demonstrate a recycling of subducted material, whereas the transience of andesite-dominated volcanism after cessation of subduction indicates that this recycled contributiion is either short-lived or insufficient to produce voluminous andesite without accompanying compression or both. That is, if volcanism simply stopped when subduction ceased, one could attribute the cessation to withdrawal of a heat or fluid source. However, volcanism often does not stop and basalts, often alkali basalts (i.e., peridotite-derived, even peridotite nodule-bearing magmas) succeed andesites as the dominant magma type within a few million years or less. Therefore, melting of subduction-altered upper mantle after subduction has ended does not continue to yield voluminous andesite. Instead it often yields the same kind of magma which is produced at the lateral edges of subducted lithosphere (i.e., at triple junctions or orthogonal plate boundaries) where the physical effects of subduction may be similar but where vertically ascending slab-derived fluids are absent.

Finally, the occurrence of volumetrically minor andesites (including orogenic andesites) in a wide variety of extensional tectonic environments (Sect. 2.8), plus their predominance in some areas arguably not underlain by subducted lithosphere (e.g., the exceptions listed above), indicate that andesites are related to basaltic volcanism. The nature of the relationship is unconstrained by tectonics; it could be by differentiation, contamination, or crustal fusion due to basalt intrusion.

Therefore, although voluminous orogenic andesites usually are underlain by subducted lithosphere, this apparently is not necessary *at the time of volcanism.* Two conditions do, however, both seem to be required for orogenic andesite to predominate. One is underlying asthenosphere which was modified via subduction not long before volcanism occurred. The other is either compressional stress or sialic crust greater than about 25 km thick. The modified asthenosphere apparently affects the composition of parental magma (Chap. 9); the compression or crustal thickness affects the extent of differentiation (Chap. 11).

Chapter 3   Geophysical Setting of Volcanism at Convergent Plate Boundaries

3.1 Topography, Gravity, Heat Flow, and Conductivity

Active volcanoes occur at topographic highs approximately 60 to 500 km from convergent plate boundaries which usually are associated with ocean trenches. Thus, variations in elevation of up to 13 km occur over short distances, resulting in some of the Earth's greatest vertical relief. The high elevations of both the volcanic arcs and the nonvolcanic ridges or coast ranges in the forearcs of island arcs and continental margins, respectively, may reflect volumetric expansion accompanying hydration of the mantle above zones of dehydration of subducted lithosphere (Fyfe and McBirney 1975). Alternatively, displacement of mantle by lower density andesite beneath volcanic arcs may cause and sustain the uplift there (Gough 1973).

Gravity profiles across island arcs typically contain a negative anomaly arcward of the trench axis, and a positive anomaly at or seaward of the volcanic front (Fig. 2.3). Neither anomaly seems related to volcanism. The negative and positive anomalies usually are attributed to effects of sediment wedges and cold, therefore dense, subducted lithosphere, respectively (Hatherton 1974). Although accurate modeling of these gravity anomalies requires lower density mantle beneath volcanic arcs and backarc basins than beneath adjacent ocean basins (e.g., Grow 1973; Segawa and Tomada 1976), available models are insufficiently unique to define boundaries between mantle of contrasting densities. On a still smaller scale, Mayfield et al. (1980) proposed that andesite volcanoes lacking large negative ($-$ 30 mgal) Bouguer anomalies also lack shallow magma chambers and have differentiation trends less influenced by plagioclase fractionation than in volcanoes with such anomalies.

Regional heat flow near orogenic andesite volcanoes has been studied primarily in the western Pacific (Watanabe et al. 1977). Typically, heat flow is low in the forearc ($\sim$ 0.7 HFU; Heat Flow Unit = 10^{-6} cal/cm^2 s), rises abruptly at the volcanic front, and remains high (2 to 4 HFU) for distances of 200 to 600 km behind the volcanic front, depending on the age of backarc crust (Fig. 3.1). The distribution of high heat flow values in northeastern Japan corresponds almost exactly to the distribution of volcanoes and hot springs (Sugimura and Uyeda 1973, Figs. 42 and 43) as in New Zealand (Studt and Thompson 1969), but in few other arcs are there sufficient published data to compare heat flow at the volcanic front with that around it.

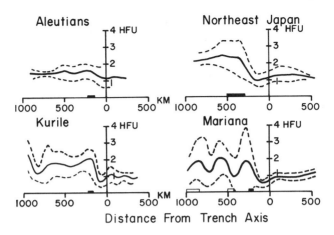

Fig. 3.1. Heat flow profile across four northwest Pacific convergent plate boundaries (after Watanabe et al. 1977). *Solid and open bars* show positions of volcanic and remnant arcs, respectively *Solid and broken lines* show means and standard deviations, respectively, of heat flow values. Arcs are arranged by the age of the crust behind them (in their backarcs):Aleutian (oldest); Japan; Kurile; Mariana (youngest)

Neither volcanism nor crustal radiogenic heat production causes this heat flow pattern. Volcanism in arcs accounts for $< 10\%$ of the regional heat flux (Oxburgh and Turcotte 1970). Radiogenic heat production is low in island arc crust because K, U, and Th contents are low. Crustal heat production is only 1.7 ± 0.2 HGU (Heat Generation Unit $= 10^{-13}$ cal/cm^3 s) in the Aleutians and Philippines (Sass and Munroe 1970), and estimates by Taylor (1969) and Jakeš and White (1971) of the average composition of arc crust predict only 1.1 and 0.5 HGU, respectively. The radiogenic contribution of arc crust 30 km thick to regional heat is, therefore, expected to be only 0.1 to 0.2 HFU. Thus, the heat flow pattern is caused by subcrustal phenomena.

Low heat flow in forearcs is attributed to effects of underthrusting cold lithosphere and endothermic metamorphic reactions within the slab (e.g., Anderson et al. 1976), and is assiciated with low temperature, high pressure metamorphism. The high heat flow values at or behind volcanic fronts cannot be explained by models of conductive heat transfer alone and require mass transfer of heated material via mantle convection or ascending magmas (McKenzie and Sclater 1968). Although the composition, amount, or vertical distribution of magma cannot be specified from heat flow data, the data do indicate that magma is unlikely beneath forearcs and that both volcanic arc and backarc regions are characterized by similarly high heat flow. Mineralogic data for peridotite xenoliths suggest lower temperatures beneath volcanic arcs than backarcs (Sect. 9.1), but the depths to which these temperatures apply are poorly constrained.

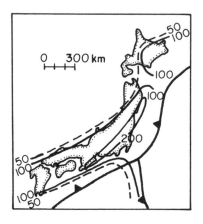

Fig. 3.2. Depth in km of highly conducting mantle, from Rikitake's (1969) interpretation of $\Delta Z/\Delta H$ value distribution, i.e., the ratio of changes in the vertical (ΔZ) and horizontal (ΔH) components of the geomagnetic field. Plate boundaries from Fig. 2.2.10

Because the amount of heat transferred by mass movements cannot be specified, and because the vertical variation in crustal radiogenic heat production is unknown, thermal gradients in arcs and continental margins are ambiguous. However, estimates for Japan by Uyeda and Horai (1964) and for a general case by Oxburgh and Turcotte (1971) agree that crustal thermal gradients in forearcs are $10°$ to $20°C/m$, while gradients at and behind the volcanic front are $35°$ to $40°C/$ km. If so, lower crustal rocks containing free water or containing amphibole but not free water will melt beneath andesite volcanoes if crust is > 20 or > 30 km thick, respectively.

High subcrustal temperatures beneath and behind the volcanic front also may be indicated by conductivity anomalies in Japan (Rikitake 1969; Honkura 1974; Honkura and Koyama 1979) and Chile (Schmucker 1969). Figure 3.2 shows the trend of steep conductivity gradients in Japan which coincide approximately with the volcanic front and the steep heat flow gradient, at least in northeastern Japan. Rikitake (1969) believed that these contours represent depths of some high-temperature isotherm whereas Waff (1974) and Honkura and Koyama (1979) argued that the high conductivity requires several percent or less of interconnected fluid (melt or water) at 30 to 100 km depths beneath the volcanic arc.

3.2 Crustal Thickness, Structure, and Age

Convergent plate boundaries involve the thrusting of oceanic crust beneath one of four kinds of more buoyant and, therefore, overlying material; mainland continental margins (e.g., Chile); peninsular or insular continental margins (e.g., Kamchatka); detached continental fragments (e.g., Japan); or island arcs (e.g., Tonga). Although this classification of subduction zones, which is modified from Dickinson (1975) and used in Fig. 3.3., apparently is geographic in character, it incorporates differences in crustal thickness, structure, and age.

Crustal thickness is poorly known in many volcanic arcs and often is predicted only from gravity data. Estimates for many volcanic arcs are collected in Table 2.1. Generally island arcs have crust < 25 cm thick, while the other three types have average thicknesses of 30 to 40 km and reach 70 km in some continental margins.

The seismic refraction data summarized in Fig. 3.3 show that most island arcs have 6 to 9 km of upper crust with P wave velocities (Vp) from 5.0 to 5.7 km/s, underlain by 10 to 15 km of lower crust with Vp from 6.5 to 7.0 km/s. These two layers are similar in velocity to layers 2 and 3, respectively, of oceanic crust but both, and especially the upper layer, are thicker in the arcs. Notably, an upper crustal layer with Vp ~ 6.1 km/s, which is typical of, though not unique to, continental interiors (e.g., Warren and Healy 1973), is missing in most arcs. Seismic refraction data for continental areas of orogenic andesite volcanism usually differ from the arcs by including a 5.8 to 6.2 km/s layer and having thicker lower crust (Fig. 3.3). In both island arcs and continental margins, crustal densities range from 2.4 to 2.9 gm/cm^3 and average about 2.7, based on the velocity-density curve of Ludwig et al. (1970) and on gravity models (e.g., Grow 1973; Kogan 1975).

In almost all areas of active orogenic andesite volcanoes, the basal crustal layer identified in refraction work has Vp ⩽ 7.0 km/s and density < 2.9 gm/cm^3 at depths ranging from around 18 km in most island arcs to 30 km in Japan to 70 km in northern Chile. Crustal layers with Vp from 7.2 to 7.7 km/s, shown by Drake and Nafe (1968) to be more characteristic of orogenic belts than shield areas, are absent except in northern Kamchatka and Colombia. Thus, gabbros or granulites probably occur but eclogites and amphibolites must be uncommon in the lower crust beneath andesite volcanoes (Manghnani et al. 1974; Ringwood 1975, p. 41). However, bulk compositions are not further constrained by the velocity data because neither mafic nor intermediate bulk compositions are expected to adopt garnet granulite or eclogite facies mineralogy at 10 kb when thermal gradients are 35° to 40°C/km (T. Green 1970; Green and Ringwood 1972). Thus, refraction data imply that feldspar or quartz are abundant within the lower crust beneath andsite volcanoes, but that large volumes of garnet-, amphibole-, or olivine-dominated cumulates or residues are not.

Implicit above is the belief that a relatively sharp boundary separates arc crust and mantle. While this expectation is consistent with P wave velocity contrasts of about 1 km/s found repeatedly by refraction studies of island arcs, it leaves uncertain whether the 7.5 to 7.9 km/s material found beneath many arcs (Fig. 3.3) is crust or mantle. Jacob and Hamada (1972) contended that the crust-mantle boundary beneath the central Aleutian arc is smeared out between 20 and 40 km with a gradual increase of Vp from 6.7 km/s at 10 km to 7.6 km/s at 23 km and 7.9 km/s below 39 km (cf. Fig. 3.3). Likewise they found a gradual increase in Rayleigh wave group velocities over the same interval. They interpret this increase

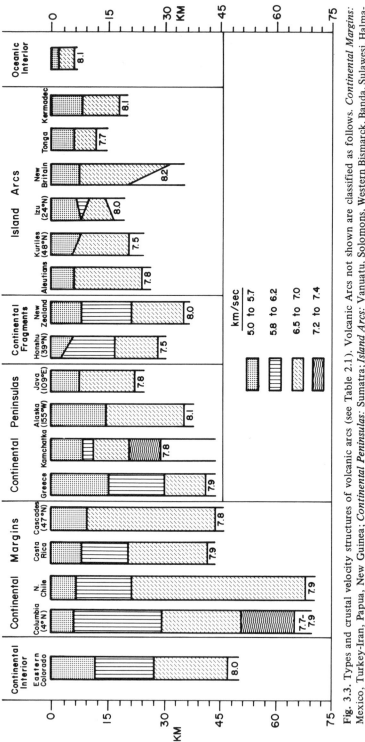

Fig. 3.3. Types and crustal velocity structures of volcanic arcs (see Table 2.1). Volcanic Arcs not shown are classified as follows. *Continental Margins*: Mexico, Turkey-Iran, Papua, New Guinea; *Continental Peninsulas*: Sumatra; *Island Arcs*: Vanuatu, Solomons, Western Bismarck, Banda, Sulawesi, Halmahera, Luzon-Taiwan, Ryuku, Mariana, South Sandwich, Eolian Aegean. Continental and oceanic crustal profiles are given for reference. Data sources as follows: E. Colorado (Warren and Healy 1973); Columbia (Ocala et al. 1975); N. Chile (James 1971); Costa Rica (Matumoto et al.1977); Cascades (Crosson 1976); Greece (Papazachos et al. 1966); Kamchatka (Balesta et al. 1977); Alaska (Kubota and Berg 1967); Java (Ben Avraham and Emery 1973); Honshu (Res. Group Explosion Seismology 1977); New Zealand (Garrick 1968); Aleutians (Grow 1973); Kuriles (Gorshkov 1970); Izu (Murauchi et al. 1968); New Britain (Wiebenga 1973); Tonga, Kermadec (Shor et al. 1971); Oceanic Interior (Christensen and Salisbury 1975)

as a gradual crust-mantle transition. The crustal thicknesses given in Table 2.1 ignore this currently unresolved issue and assume that velocities above 7.5 km/s beneath volcanic arcs represent mantle.

Pre-Cretaceous rocks are unknown at the surface of the island arcs listed in Fig. 3.3 which presumably were erected directly on oceanic crust. Pre-Cretaceous rocks also seem to be absent beneath andesite volcanoes in other environments, e.g., central Cascades or Costa Rica. All sites of active orogenic andesite volcanoes also have been regions of Mesozoic or Tertiary volcanism and orogenesis (Table 2.1). Therefore, the upper crustal layer with Vp from 5.0 to 5.7 km/s probably is < 230 m.y. old and similar in bulk composition to the modern volcanics. Precambrian crust with the elemental or isotopic composition of surface rocks in the continental shield areas (e.g., Eade and Fahrig 1971) is exposed near active andesite volcanoes only in southern Peru (Cobbing et al. 1977). As a result, effects of crustal assimilation usually will be subtle.

3.3 Upper Mantle Beneath the Forearc, Volcanic Arc, and Backarc Regions

The upper mantle beneath convergent plate boundaries contains three regions of grossly different properties, and consequently is laterally and vertically heterogenous. The most obvious region is the subducted lithosphere slab which contains a dipping seismic zone and is discussed in the following section. Overlying subducted lithosphere is a mantle wedge which is composed of two parts. The first is a lid of lithosphere which extends upward to include the crust and is characterized by high velocities (high-V) and low attenuation (high-Q) of seismic waves. This lid typically is 40 to 70 km thick (Fig. 3.4), but can be as thin as the crust or extend

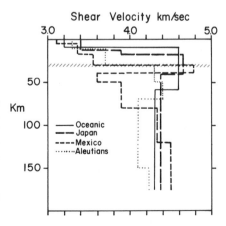

Fig. 3.4. Shear velocity of the mantle wedge above subducted lithosphere. A typical oceanic profile is given for reference (Dorman et al. 1960). Japan and Mexico profiles probably reflect the backarc mostly, whereas the Aleutian profile reflects the volcanic arc mostly. *Slashes* schematically indicate the base of the crust beneath volcanic arcs. Data sources: Japan (Evans et al. 1978); Mexico (Fix 1975); Aleutians (Jacob and Hamada 1972)

downward to the subducted slab. The second part of the wedge is a region of asthenosphere with low velocities and high attenuation which usually is sandwiched between the lithospheric slab and lid. Asthenosphere is thought to extend to the slab/wedge boundary where a sharp 5% to 7% contrast in both P and S wave velocities typically occurs over a few km, and to follow this boundary downward to about 250 to 350 km.

The mantle wedge is characterized on a regional scale by seismic velocities a few percent lower than in the Jeffreys-Bullen standard earth model. P wave velocities typically are 7.5 to 7.9 km/s and S wave velocities are 4.1 to 4.6 km/s (Utsu 1971). Minimum attenuation factors, Q, typically are about 80 and 10 for P and S waves, respectively (Barazangi et al. 1975). Estimates of viscosity within the wedge are 10^{19} to 10^{20} poise (Mavko and Nur 1975; Spence 1977) instead of 10^{22} poise as elsewhere in the mantle.

These regional velocity, attenuation, and viscosity characteristics obviously are those of the asthenospheric portion of the wedge, but what constitutes this asthenosphere is not known unequivocally and may not everywhere be the same. Asthenospheric properties most frequently are attributed to the presence of partial melt. However, these properties also may signify the presence of near-solidus but unmolten peridotite (Goetze 1977) or metasomatized ultramafic rock such as pyroxenite (Matsushima and Akeni 1977). While the coincidence of volcanism, high conductivity, and high heat flow (Sect. 3.1) with low viscosities and low P and S wave velocities and Q factors, all suggest the presence of interconnected partial melt, the interpretation is equivocal and cannot distinguish between the presence of partial melt, vapor, hot peridotite, or pyroxenite within the basal asthenosphere.

There are significant lateral heterogeneities within the mantle wedge. Beneath the forearcs of Kamchatka (Fedotov and Tokarev 1973), the southern Kurile, Japanese, Ryuku, and Izu islands (Utsu 1971), Tonga (Aggarwal et al. 1972) and New Zealand (Garrick 1968), Vp of mantle is 7.9 to 8.2 km/s and Qp is about 1000. In contrast, Vp of mantle beneath volcanic arcs often is 7.5 to 7.9 km/s when determined in seismic refraction studies (Fig. 3.3). Backarc regions of some convergent plate boundaries are characterized by high heat flow, widespread regional attenuation of seismic waves, and active crustal extension (Karig 1971; Barazangi et al. 1975). Indeed, note in Fig. 3.1 that heat flow is as high or higher in these backarc regions as in the volcanic arcs. Similarly, velocities are lower and attenuation is greater beneath backarcs than beneath volcanic arcs. Indeed, these seismic characteristics are more anomalous beneath backarcs than beneath *any* other tectonic environment (Suyehiro and Sacks 1978).

Many configurations of asthenosphere and lithosphere within the mantle wedge are possible. Three are shown in Fig. 3.5. Example 3.5 (a) is the simplest interpretation of the observations in the preceding paragraph. The sharp gradients in seismic wave velocities, heat flow, and conductivity at the volcanic front are attributed to an absence of asthenosphere beneath the forearc but its presence elsewhere.

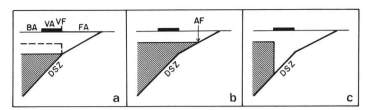

Fig. 3.5. Possible configurations of the mantle wedge above subducted lithosphere. *Ruled areas* represent asthenosphere with low velocities and high attenuation of P and S waves; *unruled areas* are lithosphere; *Heavy line marked DSZ* represents the dipping seismic zone, taken as the upper surface of subducted lithosphere. The *black bars* represent the width of the volcanic arc *(VA)*. Other abbreviations: *BA* backarc; *VF* volcanic front; *FA* forearc; *AF* aseismic front. See text for discussion. Compare with Fig. 2.3 which assumes configuration A *(dashed line)*

Consequently, upper mantle beneath volcanic arcs is similar in physical properties (and therefore possibly similar in extent of molten-ness, vapor-saturation, or chemical composition) to upper mantle beneath backarcs. The same implication would follow from a variation of example (a) in which the lithospheric lid is less thick beneath the volcanic arc and backarc than beneath the forearc, as illustrated by the dashed line in Fig. 3.5 (a), resulting in a vertical asthenosphere/lithosphere boundary beneath the volcanic front.

Example 3.5 (b) differs by invoking asthenosphere beneath part of the fore-arc. It is based on the aseismicity of the mantle wedge under both the volcanic arc and part of the forearc of the central Aleutians (Engdahl 1977) and Japan (Yoshii 1979). The trenchward boundary of asthenosphere is Yoshii's "aseismic front". The low heat flow within the forearc requires that asthenospheric characteristics beneath forearcs be due to the presence of aqueous vapor instead of partial melt (Anderson et al. 1978). Example 3.5 (c) is suggested by studies of earthquakes in Tonga (Mitronovas et al. 1969; Barazangi and Isacks 1972) and Vanuatu (Bara-zangi et al. 1974) which show that the region of low Q is absent beneath both the forearc and volcanic arc in these areas.

The different configurations illustrated in Fig. 3.5 may reflect differences in the site or source of partial melting. Consequently, differences in configuration between arcs may indicate differences in magma genesis. For example, fusion of the slab but not wedge beneath volcanic arcs is suggested by Fig. 3.5 (c). However, current resolution of three-dimensional Q structures is insufficient to confirm whether differences in configuration even exist, much less to determine whether differences have implications for magma genesis.

Vertical heterogeneities are superimposed on these lateral ones. Lithosphere underlain by asthenosphere as shown in Fig. 3.5 implies the existence of a low velocity zone in the mantle beneath some portion of the overthrust plate. Indeed, as discussed in Chapter 2, presence of asthenosphere beneath the volcanic arc seems necessary for eruption of voluminous orogenic andesite. However, demonstration of decreasing Vp or Vs with depth beneath the overthrust plate are few and weak, although the low regional P and S wave velocities cited above remain low to depths $\geqslant$ 100 km (Kanamori 1970; Kaila et al. 1971, 1974; Fedotov 1976). Beneath the East Japan and Kurile arcs, P wave velocities are most subnormal (10% to 15% below the Jeffreys-Bullen standard velocity model) in the 300 to 500 km depth range, i.e., beneath the backarc (Suyehiro and Sacks 1978).

In contrast, regional studies of surface wave velocities frequently identify a low velocity channel in the mantle wedge but do not define its lateral extent, i.e., do not determine whether it occurs beneath forearcs, volcanic arcs, or backarcs. Examples are given in Fig. 3.4. Low velocity channels in the mantle wedge generally are shallower and have lower channel velocities than beneath ocean basins, but are otherwise similar.

Even if asthenosphere beneath regions of high heat flow indicates the presence of partial melt, the proportion of such melt present is poorly constrained because the effect of magma on seismic waves is as strongly dependent on the geometry of magma distribution as on its relative abundance and viscosity. If most magma within the wedge consists of inter-connected intergranular films or thin tubules at triple grain intersections, as implied by interpretations of conductivity data cited above and by analysis of elastic rebound rates in island arcs (Mavko and Nur 1975), then the relative width to length of magma pockets (the aspect ratio) will be small and a few percent or less of magma is sufficient to explain the observed regional seismic velocities (Anderson and Sammis 1970). Melt concentrations up to 20% are inferred beneath local portions of the Kamchatka volcanic arc (Fedotov and Tokarev 1973).

The mantle wedge beneath active continental margins may differ significantly from examples discussed above. For example, the low velocity zone beneath Mexico differs from that beneath the Aleutians or Japan by having lower channel velocities, by occurring higher in the mantle, and by not extending to the inclined seismic zone (Fig. 3.4). In contrast, the lithospheric lid beneath Peru is at least 130 km thick (Sacks and Okada 1974; Barazangi and Isacks 1979a). Furthermore, whatever the thickness of subcontinental lithosphere, the chemical composition of subcontinental mantle may differ to depths exceeding 200 km, being more refractory in major elements (Jordan 1978) yet enriched in LIL-elements (Brooks et al. 1976a).

In summary, there is insufficient geophysical information to conclude whether magma within the mantle wedge forms in situ or is inherited from the slab beneath, whether melt is as widespread beneath volcanic arcs as beneath many backarc regions, whether mantle within the wedge has been extensively altered due to subduction, or whether possible differences in mantle structures between convergent plate boundaries (Fig. 3.5) have petrologic or tectonic significance.

3.4 Dipping Seismic Zones (Benioff-Wadati Zones) and Underthrust Lithosphere

Earthquake foci increase in depth away from subduction zones (Wadati 1935) and lie within a thin zone dipping away from the trench (Benioff 1954). These dipping seismic zones, known as Benioff or Wadati zones, now are considered characteristic of convergent plate boundaries (Fig. 2.3). They have been shown to occur within high-velocity, high-Q slabs which are continuous with suboceanic lithosphere of the underthrust plate at these boundaries (Isacks et al. 1968). Several aspects of these zones are significant for orogenic andesites.

Most volcanic arcs overlie earthquake foci 100 to 200 km deep (Fig. 2.2 and Table 2.1). The volcanoes parallel contours of equal focal depths just as they parallel trenches. Distance of volcanoes from the plate boundary varies inversely with dip of the seismic zone because depth to the seismic zone beneath the volcanic front is quite regular: $\bar{x} = 124 \pm 38$ km (Table 2.1). Exceptions include the Vanuatu, Solomon, and Eolian island arcs where volcanic fronts overlie deeper earthquake foci.

As methods of locating earthquake foci improve, the estimated depth and thickness of the dipping seismic zone decrease. Use of local seismic networks and correction for the effect of the lateral heterogeneities discussed in the preceding section tend to relocate foci at shallower depths than when using teleseismic data and standard earth models (e.g., Engdahl 1973; Adams and Ware 1977; Research Group Explosion Seismology 1977). Thus, the 124 km average depth cited above may need reduction as accuracy improves although Barazangi and Isacks (1979b) disagree. For example, earthquakes beneath the volcanic front of northern Honshu, Japan, may be as shallow as 80 km (cf. Hasegawa et al. 1978, with Barazangi and Isacks 1979b). Likewise, the seismic zone is only about 10 km thick when well defined (Wyss 1973; Engdahl 1977; Yoshii 1979).

Several characteristics of the dipping seismic zone change beneath volcanic fronts, implying that significant changes in the physical properties of underthrust lithosphere occur directly beneath orogenic andesite volcanoes. Most notably, a double seismic zone consisting of two 10 to 15 km thick bands has been identified beneath several volcanic arcs and attributed to slab unbending (e.g., Engdahl and Scholz 1977; Hasegawa et al. 1978). Focal mechanisms for the upper band indicate down-dip compression and such earthquakes usually are continuous from near the trench to approximately 175 km depth. Beginning at 70 to 100 km, this upper band is underlain 25 to 40 km by another which is characterized by down-dip tension; the two bands initially are parallel but eventually converge at about 175 km.

Second, the dip of the seismic zone appears to steepen beneath the volcanic front in Kyushu (Hedervari 1975), Central America (Stoiber and Carr 1973), and New Zealand (Ansell and Smith 1975). However, such changes in dip may reflect mislocation of foci due to lateral heterogeneities in mantle velocities (Adams and Ware 1977) or the effects of tectonic accretion in forearcs (Karig et al. 1976).

Third, the number of earthquakes per unit area and time decreases beneath and down-dip from the volcanic front in some areas [e.g., Kamchatka: Fedotov and Tokarev (1973); Central America: Carr and Stoiber (1973); South America: Hanus and Vanek (1978), and New Zealand], suggesting decreased shear strength in the slab or wedge or both, and possibly reflecting the presence of a fluid phase beneath and down-dip from the volcanic front. However, this pattern of strain release is not ubiquitous because no clear association with the volcanic front appears in similar data compiled earlier for numerous arcs by Isacks et al. (1968, Fig. 12). Moreover, these observations are but crude versions of the more detailed study which is summarized next.

The b-values, which describe the relationship between the frequency and magnitude of earthquakes, vary as a function of depths along the upper plane of seismicity, at least beneath northern Japan. Abrupt increases in b (decreased relative frequency of large-magnitude events) occur at 70 to 90 km depths and may indicate the depth of the amphibolite to eclogite phase transformation within subducted oceanic crust by identifying a region characterized by lowered stresses due to high pore pressure as the result of dehydration (Anderson et al. 1980). B-values then drop beneath the volcanic front, suggesting that pore fluid has been lost from the subducted slab into the overlying asthenosphere.

For Chapter 8 it will be important to know where within underthrust lithosphere these earthquakes occur. One region where the slab can be located independently of the seismic zone is the Aleutian arc, as the result of nuclear explosions there. Engdahl (1973) and Toksoz et al. (1973) found that earthquakes beneath the volcanic front occur 13 to 23 km below the top of the slab, within the colder, more brittle interior (Fig. 3.6). Similar results are obtained by utilizing ScSp waves. These are near-vertical compressional waves converted from ScS waves at an interface at most a few km thick where there is a 5% to 7% P and S wave velocity contrast (Snoke et al. 1977). If this interface is the slab-wedge boundary, then study of ScSp waves can independently locate that boundary. Beneath the northern Japanese volcanic arc the conversion boundary is found to lie about 25 km above the traditional seismic zone (Snoke et al. 1977) or to be coincident with the upper portion of the double seismic zone when resolved (Hasegawa et al. 1978). Thus, if an andesite volcano overlies earthquake foci 125 km deep, then underthrust oceanic crust probably is 100 to 125 km beneath the volcano.

The thermal effects of underthrusting cold lithosphere will produce isotherms at convergent plate boundaries which look something like those shown in Fig. 3.6. Notable features are the depression of isotherms within the slab, the temperature inversion in the mantle wedge immediately above the slab, and the occurrence of earthquake foci in the coldest portions within the slab.

Actual temperatures at the boundary between underthrust lithosphere and overlying mantle (i.e., the slab/wedge boundary), especially beneath volcanic fronts, are difficult to constrain so that published estimates are quite variable (Fig. 3.7). Uncertainties contributing significantly to the scatter are: the boundary

Fig. 3.6. Temperature and earthquake distribution within subducted lithosphere beneath the central Aleutian arc, near Amchitka (from Toksöz et al. 1973). The slab, which is outlined by *dashed lines,* has been located independently from the earthquake distribution which is indicated by the *shaded area.* Temperature estimates are in °C. The thermal gradient at the slab/mantle boundary yields line 3 in Fig. 3.7. *VF* volcanic front; *T* trench axis

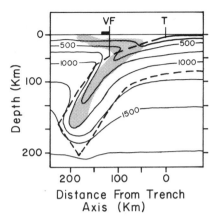

conditions adopted in numerical solutions (lines 1 and 4) or the unperturbed geotherms adopted in numerical solutions (lines 2, 3, 5); whether shear-strain heating is assumed to be high (lines 2 and 3) or low (lines 4 and 5); the effects of dehydration and convection; and what convergence rates, durations, and conductivities are adopted. For example, although lines 1 and 3 are similar, line 3 involves *de*creased temperatures (relative to an unperturbed geotherm) at the slab/wedge boundary because the slab is a heat sink, whereas line 1 involves *in*creased temperatures at this boundary due to shear-strain heating. Moreover, older analytical solutions assumed slab fusion beneath volcanic fronts. Finally, the thermal effects of shear due to the unbending cited above may be significant but have not been evaluated.

Fig. 3.7. Estimates of thermal gradients at the slab/wedge interface. Gabbro solidus and andesite liquidi are from Figs. 8.1 and 4.1. Numbered thermal gradients are from the following sources: *1* Oxburgh and Turcotte (1970); *2* Toksöz et al. (1971); *3* Toksöz et al. (1973); *4* Griggs (1972); *5* Andrews and Sleep (1974); *6* Anderson et al. (1978); *7* implicit in Fig. 8.2. The *stippled area* shows the depth range (one sigma limits) of earthquake foci beneath modern volcanic fronts, from Table 2.1

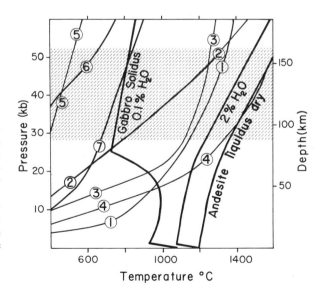

Consequently, the geotherms shown in Fig. 3.7 cannot be used with confidence to evaluate whether or at what depth melting of the slab will occur. Note, however, that thermal estimates which include the effects of dehydration of the slab or convection of overlying asthenosphere (lines 5, 6, and 7), both of which are likely, lie below the solidus of subducted basaltic crust beneath volcanic front. All estimates but one lie below the liquidus temperature of orogenic andesite containing 2 wt.% H_2O or less at these depths.

3.5 Partial Melting and Magma Ascent Beneath Volcanic Arcs

A fluid phase, probably a partial melt, exists within the mantle wedge beneath the backarc and volcanic arc regions at many convergent plate boundaries (Sects. 3.1, 3.3, 3.5). This melt probably is distributed along grain boundaries, constituting at most a few percent of the wedge (Sects. 3.1, 3.3). It is common to point out, as in Section 2.9, that the point of intersection between subducted lithosphere and this partially molten astehenosphere determines the site of the volcanic front. The more fundamental issue is why melting, hence asthenosphere and volcanism, occurs where it does.

The immediate cause of partial melting remains unknown although various possibilities were summarized by Yoder (1976, Chap. 4). The restricted spatial distribution of andesite volcanoes summarized earlier and the conclusion of Section 2.2 that melting does not occur until subducted lithosphere reaches > 70 km depths, implies that the melting point of either subducted oceanic crust or the mantle wedge consistently is reached at about 90 to 125 km. However, the slab is a heat sink because rates of underthrusting exceed rates of conductive heating and because dehydration reactions within the slab are strongly endothermic (Anderson et al. 1976). Therefore, partial melting associated with subduction must result from shear heating near the slab-wedge boundary, or from a reduction in melting point within the mantle wedge due to volatile fluxing or decompression. The energy needed to melt eclogite and peridotite at these depths is about 118 and 135 cal/gm, respectively (Yoder 1976, pp 94–95).

Shear heating is a likely source of energy at a subduction zone and, because the heat production varies directly with strain rate, which in turn increases as the temperature rises, shear heating is a self-perpetuating process (Shaw 1969). However, two observations imply that shear heating alone cannot account for the location of volcanic arcs and asthenosphere.

First, the consistent location of orogenic andesite volcanoes above earthquake foci about 125 km deep is independent of plate convergence rate, angle of dip, or angle of convergence. If depth of melting is proportional to depth of the slab, and if melting is due to down-dip shear-strain heating, then

$$d_m = \frac{\sin \theta \cdot \cos \Phi}{v} \cdot K$$

where d_m is depth of melting, θ is the dip of underthrusting, Φ is the angle of plate convergence measured from a perpendicular to the strike of the plate boundary, v is convergence rate, and K is a proportionality constant (Turcotte and Schubert 1973). No such proportional relationship exists for the data in Table 2.1, indicating that the proportionality is weak so that melting is only partially explained by shear heating. Note also that shear heating is greatest where overthrust and underthrust lithospheres abut, such as where the Nazca plate is subducted at a shallow angle but high rate beneath South America. Absence of volcanoes at these sites (Sect. 2.1) confirms that shear heating is not the principal cause of fusion beneath arcs.

Second, if shear heating alone causes melting then stresses at the slab-wedge boundary must be unrealistically high. Lines 1 to 3 in Fig. 3.7 all require a few kilobars of shear stress which is about an order or magnitude more than the stress drops calculated for intermediate depth earthquakes (e.g., Molnar and Wyss 1972), yet which is considered possible by some seismologists. However, stresses > 10 kb are necessary to cause the slab to melt if endothermic dehydration reactions are considered (Anderson et al. 1978).

Thus, the consistent and necessary depth of subducted lithosphere beneath andesite volcanoes and the insensitivity of this depth to variations in convergence geometry and, by implication, to temperature, suggests that a pressure-sensitive or structural process is more important in initiating and focusing melting beneath volcanic arcs. Instability of hydrous minerals is an example of the former (Sect. 8.3); elastic rebound of the mantle wedge as a result of underthrusting (Bischke 1974; Spence 1977), and convection within the wedge illustrate the latter.

Once formed, melt is likely to percolate upward through the mantle wedge along inter-connected grain boundaries in a few hundred to thousand years due simply to differential buoyancy (Turcotte and Ahren 1978; Walker et al. 1978). Volcano spacing within volcanic arcs may indicate that this permeable feeder region is only a few scores of meters thick (Marsh and Carmichael 1974).

Somehow melt collects by porous segregation into magma reservoir systems, such as sketched in Fig. 3.8, which feed conduits 30 to 100 km apart along and sometimes behind volcanic fronts. Rates and mechanisms of magma ascent through this system are matters of speculation. If buoyant magma blobs 1 to 10 km in radius effectively burn their way upward, then they would ascend at rates of 1000 to 10 cm/yr (10^{-7} to 10^{-9} m/s), thereby creating "conduits" or warmed paths (Fedotov 1976; Marsh 1976b, 1978). Magmas could remain molten at these slow ascent rates even if they convect, according to Marsh (1978), especially the second or subsequent time a pre-warmed pathway is used. Because such ascent is proportional to the density/viscosity ratio of magma, andesite would ascend more slowly than basalt.

If, instead, magma ascends by an elastic crack propagation, stress corrosion, or "magmafracturing" mechanism (see Weertman 1971; Yoder 1976; Anderson

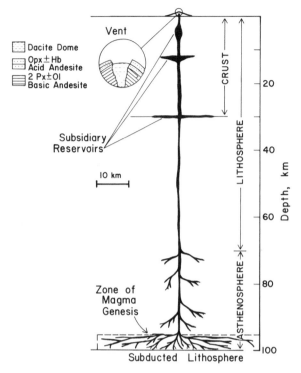

Fig. 3.8. Schematic cross-section of the reservoir system of a medium-sized stratovolcano having an 8 km diameter and 25 km^3 volume, drawn to scale except for the integrated reservoir volume which is exaggerated. Subsidiary reservoirs are shown within and at the base of the crust, as discussed in Section 3.6. Lithosphere 70 km thick is based on the Aleutian line in Fig. 3.4; note difference from Kamchatkan data discussed in Section 3.6. A zone of magma genesis is shown immediately above subducted lithosphere to explain low Q and high velocity contrasts of P and S waves there; see Sections 3.4 and 3.5. A 250 km^3 zone of magma genesis would produce 25 km^3 of ejecta if 25% fusion were followed by 80% extraction (i.e., assuming 5% melt is gravitationally stable) and by 50% crystallization or diking. The 50 km along-strike zone length assumes a 50 km volcano spacing (Sect. 4.7). The zone thickness may be only tens of meters thick, based on one interpretation of this volcano spacing (Marsh and Carmicheal 1974)

and Grew 1977), rates can be $> 10^6$ times faster, and must be $> 10^4$ faster to prevent solidification, according to Marsh (1978). Peridotite nodule-bearing alkali basalts, for example, rise through tens of km of lithosphere at rates of 0.1 to 1 m/s (Kushiro et al. 1976). Analogous calculations for hydrous andesite ($\rho = 2.4$ gm/cm^3; $\eta = 10^4$ poise) containing mafic inclusions ($\rho = 2.9$ gm/cm^3) 3 inches in diameter (7.5 cm) predict minimum ascent rates of 10^{-3} m/s, but do not similarly constrain the depth from which such ascent occurred.

A possible but unlikely test of these models comes from earthquake precursors to volcanic eruptions. For example, earthquakes preceding the 1959 eruption of tholeiitic basalt at Kilauea, Hawaii, first appeared at about 60 km depths and then ascended at about 1 km/day (10^{-2} m/s); eruption occurred when the earthquakes reached the surface (Eaton 1962). Analogous patterns preceded the 1973 eruption of basalt from Tobalchik volcano, Kamchatka, and sometimes accompany ascent of acid andesite from crustal reservoirs 5 to 6 km deep (e.g., Yasui 1963). More pertinent but also more contentious is Blot's (1972, 1976) observation that some eruptions of basic and acid orogenic andesites are preceded by patterns of earthquakes in which the earliest and deepest ones lie along the dipping seismic zone and later ones "ascend" at 0.5 to 1.8 km/day (0.5 to 1.8 $\times$ 10^{-2} m/s). As at Kilauea, eruption or at least volcanic tremor occurs when the earthquakes reach the surface. If these earthquake migration patterns accompany simultaneous magma movement, then either the erupted magma ascends rapidly from the dipping seismic zone, or the eruption of shallower, older andesite is at least triggered by a flushing of slab-derived fluid through the reservoir system. In either case, four conclusions would follow: magmafracturing is a more appropriate model than buoyant spheres, even for magma ascent through weak asthenosphere; the orientation of regional stress or the occurrence of convection within the wedge will affect ascent paths and slow ascent rates by controlling crack distributions; ascent is episodic; and something ascends from subducted lithosphere.

Evidence summarized in Sections 3.6, 4.3, and 4.4 indicates that magma is temporarily stored and most differentiation occurs in "chambers" within or at the base of the crust. These aspects of magma ascent are picked up again in Section 4.5.

3.6 Magma Chambers Beneath Orogenic Andesite Volcanoes

Several lines of evidence indicate the presence of magma chambers (i.e., portions of magma reservoir systems containing large volumes of magma; see Fig. 3.8) within the crust and within the upper mantle wedge beneath some orogenic andesite volcanoes, and their absence beneath others. Machado (1974) summarized evidence for shallow crustal magma chambers including: plutons underlying eroded volcanoes (e.g., Hopson et al. 1965); petrologic evidence for low-pressure crystal fractionation (see Chap. 11); shallow volcanic tremor at andesite volcanoes such as Asama and Kirisima, Japan (Minakami et al. 1969, 1970a); pre-eruption ground surface deformation and changes in magnetic field (e.g., Clark and Cole 1976); and interpretation of calderas such as Hakone as collapse features (Kuno et al. 1970). This evidence implies magma in chambers at most a few tens of kilometers in diameter at depths < 10 km.

Seismic studies have also identified crustal level magma chambers beneath some active orogenic andesite volcanoes. Three magma chambers about 15 km in diameter at $<$ 10 km depths, and one chamber at 20 to 30 km depth beneath the Trident and Katmai volcanoes, Alaska, were identified by studies of S-wave screening and P-wave attenuation (Matumoto 1971). Only shallow chambers were associated with Trident which has erupted acid andesite historically; both shallow and deeper chambers were associated with Katmai which has erupted rhyolitic ignimbrite historically (Fenner 1926). Minakami et al. (1970b) found P wave arrival time delays and attenuation through a zone about 10 km deep and 10 to 14 km in diameter benath Asama volcano, Japan, which has produced acid andesite in historic Vulcanian-type eruptions (Aramaki 1963). A similar zone 20 to 30 km wide with low P wave velocities (5.3 km/s) at 10 to 20 km depths has been found beneath Bezymianny volcano, Kamchatka (Utnasin et al. 1975; Balesta et al. 1977) which produced acid andesite in a Plinian-type eruption in 1956 after centuries of dormancy. These authors also cite studies indicating magma chambers 10 km in diameter at depths $<$ 8 km beneath Avachinsky (also see Steinberg and Zubin 1965) and Karymsky volcanoes in eastern Kamchatka which have produced historic eruptions of basic and acid andesite, respectively (Vlodavetz and Piip 1959).

Magma at mantle depths also has been detected seismically beneath volcanoes in Kamchatka and the Aleutians. Gorshkov (1971) reported P wave screening for paths 50 to 70 km beneath the Klyuchevskoy group of volcanoes in the central depression of Kamchatka, behind the volcanic front. Within this group, maximum screening occurs just below the crust from 40 to 60 km beneath the recently active Plosky Tolbachik volcano and at 70 to 100 km beneath Klyuchevskoy volcano (Tokarev and Zobin 1970). This screening was attributed by Utnasin et al. (1975) to a vertical magma column 2 km in diameter which can be traced from 20 to 50 km beneath Klyuchevskoy and which connects with the shallower chamber beneath Bezymianny cited above. For centuries Klyuchevskoy has erupted basic andesite at intervals of several years (Vlodavetz and Piip 1969; Ehrlich 1968). P and especially S waves also are attenuated in zones about 25 km in diameter from the base of the crust (30 to 40 km) to about 80 km depths beneath several volcanoes, including Avachinsky, at the volcanic front in eastern Kamchatka (Fedotov 1968; Faberov et al. 1973). Attenuation in this region apparently increases with depth and occurs within a wedge-shaped volume tapering upward which parallels the volcanic front, has a viscosity of about 10^{20} poise, and may contain about 20% melt by volume. In general, mantle-depth attenuating areas seem widespread in Kamchatka, lie above earthquake foci about 90 km deep in eastern Kamchatka and 150 km deep in the central depression, and occur within the region where uppermost mantle P wave velocities determined from explosion seismology are below 7.8 km/s (Fedotov and Tokarev 1973).

Grow and Qamar (1973) discovered a region of high P and S wave attenuation more than 25 km beneath Semisopochnoi volcano in the central Aleutian arc, at which the most recent cones contain mostly basalt and basic andesite (Coats 1959). Crustal level magma chambers are absent. Earthquake waves from foci between 100 and 250 km deep are strongly attenuated at stations on the volcano but not at stations > 20 km south or north. The attenuating zone could lie anywhere between 25 and 125 km depths, and might correspond to the region of decreased surface wave velocities below 70 km (Fig. 3.4).

Neither the Aleutian nor Kamchatkan data define the lower boundary of the attenuating zone, e.g., whether it extends to the dipping seismic zone. Grow and Qamar (1973) and Utnasin et al. (1975) note that their data are consistent with such an interpretation, but also may be explained by one or more isolated magma chambers at intermediate depths.

Seismic evidence for magma chambers such as that described above is difficult to obtain and often ambiguous. For example, Hedervari (1975) mapped a butterfly-shaped aseismic volume beneath Kirisima volcano, Kyushu, Japan, which is ≤ 285 km wide and ≤ 190 km deep, which he interpreted as a domain of magma generation. However, Minakami et al. (1969) found no anomalous P wave arrival times beneath Kirisima, whereas Kaminuma (1975) concluded there are small magma pockets in the mantle beneath the volcano.

Some orogenic andesite volcanoes demonstrably are not associated with attenuation zones, at least of significant size. Studies of Tofua volcano, Tonga (Mitronovas et al. 1969), two volcanoes in Vanuatu (Barazangi et al. 1974), and El Misti volcano, Peru (Sacks and Okada 1974) found no evidence of sub-volcanic seismic attenuation, although low-velocity zones at 9–12 and 36–46 km underlie the latter (Ocala and Meyer 1972).

Thus, on the basis of the limited data available, if shallow magma chambers (< 20 km deep) exist at convergent plate boundaries, they usually underlie volcanoes which have had historic eruptions of acid andesite or dacite. A deep crustal chamber (20 to 30 km) is known only at Katmai volcano, Alaska, which is a site of an historic rhyolitic ignimbrite eruption. Seismically identifiable magma reservoirs which extend into the upper mantle underlie volcanoes whose most recent eruptions are of basalt or basic andesite. These mantle-depth magma chambers underlie the volcanic front in the Aleutians and eastern Kamchatka, as well as regions behind the front in the central depression, Kamchatka. They seem to be separated from the dipping seismic zone by 10 to 35 km of aseismic mantle beneath Kamchatka, but the lower boundaries essentially are undetermined. However, the inferred presence of a 5% to 7% velocity contrast within a few kilometers of the slab/mantle boundary beneath both volcanic arcs and backarcs, and the decrease of b-values for upper plane seismicity beneath volcanic arcs (Sect. 3.4), strongly suggest the absence of a fluid phase (magma or aqueous fluid) within subducted oceanic crust but its presence in overlying asthenosphere, at least beneath northern Japan.

3.7 Conclusions

The physical processes beneath volcanic arcs are among the most complex on Earth, which compounds the problem of andesite genesis. Nevertheless, geophysical data from convergent plate boundaries strengthen the argument that orogenic andesite volcanism is intimately related to the presence of underthrust lithosphere. The parallelism between volcanic fronts and contours of the dipping seismic zone, the consistent depth of earthquake foci beneath volcanic fronts, and the changes in physical properties of underthrust lithosphere at this depth, all clearly demonstrate the important role of underthrust lithosphere in the genesis of orogenic andesites. This role would explain the conclusion in Chapter 2 that orogenic andesite usually is a dominant magma type only in areas beneath which lithosphere has been thrust to at least 70 to 100 km prior to, or at the time of, volcanism.

Volcanic fronts overlie earthquake foci 124 ± 38 km deep. These earthquakes may occur within the colder, more brittle interior of underthrust lithosphere rather than at the slab/wedge interface. Thus, underthrust oceanic crust lies about 100 to 125 km beneath volcanic fronts, at pressures between 30 and 38 kb.

Several observations imply that the depth to the dipping seismic zone beneath volcanic fronts reflects pressure-dependent processes (e.g., melting or dehydration of the slab due to negative dP/dT slopes of hydrous mineral stabilities) rather than temperature-dependent processes (e.g., shear strain heating). These observations include: the consistency of focal depths beneath volcanic fronts regardless of dip of the seismic zone; the lack of correlation between rates of underthrusting and depths to the dipping seismic zone beneath volcanic fronts; the importance of underthrust lithosphere as a heat sink; and the inability of the upper mantle to convect if its viscosity is high enough to permit significant shear strain heating.

Published estimates of geotherms at the slab/wedge interface are insufficiently unique or too model-dependent to predict whether or at what depth melting or dehydration of underthrust lithosphere will occur, although both phenomena are likely at some depth.

Differences between the mantle which is beneath forearc, volcanic arc, and backarc regions are important but poorly documented. Volcanic fronts in Japan coincide with steep heat flow and conductivity gradients. In some cases, volcanic arcs separate mantle with high density, velocity, and Q beneath forearc regions from mantle with lower density, velocity, and Q beneath backarc regions. However, the relationship between the volcanic arc and these contrasting mantle types is neither well known nor consistent between arcs. Nevertheless, in all island arcs there is evidence for at least small degrees of partial melting within the upper mantle wedge beneath either the volcanic arc or backarc region or both. For any given arc this evidence includes one or more of the following: low P or S wave velocities, a low velocity zone, high P or S wave attenuation, high conductivity, or low density. High Poisson ratios for the mantle wege have not been found yet.

Neither the spatial distribution nor the amount of magma indicated by these data is known, and evidence for the presence of magma generally is stronger for the mantle beneath backarc regions than beneath volcanic arcs. The strongest lines of evidence for magma above some slab/wedge boundaries are converted seismic waves which indicate ~5% velocity contrasts within a few km.

Once formed, magmas percolate rapidly upwards and, after aggregation, ascend at rates between 10^{-7} and 10^{-2} m/s, depending on whether they rise as buoyant spheres or via elastic crack propagation, respectively. The latter seems more likely during ascent from crustal-level reservoirs, and it may also apply to ascent within the mantle wedge if claimed patterns of earthquake precursors to andesitic eruptions are verified.

Discrete magma chambers exist beneath some orogenic andesite volcanoes. In general, shallow crustal chambers ($<$ 20 km) are associated with volcanoes having historic eruptions of acid andesite often following prolonged dormancy, and magma chambers in the mantle (40 to 100 km) are associated mostly with volcanoes having frequent eruptions of basalt or basic andesite. Magma chambers in the mantle occur beneath volcanoes both at and behind the volcanic front. Material within these chambers has been estimated to have a viscosity of about 10^{20} poise and to contain about 20% melt.

There are substantial differences in the thickness, structure, and age of crust on which active orogenic andesite volcanoes occur; however, they erupt on a Precambrian shield only in Peru. The lower crust beneath volcanoes could be basic or intermediate in composition, and beneath volcanic arcs may reach its melting point. Effects of this melting will be subtle since the wall rocks may be similar in bulk and isotopic composition to the magmas. Olivine or amphibole-dominated cumulates are uncommon in the crust, according to velocities determined in explosion seismic studies.

Thus, while geophysical data confirm the close relationship between orogenic andesites and underthrusting of lithosphere, they do not yet constrain the nature of the relationship. Consequently we must now turn to the geology and geochemistry of the andesites themselves.

Chapter 4   Andesite Magmas, Ejecta, Eruptions, and Volcanoes

Studies of the tectonic significance, petrography, and chemical composition of andesites have been more common than detailed investigations of the physical properties of andesite magma, the volcanologic characteristics of its eruption, or the geologic history of its eruptive sites. Indeed, these features sometimes are regarded as of little significance for studies of andesite genesis, as opposed to near-surface phenomenology. However, as will be shown in this chapter, the volcanologic and geologic contexts of andesite seem to favor its origin by low-pressure differentiation of basalt, and the physical properties of andesite magma probably control these differentiation processes.

4.1 Characteristics of Andesite Magma

4.1.1 Temperature

The temperature at eruption of andesite and other magmas primarily is a function of nonvolatile composition, water content, degree of crystallization, and heat liberated during eruption. Temperatures measured in the field during eruption of andesite magma are 1050 to 1100°C (Table 4.1). Presumably this underestimates liquidus temperatures slightly as the erupted lavas contained crystals. Laboratory measurements of andesite liquidi are shown in Fig. 4.1. At atmospheric pressure, liquidi are 1180° to 1250°C and solidi are ~ 1050°C. To achieve agreement between observed eruption temperatures and experimentally determined liquidi (Fig. 4.1) requires pre-eruption water contents of orogenic andesite magmas of 2 to 5 wt.% H_2O (point 5, Sect. 5.3.1).

Temperatures of equilibration can be calculated for mineral constituents in andesite if the temperature dependence of element distribution between phases is known. The mineral geothermometers which can be applied to andesites, with calibration uncertainties of about ± 50° to 100°C, utilize the composition of co-existing oxides (Sect. 6.5), co-existing augite and orthopyroxene (Wood and Banno 1973; Wells 1977); pigeonite (Ishii 1975), or crystal-melt pairs involving olivine (Bender et al. 1978; Watson 1979a; Ghiorso and Carmichael 1980), plagioclase (Kudo and Weill 1970; Mathez 1973; Ghiorso and Carmichael 1980), and hornblende (Helz 1979). Currently, the co-existing oxide and pyroxene geothermometers seem the most reliable; typical results for them are given in Table 4.1.

Table 4.1. Temperatures of andesite magmas

	% SiO$_2$	T (°C)	Reference
A. Field measurements			
1. Paricutin, 1943	55	1070–1110	Wilcox (1954), Zies (1946)
2. Akita-Komagatake, 1971	59	1090	Yagi et al. (1972)
3. Hekla, 1970	54	1050	Thorarinsson and Sigvaldason (1972)
B. Fe-Ti oxides			
1. New South Wales	61–63	1000–1050	Wilkinson (1971)
2. Lassen, USA	55	1000	Marsh and Carmichael (1974)
3. Asama, 1783	55–61 +	990–1020	Sekine et al. (1979)
4. Rabaul, PNG	57, 61 +	995, 870–970 +	Heming and Carmichael (1973)
5. Rio Grande Rift, USA	56–60 +	890–1080	Zimmerman and Kudo (1979)
	61–63 +	670–950	Zimmerman and Kudo (1979)
6. Shirouma-oike, Japan	60 +	810	Sakuyama (1978)
7. Namosi, Fiji	55–58 +	1000–1050	Gill and Till (1978)
	60–65 +	850–980	Gill and Till (1978)
8. Colima, Mexico	61 +	986	Luhr and Carmichael (1980)
C. Coexisting pyroxenes (a = Wood and Banno 1973; b = Ishii 1975)			
1. New Zealand	56–62	990–1085 [a]	Ewart (1971) *
2. Tonga	53–55	1050–1080 [a]	Ewart (1976b)
	57–63	950–1000 [a]	Ewart (1976b)
3. Rabaul	55	970–1050 [a]	Wood and Banno (1973), Heming (1974)
4. Akita-Komagatake	59	1000–1030 [a]	Aramaki and Katsura (1973) *
5. Bandai, Japan	–	970–990 [b]	Nakamura (1978)
		1064–1078 [a]	Nakamura (1978)
6. Rio Grande Rift, USA	56–62	950–1095 [a]	Zimmerman and Kudo (1979)
7. Colima	56 +	1020–1050 [a]	Luhr and Carmichael (1980)
	61 +	960–1010 [a]	Luhr and Carmichael (1980)
D. Oxygen isotopes			
1. Asama	–	900	Matsuhisa (1979)
2. Colima	56 +	1030–1140	Luhr and Carmichael (1980)
	61 +	880–1050	Luhr and Carmichael (1980)

+ Hornblende andesite. * Calculated from data therein

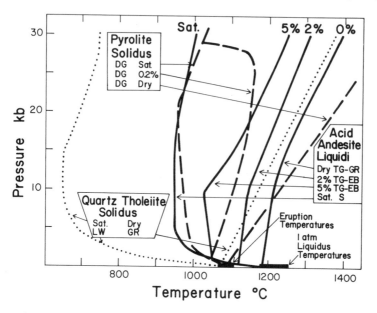

Fig. 4.1. Temperatures of andesite magmas. Eruption temperatures are from Table 4.1. One atm liquidus temperatures are from Aramaki and Katsura (1973), Sekine et al. (1979), Eggler (1972a), Brown and Schairer (1971), Kubovics (1973), and Murase and McBirney (1973). Abbreviations used for sources of higher pressure solidi and liquidi are as follows: *TG* (Green 1972); *GR* (Green and Ringwood 1968a); *EB* (Eggler and Burnham 1973); *S* (Stern et al. 1975); *DG* (Green 1973); *LW* (Lambert and Wyllie 1972). % *figures* refer to wt.% H_2O; *Sat* means water-saturated

Oxygen isotope thermometry is available for only two sites of orogenic andesite (Table 4.1), and the temperature-dependence of trace element distribution coefficients in orogenic andesite magmas is not yet well-calibrated. Temperatures implicit in the distribution coefficients of Table 6.3 are about 950° to 1250°C, using the range of available calibrations assembled by Irving (1978).

From available data, therefore, pyroxene basic andesites typically equilibrate at 1000° to 1100°C whereas hornblende acid andesites equilibrate at 900° to 1000°C.

It is unclear to which portion of the melting interval these calculated temperatures apply or whether all phases in a sample were quenched at a common temperature. The temperature calculated from coexisting oxides within the porphyritic 1970 andesites of Hekla volcano, Iceland (Baldridge et al. 1973) are 15° lower than the observed eruption temperature (Thorarinsson and Sigvaldason 1972). Differences between temperatures calculated from plagioclases and pyroxenes or oxides may reflect the actual width of the melting interval (Ewart 1976b; Heming 1977) or misjudgement of water pressure (Sect. 5.3.1), as well as calibration inaccuracy.

Too few reliable data exist to interpret clearly the apparent differences in temperature between andesite magmas, but two observations merit comment. First, internally consistent comparisons of temperature data require that andesite liquids contain modest pre-eruption water contents (point 4, Sect. 5.3.1). Anhydrous liquidus temperatures at atmospheric pressure for suites of basalt, basic and acid andesite, and dacite from Parícutin, Mexico, and from the Antilles are shown in Fig. 4.2. Unlike results from tholeiitic or alkalic basalts from Hawaii, results from the andesite-dominated suites show no regular decrease in liquidus temperature with increasing Fe/Mg ratios or silica contents. This irregularity is especially dramatic in the St. Lucia data and is attributed to the high anhydrous melting point of plagioclase. These high liquidus temperatures clearly rule out derivation of the andesite or dacite from basalt by fractional crystallization during cooling at 1 atm pressure because andesites constitute a thermal barrier between basalt and rhyolite under these conditions due to their high normative plagioclase

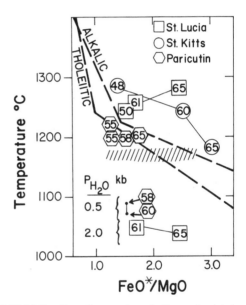

Fig. 4.2. Liquidus temperatures versus FeO*/MgO ratios of orogenic andesites and related rocks from the Antilles and Mexico. *Light lines* connect members of a suite. *Numbers inside symbols* given wt.% SiO_2 of sample. *Dashed lines* are trends for Hawaiian rock series. Note that 1 atm liquidus temperatures of volcanic arc suites correlate much less well with FeO*/MgO ratios or SiO_2 contents, as differentiation indices, than do the Hawaiian series. At 1 atm the liquidus phase in the rocks from volcanic arcs is plagioclase; the first of appearance of pyroxene in most of the samples lies in the region shown by *diagonal lines*. Also given at the *bottom of the figure* are liquidus temperatures for four of the same rocks at 0.5 or 2.0 kb under water-saturated conditions. Data are from Brown and Schairer (1971) for St. Lucia and St. Kitts, and from Tilley et al. (1968) and Eggler (1972a) for Parícutin

contents. In contrast, note in Fig. 4.2 that temperatures in the St. Lucia samples decrease with increasing Fe/Mg ratio and silica contents at P_{H_2O} = 2 kb, indicating that the samples could be cogenetic if they evolved at such water pressures.

Second, there may be consistent differences in temperature between calcalkaline and tholeiitic andesite magma. Kuno (1950a, 1968a) noted that differences in Fe-enrichment trends in Japanese andesites often were accompanied by differences in the crystal symmetry of Ca-poor pyroxenes in the groundmass which he attributed to crystallization of tholeiitic andesite at higher temperatures (see Sect. 6.2). Equilibration temperature data are insufficiently abundant or accurate to test this suggestion. For example, distribution coefficients D_{Sr}^{pl} measured for phenocryst/matrix pairs are higher (indicating lower temperatures) for calcalkaline andesite from New Zealand (Ewart et al. 1968; Ewart and Taylor 1969) that for tholeiitic andesite from Tonga (Ewart et al. 1973), but no such difference exists between a tholeiitic andesite from Osima volcano and a calcalkaline andesite from Aso volcano, Japan (Schnetzler and Philpotts 1970). Both these comparisons utilize data from the same analytical methods and laboratories. No consistent differences in temperature between New Zealand and Tongan andesites are indicated by D_{Ni}^{cpx} data or by the composition of coexisting pyroxenes (Table 4.1).

4.1.2 Density

The density (ρ) of orogenic andesite liquids within the crust is between 2.4 and 2.8 cm^{-3} and depends primarily on bulk composition, especially iron and water contents, and on pressure. Experimental determinations of ρ are about 2.45 g cm^{-3} at 1000° to 1200°C (Fig. 4.3), which is 0.1 to 0.3 g cm^{-3} lower than values calculated for average andesite compositions (including H_2O) using either Bottinga and Weill's (1970) or Nelson and Carmichael's (1979) molar volume data. Such calculated andesite densities differ amongst themselves, with tholeiitic or basic or low-K andesites being 0.1 g cm^{-3} heavier than corresponding calcalkaline or acid or high-K ones (e.g., Table 5.1). Oxidation of all iron to Fe_2O_3 reduces andesite densities by 0.15 g cm^{-3} on average, whereas addition of 5 wt.% H_2O reduces their densities by 0.3 g cm^{-3}. Increasing pressure from 5 to 15 kb increases ρ by 0.3 cm^{-3} (Fig. 4.3).

Liquid densities increase with fractionation in some tholeiitic series due to iron-enrichment. If this magma series also is characterized by low water contents, then the density of basic tholeiitic andesite magma within the crust will be close to, but still less than, that of bytownite crystals. In contrast, magma densities decrease with fractionation in the calcalkaline magma series due to coupled iron-depletion and water-enrichment. All phenocryst minerals are denser than these magmas within the crust.

At pressures greater than about 12 kb, however, labradorite may float even in calcalkaline acid andesite liquid if it contains less than approximately 3 wt.% H_2O

Fig. 4.3. Densities of orogenic andesite melt and phenocryst minerals. Note change in scale at 3 g/cc. *Solid dots* are for anhydrous melts at 1 atm from Murase and McBirney (1973). *Crosses* are for liquids containing 2.5 to 3.0 wt.% H_2O from Kushiro (1978); *adjacent numbers* give pressure in kb. All liquids contain 60 wt.% SiO_2 and have FeO*/MgO ~ 1.5. Densities of tholeiitic basic andesites (e.g., 55% SiO_2, 2.5 FeO*/MgO) are about 0.15 g/cc higher. Crustal densities estimated from seismic velocities are shown by a *shaded area* where UC and LC indicate upper and lower sialic crust, respectively; see Section 3.2. Mineral densities are from Deer et al. (1963) for 25°C; typical ranges of phenocryst compositions in orogenic andesites are shown by *bars*

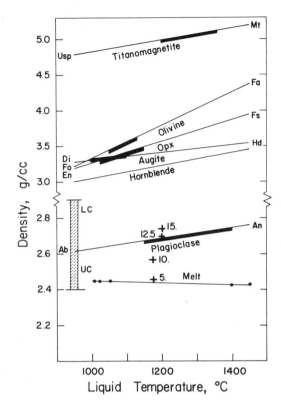

(Fig. 4.3). Consequently, permissive evidence that early ejecta during some eruptions of calcalkaline basic andesites are richer in plagioclase crystals than are later ejecta (Sect. 4.3) indicates either that the magmas in question ascended from deep reservoirs or that accumulation is accomplished by forces other than buoyancy of plagioclase, e.g., convection or vesiculation.

As noted in Section 3.2, the density of crust ranges from 2.4 to 2.9 g cm³ in the vicinity of andesite volcanoes. Thus, andesite liquid can ascend through the mantle and lower crust simply due to buoyancy, but may stagnate gravitationally in the upper crust depending on its water and iron content. The maximum height to which andesite magma ascends (i.e., summit elevations of stratovolcanoes) will depend largely on this density contrast, integrated vertically for the magma reservoir.

4.1.3 Rheology

The rheology of andesite magma is known in reconnaissance fashion (Murase and McBirney 1973; Scarfe 1973; Kushiro et al. 1976; Kushiro 1978). Like other magmas, andesite probably behaves as a Newtonian fluid above its liquidus and as a pseudoelastic substance, e.g., Bingham body, within its melting interval. In Newtonian fluids, viscosity (η_N) of a given substance at constant temperature and pressure is independent of shear stress (τ) and is given by

$$\eta_N = \frac{\tau}{d\epsilon/dt} ,\tag{4.1}$$

where ϵ is shear strain and t is time. In contrast, Bingham bodies have finite yield strength (τ_y) which must be exceeded before viscous flow occurs, so that

$$\eta_B = \frac{\tau - \tau_y}{d\epsilon/dt} .\tag{4.2}$$

When $\tau > \tau_y$, Newtonian and Bingham bodies behave similarly.

Shaw (1969) and Murase and McBirney (1973) argued that basalt magma is a Newtonian fluid above its liquidus but becomes increasingly non-Newtonian in behavior below. About 70°C below the liquidus (i.e., at about 1135°C and 30% crystallization), basalt magma may behave as a Bingham body with $\tau_y \sim 10^3$ dyne/cm^2, but at still lower temperatures the viscosity becomes a nonlinear function of ($\tau - \tau_y$) so that basalt magma mush is a plastic fluid whose rheology is more complicated than that of Bingham body. Although τ_y has not been measured experimentally for andesite magma, it has been estimated from andesite flow geometry to be 10^4 to 10^5 dyne/cm^2, to be an order of magnitude or more higher than for basalt, and to increase as temperature decreases (Hulme 1974; Moore et al. 1978; Pinkerton and Sparks 1978). Thus, if andesite magmas have yield strengths of 10^4 to 10^5 dyne/cm^2 at the low strain rate experienced by a magmatic suspension in a reservoir for 10 to 10^5 years, then this strength will be a significant barrier to gravitational separation of phenocrysts, especially plagioclase phenocrysts, from andesite because the net vertical force on the crystals is considerably less than the yield strength of the liquid.

The viscosity of andesite magma depends primarily on its temperature plus its water, crystal, and bubble content. Changing pressure from 1 atm to 20 kb decreases andesite viscosity only twofold (Kushiro et al. 1976; Kushiro 1978) whereas viscosity increases tenfold simply over the 150°C interval between precipitation of plagioclase and pyroxene from anhydrous andesite. As illustrated by Fig. 4.4, this temperature dependence is exponential and follows the relationship

$$\eta = A \cdot e^{-E_\eta/RT},\tag{4.3}$$

where A is a constant, E_η is the activation energy for viscous flow, and R is the gas content. E_η for orogenic andesites is about 68 to 75 kcal/mol (Murase and McBirney 1973; Scarfe 1973). Water is another important variable. For a highly polymerized liquid such as andesite, addition of about 3 wt.% water decreases viscosity by two orders of magnitude at about 1150°C (Fig. 4.4), a decrease which

Fig. 4.4. Viscosity of orogenic andesite magma. *Solid dots* are for an anhydrous melt at 1 atm for which the plagioclase liquidus and pyroxene-in temperatures are shown (Murase and McBirney 1973). The slope at $T < 1100°C$ yields 75 kcal for E_η. Data from Kushiro et al. (1976) and Kushiro (1978) are shown by *open symbols,* with pressure indicated. *Cross* is from Scarfe (1973)

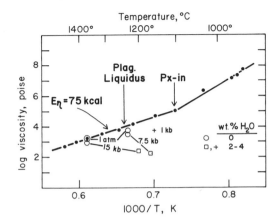

is accentuated at lower temperatures because activation energies diminish with increasing water contents (Shaw 1963). Finally, viscosity also increases exponentially as crystal and bubble contents increase through the crystallization interval, as seen below 1100°C in Fig. 4.4.

Experimentally determined near-liquidus viscosities are about 10^2 to 10^4 poise from 1 atm to 15 kb for andesite melt containing 4 to 0 wt.% H_2O, respectively (Fig. 4.4). Five published field estimates of the viscosity of andesite lava during eruption range uniformly between 10^4 and 10^8 poise at the vent, being higher for acid than basic andesites (Rose 1973; Hulme 1974). From frequency-amplitude ratios of seismic waves attenuated in shallow chambers beneath Trident volcano, Alaska, Matumoto (1971) estimated the viscosity of material in the chamber as 10^8 poise. Note that these field estimates are ambiguous and especially dependent on vesiculation.

In summary, andesite magmas containing 1 to 5 wt.% H_2O at or below their liquidi will have effective viscosities between 10^2 and 10^8 poise within the crust prior to eruption. However, near-liquidus basalt to andesite melt with 0.5 to 2 wt.% H_2O probably has an effective viscosity of only 50 to 1000 poise during ascent through the mantle wedge. The crystal-bearing magma has significant Bingham yield strength. Its rheological properties range from being essentially a Newtonian fluid to Bingham body to more complicated plastic fluid over its crystallization interval.

4.1.4 Miscellaneous Properties and Applications

The range in bulk and volatile composition of orogenic andesites leads to considerable diversity in liquid structure and, consequently, in properties such as viscosity

which are functions thereof. However, all anhydrous andesites are highly polymerized liquids in which about 90% of the oxygen atoms are shared as bridges
between adjacent $(Si,Al)O_4$ tetrahedra, resulting in fewer nonbridging oxygens
(NBO) than tetrahedrally coordinated cations (T); e.g., all anhydrous andesites
have $NBO/T \sim 0.3$.

The effect of this diversity on petrogenetically important attributes such as
chemical and thermal diffusion coefficients or rates of crystal nucleation and
growth is largely unexplored. Chemical diffusion coefficients in andesite liquid
can be estimated as between 10^{-7} and 10^{-8} cm^2/s at 1000°C (see Hofmann and
Magaritz 1976), and will be higher in more hydrous liquids for some elements, but
have not been measured.

Also, the mechanical regime and thermal or mass balance of andesite magma
reservoirs generally is unknown, primarily due to difficulties in constraining
boundary conditions and evaluating parameters. Although some success has been
achieved in studies of basalt lava lakes (e.g., Shaw et al. 1968; Wright et al. 1976),
rhyolite magma chambers (e.g., Ewart et al. 1975), and some general cases (Bartlett 1969), the only attempts to model andesitic systems are by Marsh (1978) and
Rose et al. (1978).

Preparation of such models is made difficult by inadequate knowledge of the
shape and dimensions of the magma bodies, the temperature and compositional
gradients therein, and temperature and mass flux between magma and walls, as
well as inadequate knowledge of rheologic behavior. For example, whether magma
cools as a static or convecting liquid depends on its Rayleigh number which, in
turn, depends on its viscosity, its temperature gradient, and the fourth power of
its dimension. Figure 4.5 illustrates this relationship and indicates that convection
is likely in andesite magma chambers over a few meters in thickness (Bartlett 1969).
Similar dependence on dimensions and thermal regime determines whether flow
is laminar of turbulent; laminar flow is more likely. Data are insufficiencient to
constrain realistic models of these kinds of behavior in andesite magma chambers.

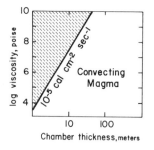

Fig. 4.5. Magma convection-stability limits in horizontal
intrusions cooled from above where the heat flux to the
surface is 10^{-5} cal cm^{-2} s^{-1}. Convection occurs only in
the *unshaded region*. If heat dissipates in all directions,
conditions become still more favorable to convection.
Yield strength will retard convection to an unknown
degree. Figure from Bartlett (1969)

Nevertheless, three general observations regarding andesite magma behavior seem pertinent.

First, if magma is static (nonconvecting), gravitational separation of roughly equidimensional crystals from a Newtonian liquid by settling or flotation will occur according to the Stokes equation. Although absolute and relative grain sizes of andesite phenocrysts vary widely, the generally inverse relationship between densitiy and grain size results in roughly similar settling velocities of about 0.3 m/ day for most minerals except hornblende, which when large can be separated more quickly. This relationship is illustrated in Fig. 4.6. (In a Bingham body the yield strength must be exceeded before the crystal separates. As noted earlier, this may be a significant qualification but the degree of significance is unknown.)

If convection occurs, gravitational separation will be more complicated and may be prevented altogether. Bartlett (1969) treated cases probably realistic for andesite magma chambers and concluded that even large (~5 mm) crystals with low density contrasts (e.g., plagioclase) would *not* separate gravitationally from convecting magma having viscosity $> 10^6$ poise, whereas even small crystals (~ 0.1 mm) with high density contrasts (e.g., magnetite) would separate effectively from convecting magma with viscosity of 10^4 poise. Clearly the efficiency of gravitational crystal-liquid fractionation will vary continuously from very low to very high within andesite magmas, depending on conditions which no one has

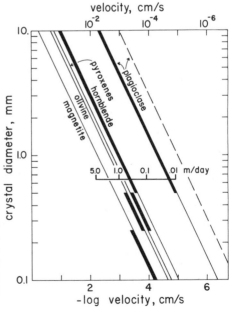

Fig. 4.6. Crystal settling rates in static orogenic andesite magma, assuming magma has negligible yield strength. *Solid lines* represent calculations using Stokes law for conditions applicable to magma at about 8 kb, 1100°C, where ρ (melt) = 2.5 g/cc, ρ (plagioclase) = 2.7, ρ (pyroxenes) = 3.3, ρ (olivine) = 3.5, ρ (hornblende) = 3.2, ρ (magnetite) = 5.0 and η (magma) = 1000 poises. *Bars* show typical phenocryst diameters in andesites. *Dashed line* shows results which are more likely to apply to plagioclase where, in addition to the preceding assumptions, allowance is made for nonsphericity of crystals and the increase in effective viscosity due to the presence of 20% crystals (see Walker et al. 1976b)

yet constrained realistically. Conversely, convection may promote crystal-liquid separation by concentrating crystals along magma chamber margins.

Second, if andesite magmas behave as Bingham bodies, then laminar flow within a cylinder (e.g., conduit) has interesting properties. Specifically, in the cylinder center where shear stress is less than yield strength, velocities will be high and relatively constant across the cylinder interior, and differential flow will not occur ("plug-flow" conditions). Under these conditions, if over $\sim 8\%$ phenocrysts are present, they will migrate toward the conduit interior (Komar 1972). Thus, while the yield strength of andesite magma may prevent gravitational differentiation, it promotes flow differentiation and may explain the putative bimodality of phenocryst concentrations in andesites (Ewart 1976a).

Finally, the cooling rate of magma is increased by convection. Although cooling rates depend significantly on parameters such as thermal and mass balances which are unknown for andesite reservoirs, the time interval between emplacement and complete solidification probably is on the order of 10^5 to 10^6 years for spherical magma bodies about 10 to 20 km in diameter, depending on whether magma is static or convecting (Jaegger 1964; Shaw 1974; Spera 1980). Shaw (1974) argued that in a magma chamber kilometers in diameter, convection will occur until the chamber is at least half crystallized, i.e., for about the first 25% of total solidification time. Increased viscosity prevents continued convection.

4.2 Andesite Rock, Eruption, and Edifice Types

The physical appearance of andesites, the manner in which they are erupted, and the edifices they constitute, all depend primarily on the physical properties of andesite magmas just described and on their gas contents. Most andesites erupt as lava flows or tephra from stratovolcanoes or cinder cones.

Andesite magma flows usually form block lavas, i.e., lava flows which are covered with angular fragments, are a few meters to scores of meters thick, have a hummocky surface, and move slowly (tens of feet per day) when over gentle slopes. Andesite block lavas frequently are characterized by flow laminations due to shear failure of the lava. This process often yields joints a few centimeters to tens of centimeters thick which crudely parallel the flow boundaries and give the rock a platey appearance. These joints sometimes provide evidence of flow direction. Andesite block lavas also frequently involve extensive auto-brecciation and resorption of fragments during magma movement, especially when lava contacts water.

Basic andesites also erupt as subaerial aa flows and submarine pillow lavas. Examples of the former include the 1974 to 1975 eruption of basic andesite at Karkar volcano, Papua New Guinea; examples of the latter have not been witnessed but pillow andesites are common in island arcs.

Many kinds of lithic and vitric andesite tephra exist in all sizes. The proportion of tephra to lava varies greatly during eruptions and from one eruption to another, but generally increases from basic to acid andesite. [However, Baker and Holland (1973) found the proportion of basic to acid andesite higher in tephra than in presumably coeval lavas from Mt. Misery volcano, Antilles, which indicates either higher explosivity of basic than acid andesite there or eolian differentiation of tephra.] Usually lavas and tephra, especially tephra carried in nuées ardentes, from the same eruption are similar in composition (e.g., Moore and Melson 1969; Melson et al. 1972). However, eolian differentiation of air-fall ash occurs such that tephra usually become more silicic with distance from the volcano.

Texturally, most orogenic andesites are porphyritic, usually containing 20 to 50 vol.% phenocrysts (Ewart 1974a, Fig. 2). Phenocrysts, particularly of plagioclase and hornblende, often are large, commonly > 5 mm long. They occur separately or in glomeroporphyritic clumps. All variations in crystal development occur although most phenocrysts are tabular and euhedral to subhedral; quench textures occur only in boninite varieties. Multiple populations of crystal grain sizes often occur, implying either multiple circumstances of nucleation or discontinuous and rapidly decreasing cooling rates (Walker et al. 1976a).

Although Eastern Europeans in particular have considerable experience in melting and recrystallizing andesites for industrial purposes (e.g., Kubovics 1973), no studies relating andesite textures to magmatic histories or to experimentally determined cooling rates have been published. Árakai (1968) inferred that his typical pyroxene basic andesites cooled at $10°$ to $10,000°C/yr$ with mineral growth rates decreasing in the order plagioclase > pyroxene > magnetite, but without experiments. For comparison, lunar basalts with plagioclase phenocrysts 5 mm in diameter apparently crystallized at about $10^2 °C/yr$. Bogoyavlenskaya and Dukik (1969) noted secular changes in mineral grain size during eruption of andesite from Bezymianny volcano, Kamchatka. After the paroxysmal eruption of March, 1956, a second, smaller generation of both plagioclase and pyroxene phenocrysts joined those already present, and mesostases became less glassy. No rock or mineral analyses were given.

The groundmass of most andesites is pilotaxitic (trachytic) or hyalopilitic, i.e., the groundmass contains parallel plagioclase microlites oriented in the direction of flow in a glassy or crystalline mesostatis, or contains brown glass in which plagioclase microlites are randomly oriented in interstices between phenocrysts. Basic andesites also may have intergranular or intersertal textures in which the proportion of glass is less than in hyalopilitic cases. Aphyric andesites are rare and seem most common in low-K suites, perhaps due to their relatively anhydrous character.

Andesitic eruptions vary considerably in character. Many basic andesites are produced by Strombolian-type eruptions, i.e., by rhythmic ejection of tephra associated with aa or block lava discharge that continues for months or years with variations in amount and proportion of effusive products. The 1943 to 1952 eruption of Parícutin volcano, Mexico, is a good example. Acid andesites also may erupt in this fashion, e.g., the 1970 eruption of Akita-Komagatake volcano, Japan. Indeed, the 1943–1952 eruption of Parícutin (U.S.G.S. Bulletin 965), the 1970 eruption of Akita-Komagatake (Bull. Volc. Soc Japan v. 16, no. 2/3), and the 1970 eruption of Hekla (Thorarinsson and Sigvaldason 1972; Sigvaldason 1974) are the best-studied andesite eruptions on record.

Other basic and acid andesite tephra and subordinate lavas are produced by violent Vulcanian, Peléean, and Plinian-type eruptions. In the former, explosive ejections of blocks and ash sometimes are accompanied by minor, stubby lava flows. Peléean- and Plinian-type eruptions are characterized additionally by nuées ardentes which accompany caldera collapse in the latter case. Many of the most violent volcanic eruptions of the past century have produced acid andesite to dacite: e.g.,Coseguina (Nicaragua),1835; Krakatoa, 1883; Pelée, 1902; Mt. Lamington, 1951; Bezymianny, 1956; Mt. St. Helens, 1980. These explosive phenomena can also accompany eruption of basic andesite. For example, the 1970 eruption of Ulawun volcano, New Britain, included several nuées ardentes of 52% to 53% SiO_2 tholeiitic basalt (Melson et al. 1972), and the 7 km long nuée during the 1968 eruption of Mayon volcano, Philippines, was of 55% SiO_2, medium-K, calcalkaline, basic andesite (Moore and Melson 1969). Frequently, nuées ardentes are accompanied by mudflows of volcanic ash (lahars) which develop due to the activity of water derived from crater lakes and streams, snow and ice on the volcano, or heavy rains during the eruption. (Mudflows also occur independently of eruptions.)

Peléean and Plinian-type eruptions are often followed by development of a resurgent dome within the volcanic caldera (i.e., endogenous dome). Usually these domes and associated lava flows are dacitic or rhyolitic in composition, though occasionally they include subordinate acid andesites (e.g., Santiaquito dome within Santa María volcano, Guatemala: Rose 1972a).

Explosive eruptions of andesite and other magmas are caused by rapid expansion of a separate gas phase with which the magma is saturated at low pressure. However, neither the proportion of magma which is vapor-saturated nor the duration of saturation is known. Some volcanologists attribute violently explosive eruptions to crystallization within a closed magma volume leading to exsolution of volatiles and build-up of gas pressures high enough to exceed the strength of surrounding rock, causing eruption. This sequence is suggested by the frequently long repose of volcanoes preceding these eruptions, the decrease in explosivity during many eruptions, and in some cases the more fractionated composition of magma initially erupted. For example, Thorarinsson (1967) noted that the longer the quiescence of the nonorogenic andesite volcano Hekla, Iceland, the higher the gas and silica content of the earliest magmas. If this mechanism is a cause of explosive eruptions, then andesite magmas, especially acid andesite magmas in the upper parts of magma reservoirs, spend some of their evolutionary history in vapor-saturated conditions.

Alternatively, explosivity may result from exsolution of gas from magma only during final ascent. In this case, eruption is not initiated by excess gas pressure but by unloading due to rock failure, by regional compression due to an independent cause such as plate movement (Sect. 4.7), by tides, or whatever. This may lead to lava fountaining or to Strombolian-type eruptions and lava vesiculation in low-viscosity magmas, but also may lead to violent eruptions if viscosity and surface tension are high enough to retard vesiculation (McBirney 1973) or due to other kinetic factors (Verhoogen 1951). A second alternative explanation is that explositity results mostly from rapid heating of meteoric water near the vent resulting in phreatic activity (McBirney 1973; Melson and Saenz 1973). If either of these alternative explanations apply, andesite magma will evolve in vapor-under-saturated conditions during all but the finale of its history.

Andesites are erupted from edifices whose form reflects the types of eruptions and ejecta described above. These edifices can be classified as monogenetic if they formed during a single, small-volume eruption, or as polygenetic if due to repeated eruptions during which a large volume of magma (1 to 100 km^3) was vented through the same conduit (Rittman 1962). Monogenetic edifices include cinder cones such as Parícutin which grew to about 350 m height in three years (mostly in one week). Cinder cones consist largely of tephra, lack sufficient strength to achieve heights above about 350 m, usually are circular in plan view to eruption from a central vent, and are produced by Strombolian or Vulcanian-type eruptions. The more common andesite edifices are the large polygenetic composite cones or stratovolcanoes which bestow at least a visual majesty on this book's otherwise prosaic topic. Summaries of their topographic dimensions are available (Pike 1978; Wood 1978). Stratovolcanoes consist of both lava and tephra in variable proportions, range in height from 1100 to 1600 m above base elevation on average, are circular to elongate in plan with basal diameters 8 to 10 km on average, are produced by Strombolian, Vulcanian, and Peléean-type eruptions, and occasionally are destroyed by Plinian-type catastrophes. They record the complex history of several tens of km^3 of magma erupted from a common conduit during 10^4 to 10^5 years.

The height of these stratovolcanoes may be significant. For example, Rittman (1962) and McBirney (1976) have noticed a correlation between magma composition or modal mineralogy and volcano summit elevation: i.e., magmas become more siliceous, olivine decreases, and hornblende and biotite increase in abundance with increasing elevation of active orogenic andesite volcanoes. McBirney attributed this to increased crystal fractionation in areas of thickened crust which, therefore, have higher base elevations due to isostacy. Similarly, Rose et al. (1977) argue that for any area there exists a critical height above which magmas do not rise. This is analogous to a hydrostatic head which, when reached by an edifice, results in dormancy or flank eruptions, and low-pressure differentiation of magma. Wherever true, these observations imply that andesites result mainly from low-pressure differentiation processes and that nucleation of hornblende is a late-stage phenomenon resulting from increased water pressure accompanying prolonged crystallization.

4.3 Variations in Magma Composition During and Between Historic Andesite Eruptions

Variations in explosivity and, consequently, lithologic character of ejecta during eruptions reflect variations in the physical properties and gas contents of magmas. As discussed above, reasons for these complex vicissitudes are ambiguous; their main potential significance for studies of andesite genesis and evolution concerns the degree of vapor-saturation of magma during its aging. Often these variations in eruption type are unaccompanied by variations in magma composition. However, variations in bulk composition or mineralogy during eruptions or between closely spaced eruptions may have broader implications.

Four kinds of such temporal variations occur. First, during some eruptions magmas become more differentiated with time, as illustrated by the 1943 to 1952 eruption of Parícutin volcano, Mexico (Foshag and Gonzalez 1954; Wilcox 1954). Medium-K, calcalkaline, basic andesite of relatively uniform composition and mineralogy was erupted during Strombolian-type activity for about three years. Thereafter, magma gradually became more acidic with time, lost first plagioclase and then olivine phenocrysts, and gained orthopyroxene phenocrysts. No major change in style or site of eruption, nor in $^{87}Sr/^{86}Sr$ ratios (Tilley et al. 1967), accompanied these changes in magma composition and mineralogy, which are shown in Figure 4.7. Wilcox (1954) attributed these variations to effects of low-pressure crystal fractionation plus crustal assimilation in convecting magma initially of basic andesite composition. However, he believed these variations existed prior to 1943 so that the eruptive sequence was fortuitous.

The 1974 eruption of Karkar volcano provides a less dramatic example (McKee et al. 1976). Strombolian-type eruption of medium-K, tholeiitic, 53% to 54% SiO_2 andesite of nearly uniform composition occurred for 135 days; after a 25-day lull, eruption resumed. Maximum eruption rates occurred after the lull when lavas contained fewer plagioclase phenocrysts and higher concentrations of incompatible elements than those erupted earlier (see comparison in Table 4.2). In both the Karkar and Parícutin examples, the later and more differentiated magmas had fewer or no plagioclase phenocrysts. This change suggests either that early eruptions tapped areas of plagioclase accumulation or that later eruptions tapped regions of higher P_{H_2O} where plagioclase was more soluble (Sect. 6.9).

Second and more common are examples in which later magmas become more mafic. For example, from 1968 to the present, lavas of Arenal volcano, Costa Rica, have become less differentiated (Table 4.2) and less explosive (Melson and Saenz 1973). A similar sequence has occurred frequently during eruption of nonorogenic basic andesite from Hekla volcano, Iceland (Thorarinsson 1967). Sometimes hornblende is a phenocryst during initial eruption of acid andesite but not in later ejecta, e.g., the Arenal eruption cited above; the 1956 eruption of Bezymianny volcano, Kamchatka (Bogoyavlenskaya and Dukik 1969); and the Mt. St. Helens example noted in the next section.

Table 4.2. Chemical and mineralogical composition of successive ejecta during single eruptions of andesite

	Karkar [a]		Arenal [b]		
	Before Lull	After Lull	1968	1970	1973
N	5	3	1	1	1
SiO_2	53.5	54.6	56.0	54.0	54.4
TiO_2	0.71	0.81	0.56	0.61	0.72
Al_2O_3	19.5	15.9	20.2	20.1	19.4
Fe_2O_3	2.6	3.5	3.8	4.4	3.0
FeO	6.1	7.6	3.1	3.3	4.7
MnO	0.16	0.20	0.14	0.20	0.16
MgO	3.1	4.1	3.1	4.2	4.8
CaO	10.5	9.0	9.1	9.6	9.3
Na_2O	2.5	2.8	3.4	3.1	3.0
K_2O	0.88	1.02	0.66	0.60	0.52
P_2O_5	0.16	0.20	0.17	0.12	0.14
H_2O	0.05	0.07	0.05	0.05	< 0.16
Phenocrysts:					
Pl	33	13	Not available		
Ol	0.5	0.5			
Cpx	1.2	1.0			
Mt	<0.5	<0.5			
Total	35	15			

a   Karkar volcano, Papua New Guinea, 1974–1975. McKee et al. (1976)
b   Arenal volcano, Costa Rica. Melson and Saenz (1973)

Likewise, changes to more mafic compositions during apparently single large-volume, but prehistoric eruptions are known, especially when accompanying collapse of large calderas. For example, massive ash-flow sheets frequently are zoned from more siliceous bases to more mafic tops, although these tops usually are rhyolite, dacite, or latite rather than andesite in composition (e.g., Hildreth 1979; Lipman et al. 1966). The paroxysmal eruption of Mt. Mazama, USA, which formed Crater Lake, Oregon, initially deposited rhyolite pumice but changed rapidly to yield basic andesite scoria at the end (Williams 1942; Ritchey 1980). All crystals present in early ejecta (from the reservoir top) occur throughout the ash column and later plagioclase, orthopyroxene, and hornblende crystals are reversely zone, consistent with crystals having settled through melt which became more mafic with depth.

Third, initial tephra are sometimes physical mixtures of disparate magma types ("banded pumice") whereas later lavas are homgeneous. An example is the 1902 eruption of La Soufrière, St. Vincent, Antilles, which initially ejected both basic andesite and dacite, but then produced basalt in the final stages (Aspinall

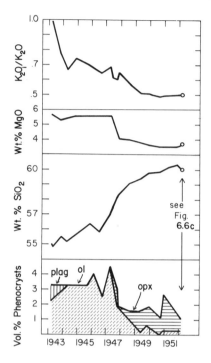

Fig. 4.7. Changes in composition and mineralogy during eruption of Paricutin volcano, Mexico, during period shown. K_2O_i/K_2O equals weight fraction of liquid remaining during fractional crystallization if $\bar{D}_K = 0$. Phase relationships for *circled sample* were determined by Eggler (1972a)

et al. 1973; Carey and Sigurdsson 1978). Additional examples are cited in Section 11.7. Presumably this temporal sequence records a magma mixing event caught in the act; indeed, the mixing itself may trigger eruption (Sparks et al. 1977).

Fourth, although analyses of samples collected during the same eruption interval usually agree to within analytical precision, there also is what seems to be random "noise", particularly in the concentration of minor elements. For example, prior to the lull during the 1974 eruption of Karkar volcano cited earlier in this section, the concentration of all oxides remained constant to within 5% of the amount present except for TiO_2 which varied randomly from 0.56 to 0.88 wt.% (McKee et al. 1976), apparently due to modal variations in oxide minerals. No analytically rigorous study of this kind of variability, especially of trace element contents, during andesite eruptions has been published.

These same four kinds of temporal trends also have been documented over longer periods, both between historic eruptions (discussed below) and over 10^3 to 10^6 years (discussed in the following section). In both instances, the trend of becoming more differentiated with time is most common.

A typical repose period for stratovolcanoes is 7 years according to Wood (1978), although this probably increases with increasing volcano age (Rose et al. 1977). Usually there is ambiguity concerning where magma is stored and what happens to it during these intervals, but increased fractionation of ejecta with time seems the most common result. For example, the 1970 eruption of Akita-Komagatake volcano, Honshu, Japan, followed 38 years of dormancy and produced low-K, tholeiitic, acid andesite lavas and subordinate tephra, whereas previous eruptions had been restricted to basalt or basic andesite (Aramaki 1971; Yagi et al. 1972). In contrast to earlier lavas, those erupted in 1970 contained more sodic plagioclase and more magnetite phenocrysts, and contained groundmass orthopyroxene. Yagi et al. (1972) interpreted these changes as indicating low-pressure crystal fractionation which produced acid from basic andesite and a calcalkaline suite from a tholeiitic one by crystallization during the repose. A second example is the increase from 60% to 66% in SiO_2 content of ejecta from Sakura-jima volcano, Kyushu, Japan, between 1956 and 1971, which apparently was accompanied by several km of magma ascent (Taneda 1977).

However, the genetic significance of trends between historic eruptions is rarely clear because cyclicity in composition of ejecta produces ambiguity about whether compositions become more or less differentiated with time, and because successive ejecta need not be consanguinous. As an example of the first point, three cycles of quiet extrusion of basic andesite ending 60 to 70 years of dormancy and followed 5 to 30 years later by explosive eruptions or more siliceous andesite, may have occurred in the last two centuries at the Soufrière of St. Vincent, Antilles (Aspinall et al. 1973), suggesting injection of new basic andesite followed by 5 to 30 years of in situ differentiation. Consequently, Aspinall interpreted magma and plagioclase compositions in the 1971 and 1972 eruption as being unrelated to those of the 1902 and 1903 eruption, thereby indicating that a new magma batch had appeared. In contrast, Roobol and Smith (1976) used the same kind of evidence from plagioclases but concluded that the younger magma was a differentiate of the older. Moreover, further back in time, ejecta which were erupted on St. Vincent following long repose periods, as indicated by paleosols, usually were basic andesites which grade upward into basalts rather than into more siliceous andesites (Rowley 1978). Such ambiguous cyclicity is common, e.g., in the 40 m thick, < 44,000 year old Mansion pyroclastics of St. Kitts, Antilles, which record a few thousand years of activity of Mt. Misery volcano (Baker and Holland 1973), or in northeastern Pacific DSDP core (Scheidegger and Kulm 1975).

Examples of successive, but unrelated, ejecta were erupted from Bagana volcano, Bougainville Island, Papua New Guinea, between 1943 and 1975. Eruptions were relatively continuous from 1943 to 1953 and again from 1959 to 1975. Although the 1959–75 ejecta are more differentiated in most respects, they are too poor in K, Rb, and Ba to have been derived from the magma which had appeared a decade earlier (Bultitude et al. 1978).

4.4 Variations in Rock Composition During Evolution
of Stratovolcanoes

Detailed geologic study of andesite stratovolcanoes is usually difficult because stratigraphy is hard to establish. The fluidity of basalt lavas or rhyolite ash flows permits
units to be widespread and therefore correlatable. However, andesite lavas are thick
and often of small areal extent, are usually limited to topographic lows such as
stream beds, and often flow in different directions than did the lavas preceding or
following them. One encounters in the field a variety of rock types and compositions
having clear spatial but ambiguous temporal relationship to each other. Products of
eruption from adjacent vents are interfingered. Such relationships require good
exposure, clever fieldwork and radiometric dating to unravel. Numerous published
papers and unpublished theses report such attempts and collectively suggest every
conceivable sequence of rock types from basalt to rhyolite. However, there are few
single andesite-producing volcanoes for which detailed stratigraphic, chronologic,
and thorough petrologic data are available. Four patterns emerge from these few, as
summarized below. These four patterns parallel those described for single or closely
spaced eruptions in the previous section. However, whereas eruptions most often
become more mafic with time, volcanoes most often become more felsic with time.

First, most andesite-producing volcanoes erupt magmas which, broadly speaking, become increasingly differentiated with time, often terminating with explosive eruption of dacite which may accompany caldera collapse or resurgent
doming or both. Examples include the recently active volcanoes of Rabaul, Papua
New Guinea (Heming 1974); Agrigan, Marianas (Stern 1979); Asama, Hakone,
Mashu, and Myoko volcanoes, Japan (Kuno 1950a,b; Aramaki 1963; Katsui et al.
1975; Hayatsu 1976, 1977); Little Sitkin, Aleutians (Snyder 1959); Mts. Hood,
Jefferson, and Shasta, USA (Wise 1969; White and McBirney 1979; Christiansen
et al. 1977); Santa María, Guatemala (Rose et al. 1977); San Salvador, El Salvador
(Fairbrothers et al. 1978); and Santorini, Aegean (Nicholls 1971a). An older
example is the Miocene volcano Banská Štiavnica in Czechoslovakia (Konečný
1971). Two case studies will illustrate the point.

The 1600 m high (3789 m above sea level) Santa Maria volcano, Guatemala,
has been studied extensively by Rose and his colleagues (Rose 1972a,b, 1973;
Rose et al. 1977) from whose work the following summary is taken. Lava samples
from the top 240 m of the volcano were exposed by a violent eruption in 1902.
These lavas have generally increasing silica contents with increasing elevation,
from 52% to 57% SiO_2. The rocks are medium-K, calcalkaline basic andesites, and
range in age from about 30,000 to a few thousand years, based on paleomagnetic
correlations. Violent eruption in 1902 ended prolonged dormancy and yielded
over 5 km^3 of dacite tephra. From 1922 to the present, a dome of acid andesite
and dacite (Santiaguito) has grown within the explosion crater.

Phenocryst minerals remained the same during evolution of the cone: plagioclase, olivine, augite, orthopyroxene, magnetite. However, the relative proportion

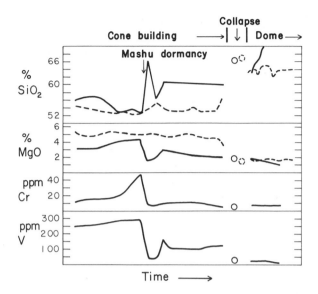

Fig. 4.8. Up-section changes of lava compositions within two stratovolcanoes. *Solid line* is Mashu volcano, Japan; *dashed line* is Santa Maria volcano, Guatemala. Time scale is approximately 25,000 years for both. *Circles* represent tephra from collapse events

of olivine decreases and of orthopyroxene increases up-section, and small resorbed hornblende crystals occur in the youngest, most silicic samples. Although Si, Na, K, Rb, and Zr contents and Fe/Mg and Rb/Sr ratios increase up-section, and Mg, Ca, Sr, Cr, and Ni contents and K/Rb ratios decrease up-section (Fig. 4.8), neither the temporal sequences nor geochemical correlations are simple. No change in the mean $^{87}Sr/^{86}Sr$ ratio of 0.7040 occurs during volcano evolution. Rose and his colleagues attribute the changes in magma composition to low pressure crystal-liquid fractionation.

A similar evolutionary history involving low-K, tholeiitic basic andesite to rhyolite occurred at approximately the same time at Mashu volcano, Hokkaido, Japan, and has been studied extensively by Katsui and his colleagues (Ando 1975; Katsui et al. 1975, 1978; Masuda et al. 1975), from whose work this summary is drawn. Mashu volcano is one of three lying within the 26 × 20 km Kutcharo caldera which formed during violent eruption of dacite ash-flow sheets and lava about 30,000 years ago. The Mashu cone is over 1000 m high and was built between 17,000 and 12,000 years ago by eruption of basic and acid andesite lavas and tephra. This was followed by formation of a 7 × 5 km caldera during eruption of acid andesite and dacite pyroclastics between 11,000 and 7000 years ago, and formation of an endogenous dome of dacite and rhyolite between 4000 and 1000 years ago. Mashu has not been active historically.

As at Santa María, magmas which were erupted during cone construction changed irregularly with time. Two periods, interrupted by dormancy of unknown duration, are identified; during each, magmas irregularly become more mafic in composition up-section. In the first period, magmas became more porphyritic with time although both the phenocryst and groundmass minerals remained the same. Si, Na, K, and Y contents decrease up-section (from 57% to 53% SiO_2) while Al, Ca, Ti, Fe, Mg, Cr, and V contents increase (Fig. 4.8). Initial products following dormancy were 66% SiO_2 dacites, followed mostly by 60% SiO_2, two-pyroxene andesites during the second period of cone formation. These define a flatter FeO^*/MgO vs SiO_2 slope than do first-period basic andesites, have only half as much V (Fig. 4.8), and have orthopyroxene in the groundmass. Tephra became much more voluminous during the following thousand years and remained so for about 4000 years during caldera formation. During this and the subsequent period of dome development, dacites with 65% to 70% SiO_2 were the most common ejecta, although minor acid andesites also were erupted. Plagioclase, two pyroxenes, quartz, and magnetite continued to be the phenocrysts. Cu, V and Y contents became significantly lower than in the acid andesites of the upper cone; all REE contents increased, and a negative Eu anomaly developed. Reported $^{87}Sr/^{86}Sr$ ratios decreased from 0.7039 during cone building to 0.7032 during caldera formation without significant change in Sr contents; δO^{18} values remained constant at $5.6 \pm 0.1\%o$.

Thus, these two volcanoes which erupted magmas of contrasting overall composition (medium-K, calcalkaline magmas with moderate K/Rb ratios and light REE-enrichment at Santa María; low-K, tholeiitic magmas with high K/Rb ratios and light REE-depletion at Mashu) had similar histories. Olivine, clinopyroxene, orthopyroxene, and magnetite were the dominant phenocrysts throughout the history of both; minor hornblende crystals occurred in the 1902 ejecta from Santa María and in subsequent dome lavas, but never formed at Mashu. Basalts with about 52% SiO_2, normative quartz, and Mg-numbers of 0.62 (Mashu) to 0.74 (Santa María) are the most mafic rocks analyzed at either. Although the chemical composition of magma changed irregularly between successive eruptions during cone building at each volcano, and although the temporal sequence differs between volcanoes in detail, generally magmas became more fractionated with time at both. Differences between the volcanoes in LIL-element contents and FeO^*/MgO ratios of the magmas erupted were retained and enhanced during volcano evolution. Although the slope of FeO^*/MgO vs SiO_2 decreases with time at both volcanoes, in neither case is there an abrupt change in slope nor any pronounced attendant change in eruption style or phenocryst mineralogy. An abrupt drop in V contents followed a period of dormancy at Mashu. Mean $^{87}Sr/^{86}Sr$ ratios remained constant at Santa María where rocks contain 400 to 600 ppm Sr, but may have decreased prior to caldera formation at Mashu where rocks contain about 100 ppm Sr.

Secondly, ejecta from andesite stratovolcanoes can become more mafic up-section, at least for a time, as during the first period of cone building at Mashu.

Similar short-term instances are numerous and include: the most recent history of the Tongariro massif, New Zealand (Cole 1978); the last 1700 years of history at Mayon volcano, Philippines (Newhall 1979); each of the three stages recognized at Bezymianny volcano, Kamchatka (Ivanov et al. 1978); the overall pattern inferred for Mt. Misery volcano from studies of the lower 2/3 of the Mansion pyroclastics at St. Kitts, Antilles (Baker and Holland 1973); Hekla volcano, Iceland (Sigvaldason 1974); and instances where flank or post-caldera eruptions are basaltic (e.g,, Fiske et al. 1963).

Hopson (1971) reported an especially interesting example of this trend during the last 35,000 years at Mt. St. Helens, USA, where there have been several cycles of explosive eruption of dacite tephra followed by emplacement of a dacite dome and then, sometimes, by extrusion of acid to basic andesite lavas. Early in the history of the present cone initial dacite tephra contained biotite, cummingtonite, and hornblende phenocrysts, whereas the more recent initial dacite tephra contains hornblende, orthopyroxene, and clinopyroxene. Initial andesite lava flows which sometimes follow dacite tephra during an eruptive cycle contain two pyroxenes (joined by subordinate hornblende in the domes), whereas later flows contain olivine as well.

However, there are no instances where this sort of pattern is sure to have prevailed throughout the history of an edifice. A possible example is Irazu volcano, Costa Rica, whose cone of high-K basalt to acid andesite overlies a dacitic ashflow sheet approximately 100,000 to 300,000 years old (Krushensky 1972; Allègre and Condomines 1976). Although detailed stratigraphy is not known, the oldest dated sample of the cone is a 110,000 year old two-pyroxene acid andesite. Rocks become increasingly mafic with decreasing age, culminating in the eruption during 1964 of basaltic andesite whose Mg-number and Ni plus Cr contents are almost potentially primary (Condomines, pers. comm.). With one exception, the initial ^{230}Th/^{232}Th activity ratios of these rocks increases regularly with time. Consequently, the simplest and most elegant interpretation is that they all are differentiates of a common parent magma rather than successive partial melts of a common source. Moreover, if the Th/U ratio of the source was 3.7, a typical value for the upper mantle but not for subducted oceanic crust, then the parent magma was produced about 140,000 years ago and has been differentiating since in an environment largely closed to U and Th. If true, then magma feeding a single cone can differentiate from the outside inward, and can remain scarcely fractionated at depth for over 10^5 years without re-stoking by new magma batches. This interpretation is consistent with the notion of differentiated, stratified magma overlying uniform, convecting magma within reservoirs (McBirney 1980). However, the timescale is so long that the entire model becomes questionable even though it is self-consistent geochemically if not thermally.

A third pattern of volcano evolution is for rock compositions to converge toward andesite with time. That is, in some volcanoes the pattern of superposed ejecta becoming increasingly differentiated up-section (as described above) is repeated cyclicly, but the range of rock compositions per cycle decreases due to the increasing silica and LIL-element concentrations of the most basic member of each cycle. The Myoko volcano group, Honshu, Japan, provides an example (Hayatsu 1976, 1977; Ishizaka et al. 1977; Yanagi and Ishizaka 1978). Four adjacent cones are aligned at 8-km intervals, decreasing in age south to north. LIL element concentrations (e.g., $K_{57.5}$ values) increase to the north whereas $^{87}Sr/^{86}Sr$ ratios generally decrease in the same direction, from 0.7047–55 to 0.7040–43. Each cone consists of stages defined by field relationships within which rocks reportedly change from olivine basalt to pyroxene basic andesite to hornblende acid andesite; all have relatively constant $FeO*/MgO$ ratios which cross the TH-CA boundary of Fig. 1.4 at about 58% SiO_2. Whether the cones are taken singlely or collectively, rock compositions generally converge toward andesite which has a restricted range of LIL-element concentrations and ratios, and lower $^{87}Sr/^{86}Sr$ ratios than in initial ejecta.

The cyclic pattern is most easily attributed to repeated influxes of basic magma followed by differentiation. The convergence may indicate mixing of previously differentiated material with subsequent basic magma, although obvious mineralogic evidence for mixing (e.g., olivine + quartz-bearing assemblages) is uncommon in the Myoko group. The isotopic diversity suggests a heterogeneous source for the basic magma (the region may be underlain by unusually radiogenic mantle; see Sect. 7.1 and Shuto 1974), or the influence of yet another factor – crustal contamination.

The fourth pattern is known in less detail than the first three upon any of which it may, in practice, be superimposed; it is for andesite volcanoes to erupt from a common vent area ejecta which apparently are unrelated genetically. That is, successively more or less differentiated rocks may differ in isotopic or trace element composition in ways which apparently preclude their consanguinity. One example from Bagana volcano in Papua New Guinea was given in the preceding section. Others have been documented from the Cascades (Condie and Swenson 1973; White and McBirney 1979), the Antilles (Gunn et al. 1974; Arculus 1976), and Japan (Masuda and Aoki 1978). This claim may be misleading; apparent lack of relationship may simply indicate misunderstanding of geochemical processes. Nevertheless, andesite stratovolcanoes are often built of rocks which, even if they become generally more silicic up-section, have elemental or isotopic compositions which are difficult to relate to one another quantitatively by known differentiation mechanisms.

These four patterns, selected to encompass the most thorough case studies currently available, do not include McBirney's (1968) "coherent" type of andesite volcano which is characterized throughout its history by andesite of monotonously uniform composition (e.g., Rainier, USA; Fiske et al. 1963). Coherency may be a first-order generalization which vanishes upon more detailed study, or may be a special case of convergence. Data from such volcanoes might lead to arguments concerning andesite genesis different from those advanced in the following section.

However, because I cannot estimate how many, if any, arc volcanoes have erupted magma which is nearly invariant in composition with time, I will not develop these arguments here.

Thus, as with individual or successive historic eruptions, many kinds of temporal variations in the composition and, occasionally, the mineralogy of andesite magmas are found during stratovolcano evolution. However, as stratovolcanoes grow they most often erupt increasingly siliceous magmas, although the evolutionary pattern usually is irregular or cyclical and occasionally reversed for a time. Long periods of dormancy (decades to centuries) often are succeeded by explosive eruptions of acid andesite, dacite, or rhyolite, by nuées ardentes, by caldera collapse, by growth of a resurgent intra-caldera dome, or various combinations thereof. This pattern may occur once, or be repeated several times during evolution of a volcano.

4.5 Conclusions About Andesite Magma Reservoirs

Is there a standard andesite stratovolcano history or andesite eruption from whose characteristics one can make inferences about magma reservoirs or genetic processes? Examples cited above clearly indicate there is not, but patterns nevertheless recur.

Changes in effusive style during an eruption, as summarized by Katsui (1963) for example, probably are linked to a decrease with time in the volatile content of magma being erupted (Sect. 5.3.1), which probably reflects evisceration of a reservoir from the top downwards. Change in the composition of magma during single eruptions can be attributed to mixing of coexisting but disparate magmas, or to existence of compositional gradients, not only of water, within the reservoir, i.e., to a gradually zoned magma chamber. Only detailed mineralogic and geochemical study can distinguish between these options.

Initial abundance of plagioclase phenocrysts in the Paricutin and Karkar examples cited above suggests crystal accumulation by flotation. Initial eruption of rocks enriched in silica and LIL-elements or containing hydrous minerals suggests that differentiated liquids as well as crystals accumulate in areas likely to be tapped first, e.g., cupolas atop the reservoir. Gravitational removal of crystals from magma during cooling is an obvious mechanism, but schemes consistent with the physical properties of the liquid and solids (Sect. 4.1), with the likely temperatures and water pressures prevailing during phenocryst precipitation (Sect. 5.3.3), and with the temporal variation of mineral species and compositions observed, have not been developed. The fluid dynamic regime is as important in this regard as it is unconstrained.

Magmatic differentiation within crustal level reservoirs is suggested even more strongly by the longer-term variations in ejecta composition and mineralogy discussed in Sections 4.3 and 4.4. These examples indicate that ejecta usually become more differentiated with time during construction of stratovolcanoes. Whether successive eruptions reflect new magma batches (possibly the 1971 eruption of the Soufrière of St. Vincent, Antilles), or sequential but incomplete eruptions from a closed-system reservoir (possibly Irazu volcano, Costa Rica), or both, is unknown in most, probably all, instances. The process can be single-stage or cyclical, usually the latter. However, acid andesites, dacites, and hornblende-bearing andesites usually are erupted late in the history of any particular cone, at high elevations, often in association with caldera collapse or resurgent doming, and often following periods of prolonged dormancy. Also, magmas sometimes change from strong to weak iron-enrichment during evolution of a cone during periods of dormancy. The converse is unknown.

Thus, the geologic history of andesite stratovolcanoes can be interpreted as recording the complex history of a continuum of magma erupted from a finite reservoir volume through a single conduit system during a limited period of time. One possible, idealised cross section of an andesite magma reservoir is sketched in Fig. 3.8. Any given edifice may result from a single pulse of magma through this plumbing system, or from a dynamic equilibrium of magma pumped through it by addition in the zone of magma genesis and eruption from the vent area. Persistence of volcanism for $> 10^6$ years from one reservoir requires restoking by magma addition, according to calculations which assume likely reservoir volumes and thermal conditions (e.g., Lachenbruch et al. 1976). How frequently and by what mechanism this restoking occurs is unknown, but it clearly bears on whether or not all magmas erupted from a common edifice are related genetically, as discussed in Chapter 5. Consistent differences in the evolutionary history of volcanoes which erupt different andesite types or which are located in different tectonic settings (e.g., at plate segment boundaries versus between them) have not been confirmed.

Several observations indicate that magma reservoirs, or portions thereof, periodically become relatively closed for $\leqslant 10^5$ years. These observations include: evidence for compositional gradients within the reservoir; increased differentiation of the initial magmas erupted as the length of repose increases; association of acid andesites and hydrous minerals with increased stratovolcano elevations and with eruptions following prolonged dormancy; and the study of $^{238}U/^{230}Th$ systematics cited earlier. What causes closure is unknown; possible reasons include episodicity of magma generation, loss of volatiles, increased regional compressive stress, or clogging of the conduit system by solidification, increased viscosity, or decreased density contrast. These periods of closure may be temporary and cyclical, resulting in the stratigraphic repetitions which occur during cone construction. That temporary closure of subsidiary reservoirs can be, perhaps often is, interrupted by influx of new magma followed by mixing and possibly triggering eruption, is indicated by the initial heterogeneity of ejecta during some eruptions (Sect. 4.3), the convergent pattern of volcano evolution (Sect. 4.4), and the mineralogic evidence for magma mixing (Sect. 11.7).

However, closure also may be terminal, sometimes leading to a theatrical finale involving caldera collapse, dome resurgence, and eruption of acid andesites and dacites, but ultimately resulting in quiescence. Closure may occur on a wide variety of scales and volcanic activity may persist for 10 to 10^5 years after closure, judging from examples cited in Sections 4.3 and 4.4. Accompanying changes in the composition or mineralogy of magmas erupted will depend on the extent of differentiation and mixing. Subsequent volcanism in the vent vicinity presumably represents magma, from elsewhere in the reservoir or from a new reservoir entirely, which utilizes a new conduit, results in a new edifice, and leads to a profound sense of déjà vu.

The depth ranges of andesite magma reservoirs are unknown but constrained by seismic observations summarized in Section 3.6 and by conclusions concerning the depth of differentiation (Sect. 6.11). These observations imply that subsidiary reservoirs at crustal levels are common beneath volcanoes which yield acid andesites, and it may be these subsidiary reservoirs which close, terminally or periodically, to produce the temporal sequences discussed. If so, differentiation occurs therein.

The temporal sequences discussed above are difficult to explain if the range of andesite compositions observed are primary partial melts of lower crust, of underthrust oceanic crust, or of the intervening upper mantle, which rise superheated into shallow chambers, precipitate some token phenocrysts, and then erupt. Instead, the sequences imply that andesites result from differentiation processes within the reservoir system as the result of its temporary closure. Where within the reservoir (e.g., at what depths), by what mechanism (e.g., crystalliquid fractionation or contamination), or at what rate differentiation occurs are not yet delimited by geologic or geophysical data; neither can the parent magma be identified. Instead, geochemical arguments bear on these problems.

4.6 Stress Fields and Volcano Spacings Within Volcanic Arcs

In two respects the specific location of currently active andesite volcanoes along convergent plate boundaries is too regular to be coincidental: their occurrence above only a portion of the region characterized by high heat flow and seismic attenuation, and their somewhat regular spacing along volcanic fronts. By my definition in Section 2.1, andesite stratovolcanoes lie only within volcanic arcs, which are 50 to 250 km wide and which overlie a restricted depth range of the dipping seismic zone (80 to 250 km maximum). However, high heat flow extends up to 1000 km behind many convergent plate boundaries (e.g., Fig. 3.1), and magmatism frequently occurs in these backarc regions but is basaltic and is less distinctive geochemically (Sect. 5.4). This spatial restriction implies either that orogenic andesite originates due to phenomena associated with subduction over a limited depth range (e.g., slab fusion or dehydration) or that magma differentiates to andesite (e.g., by crystal fractionation or by contamination) only in restricted environments. Specifically, volcanic arcs may experience horizontal compression while backarcs experience extension, and this contrast may cause predominance of andesite in volcanic arcs by slowing ascent rates or affecting ascent plaths beneath volcanic arcs.

At depth, the maximum compressional stress direction lies near the horizontal and is parallel to the direction of plate convergence within some, perhaps most, subduction complexes, but becomes oriented more vertically at some variable distance from the plate boundary (e.g., Bischke 1974). Focal mechanisms for crustal-level great earthquakes within the forearcs of several overthrust plates indicate compression axes parallel to the direction of plate convergence (see Uyeda and Kanamori 1979, for summary). Geodetic surveys also indicate that all of northern Honshu, Japan, is now compressed in this fashion (Takeuchi 1978; Mogi 1970). Finally, utilization of flank volcano alignments as an indication of the longer-term direction of maximum horizontal stress shows that some volcanic arcs (e.g., Aleutians, Japan, Chile) are under regional compression, while their corresponding backarcs are under extension (Nakamura 1977; Nakamura et al. 1977).

The issue is quite ambiguous, however. For example, forearc sediments, especially behind the trench-slope break, are not deformed by compression. Also, geodetic surveys in the forearc of New Zealand have found alternating periods of tension and compression (Walcott 1978). Some volcanic arcs (e.g., New Zealand, the Central Depression of Kamchatka, or the Nicaraguan Depression) clearly occupy grabens which indicate regional extension. The situation elsewhere is more subtle but also arguably extensional in the volcanic arc (e.g., Cascades: McBirney 1976; Guatemala: Plafker 1976).

Finally, entire arc systems (i.e., fore + volcanic + backarcs) may differ in stress regime amongst themselves. Such differences may depend on the rate of convergence and age of subducting lithosphere, and be related to the presence or absence of active backarc spreading (Uyeda and Kanamori 1979; England and Wortel 1980). That is, arc-wide extension and concurrent spreading may characterize arcs associated with rapid subduction of lithosphere which is > 40 to 70 m.y. old (Mariana-type arcs) whereas arc-wide compression may characterize the converse situation (Chilean-type arcs). Uyeda and Kanamori also propose that orogenic andesite predominates only in Chilean-type arcs, and that subducted lithosphere melts only beneath Chilean-type arcs due to high shear stress accompanying the coupling and compression.

Thus the limited data concerning the orientation, much less magnitude, of stresses at depth near convergent plate boundaries conflict, and no consensus has been reached about them. Probably both the orientations and magnitudes vary temporally within arcs and spatially along them. Voluminous orogenic andesite seems concentrated in volcanic arcs under compression, but both this conclusion and the reasons for it remain speculative.

Within the volcanic arc there is also regularity in the along-strike distance between active volcanoes, which may reflect the depth to, or thickness of, the roots of their magma reservoir systems, or tectonic control of sites of magma ascent. Estimates of the average distance between active volcanoes at convergent plate boundaries vary. They include 25 km for Central America (Carr et al. 1979), 40 km for Kamchatka (Fedotov 1976), 70 km for the Cascade and Aleutian arcs (Marsh and Carmichael 1974), 55 km for island arcs and 70 km for continental margins in general (Vogt 1974), and about 100 km for all volcanic arcs (Lingenfelter and Schubert 1976). Average values do vary between arcs but also depend heavily on the age and evidence for activity which are adopted (Baker 1974), especially in largely submarine areas. Although substantial relief can exist on the crust/mantle boundary beneath volcanic arcs, there is poor correspondence between the wave length of this relief and observed volcano spacing (e.g., Boynton et al. 1979), precluding arguments that spacing is maintained for long periods of time.

Whatever the spacing, it may approximate the depth of andesite genesis, or at least the thickness of lithosphere through which magma ascends to andesite volcanoes (Vogt 1974; Lingenfelter and Schubert 1976). The thicker the lithosphere, the greater the distance between volcanoes should be, as in East Africa for example (Mohr and Wood 1976). However, in detail andesite volcano spacings do not correlate persuasively with estimates of lithospheric thickness. For example, volcanoes in Peru and the Aleutians are about 75 and 80 km apart (Vogt 1974) whereas lithospheric lids beneath the volcanic arcs are about 70 and 150 km thick, respectively (Jacob and Hamada 1972; Sacks and Okada 1974). All four numbers are contentious.

Alternatively, variations in spacing may primarily reflect differences in the thickness and viscosity of magma at its depth of origin (Marsh and Carmichael 1974). If so, basaltic vs. andesitic volcanism, tholeiitic vs. calcalkaline volcanism, high convergence rates, or other phenomena associated with larger melt fractions should lead to closer volcano spacing. However, none of these differences alone explains the observed variability in spacing.

Finally, spacing may be related to the segmentation of convergent plate boundaries discussed in Section 2.1, but the nature of the relationship, if any, is unclear. For example, Carr et al. (1979) argued that volcanoes, especially highly explosive ones, are concentrated near arc segment boundaries, whereas Spence (1977) argued that volcanism is *less* likely at segment boundaries than within segments.

If andesite volcano spacing is controlled largely by lithospheric thickness, then andesite or its parent magma ascends from within the asthenosphere. If other factors control spacing, the spacing adds nothing to genetic questions. The latter seems more likely.

4.7 Relationships Between the Timing of Arc Volcanism and Plate Movements

Limited evidence suggests that eruption of andesite and other magmas at convergent plate boundaries is related to underthrusting on both short and long time scales. Kimura (1976) and Carr (1977) have documented a close coincidence between eruption of basalt or andesite in volcanic arcs and the occurrence of single great earthquakes, and Blot (1976) has argued that some eruptions are preceded first by deep earthquakes along the dipping seismic zone and later by earthquakes which migrate upward to the volcano over periods of a few months. On a slightly longer time-scale, the annual number of volcanoes in eruption in South America correlates with the annual energy release calculated from shallow, large-magnitude earthquakes at Nazca plate convergent boundaries (Berg and Sutton 1974). These authors also argue that volcanic activity migrates cyclically north to south from Mexico to Chile, at a time scale apparently related to the periodicity of strain release within the Nazca plate. Two similar correlations between volcanic eruptions and magnitude of strain release were observed in the New Britain arc during the 1950's and the 1970's (Cooke et al. 1976).

Neither the statistical nor physical significance of these correlations is clear, although they may reflect tectonic control of magma release from reservoirs within the overthrust lithosphere. For example, Yamashina and Nakamura (1978) suggested that Osima volcano, Izu, Japan, acts as a strain gauge such that magma level rises during periods of strain accumulation and that eruptions are associated with strain release via large earthquakes. However, these correlations also may reflect periodicity of magma generation or aggregation, as envisaged by Spence (1977) for example.

In contrast, long-term volcanic episodicity, especially if consistent along one or more arc segments, implies more about magma generation than release. Radiometric age determinations of Tertiary volcanic rocks within some of the arc segments of Fig. 2.1 frequently define discrete maxima which may coincide between arcs and may reflect periods of more rapid plate convergence. For example, four volcanic maxima beginning about 16, 10, 6, and 2 m.y. ago and lasting about 2 m.y. each, have occurred during the late Cenozoic in the central Cascades, Central American, and southwestern Pacific arcs (Kennett et al. 1977). Correlative cycles of about 2.5 m.y. duration occur both in the abundance and silica content of ash layers in DSDP core from the northeastern Pacific (Scheidegger and Kulm 1975).

This apparent episodicity has been attributed to variations in spreading and, therefore, subduction rates. Specifically, the current period of volcanic maxima which began one to two m.y. ago corresponds to about a 10% acceleration in spreading rates between eastern Pacific plates (Rea and Scheidegger 1979). Earlier periods of extensive arc volcanism, especially in the Middle Miocene, may therefore reflect periods of especially rapid plate convergence. The Middle Miocene period also coincides in time with a major reorganization of plate boundaries in the Pacific (Handschumacher 1976).

For three reasons, however, this correlation is not yet persuasive. First, the evidence for episodicity in the rate of arc volcanism remains limited and subjective. Second, no correlation between extrusion rate and plate convergence rate is apparent in modern arcs during the Pleistocene (Sect. 4.8). Finally, the episodic volcanism in the Cascades cited above can be correlated with episodic volcanism in the Snake River Plain, USA (Armstrong 1975) and along the Hawaii-Emperor chain (Vogt 1972). Thus, periodic maxima in orogenic andesite volcanism may simply reflect global oscillations in magma production.

4.8 Magma Eruption Rates at Convergent Plate Boundaries

Volcanic edifices, including those consisting predominantly of orogenic andesite, develop at geologically alarming rates. Single eruptions of andesite can produce up to 10 km^3 of ejecta, cinder cones can grow to hundreds of meters in elevation in a week, and large stratovolcanoes can develop within a few millenia. Quantitative estimates of these eruption rates are difficult to make and correspondingly scarce; those avalable for andesite are in Table 4.3.

Juvenile magma has been produced at a nearly constant rate of 0.5 to 5 X 10^{-3} km^3/yr per volcano over tens of thousands of years from Japanese and Central American volcanoes whose most recent products range from basalt to dacite (Nakamura 1974; Wood 1978). Eruption rates may decrease as volcanoes enlarge or as magmas become more differentiated. A quantitative estimate of this decreasing rate can be obtained by combining Eqs. (4) and (7) of Wood (1978) to yield the empirical expression $V = 0.4 \cdot A^{0.36} \cdot (0.02 \, A^{0.18} + 0.117)$, where V is stratovolcano volume in km^3 and A is volcano age in million year units. For comparison, ocean islands such as Hawaii and Iceland grew at rates of 2 to 10 X 10^{-2} km^3/yr (Schilling et al. 1978) and flood basalts plus andesites of the Columbia River Plateau were extruded at rates as high as 0.1 km^3/yr (Baksi and Watkins 1973).

Rate estimates in Table 4.3 for arc segments vary from 2 to 66 km^3/m.y. per km of plate boundary. If most ejecta are retained within an area 50 km wide, arc volcanism at 10 km^3/m.y. per km of plate boundary creates a kilometer-thick stratigraphic unit in only 5 m.y. This should give pause to those who casually attribute successive volcanic formations in stratigraphic columns to successive geologic periods. Again for comparison, mid-ocean ridge basalts have originated at average rates during the Tertiary of 25 to 67 km^3/m.y. per km of plate boundary (Menard 1967; Wright 1971).

Magma production rates at convergent plate boundaries could constrain genetic theories but are not yet known well enough to do so. Both subducted ocean crust and a convecting mantle wedge can supply parental material in sufficient quantity. Note, however, that rates of magma eruption, much less rates of

Table 4.3. Eruption rates of ejecta from volcanoes at convergent plate boundaries

Arc	Eruption rate $(km^3/m.y./km$ plate boundary)	Convergence rate [e] $(cm\ yr^{-1})$	Age	Reference
1. Kuriles	66	9.5	1930–1963	Markhinin (1968)
2. Northeast Japan	>1.9 [a]	9.7	Quaternary	Sugimura (1968)
3. Northeast Japan	27	9.7	Quaternary	Nakamura (1974)
4. Japan	23 [b]	–	Lower Miocene	Sugimura and Uyeda (1973, Fig. 67)
5. Cascades	9 [c]	3	0–20 m.y.	McBirney (1976)
6. Cen. America	7	8	last 400,000 yrs	Bice (1980)
7. N. Chile	4.2 [d]	10.5	Late Miocene-Holocene	Francis and Rundle (1976)

a   3825 km^3 of Quaternary ejecta; rate assumes age <1.8 m.y.
b   Assuming 1300 km arc length
c   Average value: range is from 6 to 50 at different times
d   Total erupted material of which 68% is andesite; at least twice as much batholithic rock
     is estimated to have been intruded
e   From Table 2.1

magma genesis, which exceed 25 $km^3/m.y.$ per km of plate boundary approach the rate at which ocean crust is subducted. Such high eruption rates would preclude orogenic andesite being the product of $< 40\%$ fusion of subducted crust. There is no rate problem if andesites or their parents come from the mantle wedge, so long as convection replenishes the wedge with fertile peridotite.

Vicissitudes in eruption rates may cause the periodicity in arc volcanism discussed in the preceding section. However, note in Table 4.3 the lack of correlation between eruption and convergence rates within active volcanic arcs. While the available estimates of eruption rates are few and imprecise, the proposed linkage between rates of convergence and eruption finds no support in modern arcs.

4.9 Relative Proportions of Andesite

The contention that orogenic andesite predominates only in areas recently underthrust by suboceanic lithosphere (Sect. 2.1) or that the silica mode of erupted magma increases as crust thickens (Sect. 7.2) depends on accurate estimates of relative proportions of magmas erupted. Likewise, explanations of andesite genesis which postulate differentiation of more mafic parent magmas often are embarrassed when presumed differentiates seem more voluminous than the parent, or when differentiates constitue one mode of a bimodal suite.

Unless eruptions are witnessed, relative proportions must be estimated without knowing the effects of erosion or the volume of tephra which lies outside the geographic area examined. Meaningful volume estimates of even the available rocks are difficult to make due to limited exposure, complex geometry of flows, ambiguous field distinctions between rock types, and arbitrariness concerning the temporal and spatial scales employed. For example, it is clear from Section 4.4 that the relative proportion of rock types changes considerably during evolution of stratovolcanoes and that basalts sometimes are absent from them altogether. The latter may, however, form plateaux from which the stratovolcanoes rise. Some estimates of volume relationships incorporating these ambiguities are summarized in Table 4.4. Such estimates of volume relationships understandably are few as most geologists seek refuge in qualitative measures such as "dominant rock type" rather than venture quantitative estimates.

Table 4.4. Relative proportions of Quaternary volcanic rock types at convergent plate boundaries (also see Table 7.1)

	% Basalt	% Basic Andesite	% Acid Andesite	% Dacite	% Rhyolite	Reference
Individual Volcanoes						
1. Talasea, PNG	9	23	55	9	4	Lowder and Carmicheal (1970)
2. Little Sitkin, Aleutians	0	78	4	18	0	Snyder (1959)
3. Mt. Misery, Antilles						
a) Lava	17	22	49	0	0	Baker and Holland (1973)
b) Tephra	22	42	36	0	0	Baker and Holland (1973)
4. Jefferson, USA	80	17	3	0	0	White and McBirney (1979)
Arc Segments						
1. Cascades						
a) Washington	32	-- 60 --		-- 8 --		White and McBirney (1979)
b) Oregon	80	-- 17 --		-- 3 --		White and McBirney (1979)
c) California	68	-- 30 --		-- 2 --		White and McBirney (1979)
2. Antilles	17	-- 42 --		39	2	Brown et al. (1977)
3. Japan						
a) Lava and ash fall	14	-- 85 --		2	0	Sugimura (1968)
b) Ash flows	0	-- 6 --		44	50	Sugimura (1968)

Table 4.4 shows that the apparent predominance of andesite at convergent plate boundaries certainly is true at most stratovolcanoes, from which the general predominance of andesite is inferred. In 80% of the volcanic arc segments summarized in Table 7.1., andesites constitute at least half of the analyzed rocks from at least half of the volcanoes. [Obviously, however, "volume estimates" based on relative numbers of analyses are biased. For example, basalts and basic andesites predominate in only 13% of Cascade volcanoes according to Table 7.1, whereas White and McBirney (1979) estimated that basalts alone ($< 53.5\%\ SiO_2$) constitute 70% to 85% of the Quaternary rocks of the Cascades.] Even the striking contrasts between volume estimates in Table 4.4 probably reflect methodology rather than geology. White and McBirney include data from a belt extending 300 km east of the active stratovolcanoes, incorporating flat-lying lava flows in the backarc region. In contrast, Sugimura includes only estimates of volume relations within stratovolcanoes themselves, and Brown et al. simply base their estimate on about 1500 chemical analyses.

Estimated andesite/basalt ratios of erupted lava increase as the thickness of crust traversed increases (Sect. 7.2), indicating that thick crust causes differentiation either by slowing ascent of by contaminating magma. For example, White and McBirney (1979) have argued that the andesite/basalt ratio has decreased since the Oligocene in central Oregon due to both of the above phenomea — early loss of the low-temperature melting fraction from, and speedier later ascent due to increased initial temperature of, the wall rocks — thus allowing later magmas to ascend with less differentiation en route.

No estimates of variations in the relative proportion of rock types across the width of magmatic arcs have been published. If attention is restricted to the volcanic arc, then andesites constitue 50% to 80% of the magma erupted on some continental margins or fragments. However, if backarc regions are included, andesites become a subordinate magma type at many convergent plate boundaries. Andesites definitely are a subordinate magma type at places other than convergent plate boundaries (e.g., Quaternary volcanic fields of the southwestern USA and Iceland) even though they predominate within stratovolcanoes of these fields (e.g., San Francisco Peak, Arizona, or Mt. Taylor, New Mexico, USA; and Hekla, Iceland). Finally, andesites are clearly subordinate to basalt within some oceanic volcanic arcs such as the Kermadec, Vanuatu, Mariana, and South Sandwich arcs. Andesites are subordinate to rhyolite within some continental volcanic arcs such as New Zealand, Sumatra, Guatemala, northern Chile, and Turkey.

Chapter 5 Bulk Chemical Composition of Orogenic Andesites

5.1 Rock Analyses: Significance, Averages, and Representative Samples and Suites

Bulk chemical analyses of andesites are used primarily in two ways to evaluate andesite origins. First, average elemental and isotopic compositions can be used to test genetic hypotheses and to compare volcanic rocks from different areas or ages (between-suite comparisons). Second, variation in the concentration of elements or ratio of isotopes within suites must also be consistent with the differentiation mechanisms proposed (within-suite variations).

Five factors limit straightforward utilization of andesite analyses toward these ends. First, the typically high phenocryst contents (20 to 50 vol.%), the low settling rates of some phenocrysts (Fig. 4.6), the high yield strength of andesite magma, and the possibility of magma mixing during andesite genesis (Sect. 11.7), all strongly suggest that some analyzed samples were not homogeneous liquids and that phenocrysts present did not precipitate from the liquid in which they occurred at eruption. Instead they may have accumulated in or been mixed into that liquid. Thus, an andesite analysis may not be of a liquid having andesitic composition, and variation diagrams may be mixing lines instead of lines of liquid descent. Data scatter in element-element diagrams is a common result (e.g., Fig. 5.2, solid symbols). Likewise, the positive correlations of Al_2O_3, CaO, and normative anorthite contents with the modal percent of plagioclase phenocrysts in andesites (Ewart 1976a) is significant statistically, indicating accumulation.

The infrequence of positive Eu anomalies in plagioclase-phyric andesites (Sect. 5.4.2) is sometimes cited as evidence against plagioclase accumulation. However, if D_{Eu}^{pl}/D_{Eu*}^{pl} is 4.0 as predicted by Weill and Drake (1973) for equilibrium at T = 1100°C and log f_{O_2} = -8.0 (see Sects. 4.1.1 and 5.3.5), then each 10% of accumulated plagioclase in andesite will produce only a + 3% Eu anomaly using the D values of Table 6.3. Most analyses of andesites exept by isotope dilution methods are too imprecise to identify anomalies < 10%.

Nevertheless, because most andesites contain a few wt.% water (Sect. 5.3.1), eruption *causes* precipitation (see Fig. 6.6), making aphyric rocks unlikely and perhaps atypical. Indeed, REE patterns can be more regular in appearance in strongly porphyritic than aphyric samples (e.g., Fujimaki 1975). Thus, there is no clear way to identify accumulative phenocrysts and thereby to assess their effect on whole rock compositions without thorough mineral analyses, as discussed in Chapter 6.

Second, because andesite stratovolcanoes are polygenetic (i.e., remain loci of volcanism for $\geqslant 10^4$ yr) they are potential sites of widespread alteration. However, no chemical criteria exist for determining the extent and effect of these processes on rocks with little visible alteration. Criteria were developed in studies of submarine basalts, but pre-eruption H_2O contents and Fe_2O_3/FeO ratios in andesites exceed limits set for freshness in MORB.

Third, andesite magmas become vapor-saturated during ascent and boil off constituents during eruption. Thus, even analyses of unaltered andesites will not reflect pre-eruption concentrations of some elements, especially volatiles. For example, eruption of acid andesite magma in 1976 from Sakurajima, Japan, arguably was accompanied by 50% to 60% out-gassing of Rn (Sato and Sato 1977), although such quantitative estimates usually are unavailable.

Fourth and more abstract is the problem: what is an andesite "suite" within or between which comparisons should be made? Ideally suites include rocks related genetically to each other by similar processes of magmatic differentiation or similar degrees of partial fusion of a relatively homogeneous source. Thus, a suite should have constant *ratios* of isotopes and incompatible trace elements, and concentrations should vary within suites in a manner determined by the genetic relationship.

However, variations in ratios and concentrations in excess of analytical imprecision and for which no differentiation mechanism is recognized occur even during single andesite eruptions, and successive eruptions from the same volcano can produce apparently unrelated products (Sect. 4.3). For example. Shasta volcano, USA, has developed four cones over the last 10^5 years, of which Shastina is the third (Christiansen et al. 1977). Shastina erupted magmas similar in temporal sequence and in mineralogy to acid andesites and dacites of earlier cones. However, those from Shastina are significantly depleted in K, Rb, and Ba, and enriched in Sr and Eu relative to their older equivalents (Condie and Swenson 1973) although similar in $^{87}Sr/^{86}Sr$ ratios (Peterman et al. 1970b). Similar differences are implicit in multiple K_2O-SiO_2 correlations in ejecta from single volcanoes (Sect. 5.2.2). In contrast, $^{87}Sr/^{86}Sr$ ratios increased suddenly from 0.704 to 0.709 during development of Qualibou volcano, St. Lucia, Antilles, with accompanying but unspecified changes in elemental composition (Pushkar et al. 1973).

Both kinds of variability may reflect tapping of separate, possibly coexisting magmas which have no discernible genetic relationship to one another. If the cyclical and temporary closures plus resulting differentiation episodes which I summarized in Section 4.5 are only 10's to 100's of years in duration, and if successive cycles entirely or partially result from separate extractions of partial melts, then lack of rigorous consanguinity between samples will be common unless very detailed chemical stratigraphy is available. Pooling of unrelated samples will lead to chaos or to misleading genetic hypotheses. Alternatively, unexplained variability may indicate unrecognized differentiation mechanisms. If so, the variability produces an apparently random noise level through which one might, nevertheless, identify some of the differentiation mechanisms involved; residual errors provide a guide to the unrecognized mechanisms.

Because most available studies of andesite bulk composition and mineralogy are of regional scope, or at best focus on samples of unknown age relationships from single volcanoes, the cause of variability between samples usually cannot be assessed. Thus, most "suites" discussed here probably lack actual genetic relationship and form an unreliable basis for rigorously evaluating genetic hypotheses. Nevertheless, such data are abundant, dominate what is available, and exhibit modest regularity, so the synthesis attempted here may suffice at least to identify broadly consistent trends.

Finally, and equally intractable, is the difficulty of assessing accuracy and precision of published analyses. Is the "noise" cited above magmatic or analytical? Are between-suite differences actually between-laboratory disagreements? Most published studies include insufficient information to answer these questions.

The average composition of the 2500 orogenic andesites referred to in Section 1.1 is given in Table 1.1. These analyses were then classified according to Fig. 1.2; the average compositions and sizes of the six resulting groups are given in Table 5.1. Average compositions of calcalkaline and tholeiitic members of each group differ primarily in Fe and Mg contents and are not given separately.

Table 5.1. Average composition of orogenic andesite types

	Basic andesites			Acid andesites		
	Low-K	Med-K	High-K	Low-K	Med-K	High-K
N [a]	167	648	288	171	888	338 = 2500
SiO_2	55.4	55.1	54.6	59.6	59.7	59.4
TiO_2	0.81	0.82	0.91	0.74	0.70	0.73
Al_2O_3	17.6	17.8	17.7	16.7	17.1	16.9
Fe_2O_3	3.4	3.3	3.6	3.2	2.8	2.9
FeO	6.1	4.9	4.2	4.7	3.8	3.3
MnO	0.21	0.16	0.18	0.15	0.12	0.12
MgO	4.3	4.3	3.9	3.1	3.2	3.1
CaO	9.1	8.1	7.6	7.0	6.6	6.0
Na_2O	2.7	3.1	3.3	3.2	3.3	3.3
K_2O	0.43	1.2	2.1	0.59	1.5	2.5
P_2O_5	0.16	0.21	0.30	0.15	0.19	0.24
ΣH_2O	0.78	0.93	1.2	0.79	1.0	1.3
Total	101.0	99.9	99.6	99.9	99.0	99.8
Mg#	52	55	54	49	53	55
FeO*/MgO	2.1	1.8	1.9	2.4	2.0	1.9
ρ [b]	2.78	2.72	2.66	2.66	2.61	2.56

a   N = Number of analyses in data base RKOC76; see Section 1.2

b ρ = Density (gm cm^{-3}) calculated assuming the Fe_2O_3 and H_2O contents given, and the partial molar volumes of Nelson and Carmichael (1979) at $1000°C$

Table 5.2. Representative andesite analyses

	Basic andesites						Acid andesites					
	Low-K		Medium-K		High-K		Low-K		Medium-K		High-K	
	TH	CA	TH	CA	TH	CA	TH	CA	TH	CA	TH	CA
SiO_2 (%)	55.4	55.0	55.5	55.0	54.4	55.2	60.6	58.9	58.6	59.9	59.0	59.0
TiO_2	0.57	0.52	0.81	1.1	0.73	0.92	0.57	0.57	0.89	0.69	1.0	0.72
Al_2O_3	17.6	17.3	17.7	17.4	19.4	15.9	12.9	15.6	15.4	17.1	17.2	16.5
Fe_2O_3	2.6	3.4	4.4	8.4 *	–	–	4.5	2.9	2.2	3.6	4.5	4.0
FeO	7.5	5.8	4.0	–	7.4 *	7.3 *	6.8	5.3	6.7	2.7	2.0	1.3
MnO	0.19	0.16	0.18	–	0.20	–	0.19	0.14	0.18	0.14	0.10	–
MgO	4.2	5.4	3.4	4.2	3.0	5.5	2.8	4.0	3.2	3.3	1.5	3.7
CaO	11.0	9.4	8.0	7.5	8.7	9.6	8.1	8.0	7.0	7.2	4.9	4.8
Na_2O	1.8	2.8	4.0	3.8	3.7	2.9	2.2	2.0	3.8	3.9	4.3	4.0
K_2O	0.40	0.46	1.6	1.1	2.1	2.7	0.56	0.71	1.5	1.3	2.9	2.4
P_2O_5	0.08	0.17	0.35	–	0.36	–	0.10	0.10	0.25	0.20	0.49	0.26
ΣH_2O	0.55	n.d.	–	–	0.35	–	0.56	0.72	0.37	0.72	0.97	2.9
Total	100.0	100.3	100.0	98.5	100.3	100.0	100.0	100.0	100.1	100.8	99.9	100.4
FeO*/MgO	2.3	1.6	2.4	1.8	2.4	1.3	3.8	2.0	2.7	1.8	4.0	1.3
Rb (ppm)	6	5	25	25	50	68	9	11	15	20	67	120
Sr	235	380	810	501	605	583	220	384	395	490	660	190
Ba	100	105	250	226	690	670	145	185	345	522	1200	310
K/Rb	555	735	535	365	348	328	505	536	807	544	431	164
Pb	1.8	1.0	5	–	–	12	2.5	–	–	3	33	–
Th	0.2	0.1	1.3	–	8.1	5.3	0.6	–	–	1.8	6.4	5.8
U	0.2	0.1	0.6	–	2.0	1.3	0.3	–	–	0.75	1.9	1.9
La	1.8	–	11	17	22	15	3.0	3	–	11	36	19
Ce	4.7	–	24	42	–	33	6.9	10	17	27	71	38

	1	2	3	4	5	6	7	8	9	10	11	12
Yb	1.5	—	1.7	1.9	3.1	1.6	1.3	—	3.1	1.4	1.9	1.9
Y	17	16	17	—	25	22	12	12	20	18	19	—
Zr	28	38	81	—	98	111	21	34	75	114	230	—
Hf	0.8	—	2.2	—	3.4	2.3	0.7	—	—	2.3	5	4.6
Nb	0.6	—	4	—	5	7	1.5	<1	5	2	15	—
Ni	13	34	6	20	3	45	4	23	70	9	15	60
Co	28	30	—	34	18	29	30	—	—	16	16	16
Cr	15	76	9	196	8	220	16	34	15	39	10	213
Sc	35	26	16	—	18	31	—	—	—	16	13	16
V	310	225	178	—	260	190	360	340	235	172	125	—
Cu	145	88	46	41	38	66	240	163	155	36	59	—
Zn	89	76	74	104	—	—	98	80	100	—	86	—
$^{87}Sr/^{86}Sr$	0.7038	—	0.7038	—	—	—	—	—	0.7036	0.7039	—	—
Ref.	1	2	3	4	5	6	7	8	9	10	11	12

References:

1. Tonga. Ewart et al. (1973) p. 447, L1; Oversby and Ewart (1972). Phenocrysts: plag 24%, cpx 2%, opx <1%.
2. New Britain, PNG. Blake and Ewart (1974) p 325, no. 4. Phenocrysts: plag 23%, cpx 2%, opx 3%
3. Bagana volcano, Bougainville, PNG. Bultitude et al. (1978). New Guinea "reference andesite". Mean of 31 samples, most with 20 to 30% plag, 5 to 10% cpx, <1% ol or opx, 1 to 3% hb, 3% mt
4. Mt. Rainier, Washington, USA. Condie and Swenson (1973) p. 214
5. Merapi volcano, Java, Indonesia. Whitford (1975) p. 117, no. 1071
6. Kastamonu, Turkey. Peccerillo and Taylor (1976) no. T69–159. Phenocrysts: plag 25%, cpx 15%
7. Tonga. Ewart et al. (1977) Niuatoputapu
8. New Britain, PNG. Johnson and Chappell (1979) Table E6, no. 11. Phenocrysts: plag 9%, opx 2%, cpx <1%
9. New Britain, PNG. Lowder and Carmichael (1970) no. 114; Arth (1974); Peterman et al. (1970a). Phenocrysts: plag 20%, cpx 3%, opx 3%, mt 2%
10. Fiji. Analysis is mean of four: Taylor et al. (1969) p. 21, no. X88; Gill (1970) no.66; and two unpublished. Phenocrysts: plag 40%, cpx 1%, hb 20%, mt + ilm 1%
11. Oregon. AGV-1; (Flanagan 1973; Abbey 1977). Almost aphyric with trace of pl and cpx phenocrysts
12. Ancud, Chile. Lopez-Escobar et al. (1976) p. 730, 736, no. GVO-63. Aphyric

Representative andesite analyses are provided in Table 5.2. In each case, all data are for a single sample whose phenocryst mineralogy is described and whose geologic context is available in the references cited.

Finally, broadly studied examples of each type of orogenic andesite suite are listed in Table 5.3. One example of each type is marked; analyses from these representative suites are displayed in Harker diagrams in Fig. 5.1 to 5.3, and elsewhere in this and the following chapter.

5.2 Major Elements

5.2.1 Silica Contents and Harker Variation Diagrams

Silica contents consistently and significantly correlate with other element concentrations to a greater extent in andesite than in basalt or rhyolite suites and, as a result, are used here as the primary basis for andesite nomenclature (Sect. 1.2). For the same reason, silica contents provide a convenient relative abscissa to show graphically the variation in concentration of other oxides (i.e., Harker diagrams; see Figs. 5.1 to 5.3). This utilization of silica contents as an abscissa and as an implicit measure of relative differentiation is especially effective in andesites because silica variance in them greatly exceeds the sum of other oxide variances (Chayes 1964b), silica dominates the primary factor in factor analyses of andesites (Le Maitre 1976a), and silica correlates positively and significantly with measures of differentiation. For example, the correlation coefficient between SiO_2 and the Thornton-Tuttle Differentiation Index for the orogenic andesites of Fig. 1.1 is +0.84.

Thus, Harker diagrams are utilized here to display conveniently the variations in major and trace element concentrations in andesites. However, attempts sometimes are made to infer genetic relationships between spatially related andesites whose analyses define smooth lines in these diagrams, and to determine the nature of this relationship from the linearity or curvature of the lines (e.g., Bowen 1928; Fenner 1931). Neither exercise is sound statistically, especially for those oxides correlated negatively with silica (Chayes 1960, 1964b). Finally, the usefulness of these diagrams is limited by variability attributable to crystal accumulation or "randomness" within suites of samples from the same volcano or even the same eruption.

Not all andesites of equal silica contents are equally differentiated. Differentiation indices in different suites increase relative to silica contents as potash contents increase in those suites. Also, note in Fig. 1.4 that FeO*/MgO is higher for a given silica content in tholeiitic than calcalkaline andesites and note in Fig. 5.7b that the normative ratio hy/qz increases (i.e., rocks become less silica-saturated) for a given silica content as potash contents increase. Thus, between-suite comparisons made only on the basis of silica contents may mask differentiation-dependent differences, whereas within-suite comparisons on this basis usually are profitable.

Table 5.3. Representative well-analyzed suites containing orogenic andesites

Tholeiitic		Calcalkaline	
Low-K			
1. Tonga [a]	Ewart et al. (1973, 1977) Ewart (1976b) Oversby and Ewart (1972)	1. Cape Hoskins [a] (New Britain)	Blake and Ewart (1974) Johnson and Chappell (1979)
2. Mashu (Hokkaido)	Katsui et al. (1975, 1978) Ando (1975)	2. Sarigan (Marianas)	Dixon and Batiza (1979)
3. Hakone (Honshu)	Kuno (1950a,b) Philpotts et al. (1971) Fujimaki (1975) Tatsumoto and Knight (1969)		
Medium-K			
1. Rabaul [a] (New Britain)	Heming (1974, 1977) Heming and Carmichael (1973) Peterman and Heming (1974) Heming and Rankin (1979)	1. Tongariro [a] (New Zealand)	Cole (1978) Ewart (1971) Ewart and Stipp (1968) Armstrong and Cooper (1971)
2. Talasea (New Britain)	Lowder and Carmichael (1970) Lowder (1970) Peterman et al. (1970a) Arth (1974) Johnson and Chappell (1979) De Paolo and Johnson (1979)	2. Moffett (Aleutians)	Coats (1956) Marsh (1976a) DeLong (1974) Kay (1977) Kay et al. (1978)
3. Agrigan (Marianas)	Stern (1979)	3. Medicine Lake (Cascades)	Mertzman (1977) Condie and Hayslip (1975) Smith and Carmichael (1968) Church (1976)
4. Okmok (Aleutians)	Byers (1961); DeLong (1974); Kay (1977); Kay et al. (1978)	4. Santorini (Aegean)	Nicholls (1971a) Pichler and Kussmaul (1972)
		5. Grenada (Antilles)	Arculus (1976, 1978) Brown et al. (1977) Hawkesworth et al. (1979c)
		6. Colima (Mexico)	Luhr and Carmichael (1980)
High-K			
1. Boqueron [a] (El Salvador)	Fairbrothers et al. (1978)	1. San Pedro-San Pablo [a] (Chile)	Francis et al. (1974) Roobol et al. (1976) Thorpe et al. (1976) Hawkesworth et al. (1979)
2. Bogoslof (Aleutians)	Arculus et al. (1977) Kay et al. (1978)	2. Osima-osima (Sea of Japan)	Katsui et al. (1979) Hedge and Knight (1968)
		3. Papua New Guinea Highlands	Mackenzie and Chappell (1972) Mackenzie (1976) Johnson et al. (1978b)
		4. Filicudi (Eolian)	Villari (1972) Villari and Nathan (1978)

[a] One example of each type of orogenic andesite suite which is illustrated in Figs. 5.1 to 5.3 and elsewhere

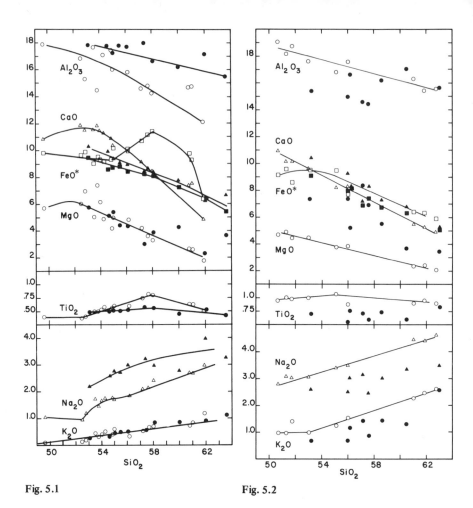

Fig. 5.1   **Fig. 5.2**

Fig. 5.1. Harker Diagram of two low-K andesitic suites on an anhydrous wt.% basis. CaO is indicated by *triangles,* FeO* by *squares.* The tholeiitic example *(open symbols)* is a composite from Tonga: Bauer (1970), Baker et al. (1971), Ewart et al. (1973), Ewart (1976b). The calc-alkaline example *(filled samples)* is from the Krummel-Hoskins area, New Britain (volcano 6.8, Appendix): Blake and Ewart (1974). Trend lines were fit by eye. FeO*/MgO ratios and CIPW norms for these suites as well as those in Fig. 5.2 and 5.3 are shown in Fig. 5.4 and 5.7. For the Tongan suite, K/Rb, Rb/Sr, REE, and V data are shown elsewhere in Chapter 5; plagioclase and pyroxene compositions are given in Chapter 6. K/Rb data are given for the Krummel-Hoskins suite in Fig. 5.10

Fig. 5.2. Harker diagram of two medium-K andesitic suites on an anhydrous wt.% basis. CaO is indicated by *triangles,* FeO* by *squares.* The tholeiitic example *(open symbols)* is from Rabaul, New Britain (volcano 6.1, Appendix): Heming (1974). Ce-anomalous samples (Heming and Rankin 1979) are included because they are not anomalous in major oxides. The calcalkaline example *(filled symbols)* is from the Tongariro center, New Zealand (volcano 1.4, Appendix): Cole (1978). Although only samples erupted since establishment of the NNE-trending arc 50,000 yrs ago are included, considerable scatter is obvious. Most samples have 7 to 10 modal% pyroxenes and the scatter was attributed to their accumulation by Cole. Trend lines are by eye and apply only to Rabaul samples. For the Rabaul suite, REE and T-fO$_2$ data are shown elsewhere in this chapter and pyroxene compositions are given in Table 6.1. For the Tongariro suite (defined more broadly than above), K/Rb, Rb/Sr, and Pb isotope data are shown elsewhere in this chapter and pyroxene compositions are in Table 6.1

Fig. 5.3. Harker diagram for two high-K andesitic suites on an anhydrous wt.% basis. The tholeiitic example *(open symbols)* is from San Salvador volcano (Boqueron), El Salvador (volcano 21.13, Appendix): Fairbrothers et al. (1978). Samples from two or three magma batches covering 300 m of stratigraphy are pooled. The calc-alkaline example *(filled symbols)* is from San Pedro volcano, northern Chile (volcano 23.15, Appendix): Francis et al. (1974), Thorpe and Francis (1976). Pyroxene compositions for the San Salvador suite are given in Table 6.1; Sr-Nd isotope data for the San Pedro samples lie within the CH field of Fig. 5.19

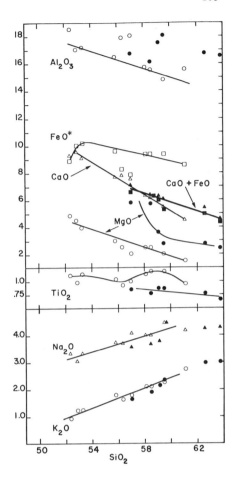

5.2.2 Alkalies

Potassium contents differ more between andesite suites than do any other oxide contents. Because these variations apparently have tectonic as well as petrogenetic significance (Sects. 7.1 and 12.4) and influence the physical properties of magma, they are used to classify andesites as low, medium, or high-K in character (Fig. 1.2).

K_2O usually varies linearly and positively with SiO_2, typically with correlation coefficients > 0.8 for samples from the same volcano (e.g., Whitford and Nicholls 1976). The positive correlation implies that K behaves as an incompatible element (i.e., with $\bar{D}_K \ll 1$), assuming that $\%SiO_2$ is proportional to the extent of differentiation. The correlation also emphasizes the need to specify silica content or range when comparing K or other LIL-element contents of separate suites, as for $K_{57.5}$ in Table 7.1 (i.e., %K_2O at 57.5% SiO_2).

Two separate K_2O-SiO_2 correlations with different slopes sometimes are observed in ejecta from single volcanoes. This has been documented most clearly by Gorton (1977) for Aoba and some other Vanuatu volcanoes where the distinction also encompasses differences in petrographic and trace element characteristics. Samples contributing to the different correlations are interbedded. Similar duality has been reported for Guntur volcano, Java (Whitford and Nicholls 1976) and Lipari volcano, Eolian arc (Kiesl et al. 1978), and may contribute to data scatter elsewhere. More subtle duality of K_2O-SiO_2 correlation which also affects petrographic and trace element characteristics but which, in addition, involves a contrast between tholeiitic and calcalkaline products of a single volcano, was reported by Masuda and Aoki (1979) for Hachimantai volcano, Honshu. In all cases, a duality of magma sources or evolutionary paths is implied for magma erupted from common or adjacent conduits.

Slopes of the K_2O-SiO_2 correlation usually increase as $K_{57.5}$ values increase (e.g., Dickinson 1968; Gill 1970). This increase is due to one or more of three phenomena associated with increasing $K_{57.5}$: decreased $\overline{D}_K$ during fractionation or fusion; increased initial potash contents of liquid at similar or decreasing silica contents; or increased silica content of fractionating solids. The third phenomenon implies either lower plagioclase/pyroxene ratios amongst fractionating solids or less calcic plagioclase in more K-rich suites. Both are true of phenocryst assemblages in orogenic andesites (Chap. 6).

As exceptions to the above, shallow or even negative K_2O-SiO_2 correlations sometimes occur at high $K_{57.5}$, although these correlations usually lack statistical significance. Examples include Sesara volcano, Papua New Guinea (Jakeš and Smith 1970) and Merabu volcano, Java (Whitford and Nicholls 1976). Involvement of hornblende or biotite is implied.

In general, soda contents are less variable between andesite suites than potash contents, although both increase together. As with potash, Na_2O-SiO_2 correlations are linear and positive in most andesite suites, though rates of increase of soda usually are less, resulting in decreasing Na_2O/K_2O ratios with increasing silica. However, the amount of decrease of Na_2O/K_2O ratios with increasing silica is inversely proportional to K_2O contents such that ratios may decrease very little in high-K suites (typically 0.25 over 5% SiO_2).

5.2.3 Iron and Magnesium

Although both FeO* and MgO correlate negatively with SiO_2 in most andesite suites (Figs. 5.1 to 5.3), their relative and absolute concentrations vary significantly. For example, 6 wt.% MgO was once proposed as a boundary between basalt and andesite (Thompson 1973) and the distinction between tholeiitic and calcalkaline series is based on differences in FeO*/MgO at a given silica content (Fig. 1.4).

Three aspects of these Fe and Mg variabilities in andesites are especially important petrogenetically: extent of Fe-enrichment during differentiation; Fe oxidation state; and molar $Mg/(Mg + Fe^{+2})$ ratios. Iron-enrichment is discussed first, from three viewpoints.

First, there are significant differences in the degree of Fe-enrichment within suites of orogenic andesites. Notice in Figs. 5.1 to 5.3 that $FeO^* \leqslant CaO$ in calc-alkaline suites but that FeO^* goes through a maximum, thereafter exceeding CaO, in tholeiitic suites. However, the extent of Fe-enrichment rarely attains the high levels observed in whole rock compositions in other tectonic environments (e.g., Skaergaard, Iceland) or even in some segregation veins within basalts at convergent plate boundaries (Fig. 5.4). Thus, to varying degrees, genesis of orogenic andesite involves suppression of Fe-enriching processes which often prevail in other tectonic environments.

Second, there is an inverse relationship between potash (and general LIL - element) contents and Fe-enrichment in some orogenic andesites even though the relationship is less consistent than implied by Jakeš and Gill (1970); see Section 1.2. Maximum FeO^* contents and FeO^*/MgO ratios decrease as potash contents increase across some volcanic arcs, and all suites with rapid Fe-enrichment (i.e., with $\Delta FeO^*/MgO > 1$ when $\Delta SiO_2 < 3$ wt.%) are low-K types occurring at volcanic fronts (Sect. 7.1 and Appendix). Also, within single volcanic arcs or single volcanoes, K_2O contents typically are lower in tholeiitic than calcalkaline andesites having similar SiO_2 contents (e.g., Miyashiro 1974, Table 3; Masuda and Aoki 1979).

However, this inverse relationship is crude at best for orogenic andesites as a whole. The strongest claim is that low-K orogenic andesites are likely to be Fe-enriched; in contrast, medium and high-K varieties, though more likely to be calc-alkaline than tholeiitic, vary widely in Fe-enrichment just as calcalkaline and tholeiitic suites do in K_2O contents. This is illustrated in Fig. 5.5 for the active volcanoes listed in the Appendix. Note the infrequence of low-K calcalkaline suites. Similarly, about 63% of the low-K orogenic andesites in the population discussed in Section 1.2 are tholeiitic according to Fig. 1.4 or Irvine and Baragar (1971, Fig. 2), whereas only 35% of the medium and high-K andesites are tholeiitic using Fig. 1.4 or only 13% using Irvine and Baragar's criterion. It is this relative consistency of Fe-enrichment in low-K arc suites but its otherwise inconsistent relationship with K_2O that led to my retention of the category "island arc tholeiitic series" in Section 1.2 but also to my use throughout the book of the adjectives tholeiitic and calcalkaline to denote only differences in FeO^*/MgO ratios *rather than* in LIL-element concentrations.

Third, although significant differences in Fe-enrichment occur, there is a continuum between the extremes (Fig. 5.4) which precludes radically different explanations for the extremes (Sect. 12.3). Moreover, diverse trends may prevail during the history of single volcanoes (Sect. 4.4). For example, at three volcanoes along the northeast Japan volcanic front, a reduction in the rate of Fe-enrichment

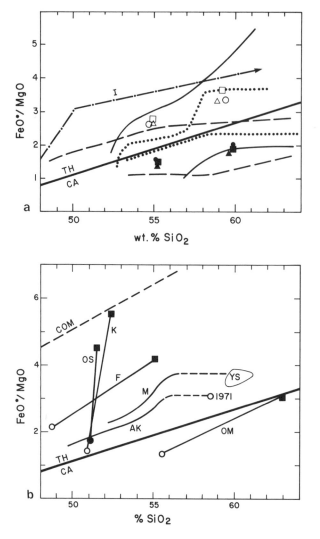

Fig. 5.4 a,b. FeO*/MgO ratios vs SiO₂. The thole-iitic/calcalkaline division is from Fig. 1.4 which also shows some similar data.

a Trends for the six suites shown in Fig. 5.1 to 5.3 are as follows: *dotted lines* low K suites; *dashed lines* medium-K; *solid lines* high-K. Data for the Rabaul, Krummel-Hoskins, and Tongariro suites scatter by up to 0.5 in FeO*/MgO ratio relative to the line fit here by eye. For comparison, also shown are the average low-K *(circles)*, medium-K *(squares)*, and high-K *(triangles)*, basic and acid, tholeiitic *(open symbols)* and calcalkaline *(closed symbols)* orogenic andesites from the population whose oxide compositions are shown in Fig. 1.1. The trend *I* is for Icelandic anorogenic andesites (Carmichael 1964)

b Lines *AK* and *M* are for Akita-Komagatake and Mashu volcanoes, Japan, respectively (Aramaki 1971; Katsui et al. 1975). The *flat dashed-line portions* represent changes in composition which occurred during repose intervals which terminated in eruption of lava in 1971 at Akita-Komagatake and in eruption of the younger somma (YS) at Mashu. Other *solid lines* connect massive lavas *(circles)* with segregation veins therein *(squares)*, thus showing in situ surficial fractionation effects (Kuno 1965). *Filled symbols* indicate the presence of magnetite phenocrysts; *open symbols* indicate their absence. All samples are from the East Japan volcanic arc except Kilauea *(K)*. Other symbols: *OS* Osima; *F* Fuji; *OM* Omuroyama. Osima lies at the volcanic front, the others behind it. The *dashed line COM* shows anorogenic andesites from the Snake River Plain, USA (Leeman et al. 1976)

Fig. 5.5. K_2O contents and FeO*/
MgO ratios at 57.5% SiO_2 for active
volcanic arc volcanoes. Data are from
the Appendix with less certain results,
e.g., those in parentheses in the
Appendix, shown as *open circles*.
Field boundaries are from Figs. 1.2
and 1.4 for 57.5% SiO_2. The *lower
right rectangle* contains "island arc
tholeiitic series". Note the relatively
even distribution of calcalkaline *(CA)*
versus tholeiitic *(TH)* character in
medium and high-K suites, but the
scarcity of low-K calcalkaline suites

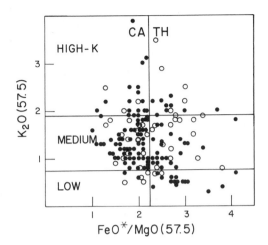

has been associated with attaining high cone elevation and with periods of dormancy during cone construction prior to caldera collapse [Hakone (Kuno 1950a,b);
Akita-Komagatake (Yagi et al. 1972); and Mashu (Katsui et al. 1975)]. This
change is shown in Fig. 5.4b. No instances of the opposite transition are known.

The oxidation state of iron in andesite magmas is poorly known. Observed
Fe_2O_3/FeO ratios in andesite lavas are variable and often high; the average ratio in
the population studied is 1.2, the median is 0.5, and the standard deviation is 4.9
(Fig. 1.1). The positively skewed distribution probably is due to surficial oxidation (Chayes 1969). There are no consistent differences in Fe_2O_3/FeO ratios
between tholeiitic and calcalkaline andesites in this population, nor between
medium and high-K suites. However, low-K suites have lower Fe_2O_3/FeO ratios.
This might indicate lower f_{O_2} but Fe_2O_3/FeO ratios also increase as the concentration increases of network-modifying cations with large ionic radii, e.g., K and
associated LIL-elements (Paul and Douglas 1965; Lauer and Morris 1977), and
there is no confirmation yet that f_{O_2} varies significantly or regularly between
andesite types (Sect. 5.3.5).

In both basic and acid medium-K andesites studied by Fudali (1965), $Fe_2O_3/$
FeO ratios by weight were about 0.25 to 0.35 when log f_{O_2} ~ −8.0 and T =
1200°C, which is considered average to low for orogenic andesites (Fig. 5.9).
Thus, a pre-eruption Fe_2O_3/FeO ratio of 0.3 was assumed when calculating Mg-
numbers and the normative mineralogy for suites shown in Fig. 5.7, although
there is no rigorous way to choose this ratio. Fe_2O_3 contents calculated by assuming this ratio often are less than those measured, or those calculated by the
method of Irvine and Baragar (1971).

The Mg-number, i.e., the molar ratio $100 \times Mg/(Mg + Fe^{2+})$, of melts and
coexisting olivines are related by the distribution coefficient (K_D), (FeO/MgO)

olivine/(FeO/MgO) melt. Consequently, if upper mantle olivines are Fo $\geqslant$ 87, then the Mg-number of a primary partial melt of the mantle must be $\geqslant$ 67 if K_D is 0.3, as determined by Roedder and Emslie (1970) and others. (A Mg-number $\geqslant$ 67 equals a FeO*/MgO ratio $\leqslant$ 1.1 if Fe_2O_3/FeO = 0.3.) This requirement that primary mantle melts have Mg-numbers $\geqslant$ 67 has been widely accepted in discussions of basalt genesis applicable to other tectonic environments. However, if fusion in the mantle wedge occurs at lower temperatures (e.g., 1100°C), then the K_D will be closer to 0.4 (Nicholls 1974; Bender et al. 1978) and the required Mg-number of primary melts will be higher, $\geqslant$ 73.

The average Mg-number of the population of orogenic andesites studied is 60. If Fe_2O_3/FeO is 0.3, the average Mg-number of orogenic andesite drops to 53. Therefore, most are not primary melts from the mantle wedge. By definition (Fig. 1.4), calcalkaline andesites have higher Mg-numbers (lower FeO*/MgO ratios) than tholeiitic andesites at any silica content and, because Mg-numbers decrease as Fe-enrichment increases, calcalkaline suites can have values which are nearly constant as well as high. Mg-numbers which are > 60 throughout suites are especially common in arcs beneath which young lithosphere is being subducted, e.g., the Cascades, Mexico, and southernmost Chile.

Only about 5% of unoxidized andesites (i.e., with Fe_2O_3/FeO < 0.5) have Mg-numbers > 67, and these magnesian andesites are of three sorts. First are the quartz-bearing but otherwise primitive "basalts" scattered throughout the western USA, including Lassen (e.g., Smith and Carmichael 1968). Some of these could have been in equilibrium with either peridotite or quartz-eclogite at high pressure (Hausel and Nash 1977). Others are hornblende andesites. Indeed, if Fe_2O_3/FeO ratios are unrestricted, most of the orogenic andesites with Mg-numbers > 67 are hornblende-bearing (see Ewart 1976a, Table 1).

The third category of andesites with high Mg-numbers consists of high-Mg andesites (> 6 wt.% MgO) usually lacking plagioclase phenocrysts. Most are pyroxene-phyric, including the "sanukites" (feldspar-free, bronzite-andesites) and "boninites" (as above but also olivine-phyric) summarized by Johannsen (1937), although unnamed high-Mg andesites containing only hornblende phenocrysts also occur. High-Mg andesites from active volcanic arcs have been described by Wilcox (1954), Noble et al. (1975), and Kay (1978); others occur in ophiolites. However, the most extensive example occurs in the present-day forearc of the Izu-Mariana arc system, cropping out in the Bonin islands (Kuroda et al. 1978), drilled in DSDP hole 458 (18°N latitude) (Meijer 1978), and dredged from the Mariana trench wall farther south (Dietrich et al. 1978; Sun and Nesbitt 1978; Bloomer et al. 1979). Most of these Mariana rocks are vesicular medium-K calcalkaline basic to acid andesites with quench phenocrysts of bronzite (En 85–90) + Cr-spinel ± clinoenstatite (En 87–92) ± olivine (Fo 85–90) ± Mg-augite. Their vesicularity despite extrusion under deep water, their lack of plagioclase, and the high H_2O contents of their glass all suggest they were water-rich. They have > 100 ppm Ni and > 500 ppm Cr so are little-fractionated. They have low normative plagioclase, very low P, Ti, Zr, Hf, and REE contents, and range from LREE-depleted to enriched.

5.2.4 Titanium

All volcanic rocks erupted at convergent plate boundaries, including andesites, have distinctively low TiO_2 contents (Chayes 1964a; Pearce and Cann 1973). Arc basalts and andesites rarely have $TiO_2 > 1.3\%$ whereas those with $FeO^*/MgO > 1.0$ in other tectonic environments rarely have $TiO_2 < 1.0\%$. Some mid-ocean ridge (especially Indian Ocean) basalts, inter-arc basin basalts, and rocks of the British Tertiary basalt province are the principal exceptions. Thus, the low TiO_2 contents in andesites reflect low contents throughout arc suites, including basalts, and the lack of sustained positive correlation between Ti and Si.

TiO_2 contents of orogenic andesites vary by a factor of five. Olivine- to quartz-normative basalts and occasional andesites of Fuji volcano, Honshu (Kuno 1962) and Slamet volcano, Java (Neumann van Padang 1951) have $TiO_2 = 1.4$ to 1.9%, and some basic andesites of Mexico have $TiO_2 \sim 1.3\%$ (Gunn and Mooser 1970), but these are exceptional. Most orogenic andesites have 0.8% to 1.0% TiO_2, although concentrations of 0.5% to 0.8% occur in the New Zealand, Tonga, New Britain, Hokkaido, and South Sandwich arcs. The lowest levels known are 0.3% to 0.4% TiO_2 in the basalts and basic andesites of Tafahi island, northernmost Tonga (Ewart 1976b) and of Manam volcano, westernmost West Bismarck arc (Morgan 1966), the acid andesites of Matthew and Hunter volcanoes, southernmost Vanuatu, and in boninites. Suggestively, all four of these instances of very-low-TiO_2 volcanism can be interpreted as products of newly established subduction (e.g., Launay et al. 1979; Meijer 1980).

TiO_2 and FeO^* contents usually correlate positively with each other and negatively with silica in orogenic andesites. However, both increase with increasing silica contents for a time in some tholeiitic basalts and basic andesites, reaching maxima between 50% and 57% SiO_2, and thereafter decreasing (Figs. 5.1 to 5.3). For example, TiO_2 increases from 0.8% to 1.2% at Miyake volcano, Izu (Kuno 1962) and from 0.5% to 0.8% at Late volcano, Tonga (Ewart et al. 1973). This parallels the Fe- and Ti-enrichment observed in tholeiitic rocks of other tectonic environments and in some groundmass glasses and basalt schlieren found at convergent boundary volcanoes. In contrast, both TiO_2 and FeO^* consistently decrease with increasing SiO_2 in calcalkaline andesites. This dissimilarity between tholeiitic and calcalkaline andesites was attributed to crystallization of magnetite throughout the calcalkaline series by Kuno (1968b) and Anderson and Gottfried (1971), and changes in slope of TiO_2 or FeO^* versus SiO_2 diagrams often coincide with the first appearance of modal magnetite phenocrysts.

5.2.5 Aluminum and Calcium

Most orogenic andesites contain 16% to 18% Al_2O_3. Low-aluminum andesites with 14% to 16% Al_2O_3 are limited to arc tholeiitic suites characterized by Fe-enrichment and low LIL-element contents (e.g., Osima and Hatizyo-zima volcanoes, Izu; Tonga; and South Sandwich arc), or to boninites. In the tholeiitic series, Al_2O_3 contents decrease modestly with increasing silica contents (Figs. 5.1 to 5.3). However, in most suites the scatter in Al_2O_3 contents exeeds any systematic variation associated with silica, probably due to plagioclase accumulation.

CaO contents vary inversely with alkalies, being highest in low-K suites but similar in tholeiitic and calcalkaline ones. In all cases, CaO decreases with increasing silica. Note in Figs. 5.1 to 5.3 that CaO < (FeO* + MgO) for orogenic andesites, which precludes their direct derivation from either an eclogitic or peridotitic source (Sects. 8.3 and 9.3).

Al_2O_3 and CaO correlate positively and significantly with plagioclase phenocryst contents in a large population of andesites having anhydrous mineralogy, indicating plagioclase accumulation (Ewart 1976a). However, aphyric andesites with 16% to 18% Al_2O_3 occur, especially in medium-K suites.

5.2.6 Phosphorous

P_2O_5 contents in orogenic andesites are between 0.05 and 0.30 wt.%, and often vary randomly within suites. However, careful analyses show contrasting behavior between tholeiitic and calcalkaline suites (Anderson and Gottfried 1971). Within tholeiitic suites, P_2O_5 contents correlate positively with silica, increasing from about 0.1% to 0.2%; P_2O_5 contents within calcalkaline suites usually remain constant or decrease within the same or an even greater P_2O_5 range. This illustrates the persistent affinity between phosphorous and iron. There is no significant correlation between P and K or Ti concentrations in the most basic andesites of the population studied.

5.2.7 CIPW Normative Mineralogy

CIPW norms represent an idealized anhydrous mineralogy into which magmas would crystallize under uniform and slow but otherwiese unspecified cooling conditions. Aspects of the normative mineralogy or the population of orogenic andesites studied are shown in Figs. 5.6 and 5.7.

Only 1% of orogenic andesites are *ol*-normative, and the average normative *qz* content of the population studied is 12.6%, which exceeds the maximum *qz* permitted andesite in some petrographic classifications (Streckeisen 1979) regardless of the iron oxidation state used when calculating the norm (Chayes 1969, 1971). High *qz* contents are most pronounced in low-K suites because *qz* contents decrease relative to silica as potash levels increase.

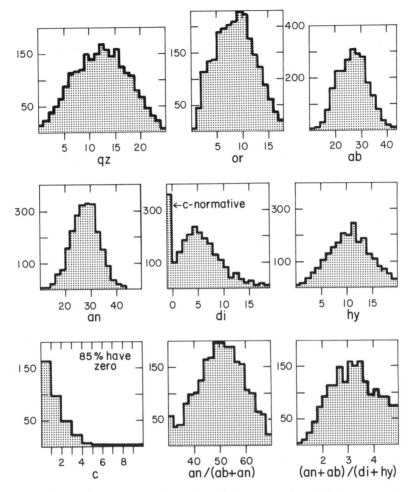

Fig. 5.6. Histogram of wt.% CIPW normative mineral compositions of the 2500 orogenic andesites whose oxide compositions are shown in Fig. 1.1. In addition to the data shown, 25 andesites have *ol* instead of *qz*, 65 have $>20\%\, hy$, and 476 have $(ab + an)/(hy + di) > 5$

The average composition of normative plagioclase in the orogenic andesite population studied is An51 ± 9, confirming the ineffectiveness of normative An50 as a useful discriminant between basalt and andesite, and confirming that normative plagioclase is more sodic than modal plagioclase phenocrysts (Fig. 6.2). The ratios $(ab + an)/(di + hy)$ and di/hy are higher in average tholeiitic than calcalkaline andesites but these differences are not reflected in phenocryst mineralogy.

About 15% of orogenic andesites in the population studied are corundum-normative. This peraluminous character is more common in acid than basic andesites, and is especially prevalent in high-K acid andesites, occurring in about 25% of these analyses. The amount of c typically is $< 2\%$ and does not increase as the percentage of corundiferous andesite increases. Unless accompanied by peraluminous minerals, this distribution of c usually indicates accentuation of analytical inaccuracy at high silica contents (Chayes 1970) rather than increasing assimilation of pelitic material (Bowen 1928) or the fractionation of pyroxene or amphibole without plagioclase (Yoder and Kushiro 1972; Ujike 1975; Cawthorn and Brown 1976).

Normative mineralogy is also useful in relating complex rocks to simple oxide systems. The basalt tetrahedron of Yoder and Tilley (1962), involving the components Ne + Ol + Qz + Cpx, includes more than 90% of the normative components of most andesites as well; in contrast, the salic tetrahedron, involving Or + Ab + Qz + An, includes only 70% to 80% of the normative components of andesites.

Trends of andesite analyses within the silica-saturated portion of the basalt tetrahedron are shown in Fig. 5.7. Neither within-suite variability nor between-suite differences in normative components are as great as within or between basalt suites, and problems resulting from phenocryst accumulation can be acute when plotting andesite analyses in this way. However, andesites define consistent trends within the tetrahedron. Orogenic andesites usually approach and occasionally reach the base of the tetrahedron (i.e., become di-free and corundum-bearing), increasing their hy/di and $(ab + an)/(di + hy)$ ratios as they become enriched in qz. In each of these aspects, most orogenic andesites seem to define an apparent cotectic trend consistent with univariant equilibrium between andesite liquids and the solid assemblage plagioclase + orthopyroxene + clinopyroxene (line ab in Fig. 5.7). The trends approximately coincide with those of intermediate members of the Skaergaard and Thingmuli suites, but lack the initial migration of liquids away from the plagioclase apex which characterizes those two suites.

No consistent difference in vector within the basalt tetrahedron distinguishes tholeiitic from calcalkaline differentiation trends or accompanies transition from tholeiitic to calcalkaline behavior within suites. Higher potash contents are accompanied by higher normative plagioclase contents in the liquid and, therefore, by reduction in the size of the plagioclase solid volume within the tetrahedron. This reduction in size may be due either to higher water or soda activities in more LIL-element-rich andesites.

There is some correspondence between modal and normative mineralogy in andesites. For example, lavas erupted from Parícutin volcano, Mexico, were distinguished by lack of clinopyroxene phenocrysts (Wilcox 1954). These lavas were notably poor in normative di as well, and the small amount of magma differentiation thoughout the three-year cone buidding stage was characterized by migration of liquids toward the di apex of the normative tetrahedron, in contrast to the trend of most andesite suites. However, subsequent differentiation at Parícutin followed the same apparent four-phase cotectic mentioned above despite the continued absence of clinopyroxene as a phenocryst, or as a near-liquidus phase at

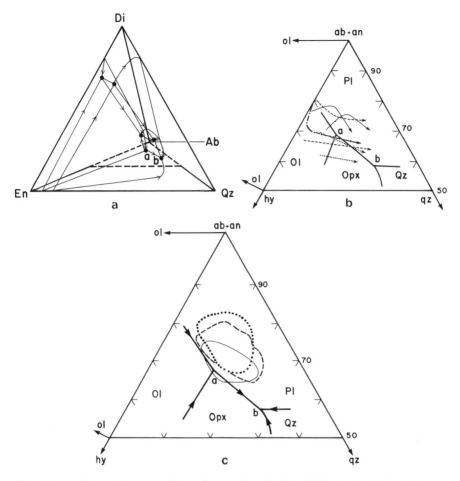

Fig. 5.7 a–c. Normative composition of orogenic andesites within the basalt tetrahedron. **a** Liquidus phase relations in the silica-saturated portion of the Ne + Di + Fo + Qz tetrahedron at 1 atm after Yoder (1976), Presnall et al. (1978), and sources therein. Field of pigeonite is omitted for simplicity. *Arrows* indicate direction of decreasing temperature. *Dashed lines* at the base show the triangle which is reproduced in **b** and **c**. **b** Analogous to a portion of the base of **a**, projected from Di. Here (*ab* + *an*) are plotted together and FeO is included in *di* and *hy*; all components amounts are CIPW normative percentages. Invariant points *a* and *b* are as in the base of **a** and apply at 1 atm to an An50 normative plagioclase composition which is the mode for orogenic andesites (Fig. 5.6). *Pl, Ol, Opx,* and *Qz* identify liquidus phases in the Fe, K, H$_2$O-free system. Trend lines are shown for the six low, medium, and high-K suites of Fig. 5.1 to 5.3. *Dotted lines* are low-K; *dashed lines* medium-K; *solid lines* high-K; and the *dash-dot line* is for the anorogenic andesites of Thingmuli, Iceland. **c** Phase boundaries as in **b**. The three lines show the location of 70% to 85% of the low *(solid line)*, medium *(dashed)*, and high-K *(dotted)* orogenic andesite populations whose average compositions are given in Table 5.1 and whose normative plagioclase compositions are shown in Fig. 6.2. Note in both **b** and **c** the displacement toward (*ab* + *an*) and away from qz of the more K-rich suites and populations.

$< 5\%$ H_2O (Fig. 6.6c; Eggler 1972a). Samples with hornblende phenocrysts lie along the same general cotectic as do hornblende-free andesites. Samples with normative hypersthene but olivine phenocrysts have low FeO^*/MgO ratios (e.g., Paricutin, Mexico or Mt. Hood, USA) or high alkali contents or both.

5.3 Volatiles

It is important to know the volatile content and especially the spatial and temporal extent of vapor-saturation of andesite magma because the latter bears on eruption prediction studies, magmatic differentiation mechanisms, the fluid dynamic regime, and relationships with hydrothermal ores. However, volatiles are lost through exsolution before or during eruption, and can be gained or lost during devitrification, weathering, or alteration after eruption. Thus, pre-eruption concentrations of volatile components in andesites are as difficult to determine as they are important, and probably are the most important unknowns in andesite petrogenesis.

5.3.1 Water

Water is the dominant constituent of andesite volcano gases (White and Waring 1963), yet informed estimates of pre-eruption water contents of andesite magmas are few. High but unspecified water contents have been assumed from the following: (1) frequency of explosive eruptions (Sect. 4.2); (2) occasional presence of hydrous minerals as phenocrysts or in inclusions or both (Sect. 11.4); (3) high anorthite content of plagioclase phenocrysts (Sect. 6.1); (4) better correlation between FeO^*/MgO ratios and liquidus temperatures when $P_{H_2O} = 2$ kb than when at 1 atm (Sect. 4.1.1); (5) observed eruption temperatures which are less than the experimentally determined 1 atm liquidus temperature of the same magmas (Sect. 4.1.1); and (6) high water contents of glass inclusions within phenocrysts. Conversely, low water contents in andesite magmas are implied by: (7) the ubiquity of plagiocalse phenocrysts (Sects. 6.1 and 6.9) and (8) reverse zoning of plagioclase phenocrysts (Sect. 6.1). Most of the observations above have alternative explanations, as noted in the sections cited, or have limited applicability as for point (2).

Quantitative estimates of water contents come from three sources: direct measurement, experimentation, and thermodynamic calculations. Direct measurements of the water contents of andesites usually do not reflect pre-eruption concentrations. However, compositions of unaltered glass on submarine pillow lavas quenched at depths sufficient to prevent vapor exsolution may. Glass on medium-K, tholeiitic, basic to acid pyroxene andesite pillows from the Mariana arc contains about 1 wt.% H_2O (Garcia et al. 1979). This might overestimate pre-eruption concentrations, based on the experience of Muenow et al. (1979) for similar data from Hawaiian samples. However, H_2O/CO_2 ratios are much higher in glass-vapor inclusions in phenocrysts from these orogenic andesites than in similar inclusions from basalts of other tectonic environments (cf. above citations with DeLaney et al. 1978), suggesting that andesitic water contents are indeed unusually high. H_2O/K_2O wt.% ratios are about 1 in the pillow andesite rims, as they also are in glassy basalt rims from other tectonic environments [see data of Moore in Anderson (1975), Fig. 9], and the ratios in andesite are constant over a 5 wt.% SiO_2 range.

Deep-water pillow andesites are uncommon. Much more ubiquitous is the presence in phenocrysts of glass inclusions whose compositions will be those of magma at the time of phenocryst precipitation (point 6 above) if two requirements are met. First, the glass must record bulk liquid compositions and not boundary layer effects, i.e., atypical concentration gradients adjacent to crystal surfaces. Second, the inclusion must not have changed in composition since entrapment due to leakage, crystallization, vapor exsolution, or diffusion. Anderson (1979), who has pioneered study of fluid inclusions in andesites, claims these requirements are met by inclusions $> 30~\mu m$ in diameter, but unfortunately such large inclusions may themselves represent atypical events (eruptions?) which accelerate crystal growth. Most large glass inclusions with andesitic compositions have summation deficiencies of 2 to 4 wt.% (Anderson 1974a,b, 1976, 1979) which may equal or overestimate water contents in andesites (Garcia et al. 1979). Similarity of glass compositions in coexisting olivines, pyroxenes, and feldspars in at least some of these instances implies that the first requirement above was met.

Furthermore, Anderson (1974b, 1975, 1979) argued that magmas represented by these glass inclusions were water-saturated during phenocryst precipitation, based on three observations. First, some inclusions contain vapor, possibly water, bubbles. Second, estimated water contents of the glass inclusions remain constant or even decrease with increasing K_2O or SiO_2, implying buffering of water contents by a vapor phase during differentiation. (Note that this constancy differs from the results for glassy rims on andesite pillow lavas cited earlier). Third, Cl contents also fail to increase as rapidly as do K_2O contents, indicating that Cl also is buffered by a water or HCl vapor.

The second way to estimate pre-eruption water contents is to compare phenocryst crystallization sequences in rocks with experimentally determined phase relationships for liquids of the same composition. This is effective because dissolved water depresses the thermal stability of plagioclase more than pyroxene, and of pyroxene more than olivine, thereby affecting liquidus phase relationships (Sect. 6.9). Several published studies of this kind are available. Eggler (1972a) concluded that a 1952 calcalkaline acid andesite from Parícutin volcano, Mexico (col. 10, Table 6.2 and the labeled point in Fig. 4.7) contained 2.2 ± 0.5 wt.% H_2O before eruption. This conclusion essentially quantifies point (7) above, as water contents must be high enough for olivine and orthopyroxene to be near-liquidus phases but low enough to retain liquidus plagioclase. Water-saturation is compatible with the observed phenocryst assemblage only at < 700 bar pressure.

Although the 1952 Parícutin andesite contained 0.1% plagioclase phenocrysts, most earlier Parícutin andesites (except earliest tephra) lack plagioclase and contain mostly olivine phenocrysts (Fig. 4.7), which weakens Eggler's particular argument (cf. his results for an older, nonplagioclase-phyric sample). However, his conclusion that most andesite magmas must contain < 3 wt.% H_2O because plagioclase phenocrysts are common, remains widely applicable.

Similar rationale led Sekine et al. (1979) to a similar conclusion concerning a 1970 tholeiitic acid andesite from Akita-Komagatake volcano, Japan. They estimated both P_{H_2O} and the temperature of phenocryst precipitation from the unique combination of these parameters at which plagioclase and pyroxene are co-liquidus phases. Their values of 0.6 to 2.8 wt.% H_2O are maxima because they assumed saturation with pure water.

A potentially more sensitive but less well-studied point involves olivine. Some andesites precipitate olivine under modest water pressures (Sect. 6.9) but lack olivine phenocrysts. For example, basic andesite sample L1 of Ewart et al. (1973) lacks olivine but nevertheless precipitates it as the sole liquidus phase between 5 and 7 kb under water-saturated conditions (Nicholls and Ringwood 1973). Determination of the minimum water content necessary for olivine precipitation in such a sample would therefore provide an estimate of the maximum pre-eruption water content.

A third approach (point 2 above) involves hornblende whose mere occurrence requires at least 3 wt.% H_2O in the liquid phase (not bulk magma) from which it precipitated (Burnham 1979). Comparison of phenocryst assemblage to phase stabilities for a hornblende-bearing calcalkaline basic andesite from the Crater Lake, USA, tephra mentioned in Section 4.4 indicates that the magma contained about 4 wt.% H_2O, and had crystallized over $100°C$ below its liquidus, before eruption (Ritchey and Eggler 1978).

The final utility of experimental petrology in constraining water contents involves comparing eruption temperatures with liquidus temperatures determined experimentally as a function of water content (point 5 above). Both the acid pyroxene andesites from Parícutin and Akita-Komagatake mentioned above erupted at about $1100°C$ (Table 4.1) and both require 1 to 2 wt.% H_2O to suppress their liquidus to this temperature (Fig. 4.1; Eggler 1972a; Aramaki and Katsura 1973; Sekine et al. 1979).

The last way to estimate pre-eruption water contents uses the composition of co-existing phases. First, one may calculate $a_{H_2O}^{fluid}$ and, from it, calculate water contents following Carmichael et al. (1977) or Burnham (1979) if P and T are known. Available solutions for orogenic andesites use co-existing hornblende + pyroxene(s) + magnetite $\pm$ plagioclase and olivine, and yield estimates of $a_{H_2O}^{fluid}$ of 01. to 1.0 or P_{H_2O} of 350 to 2800 bar (Powell 1978), $a_{H_2O}^{fluid}$ of 0.1 to 0.5 (Arculus and Sills 1980), and P_{H_2O} of 600 to 1700 bar or 2 to 3.6 Wt.% H_2O (Luhr and Carmichael 1980). The first two used data from mafic inclusions; the third from phenocrysts. The results are unreliable because of the complex composition and resulting complex crystalline solution models for hornblendes, and are useful only for relative, not absolute, water contents.

A more widely applicable approach involves calibrating the increase in anorthite content of plagioclase crystals at a given temperature and melt composition as a result of dissolved water (point 3 above). One proposed calibration is by Mathez (1973) whose equations were used in preparing Fig. 6.1 and the following discussion for illustration, but which do not seem to yield accurate results. For example, if one assumes that all phenocrysts equilibrated at a common temperature and then estimates the water pressure necessary for plagioclase-liquid pairs to yield the same temperature calculated from coexisting pyroxenes or oxides, water pressures of several kb result (e.g., Heming 1977; Ritchey 1980) which are inconsistent with plagioclase being the liquidus phase, as discussed above. Alternatively, differences in calculated temperatures between plagioclase and other minerals may reflect a crystallization interval. This argument was used by Ewart (1976b) to claim very low water contents in Tongan andesites, but efforts to quantify the argument are circular without independent information about liquidus temperatures.

Such information is available for the 1970 Akita-Komagatake andesite mentioned above whose observed eruption temperature was 1090°C (Armamaki and Katsura 1973). Plagioclase phenocrysts from this flow have An90 cores mantled by broad areas zoned in oscillatory fashion around an average composition of An65, and An50 rims similar in composition to groundmass microlites. Two pyroxenes and magnetite also occur as phenocrysts. Compositions of phenocrysts and groundmass phases indicate temperatures (at 1 bar) of 1270° to 1280°C for plagioclase core-whole rock pairs and 1110° to 1130°C for plagioclase rim-groundmass or groundmass plagioclase-groundmass glass pairs, and 1000° to 1030°C for coexisting pyroxene phenocrysts. These temperatures are similar to those calculated in the same way by Ewart (1976b) for the Tongan acid andesites mentioned above. However, laboratory experiments by Aramaki and Katsura at 1 bar found that plagioclase and pyroxene precipitated from the whole rock composition at 1200° and 1120°C, respectively. Clearly, the calculated equilibration temperatures for phenocryst cores at 1 bar are not liquidus temperatures. However, the 1 bar liquidus temperature for groundmass glass, the calculated quench temperature for groundmass plagioclase with this glass, and the observed eruption temperature all are between 1090° and 1130°C. The An65 mantles would be in equilibrium with the bulk groundmass at 1100°C when $P_{H_2O} \sim 1000$ bar (Fig. 6.1) Thus these calculations support Aramaki and Katsura's (1973) interpretation that the magma precipitated its phenocrysts at some modest water pressure (they suggest 200 to 500 bar; my calculations suggest 1000 bar), ascended along a cotectic while dehydrating, and erupted, all at ~1100°C.

Figure 5.8 shows the isothermal solubility of water in an acid andesite, indicating that pressure is the most important single limit to water solubility.

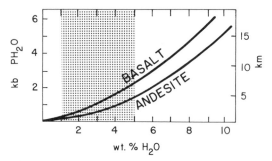

Fig. 5.8. Solubility of water in Columbia River basalt and Mount Hood acid andesite (Hamilton et al. 1964). *Stipled region* indicates the water content of most orogenic andesites which, in turn, implies that andesite magma in portions of reservoirs deeper than 0.5 to 2.5 km will be water-undersaturated. However, Sakuyama and Kushiro (1979) reported that at higher pressure 3 to 4 wt.% less water is necessary for saturation than would be inferred by extrapolating the curves shown here

This figure applies approximately to all andesite types because water solubility even on a wt.% basis is largely independent of magma composition. Note that andesite magma containing 3 wt.% water will be saturated only within 2.5 km of the Earth's surface. Saturation even at the top of shallow reservoirs 5 km deep (see Sect. 3.6) would require > 6 wt.% water and $P_{H_2O} > 2000$ bar. Thus, data summarized above suggests either that precipitation of phenocrysts occurs at $P_{H_2O} = 500$ to 1000 bar $< P_{TOTAL}$ from water-undersaturated magma within these reservoirs, or that precipitation occurs at $P_{H_2O} = 500$ to 1000 bar $= P_{TOTAL}$ within near-surface conduits above them. In either case, no separate water vapor phase coexists with andesite magma in most of the reservoir volume. In the latter case, magmatic differentiation by removal of phenocryst phases will be restricted by the limited spatial extent and temporal duration of their precipitation.

In summary, most plagioclase-phyric orogenic andesite magmas are hydrated but not extremely so, usually containing 1 to 5 wt.% H_2O before eruption, approximately equal to or greater than K_2O contents, and precipitating phenocrysts at $P_{H_2O} \leqslant 1000$ bar $< P_{TOTAL}$. Probably there are across-arc variations in H_2O contents (Sect. 7.1) and some medium to high-K hornblende-bearing acid andesite magmas may have had as much as 5 wt.% water.

Finally, three general aspects of water contents in andesite merit comment. First, if $a_{H_2O}^{melt}$ is constant due to equilibrium throughout an isothermal magma volume, then water content of the magma will decrease with depth (Kennedy 1955). Even magma conduits saturated with water at their top will be undersaturated at depth. Eruption of increasingly water-poor magma with time would explain decreasing explosivity during eruptions (Sect. 4.3) and increased thermal stability of pyroxene in younger ejecta (Sekine et al. 1979). Second, hydrous magma must become saturated in water during ascent, resulting in boiling off of water vapor and in crystallization promoted by the increase in liquidus temperature accompanying water loss. This in turn will be accompanied by significant super-saturation and viscosity increase. Also note in Fig. 4.1 that the liquidus of hydrous magma has a negative dP/dT slope below a pressure which is directly proportional to the magma's water content. Thus, water contents will affect the depth at which the ascent of andesite magma is arrested and differentiation occurs, with more water-rich magmas differentiating at greater depths. Finally, if other factors are equal, increased water contents result in decreased density, viscosity, and liquidus and solidus temperatures of magmas, increased crystal growth and nucleation rates, and increased thermal stability of nesosilicates relative to ino- and tektosilicates. Each phenomenon is associated with the role of water in depolymerizing silicate liquids (Burnham 1975). Thus, differences in temperature, texture, eruption style and mineralogy of andesites may strongly reflect differences in their water contents which are not yet known quantitatively.

5.3.2 Carbon Dioxide

Carbon gases are a distant second to water in concentration in andesite volcano fumaroles where CO_2 is the dominant species. Much less CO_2 than H_2O can dissolve in andesite magma due to the inability of CO_2 to depolymerize silicate liquid, and the solubility of CO_2 is more dependent on temperature and magma composition than in the case of H_2O. Although CO_2 solubility at crustal pressures and $900°$ to $1100°C$ is unknown, values ~ 0.5 wt.% are expected (Mysen et al. 1975). CO_2 solubility reaches a maximum at some intermediate CO_2/H_2O ratio for the entire system. [In contrast, water solubility decreases steadily as the CO_2/H_2O ratio increases (Mysen 1976), which seems to be primarily a dilution effect.] Mysen et al. (1975) argue that both CO_2 and CO_3^{2-} molecules exist as dissolved species in andesite magma, with increased fO_2 favoring CO_3^{2-}.

Although it would require little CO_2 to saturate andesite magma, it is unknown whether sufficient pre-eruption concentrations occur. The only such estimate is ~ 0.25 wt.% (Garcia et al. 1979). However, the H_2O/CO_2 molar ratio of glass inclusions in plagioclase from these Mariana arc samples is about 1:1. Consequently, CO_2 is more likely than H_2O to exsolve as a vapor phase from such andesite magmas. Indeed, Gustafson (1978) suggested that early separation of CO_2-rich vapor gradually purges intermediate magmas of sulfur and base metals, with porphyry copper deposits developing only above those magma reservoirs in which CO_2-saturation failed to occur.

5.3.3 Sulfur

Sulfur gas species (H_2S, SO, SO_2, SO_3) in fumaroles vary mostly due to temperature with SO_2 predominating at magmatic temperatures, and oxidize quickly to sulfates. Although S/Cl ratios in fumaroles consistently increase prior to eruption of andesite from several Central American volcanoes (Stoiber and Rose 1973), pre-eruption S contents of andesite magmas are little studied.

Subaerially erupted andesites typically contain < 200 ppm S (e.g., Naldrett et al. 1978). These low concentrations probably reflect exsolution of SO_2 upon eruption, but only 30 to 100 ppm S was found in glassy rims on nonvesicular pillow andesites from the Mariana arc (Garcia et al. 1979). In contrast, andesite glass inclusions in olivine commonly contain 200 to 2000 ppm S, but concentrations are less than in related basalt inclusions (Anderson 1974b; Rose et al. 1978). Values of log f_{S_2} for acid andesites from Rabaul volcano, New Britain, were estimated by Heming and Carmichael (1973) to decrease from $+ 1.0$ to $- 2.0$ with decreasing temperature, based on calculated quench temperatures and the composition of pyrrhotites with which the andesites were saturated. For comparison, submarine basalt glasses contain 500 to 2000 ppm S; differentiated samples are saturated with a condensed Fe-Ni-Cu sulfide phase (Moore and Fabbi 1971; Mathez 1976).

The solubility of S in andesite reservoirs is not yet known, although values may be less than that of CO_2 in andesite (Mysen 1977), and less than that of S in basalt. The latter is expected because S solubility decreases with increasing f_{O_2} and with decreasing temperature and Fe contents (Haughton et al. 1974). Consequently, maximum S values of 200 to 2000 ppm are expected and some orogenic andesites could be saturated with either a sulfide liquid or solid or an S-species vapor. Obviously saturation will be the case for andesites derived from S-saturated basalt by differentiation. However, lack of correlation between Ni and Cu contents in andesites (Sect. 5.4.6) indicates that silicate liquid-sulfide solid or liquid fractionation is inconsequential for most elements (Sect. 11.7), though it may help explain low S contents in andesites.

5.3.4 Halogens

Halogen contents in andesite volcano fumaroles and in andesite rocks usually decrease in the order $Cl \geqslant F > Br > I$, with Cl and F concentrations in rocks typically being similar to that of S. Cl and F contents in subaerially erupted orogenic andesites each usually are 100 to 1000 ppm; Cl/Br and Cl/I ratios are about 70 to 300 and 1000, respectively (Sugiura 1968; Yoshida et al. 1971; Heming and Carmichael 1973). These underestimate pre-eruption magma concentrations due to vaporization during eruption (Anderson 1974b, Fig. 9) or leaching during cooling, although only ~ 1000 ppm each of Cl and F occur in glassy rims of pillow andesites from the Mariana arc (Garcia et al. 1979). Strong positive correlation exists between F and K_2O, even in subaerial andesites (Ishikawa et al. 1980).

Halogen solubilites in andesite magma generally are unknown. Iwasaki and Katsura (1967) measured the solubility of HCl in acid andesite at 1 atm as about 4000 and 3000 ppm at $1200°$ and $1290°C$, respectively, and found that solubility was less than in basalt at the same temperatures. Cl contents in the andesite glass inclusions in phenocrysts discussed earlier are 500 to 2000 ppm which, like H_2O and S, remain constant or decrease despite increasing potash contents (Anderson 1974b). Anderson interpreted this behavior, as with H_2O and S, as indicating vapor-saturated conditions in which Cl is partitioned strongly into the vapor phase even if the magma is not saturated with Cl. Once an aqueous phase exsolves, Cl is rapidly lost from the silicate liquid; the resulting brine is an effective concentrator of transition metals, thereby contributing to fractionation of the andesite magma and to creation of metal-rich hydrothermal fluids (Holland 1972; Burnham 1979). In contrast, F is expected to be more soluble than Cl in andesite melts (Burnham 1979), to prefer silicate melts over coexisting vapor (Wyllie 1979), and to behave as a typical incompatible trace element, increasing in concentration as differentiation proceeds. Empirically, F behaves incompatibly in tholeiitic but not calcalkaline arc suites (Ishikawa et al. 1980).

Cl/S ratios of gases, leachates, and lavas, and Cl/K ratios of lavas and glass inclusions in phenocrysts all are higher in orogenic andesites than in MORB, ocean island basalts, or backarc basalts (Stoiber and Rose 1973; Anderson 1974b, 1979; Sigvaldason and Oskarsson 1976; Rose et al. 1978; Garcia et al. 1979). For example, Cl/K_2O ratios in orogenic andesites typically are 0.1 to 0.2 and decrease with differentiation whereas Cl/K_2O ratios in basalts from other tectonic environments are 0.025 to 0.04. Sparse data from the above sources also suggest that H_2O/Cl ratios are lower in volcanic arcs than elsewhere ($\sim$ 10 vs. 35). This Cl-enrichment could result from incorporating seawater during subduction, especially if by dehydration. However, neither the pre-eruption concentration of volatiles nor the processes which fractionate them are well enough known for this to be a strong argument. Nevertheless, volatile concentrations or ratios may eventually provide sensitive tests of the role in andesite genesis of oceanic crust or its dehydration. Andesites to check are those from the Eolian and Aegean arcs, as thick evaporite sequences are available for subduction there. However, at least F contents are no higher in Eolian than Japanese lavas (Kluger et al. 1975; Kiesl et al. 1978).

5.3.5 Oxygen

Lavas in volcanic arcs have relatively high Fe_2O_3/FeO ratios (Sect. 5.2.3) and their minerals have high Mg/Fe^{+2} ratios (Chap. 6), suggesting that the corresponding magmas are more oxidized as well as more hydrous than are the voluminous magmas of other tectonic environments. This suggestion is difficult to quantify or even confirm, but it is central to understanding andesite genesis because of the sensitive dependence of magnetite saturation on f_{O_2} (Sect. 6.9 and 11.3).

No field measurements of f_{O_2} have been made for andesite lavas, but laboratory experiments: (1) have reproduced the observed Fe_2O_3/FeO ratio of basic and acid andesites when log f_{O_2} = -7.0 to -8.2 at 1200°C (Fudali 1965); (2) have found that liquidus temperatures were consistent with an observed eruption temperature of $\sim$ 1100°C for acid andesite when log f_{O_2} $\leqslant$ $-$ 9 to 10 (Aramaki and Katsura 1973); and (3) have indicated log f_{O_2} = $-$ 7.5 to 8.2 at 1200°C using the ZrO_2 electrolytic cell technique (Drory and Ulmer 1974). Quench oxygen fugacities can be calculated to ± 1 log unit from the composition of coexisting Fe-Ti oxides using the calibrations of Buddington and Lindsley (1964), Katsura

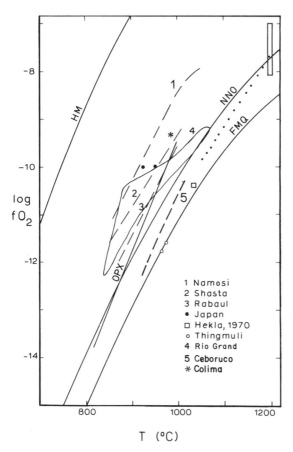

Fig. 5.9. Quench temperatures and oxygen fugacities of orogenic and other andesites estimated from compositions of co-existing oxide phenocrysts. Data for other rock types are omitted. *Rectangle* at 1200°C encloses experimental results of Fudali (1965) for an orogenic andesite at 1 atm. Data sources: Namosi, Fiji (Gill and Till 1978); Shasta, USA (Anderson 1975); Rabaul, PNG (Heming and Carmichael 1973); Japan (Buddington and Lindsley 1964); OPX (Carmichael 1967a); Hekla and Thingmuli, Iceland (Baldridge et al. 1973; Carmichael 1967b); Rio Grande, USA (Zimmerman and Kudo 1979); Ceboruco and Colima, Mexico (Luhr and Carmichael 1980). Also shown by the *dotted line* are average results obtained for three orogenic andesites at 1 atm by the ZrO_2 electrolytic cell method (Drory and Ulmer 1974)

et al. (1976), and Powell and Powell (1977), subject to the difficulties discussed for this method in Section 6.5. The few data available for orogenic andesites are shown in Fig. 5.9. Oxygen fugacity does not remain constant. The trends are parallel to but usually higher than the fayalite + magnetite + quartz (FMQ) oxygen buffer followed by many basalts and icelandites. Despite containing the same orthopyroxene + clinopyroxene + magnetite assemblage which helps keep f_{O_2} near the nickel + nickel oxide (NNO) buffer in many dacites and rhyolites, most orogenic andesites lie up to one log unit above that buffer curve also, possibly due to the lower silica activities of the andesites (Heming and Carmichael 1973). Estimates of log f_{O_2} in andesites from the magnitude of Eu anomalies in plagioclase phenocrysts are $- 8.1 \pm 5.7$ (Drake 1975), but this method can yield discrepant results (Ewart 1976b; Mysen et al. 1978b).

Available data are insufficient to test whether significant and consistent differences in f_{O_2} exist between low and high-K andesites, calcalkaline or tholeiitic andesites, or andesites with hydrous instead of anhydrous mineral assemblages. Consequently the relative roles of minerals, liquid composition, and volatiles in buffering f_{O_2} remain unknown. The important observation is that f_{O_2} is unexpectedly high.

5.4 Trace Elements

Most trace element concentrations in andesites have been determined since 1965. The following discussion is limited by frequent inability to assess accuracy of these data, by infrequency of thorough analyses of single samples, and by lack of information about the mineralogy or stratigraphic relationships of analyzed samples. Thus, most studies have been regional in scope, descriptive in style, and limited to a few elements. However, these studies are adequate to show that andesites are sufficiently diverse to require separate discussion of different types and of discrete suites, rather than of andesites in general. Thus, my emphasis here is on similarities and differences in overall concentrations and on within-suite variability between low, medium and high-K andesites, between calcalkaline versus tholeiitic andesites, and between orogenic andesites and oceanic basalts. For the latter comparison I will follow Sun and Nesbitt (1978) and Wood et al. (1979a,b) in distinguishing between N-type MORB (normal) and E-type MORB (enriched).

I did not calculate average trace element concentrations for all orogenic andesites or for sub-types thereof; consult Taylor (1969), Jakeš and White (1972a), or Ewart (1979, 1980) for such estimates. However, trace element data are included in Table 5.2 for representative examples of different andesite types, and are available for entire suites in the references cited in Table 5.3.

5.4.1 The K-Group: Rb, Cs, Ba, and Sr

These elements form large cations whose concentrations generally correlate positively with K_2O and SiO_2 in andesites. Thus, Rb, Cs, and Ba contents are higher in high-K than low-K andesites at a given silica content and usually do not reflect differences in iron-enrichment between suites, although K_2O is higher relative to SiO_2 or MgO in calcalkaline than tholeiitic suites in northeastern Japan (Masuda and Aoki 1979). Rb and Cs typically increase twofold or more over the range 53% to 63% SiO_2, regardless of absolute concentration. Rb contents range from 1 to 5 ppm in low-K basic andesites (e.g., Blake and Ewart 1974) to $\sim$ 100 ppm in high-K acid andesites (e.g., Dupuy and Lefèvre 1974). Cs data are less abundant but contents vary from $<$ 1 to 10 ppm (e.g., Condie and Swenson 1973; Kiesl et al. 1978). Reported Ba contents often vary erratically by $\pm$ 50% in orogenic andesites.

K/Rb ratios decrease with increasing K_2O contents (Fig. 5.10) such that low-K andesites sometimes have K/Rb $\sim$ 1000, as high as in N-type MORB, while high-K andesites have K/Rb $\sim$ 250, as low as in average continental crust (Jakeš and White 1970). Variations within suites and differences between low, medium, and high-K suites usually lie above but parallel to Shaw's (1968) main trend,

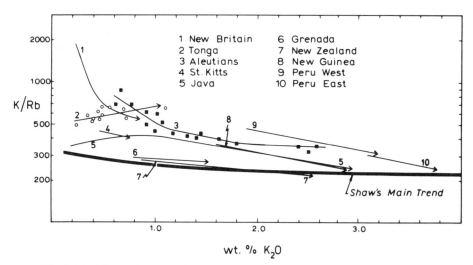

Fig. 5.10. K/Rb ratios vs. K_2O contents in selected orogenic andesites. *Lines* are informal best-fits to data, examples of which are shown for two suites. *Arrows* point in the direction of increasing SiO_2 contents. *Heavy line* is from Shaw (1968). Other data sources are: *1* Blake and Ewart (1974); *2* Ewart et al. (1973); *3* DeLong (1974), Semisopochnoi; *4* and *6* Brown et al. (1977); *5* Whitford (1975); *7* Ewart and Stipp (1968); *8* Mackenzie (1976); *9* and *10* Dupuy and Lefèvre (1974). Lines *9* and *10* refer to volcanoes closer to and farther from, respectively, the Peru-Chile trench

although two kinds of exceptions occur. First are apparent regional variations. For example, andesites of the Antilles and New Zealand seem to have K/Rb ratios about 50% lower relative to K_2O contents than in other arcs (Ewart and Stipp 1968; Hedge and Lewis 1971; Brown et al. 1977). Second, the decrease in K/Rb within andesite suites often is less than as summarized by Shaw (1968) and in some suites K/Rb ratios even increase with silica contents (Fig. 5.10).

Only in some low-K andesites do Sr contents correlate positively with silica (Ewart et al. 1973; Ando 1975; Baker 1978). Instead, Sr contents typically remain quite constant, sometimes decreasing in acid andesites of high-K suites (Fig. 5.11). The range in Sr contents between suites is considerable, from about 100 ppm which is typical of MORB (e.g., Ando 1975; Baker 1978) to ~ 1000 ppm. Sr contents usually increase between suites as K_2O increases, accounting for some of this variability. The inverse correlation between Sr and K suggested by Gunn (1974) seems atypical and limited to a few anomalous instances of local and as yet inexplicable Sr enrichment. For example, the Shasta and Lassen volcanoes, USA, inconsistently have erupted medium-K basic and acid andesites with 1100 to 1500 ppm Sr (Peterman et al. 1970b; Condie and Swenson 1973), and high concentrations also characterize medium-K basic andesites of Grenada, Antilles

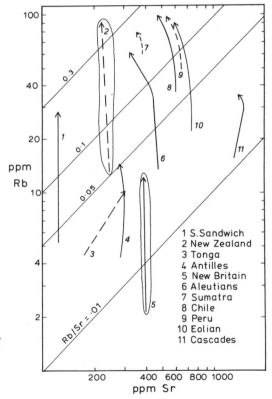

Fig. 5.11. Rb vs. Sr contents in selected orogenic andesites. *Lines* and *arrows* are as in Fig. 5.10; size of data fields are shown for two examples. *Dashed lines* indicate that the data populations yield pseudoisochrons, e.g., Fig. 10.2. Note that there is no difference in elemental behavior between suites with or without pseudoisochrons. Data sources: *1* Baker (1978); *2* Ewart and Stipp (1968); *3* Ewart et al. (1973); *4* Brown et al. (1977), St. Kitts; *5* Blake and Ewart (1974); *6* DeLong (1974), Semisopochnoi; *7* Whitford (1975), Merapi; *8* Roobol et al. (1976); *9* James et al. (1976); *10* Keller (1974); *11* Church and Tilton (1973)

1 S.Sandwich
2 New Zealand
3 Tonga
4 Antilles
5 New Britain
6 Aleutians
7 Sumatra
8 Chile
9 Peru
10 Eolian
11 Cascades

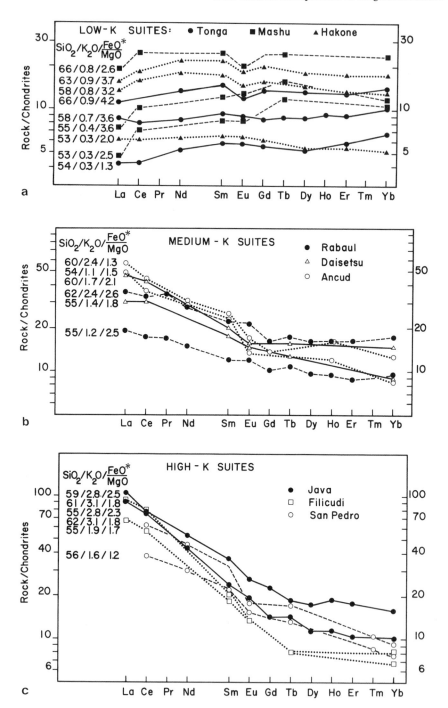

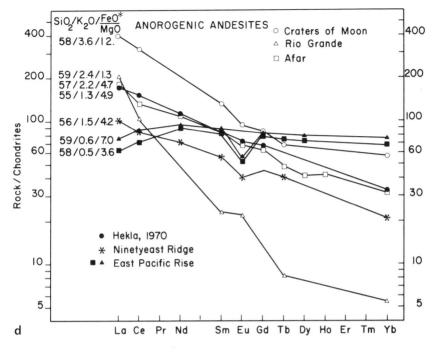

Fig. 5.12 a–d. REE patterns of andesites. Representative basic and acid andesites or dacites from single suites, in most cases representing superposed, possibly consanguinous lavas. All data are normalized to the chondritic average of Nakamura (1974). a Low-K orogenic andesites all of which are tholeiitic according to Fig. 1.4. Note the large increase in concentration of all REE which accompanies differentiation, the development of a negative Eu anomaly, and the light-REE depletions. Data sources: Tonga (Ewart et al. 1973; Kay and Hubbard 1978); Mashu, Japan (Masuda et al. 1975); Hakone, Japan (Fujimaki 1975). b Medium-K orogenic andesites. Note more restricted increases in concentration and less development of Eu anomalies accompanying differentiation. The Ancud patterns which cross may not represent consanguinous material given their Miocene age and lack of field control. *Closed symbols* are tholeiitic; *open symbols* are calcalkaline according to Fig. 1.4. Data sources: Rabaul, PNG (Heming and Rankin 1979; non-Ce-anomalous lavas); Daisetsu, Japan (Masuda et al. 1975); Ancud, Chile (Lopez-Escobar et al. 1976). c High-K suites. *Closed versus open symbols* as in b. Javanese samples are averages of samples from several volcanoes. Data sources: Java, Indonesia (Whitford et al. 1979a); Filicudi, Eolian arc (Villari and Nathan 1978); San Pedro, Chile (Thorpe et al. 1976). d Anorogenic andesites. *Open symbols,* continental; *closed symbols,* oceanic. All tholeiitic according to Fig. 1.4 except Rio Grande sample. Compositions of all samples except solid square are given in Table 5.5. Data sources: Craters of the Moon, USA (Leeman et al. 1976); Rio Grande (Taos Plateau), USA (Zimmerman and Kudo 1979); Afar (Boina Centre), Ethiopia (Barberi et al. 1975); Hekla (1970 eruption), Iceland, and Ninetyeast ridge, Indian Ocean (Thompson et al. 1974); East Pacific Rise [*solid triangle* is V2140 from Kay et al. (1970); *solid square* is Amphrite 3 andesite from Hart (1971), and Haskin et al. (1970)]

(Arculus 1976) and southeastern Iran (Dupuy and Dostal 1978). Still higher Sr contents were found by Kay (1978) in two high-Mg Aleutian andesites.

Rb/Sr ratios increase regularly with silica within andesite suites and with K_2O between suites (Fig. 5.11), as do Rb/Ba and Ba/Sr ratios.

K-group elements are more enriched in orogenic andesites and related basalts than in oceanic basalts when samples with similar concentrations of rare earth elements are compared. For example, Ba/La ratios exceed 15 in most volcanic rocks from convergent plate boundaries (Fig. 5.16a), in contrast to values of 4 to 10 in N-MORB or 10 to 15 in E-MORB and most within-plate basalts (e.g., Wood et al. 1980). Ba/La ratios often increase with differentiation but also are consistently higher in low-K than high-K andesites, are higher in island arcs as Tonga-Kermadec and Izu than in their extensions onto the thicker crusts of New Zealand and Japan, respectively, and are higher in volcanic arcs than in most adjacent backarcs (Table 5.4). Also, Sr/Nd or Sr/Ce ratios from volcanic arcs are higher than elsewhere (DePaolo and Johnson 1979; Pearce 1980). This enrichment of K-group elements reflects either the unique initial composition or refractory mineralogy of the magma source beneath volcanic arcs or crustal contamination during ascent.

5.4.2 REE Group: Rare Earth Elements Plus Y

Generally speaking, patterns of orogenic andesites are similar to those of oceanic basalts both in shape and concentration (Fig. 5.12). That is, there are both enrichments and depletions in light REE, heavy REE enrichments are 5 to 20 times chondritic, and light-enriched patterns may have Tb_N/Yb_N ratios of 1 (flat heavy REE pattern) to 2 (depleted heavy REE pattern).

Low-K basic andesites have negative or flat slopes in REE diagrams and among the lowest total REE concentrations of any terrestrial volcanic rocks (e.g., Jakeš and Gill 1970; Ewart et al. 1973, 1977; Fujimaki 1975; Masuda et al. 1975; Hawkesworth et al. 1977; Kay and Hubbard 1978; DePaolo and Johnson 1979; Masuda and Aoki 1979; Wood et al. 1980). Light REE contents and the slopes of REE patterns increase steadily from low to high-K suites (Figs. 5.12 and 5.13a). In contrast, Y and heavy REE contents change little. Most commonly, concentrations of all REE increase with silica within andesite suites, although in some instances heavy REE or Y contents remain constant or decrease (Fig. 5.13b). To date, most published reports of constant or decreasing heavy REE or Y contents are restricted to but not typical of calcalkaline suites of continental margins; exceptions include Oshima-oshima and Rishiri volcanoes in the Japan Sea (Masuda et al. 1975), and Dominica, Antilles (Brown et al. 1977). Light REE contents usually increase more rapidly than or to the same extent as heavy REE with increasing silica (Fig. 5.13a), although the opposite sometimes occurs (e.g., Gill 1976a; Kay 1977). REE patterns are roughly similar in tholeiitic and calcalkaline suites of comparable K-enrichment (e.g., Masuda and Aoki 1979).

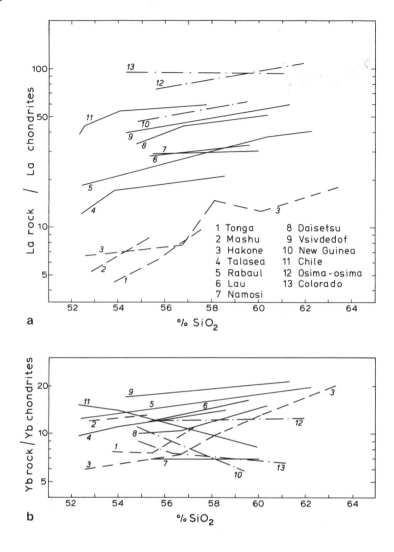

Fig. 5.13 a,b. La (a) and Yb (b) enrichments (i.e., La$_N$ and Yb$_N$) vs. SiO$_2$ contents in selected orogenic andesites. *Lines* drawn as in Fig. 5.10. *Dashed lines* are low-K andesites according to Fig. 1.2; *solid lines* are medium-K; *dash-dot lines* are high-K. Chondritic La and Yb values used are 0.33 and 0.22 ppm, respectively. Data sources are: *1* Ewart et al. (1973); *2, 8,* and *12* Masuda et al. (1975); *3* Fujimaki (1975); *4* and *5* Arth (1974); *6* Gill (1976a); *7* Gill and Till (1978); *9* Kay (1977); *10* Jakeš and Gill (1970); *11* Lopez-Escobar et al. (1976), Ancud; *13* Zielinski and Lipman (1976)

Unlike other REE, Eu may be partially divalent in magmas resulting in increased ionic radius and difference in behavior. In lunar and some terrestrial basalts, for example, Eu is highly concentrated relative to other REE in plagioclase and consequently depleted in liquids. However, Eu anomalies $> 10\%$ are uncommon in orogenic andesites despite evidence of plagioclase accumulation and fractionation. Positive Eu anomalies occur in some samples, including aphyric ones (e.g., Yajima et al. 1972; Condie and Swenson 1973; Arth 1974). Negative Eu anomalies are less common and occur primarily in acid andesites (e.g., Fujimaki 1975; Masuda et al. 1975; Lopez-Escobar et al. 1976). This similarity in behavior between Eu and other REE indicates high Eu^{+3}/Eu^{+2} as well as Fe^{+3}/Fe^{+2} ratios in orogenic andesites for reasons given in Section 5.2.3.

In contrast, Ce is partially tetravalent in highly oxidized environments such as seawater, resulting in decreased ionic radius. Ce anomalies, although uncommon and dependent on analytical accuracy, sometimes occur in orogenic andesites and usually are negative. Typically they appear erratically within suites (e.g., Woodruff et al. 1979; Wood et al. 1980) but other times characterize entire sets of spatially or temporally related rocks (Dixon and Batiza 1979; Heming and Rankin 1979). While negative Ce anomalies might indicate incorporation of seafloor material during genesis (Sect. 8.3), both pre- and post-eruption near-surface oxidation processes need evaluation before this argument is compelling. Positive Ce anomalies are more difficult to identify because some low or medium-K arc basalts and andesites have La_N/Ce_N ratios < 1. La_N/Ce_N ratios < 1 indicate derivation from a light REE depleted source.

Two other features of the shape of REE patterns in andesites occasionally have been observed. One is a concave downward hump at Nd to Sm (e.g., Kay 1977). The other is a concave upward dip at Dy to Er (e.g., Arth 1974). Either could be important but neither is common.

Y contents, like heavy REE contents, are quite similar in most orogenic andesites, the range 20 to 25 ppm being considered "standard" by Lambert and Holland (1974). Atypically higher Y contents occur in some western Pacific low and medium-K andesites (Ewart and Bryan 1972; Gill 1976a), whereas Y contents < 15 ppm usually are restricted to andesites erupted through crust > 30 cm thick as in New Zealand and Chile (Ewart and Taylor 1969; Roobol et al. 1976).

5.4.3 The Th Group: Th, U, and Pb

Th group elemental concentrations increase by an order of magnitude from low to high-K andesites (Table 5.2). In low-K basic andesites, concentrations of all three elements approach the low levels of MORB, whereas concentrations in high-K acid andesites approach those of granites. Typical Th contents are < 1, 1 to 5, and > 5 ppm in low, medium, and high-K suites, respectively; U and Pb contents are less clearly separated. Concentrations of all three elements increase with silica in most suites and U concentrations seem especially sensitive to the thickness of crust through which magmas ascend (DuPuy et al. 1976; Zentilli and Dostal 1977).

Th/U ratios likewise increase with K_2O level, typically being < 2, 2 to 4, and > 4 in low, medium, and high-K suites, respectively. Reported Th/U ratios usually vary randomly with silica but can either increase [e.g., Tatsumoto and Knight 1969 (Hakone); Katsui et al. 1978 (Mashu)] or decrease [e.g., Tatsumoto and Knight 1969 (Omuro-yama); Gill 1974 (Fiji)]. Significantly, the twofold increase in Th/U ratios of rocks from Irazu (Allègre and Condomines 1976) requires fractionation of U into vapor as well as crystals in order to suppress U-enrichment in residual andesitic melt.

Few high precision measurements of U/Pb ratios in andesites have been published but the available data indicate no consistent correlation of U/Pb ratios with silica within suites but positive correlation with K-enrichment between suites (Hedge and Knight 1969; Tatsumoto and Knight 1969; Oversby and Ewart 1972; Lipman et al. 1978). Observed $^{238}U/^{204}Pb$ ratios (μ) in andesites studied by these authors range from 4 to 19; Doe's (1970) median value for andesites is 10.

It is not yet clear whether Th-group elements share with the K-group elements the anomalous enrichment in rocks of volcanic arcs which I cited in Section 5.4.1. Suggestively, Th/La ratios, like Ba/La ratios, usually are higher (i.e., La/Th ratios are < 7) in most orogenic andesites and related basalts than in oceanic basalts (Fig. 5.16b), but such ratios also occur in some ocean island basalts (e.g., Joron and Treuil 1977).

5.4.4 The Ti Group: Zr, Hf, Nb, and Ta

Ti-group elements have high field strengths (charge/ionic radius) and consequently are not incorporated appreciably in common minerals except for Ti in magnetite and ilmenite. As a result, all of these elements correlate positively with each other and with indices of differentiation in silica-saturated magmas, especially those not precipitating Fe-Ti oxides or amphibole. Ratios of these elements to one another remain quite constant in basalts of all tectonic environments: Ti/Zr = 80 to 100; Zr/Hf = 39; Nb/Ta = 16.

A characteristic of orogenic andesites and related rocks at convergent plate boundaries is their lower TiO_2 contents than in volcanic rocks from intra-plate locales (Sect. 5.2.4). This is just the most obvious example of a general impoverishment of arc magmas in all the Ti-group elements, an impoverishment which has been used to distinguish arc-derived rocks from others by Pearce and Cann (1973) and Wood et al. (1979b). This impoverishment is both in absolute and relative terms. Agate-crushed low or medium-K orogenic andesites have less Nb and Ta than do basalts, even little-fractionated basalts, from other tectonic environments. However, the depletion is especially evident in ratios involving members of different element groups. For example, La/Ta or La/Nb ratios in orogenic andesites and related basalts exceed 30 or 2, respectively (Fig. 5.16c), which are the bulk earth values of Anders (1977) and the upper limits for oceanic basalts (Wood et al. 1979b).

Within suites, Ti differs in behavior from other members of the group. As noted in Section 5.2.4, TiO_2 either correlates negatively with silica in orogenic andesites or goes through a maximum in basic andesites characterized by rapid iron-enrichment. In contrast, other group members correlate positively with silica or potash contents throughout basalt to andesite suites whether or not iron-enrichment occurs, although an irregular increase in Nb accompanies increasing K_2O in the calcalkaline andesites of Mt. Jefferson, USA (Anderson and Gottfried 1971). Zr/Nb ratios apparently increase with differentiation in andesites but no example of changing Zr/Hf or Nb/Ta within andesites has yet been confirmed by high-quality analyses, as was done for an ocean island differentiation series by Ehmann et al. (1979).

Most orogenic andesites have 50 to 150 ppm Zr. Concentrations of Zr, Hf, Nb, and Ta at any given silica content increase from low to high-K suites. So also do Zr/Nb ratios which are 40 or more in low-K suites from Tonga, New Britain, and Japan, but 10 to 20 in high-K suites from both island arcs (e.g., Taylor et al. 1969a; Kiesl et al. 1978) and continental margins (AGV-1; Noble et al. 1975; Dostal and Zerbi 1978). This inverse correlation between LIL-element enrichment and Zr/Nb ratios is also observed in oceanic basalts (Erlank and Kable 1976; Wood et al. 1979a).

5.4.5 The Compatible Group: Ni, Co, Cr, V, and Sc

These elements differ from the ones discussed in preceding sections because the compatible elements are concentrated more in ferromagnesian minerals than in coexisting melt (Sect. 6.10) and consequently correlate negatively with silica.

Ni concentrations in andesites correlate positively with MgO, most lying in the stipled field of Fig. 5.14 which is approximately along the projection of basalt suites in other tectonic environments. Nevertheless, Ni contents can vary tenfold in orogenic andesites of equal MgO. Although no single generalization adequately describes this diversity, three seem apt. First, Ni contents are lower relative to MgO in tholeiite than calcalkaline suites; i.e., most samples which plot in region A of Fig. 5.14 are tholeiitic according to Fig. 1.4 whereas most in area B are calc-alkaline. Certainly this explains why Ni contents are lower relative to SiO_2 in tho-leiitic than calcalkaline andesites (Jakeš and Gill 1970) because Ni correlates nega-tively with FeO*/MgO ratios which are by definition (i.e., Fig. 1.4) higher in tho-leiitic than calcalkaline andesites at any SiO_2 (Miyashiro and Shido 1975, 1976). This contrast between magma series is especially striking along the volcanic front in northeastern Japan where calcalkaline andesites contain more Ni and Cr relative to MgO or SiO_2 than do tholeiitic andesites from adjacent volcanoes (Katsui et al. 1978; Masuda and Aoki 1979). However, there is no difference between calcakaline and tholeiitic andesites of New Britain, for example, where both lie in region A of Fig. 5.14 (Johnson and Chappell 1979). Also, the tholeiitic andesites of Mashu volcano, Hokkaido, lie in region B (Ando 1975).

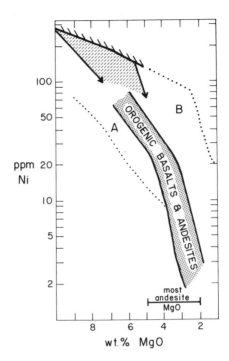

Fig. 5.14. Ni-MgO relationships for orogenic basalts and andesites. The *stippled region* contains most analyses. Analyses also lie in regions *A* and *B* which are discussed in the text. The *bar in lower right* shows the MgO contents of most orogenic andesites; see Fig. 1.1. The *dashed-line region bordered by arrows* shows Ni-MgO relationships of MORB and within-plate basalts from Hart and Davis (1978, Fig. 10) and other sources. The *heavy line* with *hatchures at the top* shows compositions of liquids in equilibrium with mantle olivine according to Hart and Davis (1978, Fig. 11)

Second, Ni contents in andesites sometimes remain remarkably constant or vary randomly over a constant range with decreasing MgO or increasing SiO_2 (Gill 1978, Fig. 6), thus causing some intitially ordinary suites to enter region B of Fig. 5.14. Constant Ni contents could result from varying degrees of partial fusion rather than fractional crystallization (Sect. 5.4.7) or, more prosaically, from uniform Ni contamination during grinding or from analytical imprecision. $\overline{D}_{Ni}$ values calculated as discussed in Section 5.4.7 range from 1 to 5 within orogenic andesites (e.g., Fig. 5.17b and Table 11.1), typically being higher in calcalkaline than tholeiitic suites (e.g., Masuda and Aoki 1979, Fig. 7) but often being lost in data scatter.

Finally, andesites erupted through thick crust tend to have higher Ni contents than do andesites with similar MgO or SiO_2 contents in island arcs (Andriambololona et al. 1974). For example, most andesites from Mexico, Peru, Chile, Turkey, northwest Iran, and the USA including the Cascades lie in region B of Fig. 5.14. Also, andesites from the New Guinea Highlands have higher Ni relative to MgO than do those from New Britain. This may be a consequence of sub-continental peridotite being more refractory (Jordan 1978) but many Antillean andesites likewise plot in region B.

To what extent the diversity of Ni contents in orogenic andesites reflects significant differences in source compositions versus differentiation processes remains unknown and will continue to remain so until reliable data are available for a spectrum of suites. However, it is already clear that Ni contents of andesites are characteristically low: rarely > 100 ppm, usually < 40 ppm and sometimes < 5 ppm. These concentrations are too low for liquids in equilibrium with mantle peridotite (Fig. 5.14).

Co contents are less variable than Ni, usually decreasing from about 40 to 10 ppm between basic and acid andesites of any type. Thus Co contents usually equal or exceed Ni contents, except in basic high-Mg andesites (e.g., Taylor and White 1966; Hormann et al. 1973; Condie and Hayslip 1975; Peccerillo and Taylor 1976). Ni/Co ratios increase or decrease with silica within andesite suites depending on the rate of Ni decrease, but usually are < 1 throughout.

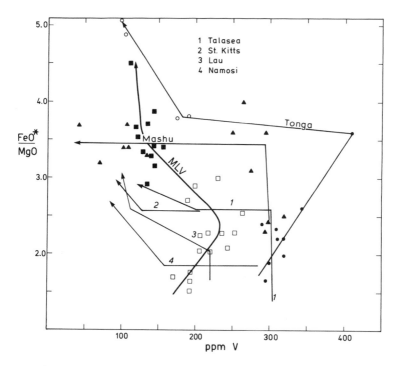

Fig. 5.15. V contents vs. FeO*/MgO ratios in selected orogenic andesites. *Solid triangles* are data for Mashu volcano, Japan; *filled* and *open circles* are for Tongan andesites which lack and bear, respectively, magnetite phenocrysts; *filled* and *open squares* make the opposite distinction for Medicine Lake volcano *(MLV)*, USA, samples. At MLV, magnetite-phyric lavas overlie magnetite-free ones, and mostly post-date caldera collapse. *Lines* show trends of data which, for examples 1 to 4, are omitted for clarity; *arrows* point in the direction of increasing SiO$_2$. Data sources: *1* Lowder and Carmichael (1970); *2* Brown et al. (1977); *3* Gill (1976a); *4* Gill and Till (1978); Tonga, Ewart et al. (1973); Mashu, Ando (1975); MLV, Mertzman (1977)

Cr behaves much like Ni in andesites (Fig. 5.17d), decreasing with increasing silica contents or FeO*/MgO ratios or decreasing MgO contents (Miyashiro and Shido 1975). Cr/Ni ratios usually fluctuate between 1 and 3 (e.g., Gunn and Mooser 1970), although Cr contents rarely remain as consistent throughout a suite as Ni contents sometimes do. Available $\bar{D}_{Cr}$ values within orogenic andesites are 2 to 4.

V contents generally follow TiO_2 and FeO*. All three decrease regularly with increasing silica in suites characterized by modest or no iron-enrichment ($\bar{D}_V$ values are 3 to 4), but all remain constant or go through a maximum in suites characterized by rapidly increasing FeO*/MgO ratios in basic andesites (Miyashiro and Shido 1975). This is illustrated for V in Fig. 5.15. The elbow shown for the Tonga, Mashu, Talasea, and Medicine Lake suites corresponds to the elbow seen in plots of FeO*/MgO versus SiO_2 for the same suites. That is, cessation of the rapid increase in FeO*/MgO ratios is accompanied by initiation of a rapid decrease in V contents. In some instances the elbow also corresponds approximately to the first modal appearance of magnetite microphenocrysts which are missing in more basic samples of the suite.

Sc contents usually decrease from about 40 to 10 ppm between basic and acid andesites without apparent relation to alkali or iron-enrichment, or to geographic region.

5.4.6 The Chalcophile Group: Cu, Zn, and Mo

Frequently, andesite volcanic provinces also are Cu-Pb-Zn-Mo-Ag-Au metallogenic provinces. Although most of these ore deposits occur in or near plutonic rocks whose relationships to synchronous volcanism is ambiguous, some ores clearly have intimate spatial and temporal association with andesite to dacite stratovolcanoes and rhyolite domes (e.g., Gustafson and Hunt 1975; Sillitoe 1977). Thus, there is debate whether these metals were derived from andesite magma during its evolution thereby requiring vapor fractionation of magma (Sect. 11.7), or were derived from subsolidus andesite or surrounding rock by hydrothermal convection cells driven by the magma chamber. Cu contents tend to be higher in andesites at convergent plate boundaries than elsewhere (Sect. 5.6).

Cu contents have been determined in over 30 andesite suites, Zn in about half that number, and Mo in only a few. Cu contents range from about 150 to 10 ppm, the median being about 60 ppm. In most suites Cu contents decrease irregularly with increasing silica (e.g., Siegers et al. 1969; Taylor et al. 1969a; Gill 1976a; Cole 1978) but remain relatively consistent in others (e.g., Taylor and White 1966; Gunn and Mooser 1970; Condie and Swenson 1973). Occasionally an apparent maximum occurs between 55% and 60% SiO_2 (e.g., Ewart et al. 1973; Blake and Ewart 1974; Heming 1974; Ando 1975; Brown et al. 1977).

Zn contents in andesites vary less than Cu, ranging from 50 to 100 ppm. Most often, variation of ± 20 ppm occurs irregularly but in some suites Zn, like Cu, decreases with increasing silica [Brown et al. 1977 (Grenada); Andriambololona 1976]. Occasionally this decrease only commences between 58% and 60% SiO_2. In at least two instances, Zn contents increase with silica throughout the andesite range of silica contents (Ewart et al. 1973; Ando 1975).

The few available Mo data vary irregularly between 0.5 and 2.5 ppm (Taylor and White 1966; Taylor et al. 1969a; Peccerillo and Taylor 1976).

Although data are inadequate for confidence, Cu and Zn seem to increase with silica only in andesitic suites characterized by rapidly increasing FeO*/MgO ratios; both elements decrease with silica or vary irregularly otherwise. Both Cu and Mo concentrations are higher in high-K than medium-K andesites of Bougainville (Taylor et al. 1969a) and, similarly, Cu contents are higher in K-rich granodiorites than in quartz diorites in the northern Caribbean (Kesler et al. 1975). However, there is no consistent relationship between Cu, Zn, or Mo contents of andesites and their K-enrichment. Cu and Ni correlate poorly in andesite suites, implying little effect of S-saturation on the distribution of transition metals (Sect. 5.3.4).

5.4.7 Trace Element Systematics

Trace elements can provide information both about the source and the differentiation history of magmas. Ratios of highly incompatible trace elements, like isotopic ratios, are indicative of source compositions because they do not change appreciably during either melting or fractionation as long as the weight fraction of liquid exeeds 10%. Consequently, the high Ba/La, Th/La, and La/Nb ratios of orogenic andesites and related basalts indicate that their sources differ significantly from those of oceanic basalts (Fig. 5.16). Relative to the REE (which are similar in orogenic andesites and oceanic basalts), the source regions for volcanic arcs are enriched in Rb, Cs, Ba, Sr, Th, U, and Pb but depleted in Nb and Ta and, to a lesser extent, Ti, Zr, and Hf. Consequently, a Ba/Ta ratio greater than 450 is the single most diagnostic geochemical characteristic of arc magma.

Moreover, this geochemical signature occurs only within the volcanic arc, not the backarc. Data compiled in Table 5.4 show consistent differences in elemental and isotopic composition between volcanic arcs and adjacent backarcs, differences which reflect source compositions.

Secondly, concentrations of trace elements can be used to test specific models of andesite genesis. Examples are given in Figure 5.17a and discussed in

---

Fig. 5.16 a–c. Schematic summary of incompatible trace element ratios in volcanic rocks ▶ from different tectonic settings: a La vs. Ba; b La vs. Th; c La vs. Nb. In each, bulk earth ratios are from Anders (1977); N-type and E-type MORB are from Wood et al. (1979a,b) and other sources; orogenic andesite data are from many sources. Ocean island and continental basalts typically lie in the E-MORB field. La/Th and La/Nb ratios both decrease during fractionation; Ba/La ratios increase

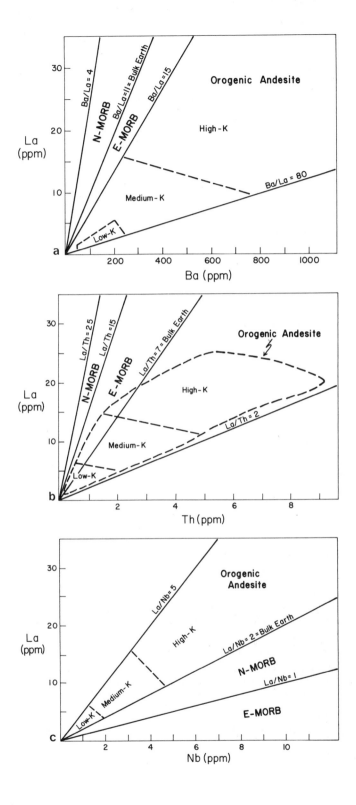

Table 5.4. Geochemical comparison of volcanic rocks from volcanic arcs (VA) and back arcs (BA)

	New Zealand VA	BA[3]	Tonga VA	BA[2]	Mariana VA	BA[2]
Ba/La	12–25	5	70	8	35	4–13
La/Th	3–9	–	2–7	20	10	–
La/Nb	2–4	7	3–5	1	–	–
$^{87}Sr/^{86}Sr$	0.704– 0.706	0.7027– 0.7036	0.7036– 0.7043	0.7036	0.7032– 0.7038	0.7027– 0.7030
$^{206}Pb/^{204}Pb$	18.8	19.1	18.6	18.3	18.8	18.2
$^{207}Pb/^{204}Pb$	15.60	15.58	15.55	15.52	15.53	15.44
References	1		2		3	

a Assuming Nb = Ta x 16

1 Comparison possibly invalid because VA data are from NE Japan arc whereas BA data from SW Japan arc where Sr is more radiogenic (Shuto 1974)

2 Tholeiites in inter-arc basins

3 Transitional to alkali basalts in monogenetic cones

References:

1. Cole (1978); Ewart and Stipp (1968); Ewart et al. (1977); Armstrong and Cooper (1971); Rafferty and Heming (1979)

2. Ewart et al. (1973, 1977); Oversby and Ewart (1972); Gill (1976a); Hamelin et al. (1979)

3. Hart et al. (1972); Meijer (1976); Dixon and Batiza (1980); Stern (1979)

4. Hedge and Knight (1969); Kurasawa (1968); Masuda et al. (1975); Masuda and Aoki (1979); Wood et al. (1980)

5. Kay (1977); Kay et al. (1978)

6. Tarney et al. (1977); Saunders and Tarney (1979); Hawkesworth et al. (1977)

7. Kiesl et al. (1978); Villari and Nathan (1978); Klerkx et al. (1974); Hamelin et al. (1979)

8. Wood et al. (1979a,b)

Chapters 8 to 11. However, most models usually fail such tests, probably indicating that the tests themselves are inappropriately simple but nevertheless ruling out some otherwise viable hypotheses (Gill 1978).

However, within-suite variations in composition also can be used to distinguish between general petrogenetic processes such as fractional crystallization, batch partial melting, and mixing, as discussed by Allègre and Minster (1978) and others. This process determination can be done by inspection of correlation diagrams such as those shown in Fig. 5.17 and discussed below. Symbols used below are defined as follows: $\bar{D}$ = bulk solid/liquid distribution coefficient [see Eq. (6.1)]; c = element concentration in residual liquid; c^o = initial concentration in liquid; H = highly incompatible element with $\bar{D} < 0.1$; M = somewhat incompatible element with $0.1 < \bar{D} < 1.0$; T = compatible element with $\bar{D} > 1$; f_ϱ = wt. fraction of liquid (percent fusion); and $f_s = (1 - f_\varrho)$ or degree of crystallization.

pan[1] A	BA[3]	Aleutians VA	BA[3]	Scotia VA	BA[3]	Eolian VA	BA[2]	N-MORB	E-MORB
−80	12−21	30−40	−	50−100	10−20	10−20	8−12	4−11	11−20
	10	−	−	−	3−9	1−5	10	15−25	7−15
−5	1	−	−	−	1−2	1.5−3	1	1−2	<1 [a]
703−	0.705	0.7030−	0.7028	0.7038−	0.7030	0.704−	0.7030	0.7023−	0.703−
704		0.7035		0.7042		0.706		0.7028	0.706
.3−	18.2	18.8	18.9	−	−	−	18.8	17.5−	18.0−
.7								18.5	19.5
.5−	15.4	15.57	15.49	−	−	−	15.60		
.6								See Fig. 5.18	
		5		6		7		8	

Log c_H − log c_T diagrams: Fig. 5.17b. Fractional crystallization is the only differentiation process which causes trace elements to follow a Rayleigh distillation pattern, resulting in exponential changes in the concentration of elements in residual liquids. Consequently, concentrations determined by fractional crystallization define straight lines on log-log diagrams whereas concentrations determined by other processes do not, and straight lines with steep slopes on such diagrams unambiguously identify fractional crystallization as the cause.

In contrast, neither batch melting nor mixing are exponential processes and therefore lead to curves on log c_H versus log c_T diagrams. The two processes can be distinguished because concave-up parabola are due to fusion whereas concave-down parabola are due to mixing. Also, note that mixing always results in higher c_H than does fractional crystallization for a liquid equally part-way between $c°$ and some more differentiated composition. Even higher c_T due to mixing occurs if $\bar{D}_T \gg 1$. For example, both a and a′ in Fig. 5.17b are exactly mid-way between liquids B and D and could be, for example, basic andesites derived from basalt (B) by fractional crystallization or by mixing with dacite (D), respectively. The andesite-by-mixing (a′) has nearly twice the c_H and the c_T as does the andesite-by-fractionation.

Moreover, because $c_H°/c_H = f_\ell$ if $\bar{D}_H = 0$, the slope of a straight line in a log c_T versus log c_H diagram equals $(\bar{D}_T - 1)$ if $\bar{D}_H = 0$. [The slope equals $(1 + \bar{D}_H) \cdot (\bar{D}_T - 1)$ in the general case]. Typical $\bar{D}$'s calculated in this fasion for suites of orogenic andesites are as follows: Rb, Cs, Ba, Th, light REE < 0.1; Zr, Y, heavy REE = 0.5 to 1.0; Sr = 1 to 2; Ni, Co, Sc, V, Cr = 1 to 4. Comparison of these $\bar{D}$'s with those given in Table 6.3 can identify or preclude the involvement of some minerals if crystal fractionation is the appropriate mechanism. As examples, the weight fraction of magnetite or hornblende amongst crystallizing phases must be 0.06 to 0.10 if $\bar{D}_V = 3$, whereas the weight fraction of garnet must be < 0.02 if

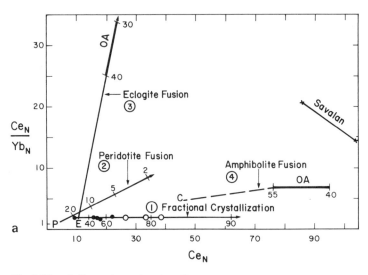

Fig. 5.17 a–d. Trace element systematics

a Testing specific models: Ce_N/Yb_N vs. Ce_N for samples from Rabaul volcano, PNG (volcano 6.1 in Appendix: Heming and Rankin 1979). *Filled circles* are basalts and basic andesites; *open circles* are acid andesites and dacite. Ce-anomalous samples are omitted. The subcript N means concentrations are normalized to chondritic values (Ce = 0.865 ppm, Yb = 0.22 ppm). *Line 1* shows effects of fractional crystallization with constant mineralogy (plagioclase:augite:olivine or orthopyroxene:magnetite of 67:13:13:7), assuming D's of Table 6.3. *Numbered tickmarks* indicate percent solidification of the least-differentiated composition. The more complex model of Heming (1974, Table 5), which explains most major element variations, defines a very similar line. *Line 2* shows compositions of partial melts of garnet peridotite P (2 times chondritic compositions), assuming $\bar{D}_{Ce}/\bar{D}_{Yb}$ of 0.01/0.01, 0.03/0.45, 0.15/0.61, and 0.01/9.0 for olivine, orthopyroxene, augite, and garnet, respectively, and assuming olivine:orthopyroxene:augite: garnet of 54:25:15:6 in both the subsolidus and melting proportions. *Numbered tickmarks* indicate percent fusion. *Line 3* shows compositions of partial melts of eclogite E (10 times chondritic – possible ocean floor basalt), assuming D's of Table 6.3. The refractory mineralogy (garnet:clinopyroxene $\sim$ 1:2) is taken from Stern and Wyllie (1978, Table 8) for estimated modes of experiments at 30 kb and 5 wt.% H_2O. Segment *OA* represents melts whose major element composition is andesitic according to Stern and Wyllie. *Line 4* shows compositions of partial melts of average continental crust, assuming initial composition C (Ce_N = 45, Yb_N = 10), D's from Table 6.3, and melting conditions appropriate to production of orogenic andesite from a basaltic parent at PH_2O = 5 kb. *Numbered tickmarks* and *segment OA* indicate percent fusion where melts are andesitic in major element compositions (Helz 1976, Table 7b, runs 15a and 16a). The segment is flat because the proportion of amphibole and pyroxene decrease as percent fusion increases. Note differences from datapoints for Savalan volcano, Iran (see Sect. 10.2). The constancy of Ce_N/Yb_N ratios throughout the Rabaul suite clearly favors origin by fractional crystallization rather than by partial melting of any of the sources considered

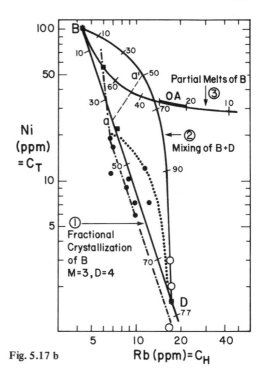

Fig. 5.17 b

b Log C_T vs. C_H diagram: ln Ni vs. ln Rb samples from Hargy-Galloseulo volcano, PNG (volcano 6.4 in Appendix: Johnson and Chappell 1979). *Solid symbols* are orogenic andesites; *open symbols* are dacites. *Squares* indicate samples with olivine phenocrysts ($<$ 2%); others lack olivine. The three highest-Ni samples lie in region A of Fig. 5.14. *Line 1* is a best-fit to the data with a slope (m) of 2.97, indicating $\bar{D}_{Ni}$ = 4 if $\bar{D}_{Rb}$ = 0. *Asterisks B and D on line 1* schematically illustrate potentially primitive basalt *(B)* and derivative dacite *(D); numbered tickmarks* indicate the percent solidification of B. *Line 2* is a mixing line showing effects of mixing basalt *B* and dacite *D; numbered tickmarks* indicate the percentage of D. Points *a* and *a'* connect midpoints between *B* and *D* for fractional crystallization and mixing, respectively, showing that both Rb and Ni contents are higher in the mix than in the equivalent fractionate. *Line 3* shows compositions of partial melts of eclogite with initial composition B, refractory mineralogy as for line 3 in a, and D_{Ni} values from Table 6.3. These parameters should be applicable to partial melts of subducted ocean floor basalts. *Numbered tickmarks* indicate percent fusion; the line segment labeled *OA* is as in a. Although the absolute Ni contents are only 20 to 25 ppm below those expected in melts from eclogite *B*, note that the data points are generally consistent with fractional crystallization but will be inconsistent with the *shape* of the mixing or fusion curves, *regardless of choice of reasonable D's or starting materials.* A more complex fractionation model showing decreased $\bar{D}_{Ni}$ upon loss of modal olivine is shown by the *dash-dot line,* and the effects of back-mixing dacite with andesite to explain some samples is shown by the *dotted line.* In either fractionation case, ferromagnesian minerals must constitute about 50% of the solid phases if the D values of Table 6.3 are appropriate

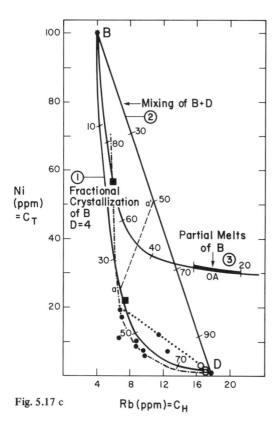

Fig. 5.17 c   Rb (ppm)=C_H

c Rectilinear C_T vs. C_H diagrams: Ni vs. Rb data as in **b**. All *symbols, lines,* and *numbered tickmarks* are as in **b**. Conclusions also are the same

$\bar{D}_{Yb} < 1.0$. More elegantly, inverse theory can be used, in principle, to determine in a least squares sense both the nature and weight fraction of fractionating phases in specific suites (Minster et al. 1977). This procedure has not yet been applied successfully to suites dominated by orogenic andesite.

Rectilinear c_H versus c_T Diagrams: Fig. 5.17c. Both fractional crystallization and batch melting with constant $\bar{D}$ lead to parabolic curves on this diagram. However, because c_T changes little for f_ϱ less than 0.3 during batch melting, whereas c_T changes considerably for f_s less than 0.3 during fractional crystallization, the two processes sometimes can be distinguished. Distinguishability increases as $(\bar{D}_T - \bar{D}_H)$ increases.

In contrast, binary mixing is the only differentiation process leading to straight lines on this diagram. In practice, however, straightness is not diagnostic

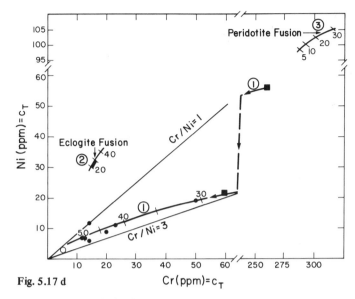

Fig. 5.17 d $Cr(ppm)=c_T$

d Rectilinear C_T vs. C_T diagram: Ni vs. Cr for the same samples as in **b**. Note changes in scales. *Symbols* also as in **b**. *Line 1* shows effects of fractional crystallization of the least differentiated sample when $\overline{D}_{Ni} = 4$ and $\overline{D}_{Cr} = 5.5$; *numbered tickmarks* indicate percent solidification. *Lines 2 and 3* show the much smaller variation of C_T during equilibrium partial fusion, a process which clearly cannot account for the range in concentration of the data plotted. *Line 2* should be applicable to partial melts of subducted ocean floor basalts, assuming $C_0^{Ni} =$ 100 ppm, $C_0^{Cr} = 260$ ppm, D's from Table 6.3, and melting conditions as for line 3 in **a**. Note that if $Cr/Ni = 1 - 3$ in orogenic andesites (Sect. 5.4.5), partial fusion of eclogitic MORB is an unlikely genetic process. *Line 3* roughly illustrates the composition of primary melts of mantle peridotite under hydrous conditions when D_{Ni}/D_{Cr} are 30/1, 2/10, and 3/10 for olivine, orthopyroxene, and augite, respectively, and when olivine:orthopyroxene:augite proportions are 66:23:11 in the subsolidus and 14:29:57 as melting proportions. The orogenic andesites which are plotted clearly must be fractionates

of mixing because straightness also results from partial melting when $f_\varrho < 0.3$ and from fractional crystallization accompanied by increasing $\overline{D}$.

Rectlinear $c_T{}^1$ versus $c_T{}^2$ Diagrams: Fig. 5.17d. In this diagram fractional crystallization yields a parabolic curve, mixing yields a straight line, and small degrees of batch fusion yield a nearly constant (eutectic) composition.

Despite the potential utility of these diagrams for discriminating between petrogenetic processes, few suites of orogenic andesites yield diagnostic results. In part this is because Δf_ϱ and Δc_T's are small amongst andesites. More fundamentally, however, determinative diagrams for orogenic andesites typically show wide data scatter or flat slopes or both. Wide data scatter may indicate the pooling of nonconsanguinous rocks, analytical imprecision, or crystal accumulation (Sect. 5.1). Flat slopes are nondiagnostic, possibly indicating that $\overline{D} \sim 1$ for one of the elements for the mineral assemblage controlling andesite genesis, or that partial melting is the responsible process, or that mixing has occurred. For example, one of these options presumably explains the relative constancy of Ni, Y, and heavy REE in some orogenic andesites (Sects. 5.4.2 and 5.4.5).

Commonly, Cr and V are the most diagnostic T elements in orogenic andesites and sometimes they clearly demonstrate that fractional crystallization has occurred. This is shown using Ni in Fig. 5.17b. Other clear examples of this kind occur in the data of Masuda and Aoki (1979), Mertzman (1978), Ewart et al. (1973), Ando (1975), Cole (1978), and Chappell and Johnson (1979).

5.5 Isotopes

Isotopic studies of orogenic andesites are used primarily to evaluate the extent to which certain elements in andesite come from the upper mantle versus other sources, such as sialic or subducted crust, and to investigate the history of these source regions.

5.5.1 Strontium

Between 1966 and 1980 over 800 measurements of $^{87}Sr/^{86}Sr$ ratios of Quaternary andesites and related rocks at convergent plate boundaries were published. Those discussed below have been normalized to 0.7080 for E & A $SrCO_3$ or to 0.7102 for NBS $SrCO_3$.

The average $^{87}Sr/^{86}Sr$ ratio of most suites containing orogenic andesites at modern convergent plate boundaries lies between 0.7030 and 0.7040. Within this range are most andesites from Tonga, Fiji, and Vanuatu (Gill and Compston 1973; Gill 1976b; Gorton 1977); New Britain (Peterman et al. 1970a; Page and Johnson 1974; Peterman and Heming 1974; DePaolo and Johnson 1979); the Mariana and Izu islands (Pushkar 1968; Hart et al. 1970; Meijer 1976; Matsuda et al. 1976; Stern 1979); northern Honshu and Hokkaido (Hedge and Knight 1969; Katsui et al. 1978); the Kuriles and Kamchatka (Hedge and Gorshkov 1978); the Aleutians (Kay et al. 1978); the Cascades (Hedge et al. 1970; Peterman et al. 1970b; Church and Tilton 1973); Mexico (Whitford and Bloomfield 1976; Moorbath et al. 1978); Central America (Pushkar 1968; Montigny et al. 1969; Rose et al. 1977; Thorpe et al. 1979); the northern Antilles (Hedge and Lewis 1971; Rea 1974; Hawkesworth and Powell 1980); and the South Sandwich islands (Gledhill and Baker 1973; Hawkesworth et al. 1977).

These data are notable for several reasons besides their striking consistency. First, ratios of 0.703 to 0.704 lie within the range of ocean island basalts but above that of N-type MORB. Thus, andesites from the arcs listed above are similar in $^{87}Sr/^{86}Sr$ ratios to many mantle-derived magmas, but dissimilar to the majority of fresh oceanic basalts being subducted beneath those arcs. However, these andesites have the same $^{87}Sr/^{86}Sr$ ratio as thought to be average for oceanic crust being subducted due to its alteration by sea water between creation and subduction (0.7039: Spooner 1976). Second, these andesites are little if at all affected by Sr in pre-Mesozoic sialic upper crust (average $^{87}Sr/^{86}Sr \sim 0.72$) regardless of whether such crust might have been encountered during ascent (e.g., Alaska, northern Japan, northern Cascades, Guatemala) or not.

Third, ratios are similar in basalts and andesites from the same volcano or suite to within analytical uncertainty (usually ± 0.0002 at 2 σ limits) in most of the suites studied. Thus, these basalts and andesites could be related genetically to a common, isotopically homogeneous magma body or source region. There are, of course, exceptions such as Shasta, USA, where a single stratovolcano is heterogeneous isotopically and in trace elements, including Sr, and where the isotopic heterogeneity is unrelated to Rb/Sr ratios or Sr concentrations (Peterman et al. 1970b; Condie and Swenson 1973). This indicates eruption from a common edifice of genetically diverse material all having normal $^{87}Sr/^{86}Sr$ ratios compatible with mantle derivation. No study yet has documented clear temporal patterns within such polygenetic fields, or found a relationship with stratovolcano evolutionary history. For example, it is unclear whether the decrease in $^{87}Sr/^{86}Sr$ ratios from 0.7037–39 to 0.7032–34 associated with caldera collapse at Mashu volcano, Hokkaido, is analytically significant or not. More precise measurements from Santa María, Guatemala, show no change during a similar edifice evolution (Sect. 4.4).

Although published $^{87}Sr/^{86}Sr$ ratios of andesites most often lie between 0.703 and 0.704, many (about 35%) lie outside this range, generally between 0.704 and 0.708. One high-K basic andesite from Achilleion, Greece (beneath which subducted crust apparently is absent: LePichon and Angelier 1979) reaches 0.71 (Pe 1975). These higher ratios usually occur in volcanic arcs overlying crust which is > 30 km thick and pre-Meoszoic in age: New Zealand (Ewart and Stipp 1968); Papua New Guinea Highlands (Page and Johnson 1974); Sumatra and Java (Whitford 1975); central and southwest Japan (Shuto 1974; Ishizaka et al. 1977); Ceboruco volcano, western Mexico (Moorbath et al. 1978); Ecuador (Francis et al. 1977); Peru and Chile (Pichler and Zeil 1972; McNutt et al. 1975; James et al. 1976; Klerkx et al. 1977; Magaritz et al. 1978; Briqueu and Lancelot 1979); Sardinia (Dupuy et al. 1974); Greece (Pe 1975); Turkey (Lambert et al. 1974); and the Absaroka and San Juan fields, USA (Peterman et al. 1970c; Lipman et al. 1978). Again, there are exceptions to this generalization although information about crustal thickness and age is limited in several cases. These exceptional instances of high ratios depsite thin crust include: eastern Indonesia (Whitford et al. 1977; Magaritz et al. 1978); the Antilles from Dominica south to Grenada (Hedge and Lewis 1971; Pushkar et al. 1973; Hawkesworth et al. 1970c; Hawkesworth and

Powell 1980); the Eolian arc (Barberi et al. 1974; Klerkx et al. 1974); and Santorini, Aegean (Pe and Gledhill 1975).

Ratios in all of the areas listed above usually are more variable as well as higher than those in the first group; in only two of the areas with ratios > 0.705 (Carriacou, Antilles; several Banda arc volcanoes) does isotopic variability at a given volcano lie within or near limits of analytical uncertainty. Often $^{87}Sr/^{86}Sr$ ratios vary without apparent relation to anything (e.g., Whitford 1975), although sometimes ratios correlate negatively with Sr concentrations (e.g., Page and Johnson 1974; James et al. 1976; Briqueu and Lancelot 1979) or even Rb/Sr ratios (e.g., Ishizaka et al. 1977).

However, most dramatic of these isotopically heterogeneous suites are those in which $^{87}Sr/^{86}Sr$ ratios correlate positively with Rb/Sr ratios, yielding pseudo-isochrons which imply ages of 100 m.y. or older for andesites which are Pliocene or younger in age (see Figs. 5.11 and 10.2). Examples include New Zealand (Ewart and Stipp 1968), Tonga (Ewart et al. 1973). the New Guinea Highlands (Johnson et al. 1978b), Merapi, Sumatra (Whitford 1975), Absaroka field, USA (Peterman et al. 1970c), Nevado de Toluca, Mexico (Whitford and Bloomfield 1976), and the central Andes (James et al. 1976; Klerkx et al. 1977). These positive correlations are not good statistically (cf. James et al. 1976, with James 1978), suggesting superimposed effects of fractionation and mixing.

Unsupported Sr describes Sr in a rock whose Rb/Sr ratio is too low to have led to the rock's observed $^{87}Sr/^{86}Sr$ ratio from primordial Earth Sr ($^{87}Sr/^{86}Sr =$ 0.6990) within the age of the Earth (4.5×10^9 yr). Because partial melting of peridotite produces Rb/Sr ratios in melt which equal or exceed that in the source, unsupported Sr usually is interpreted as indicating prior depletion of this source in Rb relative to Sr (Gast 1968). Alternatively, the source may have become relatively enriched in ^{87}Sr, for example by incorporation of seawater Sr. At convergent plate boundaries, unsupported Sr occurs in basalts from several volcanoes (e.g., throughout New Britain, PNG; Krakatoa, Indonesia; St. Kitts and Carriacou, Antilles). However, the only orogenic andesites with unsupported Sr are the anomalously high-Sr andesites from Shasta and the Aleutians cited in Section 5.4.1, and low-K basic andesites from Tonga and from Cape Hoskins, New Britain (Ewart et al. 1973; Page and Johnson 1974; DePaolo and Johnson 1979). Thus, the difference in Rb/Sr ratios between basalts and andesites in the suites listed above reflects magmatic differentiation processes; magma source regions for some low and even medium-K suites contain unsupported Sr.

5.5.2 Lead

By 1980, Pb isotope data were available for orogenic andesites and related arcs from nine active volcanic arcs plus mid-Tertiary examples from the western USA (Fig. 5.18). Pb isotope ratios of Pliocene to Holocene andesites are restricted to a portion of the range of compositions found in ocean basin basalts of similar age;

$^{206}Pb/^{204}Pb$ ratios are lower (more "planetary") than those of most oceanic islands. Eocene andesites from Wyoming, USA, contain Pb which is still less radiogenic than this range, as do young basalts from the USA interior (Peterman et al. 1970c). Thus, as in the case of Sr, the average isotopic composition of Pb in all analyzed andesites could be derived from the upper mantle.

In detail, however, only the Vanuata and Mariana arcs have Pb which is thought characteristic of the mantle beneath ocean basins (i.e., which lie in the PUM line in Fig. 5.18). Some or all andesites from other arcs are more enriched in ^{207}Pb and sometimes in ^{208}Pb relative to ^{206}Pb than in most MORB or ocean island basalts, indicating the presence of a component from an environment with higher μ and K ratios than are typical of the upper mantle. Moreover, Pb isotope ratios of andesites and related rocks usually form linear arrays with steep slopes in $^{207}Pb/^{204}Pb-^{206}Pb/^{204}Pb$ diagrams (Fig. 5.15a). These slopes are steeper(~ 0.3) than observed in MORB or ocean island basalts (~ 0.1) and yield ages of ~ 3500 m.y. which are too old to be interpreted reasonably as secondary isochrons or as mixing lines for ordinary mantle reservoirs. Thus, the arrays probably reflect mixing of mantle Pb with a ^{207}Pb-enriched component. Old sialic crustal rocks and material transported from them to the oceans by erosion are typically enriched in ^{207}Pb as in ^{87}Sr, and constitute the most likely second component.

Because magmatic differentiation processes do not result in Pb isotopic fractionation, all rocks derived from an isotopically homogeneous magma should lie at one point in Fig. 5.18, within analytical precision. Analyses of multiple samples from several andesite stratovolcanoes are not coincident and define apparent mixing lines. In some cases this heterogeneity may be inherited from the source. For example, $^{238}U/^{204}Pb$ and $^{206}Pb/^{204}Pb$ ratios are positively correlated in basalts to rhyolites from Kermadec (Oversby and Ewart 1972). The resulting pseudoisochron may be a mixing line reflecting variable U/Pb ratios in the upper mantle, as U/Pb ratios change little during differentiation of magma (Sect. 5.4.3). However, other suites show no correlation between $^{206}Pb/^{204}Pb$ and U/Pb ratios (see discussion of Japanese data by Oversby and Ewart 1972) and the Pb-enrichment probably reflects mixing with crustal Pb. In the light of this, it is important that an analytically significant positive correlation between $^{207}Pb/^{204}Pb$ and $^{87}Sr/^{86}Sr$ ratios occurs in samples from two Aleutian and one Cascade volcano (Church 1976; Kay 1977), although only two to four samples per volcano were analyzed. This correlation implies a common source for both ^{207}Pb and ^{87}Sr.

Pb isotope ratios in andesites imply time-averaged $^{238}U/^{204}Pb$ (μ) and $^{232}Th/^{238}U$ (K) ratios of about 8.2 to 8.4 and 3.6 to 3.9 respectively, in their source regions (Fig. 5.18). Wyoming and Peruvian andesite Pb came from a source with a higher Th/U ratio of 4.0 to 4.4. Comparison with the limited available data summarized in Section 5.4.3 indicates that μ and K are lower in most low-K andesites of Tonga and Japan than in their source regions, whereas μ in high-K andesites of northern Honshu is higher.

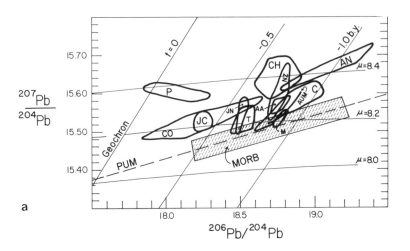

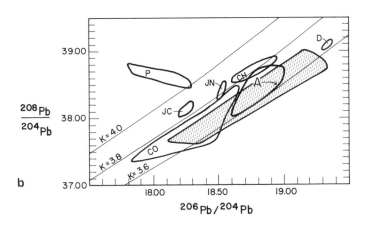

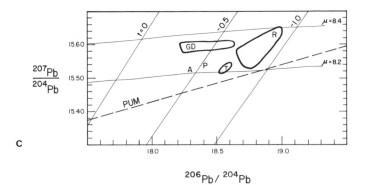

5.5.3 Neodymium

Study of Nd isotopes is in its infancy; only about 70 analyses of volcanic rocks from convergent plate boundaries, including 25 orogenic andesites, had been published before 1980 (Fig. 5.19; Tilton 1979). The range of $^{143}Nd/^{144}Nd$ ratios in most orogenic andesites studied thus far lies within that of oceanic basalts and therefore is superficially consistent with mantle derivation of andesites. The exceptions are andesites from northern Chile and Peru whose ratios are lower, more typical of sialic crust (Fig. 5.19). Few rocks from volcanic arcs have ratios as high as in N-type MORB (i.e., $\epsilon_{Nd} \geqslant + 9$); their source is more like that of ocean islands ($\epsilon_{Nd} = 0$ to 8). Like most ocean basalts, modern orogenic andesites have higher $^{143}Nd/^{144}Nd$ ratios than would a chondritic source, indicating time-averaged light REE depletion of the source despite the light REE enrichment of most of the andesites in question. Finally, pseudoisochrons occur in the Sm-Nd as well as the Rb-Sr systematics of Chilean andesites, with the apparent age being similar in both instances (Hawkesworth et al. 1979a).

◄ Fig. 5.18 a–c. Pb isotopic compositions of orogenic andesites and related rocks.

a $^{207}/Pb/^{204}Pb$ vs. $^{206}Pb/^{204}Pb$. Typical upper mantle Pb as defined by oceanic basalts lies along the *PUM* line with a slope of 0.11; data for most *MORB* as shown in the *dashed-line rectangle* (Tatsumoto, 1978; Cohen et al. 1980). Arc data are from these sources: *P* Peru (Tilton 1979); *CO* Colorado, andesites only (Lipman et al. 1978); *JC* central Honshu, Japan (Tatsumoto and Knight 1969, corrected data); *JN* northern Honshu, Japan (Hedge and Knight 1969, corrected data); *CH* northern to central Chile (Tilton 1979; McNutt et al. 1979); *NZ* Taupo zone, New Zealand (Armstrong and Cooper 1971); *AA* Adak volcano, central Aleutians, and *A UM* Okmok volcano, eastern Aleutians (Kay et al. 1978); *C* Cascades (Church and Tilton 1973; Church 1976); *M* Mariana active volcanoes (Meijer 1976); *T* Tonga (Oversby and Ewart 1972); *AN* Antilles active volcanoes (Armstrong and Cooper 1971). Data for Vanuatu, Taiwan, and Mexico are also available (Lancelot et al. 1978; Sun 1980).

b $^{208}Pb/^{204}Pb$ vs. ^{206}Pb vs. ^{206}Pb. The composition of oceanic basalts lies in the *dashed-line region*. Arc data are identified as in a except that, for simplicity, region A includes all not-otherwise-identified data and *D* = Dominica, Antilles (Armstrong and Cooper 1971). Missing from both (a and b) diagrams are data for Eocene andesites of Wyoming which, after correction, have a = 16.3–16.6, b = 15.33–15.44, c = 36.7–37.7 (Peterman et al. 1970c). For reference, single-stage growth curves and isochrons are shown, calculated from the following parameters: $T_0 = 4.55$ b.y.; $a_0 = 9.307$; $b_0 = 10.294$; $c_0 = 29.476$; $\lambda 238 = 0.155125 \times 10^{-9}$; $\lambda 235 = 0.98485 \times 10^{-9}$; $\lambda 232 = 0.49475 \times 10^{-9}$; $\mu = 8.2$ for b.

c $^{207}Pb/^{204}Pb$ vs. $^{206}Pb/^{204}Pb$ for atypical MORB and ocean island basalts, showing that mantle not in wedges above subducted lithosphere also can have arc-like Pb isotopic compositions. Growth curves and isochrons are reproduced from a for reference. Data sources are: *GD* Gough and Discovery Tablemount, and *T* Tristan de Cunha islands (Sun 1980) and *R* Reunion (Oversby 1972). Points *A* and *P* are single MORB's from Atlantic and Pacific oceans (Cohen et al. 1980; Bass et al. 1973)

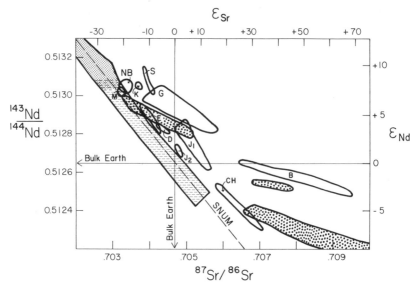

Fig. 5.19. Sr-Nd isotopic compositions of orogenic andesites. Nd isotopes have been normalized to $^{146}Nd/^{144}Nd$ = 0.72190. The Sr-Nd correlation line for the upper mantle *(SNUM)* is defined by the compositions of MORB which lie in the *lightly dotted area in the upper left,* by results for oceanic islands which lie in the partially overlapping *dashed area,* and by results for peridotite nodules from kimberlites which complete the SNUM line (e.g., O'Nions et al. 1979). The *horizontal line* shows the present-day $^{143}Nd/^{144}Nd$ ratio of chondritic meteorites, taken as that of bulk Earth. Its intersection with the SNUM line, therefore, gives the average $^{87}Sr/^{86}Sr$ ratio of bulk Earth, 0.7047. ϵ_{Nd} and ϵ_{Sr} values also are shown and are calculated relative to these bulk Earth ratios as described by DePaolo and Wasserburg (1977). Magma sources in the *upper left quadrant* have been depleted in Rb and light REE relative to the bulk Earth by amounts or for times which increase away from the quadrant intersection. Atypical mantle whose composition lies to the right of the SNUM line occurs beneath the Azores and the Campanian province of Italy (Hawkesworth and Vollmer 1979; Hawkesworth et al. 1979b) and is shown by the *heavily dotted areas.* Compositions of Quaternary basalts and andesites (not dacites or rhyolites) from volcanic arcs are shown in the *open, lettered areas.* Data sources are: *M* Mariana (DePaolo and Wasserburg 1977); *NB* New Britain (DePaolo and Johnson 1979); *D* Dominica and *K* St. Kitts, Antilles (Hawkesworth and Powell 1980); *G* Grenada, Antilles (Hawkesworth et al. 1979c); *E* Ecuador, and *CH* northern Chile (Hawkesworth et al. 1979a); *J1* Java calcalkaline, *J2* Java high-K calcalkaline, and *B* Banda (Whitford et al. 1979b); *S* South Sandwich (Hawkesworth et al. 1977)

Collectively, arc rocks display a negative correlation between $^{143}Nd/^{144}Nd$ and $^{87}Sr/^{86}Sr$ ratios. Some lie on the upper mantle Sr-Nd correlation line (SNUM line of Fig. 5.19) and therefore could be totally unaffected by subduction. However, these rocks could also represent mixtures between old sial and unaltered MORB or its source. This latter alternative is favored when the samples also have atypically high Sr/Nd ratios (> 20) or have $^{207}Pb/^{204}Pb$ ratios above the PUM line of Fig. 5.18.

Most arc volcanics are displaced by about 0.005 to the high Sr side of the SNUM line and define correlations with shallower slopes (Fig. 5.19). Significantly, of samples from active volcanoes in the New Britain, Mariana, and Peruvian arcs, only dacites and rhyolites are displaced in this fashion, implying that upper crustal-level contamination, not source characteristics, causes the increased displacement of the more differentiated rocks.

5.5.4 Inert Gases

Inert gas isotope geochemistry of orogenic andesites is even more in its infancy than Nd isotope geochemistry. Available data are restricted principally to $^3He/$ ^4He ratios of gases from volcanoes in northwestern Pacific volcanic arcs plus Lassen, USA (Baskov et al. 1973; Wakita et al. 1978; Craig et al. 1978a,b). However, these examples include volcanoes whose andesites span a wide range in $K_{57.5}$ (0.6 to 1.8) and $FeO*/MgO$ (1.4 to 4.1) and whose situations differ significantly in age of subducted crust and in crustal thickness.

Oceanic basalts and continental crust differ in $^3He/^4He$ ratios by at least 100 times, with the basalts having ratios 8 to 15 times higher than atmospheric while continental crust has ratios about 10 times lower than atmospheric (Craig et al. 1978b). Ratios in subducted oceanic crust will be lower than in fresh oceanic basalt due to in situ formation of ^4He from uranium and thorium decay prior to subduction. Although these lower ratios are unknown empirically, they should be two to seven times atmospheric, depending on the age and alteration (U content) of subducted crust (Craig et al. 1978a).

Available $^3He/^4He$ ratios for gases from convergent plate boundaries all are five to nine times atmospheric whereas boninite lava from the Mariana forearc (see Sect. 5.2.3) has a $^3He/^4He$ ratio only 1.5 times atmospheric (Bloomer et al. 1979). To date there is no correlation with crustal thickness. Perhaps fortuitously, the highest ratios for volcanic gases are from Lassen beneath which subducted crust is youngest.

High $^3He/^4He$ ratios in northwestern Pacific arcs require that most He and, by implication, other magmatic volatiles in these arcs come from typical upper mantle peridotite rather than from subducted crust via dehydration, or from continental crust. Conversely, because $^3He/^4He$ ratios in volcanic arc gases apparently are less than for spreading centers or plumes, the incorporation of some volatiles from either slab or continental crust is suggested, especially for boninite. However, more data are needed to assess roles of meteoric water, the atmosphere itself, and vapor-phase isotopic fractionation as well as regional variability.

5.5.5 U-Disequilibrium

The four preceding sections discussed variations in isotopic composition due to decay of long-lived radioisotopes with half-lives of $> 10^9$ yr. In contrast, several members of the ^{238}U decay series have half-lives of days to 10^5 yr, which make them useful monitors of events occurring on the time scale of individual magmatic differentiation episodes or eruptions. Three members have been studied in connection with orogenic andesites: ^{230}Th, ^{226}Ra, and ^{222}Rn. Parentheses in this section denote activities or activity ratios where (i) is the activity of i, which is the concentration of i times its decay rate.

In orogenic andesites $< 10^6$ yr old, $(^{230}Th/^{238}U)$ ratios frequently differ from 1.0, thereby indicating radioactive disequilibrium (Nishimura 1970; Cherdyntzev 1973). This activity ratio is more likely to be < 1.0 in magmas from volcanic arcs than in magmas from elsewhere, suggesting either U enrichment in melts via vapor fractionation, or U enrichment in the source due to slab recycling $\leqslant 4 \times 10^5$ yr before magma genesis.

In addition, $(^{230}Th/^{238}U)$ and $(^{230}Th/^{232}Th)$ ratios increased with time at Irazu volcano, Costa Rica, which indicates in situ differentiation during about 10^5 yr rather than episodic partial fusion of a common source (Sect. 4.4) (Allègre and Condomines, 1976). A third implication of these Irazu data, though not discussed by Allègre and Condomines, is presented below.

Th in Irazu andesites is more likely to have come from the upper mantle than from subducted ocean crust for the following reason. One may safely assume that, at the time of fusion, $(^{230}Th/^{232}Th)$ ratios of magma and source were equal and reflected the U/Th ratio of the source. At Irazu, $(^{230}Th/^{232}Th)$ ratios in the magma increased with time from $\leqslant 1.0$ to 1.3 due to progress toward radioactive equilibrium with a magma having a $(^{238}U/^{232}Th)$ ratio $\leqslant 1.6$ but greater than the source. Thus, the source must have had a $(^{230}Th/^{232}Th)$ ratio $\leqslant 1.0$, or a Th/U ratio $\geqslant$ to 3. Fresh and altered MORB have Th/U = 1 to 3 (Tatsumoto 1978), and in general weathering of oceanic crust increases its U content, decreasing its Th/U ratio. Thus, Irazu andesites or their more mafic parents are more likely to be partial melts of peridotite with Th/U = 3 to 4 (which also is implied by andesite $^{208}Pb/^{204}Pb$ ratios; Sect. 5.5.2), than to be partial melts of subducted MORB.

A second example of isotopic disequilibrium in orogenic andesite is excess ^{226}Ra and ^{228}Ra, unsupported by their U and Th ancestors. This is common in andesites from Japan (Nishimura 1970) and from Stromboli volcano, Eolian arc (Capaldi et al. 1976), where the disequilibrium was attributed to multiple episodic or continuous enrichment of lava in Ra, an alkaline earth, through vapor fractionation.

The final example of disequilibrium involves an initial ^{214}Bi depletion in pumice from the 1975 and 1976 eruptions of Sakurajimi volcano, Kyushu, Japan, resulting in apparent U/Th ratios lower than in older ejecta from that volcano (Sato and Sato 1977). Because this apparent depletion disappeared within one month of eruption, the depletion was attributed to loss during eruption of ^{222}Rn, which is the grandfather of ^{214}Bi and has a 3.8 day half-life. Degassing efficiencies of 50% to 60% were deduced from the degree of depletion. Conversely, disequilibrium enrichments of ^{210}Pb in andesite lavas (Cherdyntzev 1973) may reflect enrichment of lava in Rn through vapor fractionation much as decribed above for Ra.

5.5.6 Oxygen

Most analyses of δO^{18} in orogenic andesites lie between $+ 5.5\%o$ and $7.0\%o$ as they also do for most fresh basalts (Taylor 1968). Thus there is no evidence for ubiquitous or necessary involvement in andesite genesis of sialic crust of any age or of ocean floor rocks, both of which are enriched in O^{18}. Variations within suites of up to 1‰ in δO^{18} can be explained by up to 90% fractionation of plagio-clase + pyroxenes + magnetite (Matsuhisa 1979), but this mineral assemblage is not unique in its ability to explain the data. Somewhat greater variations in δO^{18} within suites due to crystal fractionation are expected in island arcs than in mid-ocean regions due to lower magma temperatures and a larger weight fraction of magnetite in the arcs.

Some orogenic andesites have $^{18}O/^{16}O$ ratios higher than attributable to frac-tionation of conventional mantle-derived magma. For example, rocks from Irazu volcano, Costa Rica, have δO^{18} values of about 7.5‰ (Montigny et al. 1969). Andesites from central Honshu and Hokkaido volcanoes have similar δO^{18} values which are 1‰ to 2‰ higher than in andesites from the Izu Peninsula or Islands (Matsuhisa et al. 1973; Muehlenbachs and Hoering 1972; Matsuhisa 1979) or the Ryuku arc (Matsubaya et al. 1975) where crust is thinner or younger or both. These same central Honshu volcanoes have erupted andesites whose $^{87}Sr/^{86}Sr$ ratios exceed 0.705 (Shuto 1974). Also, late-stage acid andesite flows from Asama volcano, which carry cordierite and sillimanite-bearing xenoliths, have δO^{18} values of 7.0‰ to 7.5‰, which exceed those of other Asama andesites. Finally, the Peruvian acid andesites which display a Rb-Sr pseudoisochron (Sect. 5.5.1) have δO^{18} values of $+ 7.5\%o$ to 9‰, but without correlation between $^{87}Sr/^{86}Sr$ and δO^{18} (Magaritz et al. 1978). These three observations suggest that assimilation of sialic rocks or partial melts derived therefrom has occurred, resulting in heavier oxygen (and Sr), but detailed consideration precludes bulk assimilation processes (Taylor 1980).

A more puzzling discovery is of δO^{18} values of $+ 7.0\%o$ to 9.2‰ in orogenic andesites from Serua and Damar islands, Banda arc, Indonesia, beneath which sialic crust is not expected (Magaritz et al. 1978). Incorporation of subducted material is suggested.

Too few data are available to assess whether there are consistent differences in δO^{18} between low, medium, and high-K andesites or tholeiitic versus calcalka-line andesites. Differences could exist due to differences in degree of crustal assi-milation, vapor fractionation, or magnetite fractionation, for example. However, there are no differences between tholeiitic and calcalkaline suites at Hakone vol-cano, Japan (Matsuhisa et al. 1973; Matsuhisa 1979).

5.5.7 Synthesis of Isotope Data

The isotope data presented above constrain the source region(s) of certain elements in orogenic andesites, but do not yet definitively identify that source. Four isotopically different possibilites exist: the MORB-source, i.e., sub-oceanic upper mantle; heterogeneous mantle unaffected by subduction; seawater; or sialic crust. Subducted ocean crust, or mantle which is modified isotopically by mass transfer from subducted lithosphere, will combine characteristics of two or more of these sources.

As mentioned above, the Sr, Nd, He, Th, and O in most orogenic andesites could, in a general sense, come from the mantle. In detail, however, only the Mariana arc has basalts and andesites (not dacites) all of which are indistinguishable isotopically (in Sr, Pb, Nd, and He) from ocean basin basalts, and even then the ϵ_{Nd} values for the Marianas are atypically low for MORB or its mantle source. Four differences from MORB, and its source, recur in other arcs. First, most $^{87}Sr/^{86}Sr$ ratios are displaced to the right of the mantle Sr-Nd correlation line (Fig. 5.19) by $\leqslant 0.001$, and when $^{87}Sr/^{86}Sr$ ratios exceed 0.704 they often become variable within suites. Second, most $^{207}Pb/^{204}Pb$ ratios lie above the mantle Pb line (Fig. 5.18), defining steeper slopes than are observed even in oceanic islands. Third, most ϵ_{Nd} values are less than in MORB, and some are negative. Fourth, some δO^{18} values exceed $+ 7.0\%$. Of these differences, only those in Nd are compatible with an uncontaminated but nonMORB mantle source. Another component is present.

The difference in Sr may result from incorporation of seawater which could occur in one of three ways: submarine alteration after eruption; mixing in near-surface magma chambers; or mass transfer from altered ocean floor basalt during subduction. Both near-surface processes are possible in oceanic arcs such as Tonga, Scotia, or the Eolian, and have not been adequately tested. Phenocrysts as well as rocks need analysis to assess submarine alteration; Th activity measurements may help assess pre-eruption interaction with seawater (see Condomines et al. 1976). In any case, near-surface seawater incorporation requires water/rock ratios much larger than seem likely, will reduce δO^{18} values, and applies only to oceanic arcs.

A more widely applicable means of introducing seawater or sial is via mass transfer from subducted ocean crust. Indeed, the isotopic differences between oceanic basalts and most orogenic andesites seem to indicate elemental recycling beneath most volcanic arcs due to subduction. Circumstantially, this argument is strongest in the Banda, Aleutian, southern Antilles, and Eolian arcs whose magmas are enriched in radiogenic isotopes (plus ^{18}O in the Banda case) without having ascended through thick or old sialic crust.

Moreover, the isotopic evidence constrains the mechanism of recycling. Isotopically, the Sr in most orogenic andesites could be from altered ocean floor basalts in so far as the $^{87}Sr/^{86}Sr$ ratio of this source lies between 0.7028 and 0.709 (the range of andesites), is about 0.7039 on average (Spooner 1976), and is displaced to the right of the mantle Sr-Nd correlation line. However, this possibility is ineffectual for other isotopic systems. Neither hydrothermal nor seafloor alteration changes the Pb or distinctive Nd isotopic composition of MORB (Church 1973; Meijer 1976; O'Nions et al. 1978). Low-temperature seawater alteration increases δO^{18} values whereas high temperature ($> 300°C$) hydrothermal alteration has the opposite effect (Magaritz and Taylor 1976). Submarine alteration also increases U concentrations which has two isotopic consequences. First, Th/U ratios decrease from their already low values and, therefore, $(^{230}Th/^{232}Th)$ activity ratios in altered crust will be higher (probably greater than 2) than typical of the mantle. Second, decay of Th and U leads to $^3He/^4He$ ratios which are less than typical of the mantle. Consequently, altered MORB lacks the ^{207}Pb enrichment observed in most orogenic andesites, has higher $^{143}Nd/^{144}Nd$ and different δO^{18} than almost all andesites, and should have lower $^3He/^4He$ and higher $(^{230}Th/^{232}Th)$ ratios than are known for andesites. Isotopically, therefore, orogenic andesites cannot be fused ocean floor basalt. This issue is discussed further in Section 8.3.

Sediments may be subducted and thereby alter the compostion of oceanic crust which could be fused to yield orogenic andesites. Specifically, subducted sediments could alleviate the Sr, Pb, Nd, and Th problems discussed above, leaving only the O and He as isotopic dilemmas.

Pb has been studied most in this regard, allowing a distinction between types of sediment. Illite-rich pelagic sediments have high Pb contents and high $^{207}Pb/^{204}Pb$ ratios, so are promising Pb sources. Moreover, note in Fig. 5.18 that average $^{206}Pb/^{204}Pb$ ratios in andesites increase in the order: northwest Pacific arcs (Japan) $<$ northeast Pacific arcs (Aleutians, Cascades) $<$ Atlantic arcs (Antilles). Regional averages of $^{206}Pb/^{204}Pb$ ratios in surface pelagic sediments increase in the same order (Chow and Patterson 1962). Thus, Pb in andesites may be a mixture between Pb from pelagic sediments and Pb from the upper mantle or melted MORB.

However, detailed arguments mitigate this hypothesis. First, published Pb isotope ratios for pelagic sediments from the seafloor adjacent to Japan, the Cascades, Peru, and New Zealand do not lie along linear projections of the arrays characteristic of volcanic rocks from these arcs. Thus, only central Chile and the Antilles are examples in which pelagic sediment is even a possible end-member component of volcanic rock Pb in adjacent arcs. Second, although Pb, U, and Th concentrations in pelagic sediments and the upper mantle are variable, quantitative mixing models consistent both with elemental concentrations and Pb isotopic compositions of arc magmas have not been successful, especially if results from the two Japanese arcs are not combined [cf. Tatsumoto (1969) with Oversby and Ewart (1972) and Church (1973)]. Finally, ^{207}Pb-enrichment is least in the Tonga and Mariana arcs where the ratio of pelagic to terrigenous sediment in the trench is highest and where pelatic sediment is being unequivocally subducted (Sect. 7.4).

Although terrigenous sediment is less likely to be subducted than is pelagic sediment (Sect. 7.4), it would provide an isotopically more appropriate source of crustal Pb during magma genesis. Addition of a few percent of terrigenous sediment could account for the Pb isotopic arrays of andesites from the Aleutians, and the Cascades (Kay et al. 1978; Church 1976). Because terrigenous sediments also have relatively high $^{87}Sr/^{86}Sr$, δO^{18}, and Th/U ratios but low $^{143}Nd/^{144}Nd$ ratios, their presence could reduce or eliminate the isotopic differences between altered oceanic crust and orogenic andesites. Even if andesites are not primary partial melts of such basalt-plus-terrigenous sediment mixtures (Sect. 8.4) the isotopic signature of andesites at least shows the influence of the mixture (Sect. 9.4).

Just as seawater can be introduced into andesite either near the surface or via subduction, so also can sialic crust. Geochemically speaking, incorporation of subducted sediment, especially terrigenous sediment, is just one mechanism of crustal contamination. More straightforward, of course, is appeal to such contamination during ascent of magma through, or storage within, the crust. Qualitatively, this explains the Sr, Pb, Nd, and O (as noted above) and is consistent with the location of most isotopically deviant andesites above thick crust. Indeed, pseudoisochrons in the Quaternary andesites of New Zealand and central Chile give the same age as basement rocks (Fig. 10.1), and both Rb-Sr and Sm-Nd pseudoisochrons give the same (inexplicable) age in northern Chile (Hawkesworth et al. 1979a), consistent with mixing between magma and Rb-Nd-rich crust. Rock Pb which is an isotopically appropriate local contaminant has been found in upper crustal rocks of the Cascades, New Zealand, and Japanese volcanic arcs (Shimizu 1969; Armstrong and Cooper 1971; Church 1976) and in lower crustal rocks of Colorado and Peru (Lipman et al. 1978; Tilton 1979).

Contamination within the crust rather than incorporation of subducted sediment is the simplest explanation of isotopic heterogeneity within suites whose elemental concentrations are generally explicable by fractional crystallization (e.g., Ewart et al. 1973; Serua volcano in Whitford and Jezek 1979) including instances where dacite and rhyolite but not co-genetic basalt are displaced to the right of the SNUM line of Fig. 5.19 (DePaolo and Wasserburg 1977; DePaolo and Johnson 1979), and is the simplest explanation of Ishizaki et al.'s (1977) observation of highest Sr isotope ratios and greatest variability in early ejecta from andesite volcanoes. Crustal-level contamination also would explain instances of atypically high alkali concentrations in andesites whose $^{87}Sr/^{86}Sr$ ratios exceed the regional average (Whitford 1975; Moorbath et al. 1978). Finally, crustal-level contamination alone can explain instances of very low $^{206}Pb/^{204}Pb$ ratios despite high $^{207}Pb/^{204}Pb$ and $^{208}Pb/^{204}Pb$ ratios in Wyoming and Peruvian andesites because these characteristics uniquely identify old granulitic rocks which long ago lost U and Pb relative to Th during high pressure metamorphism or anatexis. Such rocks are expected in old lower crust but not in continental margin sediments.

However, no claimed example of either contamination within the crust or sediment subduction quantitatively explains multiple isotope systems or elemental concentrations. While internally consistent patterns for Sr alone may exist (e.g., Ewart and Stipp 1968; Briqueu and Lancelot 1979) no instance of contamination in arcs is yet as clear as the contamination of Tertiary basalts in Scotland described by Carter et al. (1978). Whatever contamination occurs apparently happens selectively, not by bulk assimilation (Sect. 10.3). Even pseudoisochrons involving orogenic andesites often have intercepts too radiogenic for typical mantle, implying that crustal contamination need not even be invoked.

It is this quantitative failure of mixing models which currently limits interpretation of andesite isotopery. Circumstantial arguments therefore prevail. Sediment subduction is favored by the occurrence of andesites which apparently have been contaminated by sialic upper crust without ascending through sial. Crustal-level contamination is favored by isotopic heterogeneity within suites which seem consanguinous otherwise, or by isotopic evidence for contamination within the lower crust. However, both heterogeneity and absence of sial apply to some suites, e.g., at Serua volcano, Banda arc, or Grenada volcano, Antilles arc. This can be attributed to mantle heterogeneity, as is currently fashionable in studies of oceanic basalts.

The final explanation of why most andesites differ isotopically from most oceanic basalts is that the mantle underlying them differed before subduction; i.e., that the mantle is sufficiently heterogeneous isotopically to produce andesites without help from seawater or sial. Support comes from atypical ocean island basalts. Ocean island tholeiite basalts in general, and the basalts of several ocean island groups in particular (e.g., Samoa, Kerguelen, French Polynesia) also have high $^{87}Sr/^{86}Sr$ ratios extending to 0.706 and have pseudoisochrons, but lack opportunity for contamination by radiogenic Sr from sialic crust (Brooks et al. 1976b; Duncan and Compston 1976). (Mantle rocks can have $^{87}Sr/^{86}Sr$ ratios as high as 0.71: Kramers 1977.) Enrichments in ^{207}Pb and ^{208}Pb somewhat similar to those in andesites also occur in some ocean island (e.g., Reunion, Gough, Tristan da Cunha) and some ocean floor basalts (Fig. 5.17c). Sr-Nd isotope correlations which are less steep than, and displaced to the right of, the typical mantle correlation line have already been found in ocean islands (Fig. 5.18).

In addition, subcontinental mantle apparently differs isotopically from suboceanic mantle. Some subcontinental mantle is isotopically "planetary" with $\epsilon_{Nd} = 0$ and $^{87}Sr/^{86}Sr = 0.7047$; some is even enriched relative to these values and in ^{207}Pb (DePaolo and Wasserburg 1977; Leeman 1977; Basu and Tatsumoto 1980). Because many volcanic arcs rim continents, these differences between mantles may cause isotopic differences between, across, and along arc segments (Sects. 7.1 and 7.2). Because andesite volcanoes provide the greatest available sampling of subcontinental mantle, they may simply be extending our knowledge of its heterogeneity.

While some of the isotopic characteristics of orogenic andesites may be independent of subduction, pre-existing mantle heterogeneity alone is not the answer for two isotopic reasons. First, atypical mantle does not fortuitously underlie all areas where plates converge. Today, for example, isotopically typical mantle underlies the backarcs of New Zealand, southwest Japan, and the Aleutians, in contrast to which andesites (and other rocks) within the volcanic arc contain more radiogenic Pb, Nd, and Sr (Table 5.4). Second, δO^{18} values above + 7.0 in andesites are incompatible with a mantle source, however heterogeneous in radiogenic isotopes.

Thus, the existence of some oceanic basalts which are isotopically similar to orogenic andesites does not prove that andesites are derived from heterogeneous mantle unaffected by subduction. Indeed, the converse may apply: the mantle currently beneath some intra-plate volcanoes may have resided earlier in a mantle wedge above the subducting slab.

In summary, the Sr, Pb, and perhaps Nd of almost all orogenic andesites is drawn from *both* mantle and sialic crustal sources, whereas the O, He, and Th seems primarily of mantle origin. The sialic component is most abundant in the Banda and Chile arcs; least in the Vanuatu and Mariana arcs. The sialic component may be added either as subducted sediment or during ascent. The mantle source(s) (MORB, ocean island, subcontinental) is not presently identifiable because Nd in orogenic andesites seems more "planetary" than in the MORB source whereas Pb in orogenic andesites seems more "planetary" than in the ocean island basalt source. Isotopically, the "mantle component" could be either mafic subducted crust, or the overlying mantle wedge.

5.6 Comparison with Andesites at Not Convergent Plate Boundaries

As noted in Section 2.8, orogenic andesites as defined in Section 1.2 sometimes occur in situations clearly unrelated to subduction, but are subordinate volumetrically to basalt in these instances. Representative analyses are given in Table 5.5. Some of these anorogenic andesites have compositions similar to ones described previously in this chapter. Most, however, differ and fall into two broad categories: icelandites or trachytes.

The icelandite category includes rocks with 53% to 63% SiO_2 and with *hy*, and described in their primary references as icelandites, ferrobasalts, or ferrolatites, as well as simply basalts or andesites. Well-studied examples occur in Iceland (Carmichael 1964, 1967b; Baldridge et al. 1973; Sigvaldason 1974) and other mid-ocean ridges and islands (Kay et al. 1970; Hart 1971; Thompson et al. 1974; Byer et al. 1976), the Snake River Plain, USA (Leeman et al. 1976), and Scotland (Holland and Brown 1972). Also included are the "tholeiitic anorogenic andesites" which constitute significant portions of some continental flood basalt provinces, such as the Grande Ronde "Basalt" which constitutes over 80% of the Columbia River Basalt Group.. Whether or not these andesites are technically "orogenic" or not depends on their TiO_2 contents which decrease from $\leqslant 1.7\%$ within the andesite silica range; they are medium to high-K and tholeiitic in character.

Table 5.5. Representatative andesites not from convergent plate boundaries

	A. Icelandites									B. Trychyte and andesite		C. Impact melt
	Ocean floor		Ocean islands			Continents						
SiO_2	59.0	55.7	54.2	55.3	55.4	54.8	58.1	57.9	57.7	56.8	59.2	57.8
TiO_2	1.8	1.4	2.2	1.9	2.5	2.0	1.3	1.8	1.7	1.8	0.7	0.8
Al_2O_3	12.6	15.3	14.6	14.0	13.5	14.2	14.4	11.3	15.2	13.9	15.9	16.5
Fe_2O_3	–	2.9	1.5	3.3	2.2	–	–	2.4	4.2	0.7	2.5	4.0
FeO	12.0	7.0	10.3	8.7	8.2	11.6	12.3	11.9	5.4	9.4	2.7	2.3
MgO	1.7	2.3	3.2	2.7	5.6	4.3	1.0	2.2	2.1	2.1	3.7	3.5
CaO	5.6	5.7	7.0	6.5	7.4	8.0	4.4	5.6	5.5	5.0	5.8	5.9
Na_2O	4.2	3.8	4.0	4.5	3.0	3.0	3.6	3.1	4.1	5.0	4.1	3.8
K_2O	0.6	1.5	1.2	1.3	1.2	1.5	3.3	2.2	3.1	2.1	2.4	3.0
P_2O_5	–	0.7	0.8	0.6	0.3	0.4	0.6	0.7	0.8	0.7	0.6	0.2
other [a]	1.8	1.4	0.5	0.8	0.5	–	–	1.2	–	1.3	1.3	1.7
Sr	105	580	379	250	–	290	217	180	535	360	–	–
La	25	33	48	–	–	22	134	63	24	60	68	38
Yb	17	4.6	7	–	–	3.3	13	–	–	7	1.2	2.4
Zr	–	235	390	–	–	190	2250	290	437	405	178	–
Ni	<10	<5	0	14	–	18	–	7	46	6	3	–
V	–	40	40	98	–	350	–	11	120	–	–	–
Ba/La	2.2	16	10	–	–	25	16	10	22	7	18	32
La/Th	–	7	11	–	–	5	13	–	–	13	11	13
La/Ta	–	3.2	11	–	–	25	16	–	10 [b]	14	–	20
References	1	2	3	4	5	6	7	8	9	10	11	12

a　$(H_2O + CO_2)$ or LOI
b　Assuming $Ta = Nb \div 16$

References:

1. East Pacific Rise: Kay et al. (1970), sample V2140
2. Ninety-east Ridge: Thompson et al. (1970), sample J40
3. Hekla, Iceland, 1970 lava. Wood et al. (1979a), sample ISL83 Zr, V from Sigvaldason (1974)
4. Galapagos. McBirney and Williams (1969), sample 71
5. Kilauea, Hawaii. Wright and Fiske (1971), sample TLW67-126a
6. Columbia River Basalt, Grande Ronde formation. Wright et al. (1979) and McDougall (1976), averages
7. Snake River Plain. Leeman et al. (1976), sample 69-19. K_2O in excess of andesite range
8. Palisades sill, NJ. Waker (1969), sample W-WU3-61
9. Queensland, Australia. Ewart (1980), average
10. Boina Center, Afar, Ethiopia. Barberi et al. (1975), sample D237 except La/Ta
11. Rio Grande Rift, NM. Zimmerman and Kudo (1979), sample 8 except La/Th
12. Manicouagan Impact Melt, Quebec. Floran et al. (1978), average

These icelandites differ most consistently from orogenic andesites at convergent plate boundaries by being enriched in Fe, Mn, P, and Ti but depleted in Al, Ca, and Mg at any given silica content (e.g., Fig. 5.4). This is especially true when comparing analyses with similar potash contents. Some andesites at convergent plate boundaries are similarly rich in Fe and Mn and depleted in Al, but these are always richer in Ca and Mg and poorer in K. Such differences lead to different normative and modal mineralogies. Andesites not from convergent plate boundaries have lower an contents and $(an + ab)/(di + hy)$ ratios, more albitic normative plagioclase compositions, di/hy ratios higher than the 1/2 average of orogenic andesites, and higher differentiation indices. However, trends of icelandites within the basalt tetrahedron are similar to those of orogenic andesites (Fig. 5.7).

Ti-group trace element concentrations are roughly twice as high or more in the icelandites, and go through maxima or uniformly decrease with increasing silica in the icelandites or related basalts. Ba/Sr ratios are higher relative to silica than in orogenic andesites at convergent plate boundaries. Ni contents usually are low relative to silica for the same reason as in arc tholeiitic andesites, but have a similar range relative to MgO in both groups. V contents are very low in the icelandites. Although observed and calculated magmatic temperatures are similar, calculated fO_2 values are lower for icelandite and near the FMQ buffer. Both light and heavy REE and Y contents are higher in icelandites, especially relative to potash, and consistently increase with silica throughout suites. Acid andesites from the East Pacific Rise have flat REE patterns enriched about 100 times relative to chondrites except for large negative Eu anomalies (Fig. 5.12d). REE patterns for other icelandites are similar in shape to those in Fig. 5.12a although at higher overall concentrations. Concentrations of elements such as Rb, Th, Sc, Cu, and Zn, and ratios such as K/Rb and Zr/Hf are similar in both groups although only orogenic andesites at convergent plate boundaries contain > 70 ppm Cu and have Cu contents which remain constant or increase with silica. Cl/K_2O ratios in Hekla icelandites are 50 times lower than in orogenic andesites at convergent plate boundaries (Sigvaldason and Oskarsson 1976), but plagioclase composition-quench temperature relationships predict similar water contents (2 to 6 wt.%: Baldridge et al. 1973).

A second useful comparison is with the andesine basalts and trachytes of continental rifts (e.g., Weaver et al. 1972; Barberi et al. 1975). Those with 53% to 63% SiO_2 and hy also have high K_2O contents and high Feo*/MgO ratios which makes them calcalkaline on AFM diagrams but often tholeiitic in Fig. 1.4. They resemble and differ from orogenic andesites at convergent plate boundaries in many of the ways described above. However, in these more alkaline suites, rocks with 53% to 63% SiO_2 approach peralkalinity and are strongly differentiated, which is reflected in their normative mineralogy and in much higher concentrations of incompatible trace elements. The trachytes have larger negative Eu anomalies relative to silica than do orogenic andesites at convergent plate boundaries or icelandites.

Both the icelandite and trachyte varieties of andesites not from convergent plate boundaries also lack the distinctive trace element enrichments and depletions of orogenic andesites which I presented in Fig. 5.16 and Table 5.4. That is, the anorogenic andesites cited above have Ba/La ratios < 20, La/Th ratios > 7, and La/Ta ratios < 30 (e.g., Table 5.5). Also note that calcalkaline rocks (i.e., with low as well as relatively constant FeO^*/MgO ratios) are less common in places other than convergent plate boundaries.

A third category of anorogenic andesites includes Pliocene varieties from the southwestern USA which post-date subduction there and at times occupy a tectonic niche like that of the East African trachytes discussed above (e.g., the Taos plateau of the Rio Grande Rift), but which differ geochemically from the icelandites and trachytes in two respects. They are calcalkaline according to Fig. 1.4 as well as subalkaline according to Irvine and Baragar (1971), and their oxide minerals equilibrated at or above the NNO fO_2 buffer (Zimmerman and Kudo 1979). However, even these rocks apparently lack the characteristic trace element signature of orogenic andesites at convergent plate boundaries; i.e., although their Ba/La ratios are in the 15 to 20 range where tectonic distinctions are impossible, both their La/Th and La/Nb ratios are about 12, assuming Hf and Nb concentrations are the same.

Note that both the icelandite and trachyte varieties of orogenic andesite were interpreted quite rigorously as differentiates of basalt in the references cited. Thus, similarities between these two rock types and orogenic andesites at convergent plate boundaries imply the latter also can result from basalt differentiation. Indeed, the greatest difference between orogenic andesites associated with subduction and those not similarly associated is in relative volume, not composition (Sect. 2.8). However, dissimilarities exist, especially in trace element signatures, which indicate differences in source composition or differentiation mechanisms or both, as discussed further in Section 11.9.

5.7 Geochemical Distinctiveness of Volcanism at Convergent Plate Boundaries

This chapter has shown that magmas erupted at convergent plate boundaries differ in composition from magmas erupted in other tectonic environments in five principal ways: silica mode, iron-enrichment, volatile composition, trace element enrichment and depletion patterns, and Sr and Pb isotopic composition. The higher silica mode (i.e., the predominance of andesite) is obvious and the reason for this book; the other points are less widely appreciated.

Subalkaline andesites with FeO^*/MgO ratios less than 2 are rare elsewhere than at convergent plate boundaries regardless of $K_{57.5}$ values. Thus, processes which suppress Fe-enrichment in subalkaline melts are mostly restricted to, but not necessarily operative within, sites of volcanism associated with subduction.

Volatiles likewise appear distinctive. The limited available data suggest that although H_2O contents are broadly similar in orogenic and anorogenic andesites, fO_2 as well as H_2O/CO_2, Cl/K, Cl/S, and possibly even Cl/H_2O ratios are higher in andesites from convergent plate boundaries than elsewhere, including backarcs (Sects. 5.3 and 5.6). The clearest exceptions are Pliocene andesites from the Rio Grande Rift, USA, in which at least fO_2 also was high and the andesites also are calcalkaline.

Next, there is a characteristic trace element signature of rocks, including orogenic andesites, from convergent plate boundaries. In them, K-group and Th-group trace elements are enriched relative to REE whereas Ti-group elements are depleted (Sect. 5.4). Consequently, Ba/La and La/Nb ratios are higher ($>$ 15 and 2, respectively) in basalts and andesites of volcanic arcs than in similar rocks from elsewhere, including backarcs, whereas La/Th ratios are lower ($<$ 7).

Finally, although Quaternary basalts and andesites of the Mariana arc are indistinguishable from oceanic basalts, most arc rocks have higher $^{87}Sr/^{86}Sr$ relative to $^{143}Nd/^{144}Nd$ ratios, $\epsilon_{Nd} < + 9$, higher $^{207}Pb/^{204}Pb$ relative to $^{206}Pb/^{204}Pb$ ratios, and occasionally higher $^{208}Pb/^{204}Pb$ relative to $^{206}Pb/^{204}Pb$ ratios (Figs. 5.18. 5.19).

All of the geochemical comparisons listed above not only distinguish volcanic arcs from sites of basaltic or andesitic volcanism not at convergent plate boundaries, but also separate volcanic arcs from backarcs (Table 5.4). This latter distinction is remarkable both because of the close proximity involved and because some backarc basalts themselves can be distinguished from N-type MORB by several of these same criteria (e.g., Saunders and Tarney 1979).

Although it is less well known, arc rocks also seem to have $^3He/^4He$ ratios and $(^{230}Th/^{238}U)$ activity ratios which are lower than in oceanic basalts.

It is not yet known whether components of this geochemical signature occur in the andesites discussed in Chapter 2 which are "associated with" subduction in the sense that they were erupted within several hundred kilometers of a subduction zone within, say, 10 million years of the subduction, but which clearly, or at least arguably, did not overlie subducted oceanic crust at the time of volcanism. Similarly, the time interval during which the signature persists after subduction ceases is unknown. Both the spatial and temporal extent of this distinctiveness obviously is important to determine.

5.8 Conclusions: Chemical Diversity of Orogenic Andesites

The preceding section summarized geochemical characteristics common to and distinctive of all orogenic andesites associated with convergent plate boundaries. This section summarizes the ways in which orogenic andesites differ among themselves depending on whether they are low, medium, or high-K, calcalkaline or tholeiitic, erupted through thin or thick crust, or simply located in different regions.

Increasing potash levels ($K_{57.5}$) of andesites are accompanied by different behavior of alkalies within andesite suites. K_2O-SiO_2 slopes increase more, but Na_2O/K_2O ratios decrease less relative to increasing silica, as potash levels increase. CaO contents vary inversely with potash between suites. One result of all this is that, at any given silica content, normative plagioclase contents are higher in high than low-K andesites. Also, hy/qz ratios are higher, and differentiation indices are lower in high-K andesites; i.e., they are less silica-saturated and less differentiated at any given silica content.

Predictably, concentrations of Rb, Cs, Ba, Sr, light REE, Th, U, Zr, Nb, and Hf increase with potash level in andesites. Rb concentrations increase more than K or Sr so that K/Rb ratios decrease while Rb/Sr ratios increase with increasing $K_{57.5}$. Zr/Nb ratios decrease in the same direction. Heavy element concentrations increase in the order Th > U > Pb so that Th/U and U/Pb ratios increase with increasing $K_{57.5}$.

The elements cited above correlate positively with SiO_2 to a similar extent within andesite suites regardless of potash level, i.e., bulk distribution coefficients remain similar. However, heavy REE and Y contents sometimes increase less or even vary inversely with silica as potash levels increase. (The inverse correlation is not always limited to high-K suites, however). Negative Eu anomalies generally are limited to low-K andesites, and even then are uncommon and small. Low-K andesites often are similar to N-type MORB in ratios of incompatible elements belonging to the same element group, e.g., K/Rb, La/Yb, Th/U, and Zr/Hf but not Ba/Sr. These ratios in medium to high-K andesites are intermediate between those of E-Type MORB and alkali basalts. No consistent differences in other minor or trace element contents or ratios seem related to potash levels.

No consistent correlation exists between absolute iron-enrichment (e.g, FeO*/MgO ratios) and the concentration of K_2O or related elements in andesites at given silica contents, although this distinction is striking in northern Japan. While many low-K andesites have high FeO*/MgO ratios (i.e., are tholeiitic as defined in Fig. 1.4 and thereby constitute the "island arc tholeiitic series" of Jakeš and Gill, 1970), no other combination of potash and iron-enrichment variables is consistent. Thus, tholeiitic and calcalkaline andesites each usually have a similar range of concentrations of potash and the related elements discussed above. There are some differences in normative mineralogy, however, with $(an + ab)/(di + hy)$ and di/hy ratios usually being higher in thoeliitic than in calc-alkaline andesites of similar silica contents. At any given silica content, some differences in the concentrations of trace elements strongly correlated with FeO*/MgO ratios also may occur between tholeiitic and calcalkaline andesites. Specifically, Ni and Cr correlate negatively with FeO*/MgO ratios and often are higher in calcalkaline than tholeiitic andesites of similar SiO_2 or even MgO contents.

Other differences between tholeiitic and calcalkaline andesites seem more related to the rate of iron-enrichment relative to silica than to absolute FeO*/MgO ratios. Some tholeiitic andesites experience rapid Fe-enrichment although this is not prerequisite according to the definition adopted. Such andesites differ from calcalkaline and other tholeiitic andesites in the behavior of Ti, P, V and possibly Cu, and Zn. Their Ti and V contents follow Fe in rapid enrichment, going through a maximum concentration before decreasing with increasing silica. In contrast, these elements correlate negatively with silica throughout calcalkaline and other tholeiitic suites. P and possibly Cu and Zn increase with silica in rapidly iron-enriched andesites but decrease with silica throughout all or part of other suites. Kuno (1968a) and others have suggested that tholeiitic andesites are characterized by higher eruption temperatures, lower water contents, and lower fO_2 than calc-alkaline andesites of similar silica contents (or some other index of differentiation). One might also expect tholeiitic andesites to be characterized by lower halogen contents, lower Fe_2O_3/FeO ratios, and a smaller range of δO^{18} values within suites for related reasons. Available data are insufficient to demonstrate any of the above.

Several other aspects of bulk composition should be noted which are common to all or many andesites but which are related inconsistently if at all potash or iron enrichment. Some have to do with general concentration levels. Ni contents and Mg-numbers are low, usually 10 to 50 ppm and about 53, respectively. Ni contents vary by an order of magnitude relative to MgO. Heavy REE and Y concentrations usually are about ten times chondritic levels and relatively constant; these concentrations are less than in MORB but similar to those in ocean island tholeiitic basalts. Positive or negative Ce or Eu anomalies > 10% are uncommon. Measured Th/U and U/Pb ratios can be greater (in high-K) or less (in low-K) andesites) than the calculated, time-averaged ratios of andesite source regions. Unsupported Sr is uncommon and limited to low-K basic andesites or regions of anomalous Sr-enrichment. Nd/Sm ratios in medium and high-K andesites always exceed those of their source during most of its history. About 2/3 of all analyzed andesites, including most of those from 12 out of 20 of the arcs for which data are available, have $^{87}Sr/^{86}Sr$ ratios between 0.703 and 0.704. All andesites except those from continental interiors have $^{206}Pb/^{204}Pb$ ratios similar to those of MORB. Most andesites resemble basalts in δO^{18} values.

The second group of notable features relates to within-suite variations. Most orogenic andesites, regardless of potash or iron-enrichment, crudely define an apparent cotectic trend between the plagioclase, orthopyroxene, and clinopyroxene volumes in the basalt tetrahedron. Nevertheless, K/Rb ratios and Ni contents frequently remain anomalously constant in some of these suites. It is not known whether this constancy reflects inaccurate Rb or Ni data, or an unexpected differentiation mechanism.

The thickness of crust encountered increases the proportion of andesite relative to basalt erupted at convergent plate boundaries and also influences andesite compositions (Sect. 7.2). Heavy REE and Y contents are anomalously low or decrease with increasing silica only in andesites erupted on continental margins or interiors. Ni contents sometimes are high relative to MgO, and U contents are especially high relative to Si and K in andesites erupted through thick crust. $^{87}Sr/^{86}Sr$, $^{207}Pb/^{204}Pb$, and possibly $^{18}O/^{16}O$ ratios are higher and more heterogeneous, whereas $^{143}Nd/^{144}Nd$ ratios are lower in andesites ascending through thicker crust. At times the heterogeneity in $^{87}Sr/^{86}Sr$ correlates positively with Rb/Sr ratios, forming pseudoisochrons. However, it is not known whether these higher isotopic ratios are due to incorporation during magma genesis of subducted terrigenous sediment (which is most abundant adjacent to regions of thick crust) or due to contamination of magma by crustal rocks during magma ascent. However, incorporation of sea water or subducted sediment during magma genesis is suggested by the high Cl/S, Cl/K, Ba/La, and Th/La ratios of all orogenic andesites, and by the presence of isotopically unusual Sr or Pb or both in arcs where radiogenic crustal rocks apparently are absent within the arc but present on subducting lithosphere. The presence of normative corundum in many andesites is attributed to analytical error rather than to incorporation of pelitic material.

Finally, some regional heterogeneities in source composition may exist, such as regionally high FeO*/MgO ratios in Indonesia and New Britain, locally high Sr in northern California and the southern Antilles, regionally high Rb relative to K in New Zealand and much of the Antilles, and high Zr and Nb relative to Ti in Turkey.

Chapter 6 Mineralogy and Mineral Stabilities

As noted in Section 4.2, andesites are usually porphyritic in texture, thereby providing a rich mineralogic record of their history, as well as sometimes providing Earth's finest examples of anorthite and of coexisting pigeonite and hypersthene. However, most studies of andesite report only the identity or, at most, the proportion of phenocrysts present. Indeed, the isotopic composition of Sr is known for more andesites than is the composition of their conspicuous pyroxene phenocrysts! The data summarized below broadly constrain the physical conditions of phenocryst crystallization but are insufficient to make possible a systematic investigation of relationships between mineralogy and andesite type, eruption history, or the tectonic environment of eruption. In addition, Ewart (1976a, 1979, 1980) has summarized relationships between modal proportions of phenocrysts, bulk compositions, and crustal environments.

In general, phenocrysts record one or more of three end-member phenomena: the polybaric, polythermal, closed-system precipitation of crystals from a liquid whose residue was host to the crystals upon eruption and which chilled to form the surrounding groundmass; an open-system accumulation of crystals precipitated from genetically related melts of diverse composition within a zoned reservoir; or the mixing of unrelated magmas pumped through common conduits. Each phenomenon apparently is respresented in andesites, but their relative significance cannot yet be assessed. A given lava or ash may contain crystals originating in two or more ways, i.e., the phenocrysts may be a polygenetic assemblage.

6.1 Plagioclase

Plagioclase crystals are the most ubiquitous and usually the most abundant phenocrysts in andesite, typically constituting 50% to 70% of the phenocrysts by volume. Indeed, porphyritic andesites lacking plagioclase phenocrysts are very uncommon except in high-Mg varieties (e.g., Hormann et al. 1973; Kuroda et al. 1978). This ubiquity indicates that plagioclase is the liquidus phase in most andesites prior to eruption which in turn implies less than 2 to 5 wt.% water in the magma (Sect. 5.3.1). Usually plagioclase is the only feldspar present as a phenocryst or in the groundmass, although microphenocrysts and groundmass crystals of sanidine occur in some high-K acid andesites (e.g., Jakeš and Smith 1970).

Plagioclase phenocrysts are the most striking feature of andesites in thin section. They are highly variable in composition due to pronounced and complex zoning; they frequently contain inclusions in certain zones and appear moth-eaten, frittered, or sieve-like in texture; sometimes they occur as aggregates enclosing clear glass. Because the density contrast between plagioclase crystals and surrounding melt is slight (Sect. 4.1.2), these diverse characteristics of plagioclase phenocrysts are especially likely to reflect crystal accumulation (Sect. 5.2.5) and a complex polygenetic history.

Plagioclase phenocrysts in andesite range in composition from An_{15} to An_{99} with modes of about An_{75} and An_{60} for basic and acid andesites, respectively (Ewart 1976a). Ironically, compositions $< An_{50}$, once regarded as characteristic of orogenic andesite, are common only in anorogenic andesites or in the acid andesites of restricted areas such as the central Andes or Papua.

One cannot tell a priori whether or not specific plagioclase phenocrysts precipitated from a liquid whose composition lies between that of their host rock and host groundmass due to uncertainties regarding the thermodynamics of plagioclase-melt equilibria and uncertainties regarding the pre-eruption water pressure of andesite magmas. However, application of equations from Mathez (1973) and Kudo and Weill (1970) yields three conclusions premised on data summarized in Fig. 6.1. First, precipitation of $\geqslant An_{90}$ plagioclase from andesite magma at $1000°$ to $1100°C$ (probable liquidus temperatures: Sect. 4.1.1) is likely only if $P_{H_2O} \geq 5\,kb$, which is unrealistic (Sect. 5.3.1). Thus, highly calcic plagioclase crystals (e.g., Ishikawa 1951) either accumulated after cognate precipitation at unexpectedly high temperatures ($> 1300°C$) or are xenocrysts. Second, $P_{H_2O} > 1\,kb$

Fig. 6.1. Calculated compositions of plagioclase in equilibrium with average orogenic andesites at various water pressures. Calculations use the equations of Mathez (1973) for 0.5 and 1.0 kb P_{H_2O}, and Kudo and Weill (1970) for 5 kb. Melt compositions used are the range of average andesite compostions from Table 5.1. Within each field, low-K acid andesite melt yields the most calcic plagioclase for any given P-T condition; high-K basic andesite yields the least calcic

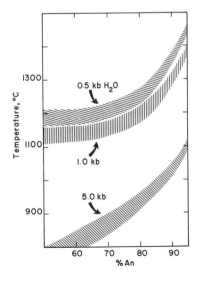

is necessary for plagioclase $> An_{50}$ to precipitate at $T < 1100°C$ from average basic or acid andesite magmas. This is the basis for point (3) in Section 5.3.1. Third, differences in nonvolatile composition between low and high-K andesites can result in differences of > 20 mol% An at constant temperature, with An contents decreasing as potash level increases.

Although not apparent in the data chosen for Fig. 6.2, average An contents of phenocrysts decrease from low- to high-K andesites, as predicted above, and as was noted earlier for normative plagioclase compositions. Also, average An contents usually decrease from basic to acid andesites of consanguinous suites (Ewart 1980).

Compositions within thin sections often are as variable as within suites, commonly covering a range of 20 to 50 mol% An (Fig. 6.2). Compositions within individual crystals frequently vary by 10 to 40 mol% An, but can be homogeneous.

Because plagioclase usually is the sole feldspar present, andesite liquids rarely reach the natural equivalent of the two-feldspar surface, whether or not normative rock compositions lie within the projection of the two-feldspar field on the Ab-An-Or plane. Thus, Or contents of andesite plagioclases increase with whole rock K_2O contents, from about 0.1 mol% Or in low-K suites (e.g., Aramaki and Katsura 1973) to 5 mol% in some high-K suites (e.g., Mackenzie 1976). Characteristically, Or contents are lower in phenocrysts or groundmass plagioclases than in normative feldspars (Fig. 6.2), indicating concentration of K in glass.

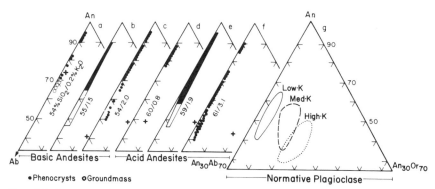

Fig. 6.2. Representative compositions of plagioclase phenocrysts *(solid circles* or *bar)* and groundmass crystals *(open circles* or *bar)* in single specimens of orogenic andesite. One basic and acid andesite sample is shown for each of three arcs and illustrate the low, medium, and high-K andesite types. SiO_2 and K_2O contents of the host rock are given. The normative feldspar compositions of the host rocks are shown as *crosses*. The low-K suite (samples *a* and *d*) is from Tonga (Ewart 1976b, samples HH1 and F31); the medium-K suite (samples *b* and *e*) is from the Aleutians (Marsh 1976, samples AD44 and AMK6; individual plagioclase analyses were not reported); the high-K suite (samples *c* and *f*) is from Filicudi volcano, Eolian arc (Villari and Nathan 1978, samples F30 and F1). *g* shows the CIPW normative plagioclase composition of 85% of the 2500 low, medium, and high-K orogenic andesites for which data are compiled in Fig. 1.1 and 5.6

Plagioclase phenocrysts commonly contain 0.5 to 1.0 wt.% FeO* in solid solution, regardles of the FeO*/MgO ratio of the host rock (cf. Smith and Carmichael 1968; Lowder 1970; Aramaki and Katsura 1973). The Fe oxidation state is not known. Both Iida (1961) and Heming (1977) reported maximum Fe contents at around An_{70} in andesite plagioclases whereas plagioclase from many MORB and lunar basalts have a continuous inverse relationship between Fe and An contents (Smith 1974; Ayuso et al. 1976) or a maximum around An_{50} (Mazzulo and Bence 1976). This difference may indicate a wider interval of co-precipitation of plagioclase and magnetite in andesites.

The zoning of andesite plagioclase phenocrysts is even more striking than their overall calcic compositions. Various categorizations of this zoning exist (e.g., Homma 1936) but are followed here only informally. Trends in composition can be for interior regions to be more calcic than exterior regions (normal zoning), or more sodic (reverse zoning). Changes in composition can be abrupt or gradual, oscillatory or continuous. Crystals usually include one or more of the following features in any combination: a homogeneous internal region, frequently the core; patchy-zoned, inclusion-rich (frittered) regions; regions of oscillatory zoning; abrupt, nonoscillatory changes in compositions of 10 to 30 mol% An, often accompanied by resorption of the inner feldspar; a clear, normally zoned mantle; and a thin ($< 20 \mu m$) rim, usually similar in composition to groundmass microlites. Correlations of zoning patterns between different crystals in the same ejecta, much less in successive ejecta, are notoriously tedious and difficult; no definitive results for andesitic hosts have been published.

Oscillatory zoning is characteristic of plagioclase phenocrysts in orogenic andesite. Oscillations are 1 to $10 \mu m$ thick and typically fluctuate by 1 to 10 mol% An, returning repeatedly to a constant maximum An content (e.g., Sibley et al. 1976) or gradually decreasing in maximum An content outward (e.g. Ewart 1976b). While possibly related to rhythmic changes in environmental variables such as P or P_{H_2O}, oscillatory zoning may also result from diffusion rate-controlled compositional gradients at crystal-liquid surfaces (Lofgren 1974; Haase et al. 1980). Smith (1974) reviewed this topic. Because neither magmatic convection nor fluctuating P_{H_2O} needs to be invoked to explain oscillatory zoning, its ubiquity in andesite plagioclases may only reflect the structural and thermal characteristics of andesite magmas and, therefore, elucidate the kinetics of crystal precipitation but not the genesis of andesites. However, abrupt nonoscillatory changes in composition probably do reflect abrupt and widespread changes in magmatic environment, but few attempts to correlate these changes throughout flow units or to relate them to a stratigraphic context have been successful. One example is cited on the next page.

Normal zoning seems to predominate throughout most of the volume of most plagioclase phenocrysts in andesites. Although this stagement has never been rigorously quantified, typical examples are described in papers listed in Table 5.3. Rims are usually abruptly more sodic than mantle regions. Normal zoning can result from fractional crystallization during isobaric cooling, or during iso-thermal ascent of hydrous magma, or any combination thereof. Both phenomena, therefore, seem common in andesites. For example, a decrease by 15 to 20 mol% An can result from an isothermal drop of about 1000 bar water pressure during ascent of andesite magma (Fig. 6.1 and Sect. 5.3.1). In the simpler system Ab-An-H_2O, the 1 mol% An decrease in calcic plagioclase for each 10 bar of decompres-sion which was reported by Yoder et al. (1957) is an underestimate (Johannes 1978).

However, nonoscillatory reversed zoning also is common, either as one or more narrow interruptions (spikes) in an otherwise normally or oscillatorally zoned crystal, or as an overlay of a broad calcic sheath, usually including rim, on a sodic core. The narrow calcic spikes may be the most calcic portions of the crystal. Crystals with such spikes have been interpreted most often as due to the isothermal ascent of nearly anhydrous magma (e.g., Vance 1965; Ewart 1976b), although they may also be due to inward as well as outward growth of originally skeletal crystals during isobaric, isothermal crystallization (Lofgren 1974) or due to local rise in temperature. Generally, the stratigraphic context of such crystals in andesite is not given. However, Wise (1969, p. 987) described an example in which basal flows contained thin, abrupt An_{70} spikes overlying more sodic cores, whereas crystals with these spikes became proportionally less common in later flows. He interpreted the spikes as having precipitated during ascent to a shallower sub-sidiary reservoir and the subsequent flows as containing in addition crystals nuc-leated therein.

Examples of reverse zoning involving broad calcic sheaths surrounding sodic cores are described by Larsen et al. (1938), Pe (1973, 1974), and Eichelberger and Gooley (1977), for example, who estimate that such phenocrysts occur about 10% of the time, commonly, or abundantly in every thin section, respectively, in the areas they studied. Often the sodic cores are resorbed, the contact area is rich in inclusions, and the calcic rim is similar in composition to groundmass microlites or is normally zoned to such a composition. Although these phenomena some-times can be attributed to ascent of anhydrous magma as discussed above, ascent from 30 km depths is likely to increase An contents by at most 10 mol%, based on experiments in the simple Ab-An system (see Pringle et al. 1974) or in liquid of andesitic composition (Green 1968); Drake (1976) concluded even less varia-tion would occur. Thus, greater differences in plagioclase compositions, especially where sodic cores contain rhyolite glass inclusions, probably reflect magma mixing or assimilation. If so, the sodic cores are xenocrysts from dacitic to rhyolitic magmas or are inclusions from silicic wall rocks.

Criteria for choosing in any specific instance between the interpretations of reverse zoning summarized above, and estimates of the relative importance or stratigraphic context of such crystals, are difficult and not yet available for any andesite. Thus, the tantalizing implications of plagioclase zoning for the water contents or mixing history of andesite magma, for example, remain ambiguous.

Inclusions in plagioclase phenocrysts usually are of glass or its devitrification products or other phenocryst minerals, are restricted to certain zones but often result in widespread sieve-like textures, and can be very abundant.

Notably, inclusions of minerals which are not coexisting phenocrysts are uncommon; e.g., hornblende inclusions in plagioclase phenocrysts of pyroxene andesites are rare. Often zones especially rich in inclusions are resorbed on their exterior margin, or overlie resorbed cores, or both; typically they are rimmed by clear, normally zoned plagioclase. An contents of inclusion-rich zones can be higher (e.g., Lowder 1970) or lower (e.g., Pe 1974) than in the clear outer rim; sometimes both possibilities occur in the same flow (e.g., Larson et al. 1938). Often the region immediately overlying the sieve zone is the most calcic portion of the crystal.

The glass or mineral inclusions are interpreted as due to one or more of four phenomena: localized nucleation of minerals due to compositional gradients at the interface between plagioclase rims and melt (Bottinga et al. 1966); melt entrapment due to changes in crystal morphology during rapid growth and enhanced supercooling (Lofgren 1974); resorption during ascent of anhydrous magma (Mac-Donald and Katsura 1965); or resorption during magma mixing (Eichelberger and Gooley 1977; Anderson 1976; Dungan and Rhodes 1978). As with explanations for zoning, each is possible, but the important questions of relative significance and stratigraphic context are unanswered. However, at least two conclusions need emphasis. First, large glass inclusions frequently have compositions which lie outside the range represented by the bulk rock and groundmass. Reversely zoned plagiocalse can include rhyolitic glass in their sodic cores; highly calcic cores of normally zoned plagioclases sometimes include basaltic glass (Anderson 1976). Thus, magma mixing and crystal accumulation, respectively, are implied. Second, minerals included in plagioclase do not indicate precipitation of minerals other than coexisting phenocrysts. Thus, there is no indication of ubiquitous polybaric fractionation involving nonphenocryst phases such as amphibole, quartz, or garnet at depth.

In summary, the zoning, inclusions, and resorptions of plagioclase phenocrysts may provide the fullest available record of andesite magma history, but their ubiquity and bewildering complexity make their interpretation formidable.

6.2 Pyroxenes

Five pyroxenes occur in andesite and will be discussed in the order: augite, clinoenstatite, pigeonite, subcalcic augite, and orthopyroxene. *Augite* is second only to plagioclase in its abundance and ubiquity as a phenocryst in andesites, averaging about 3 modal % (Ewart 1976a) and occurring in almost all porphyritic andesites except those with very low normative *di* contents [e.g., Parícutin, Mexico (Wilcox 1954); Ancud, Chile (Lopez-Escobar et al. 1976); L'Esperance, Kermadec (Ewart et al. 1977)]. The most striking feature of augite phenocrysts is the restricted range in composition of their cores, which mostly lie between Wo 38–50 En 40–55 Fs 7–20. Variations within single thin sections or even single crystals often are as great as variations between augites from basalts to rhyolites.

Typical compositions are given in Table 6.1 and shown in Figure 6.3, and illustrate two differences amongst augites from various orogenic andesite types. First, variations in FeO*/MgO ratios of the whole rocks are reflected in augite phenocrysts as expected. Thus, Fs contents are higher in augites from acid tholeiitic andesites than acid calcalkaline andesites, and are lowest in suites with unusually low FeO*/MgO ratios such as at Shasta, USA (Smith and Carmichael 1968). However, just as orogenic andesites are less Fe-rich than anorogenic icelandites (Sect. 5.6), so ferroaugite phenocrysts are uncommon in orogenic andesites and Fs contents rarely exceed 20 mol% except on phenocryst rims. Exceptions exist (e.g., Kuno 1969; Fairbrothers et al. 1978) but are restricted to tholeiitic acid andesites. Thus, whatever usually restricts whole rock Fe-enrichment in orogenic andesites also prevents their augites from becoming as Fe-enriched as in differentiated tholeiitic or alkalic suites elsewhere. The cause may be high Fe_2O_3/FeO ratios and coprecipitation from orogenic andesite of magnetite or amphibole or both with augite.

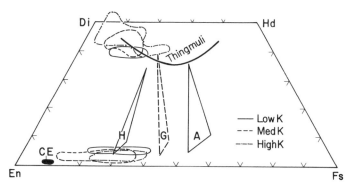

Fig. 6.3. Summary of phenocryst pyroxene compositions in orogenic andesites. Data are restricted to phenocryst cores from andesite hosts, and are generalized from the following suites. *Low-K* Tonga, Hakone, Akita-Komagatake; *Med-K* Tongariro, Talasea, Rabaul, Marianas, Aleutians, Cascades, Colima, Santorini; *High-K* Papua, New Guinea Highlands, Shiroumeaoike, Osima-osima, Bogoslof, San Salvador, Filicudi, Aegina, Methana. See Appendix for references. The *triangles* connect co-existing phases in orogenic andesites from Hakone *(H)* and Asia *(A)*, Japan, and Germany *(G)*; see text for references. Field *CE* shows clinoenstatites in high-Mg orogenic andesites (boninites) from Shiraki et al. (1980). *Thingmuli* trend shows augites from anorogenic andesites (Carmichael 1967b)

A second feature of Fig. 6.3 is the general though inconsistent increase in Wo content of augite as K-enrichment of the whole rock increases. This phenomenon has been recognized in basalts for some time (Le Bas 1962) and can be attributed to lower equilibration temperature (hence, greater pyroxene immiscibility), to lower a_{SiO_2} (hence, higher $CaAl_2SiO_6$ component in the autite), or to coexistence with amphibole in the alkali-enriched rock. Where the latter occurs without orthopyroxene, compositions usually lie near the Di-Hd join (e.g., Lewis 1973; Arculus et al. 1977), as also noted experimentally by Helz (1973).

Augite is the most common groundmass pyroxene in andesites. Groundmass augites either are similar to phenocrysts in composition, or are more ferrous or less calcic or both.

Clinoenstatite, a mineralogic rarity, occurs as phenocryst mostly in atypical orogenic andesites known as boninites (Sect. 5.2.3), and indeed may owe its existence to the unusual combination of high temperature, MgO, SiO_2, and H_2O in these magmas. Examples are described by Dallwitz et al. (1966), Dietrich et al. (1978), and Shiraki et al. (1980). The clinoenstatites in andesites have En contents of 87 to 92, < 1 mol% Wo, and < 0.5wt.% Al_2O_3 (Fig. 6.3).

Pigeonite occasionally occurs as phenocrysts in tholeiitic andesites which lack olivine phenocrysts (e.g., Kuno and Nagashima 1952; Ginzburg et al. 1964; Oba 1966; Kuno 1969; Nakamura and Kushiro 1970a,b; Virgo and Ross 1973; Ewart 1976b). The pigeonites are more Fe-rich than Fs30. Those coexisting with augite phenocrysts typically have 10 mol% Wo; those existing only with hypersthene phenocrysts are less calcic in one case (Wo4 in a Mull andesite; Virgo and Ross 1973) but not in another (Wo10 in a Pleistocene Carpathian andesite; Ginsburg et al 1964). Pigeonite phenocrysts have been reported in calcalkaline andesites (e.g., Taylor 1958; Crowe et al. 1976; Arculus 1978), but usually not confirmed by analyses.

In contrast, pigeonite occurs quite frequently as rims on olivine or other pyroxene phenocrysts or in the groundmass of andesites which are tholeiitic according to Fig. 1.4 (e.g., Kuno 1950a; Kawano et al. 1961), and occasionally in the groundmass of andesites which are calcalkaline according to that criterion (e.g., Akrotiri-Thira Series, Santorini; Nicholls 1971a). Groundmass pigeonite in andesites may coexist with subcalcic augite only (e.g., Lowder 1970; Ewart 1976b), with hypersthene and augite (e.g., Nakamura and Kushiro 1970a), or with augite only (e.g., Nakamura and Kushiro 1970b; Nicholls 1971a).

Subcalcic augite is restricted to the rims of other pyroxene phenocrysts and to the groundmass where it merges with pigeonite in composition (Lowder 1970; Aramaki and Katsura 1973; Ewart 1976b). Subcalcic augites display wide compositional ranges between grains and, sectorally, within grains, indicating disequilibrium. Thus, they probably reflect quenching as suggested by Muir and Tilley (1964). However, glassy calcalkaline andesites usually lack subcalcic augites, and no correlation with texture or eruption style has been identified.

Most suites of orogenic andesites contain primary *orthopyroxene* phenocrysts but the range in silica contents of rocks in which they occur varies considerably.

Generally, orthopyroxene and olivine phenocrysts are antipathic, although the two coexist in some calcalkaline basic andesites. The in-coming of orthopyroxene instead of olivine is a function of the extent of magma polymerization (Kushiro 1975) and a_{SiO_2} and occurs at higher SiO_2 contents in high-K than low-K andesites. Orthopyoroxene also seems antipathic with hornblende and biotite (Bowen 1928) but where these hydrous minerals are absent, orthopyroxene persists into dacites and rhyolites. Predictably, the minimum silica contents of rocks in which hornblende or biotite occurs instead of orthopyroxene decreases as the alkali content of the rocks increases. Thus, orthopyroxene phenocrysts can be absent altogether in high-K andesites, which contain olivine, hornblende, or biotite instead (e.g., Arculus et al. 1977; Pe 1974).

Orthopyroxene phenocrysts have more variable and higher Fe/Mg ratios than coexisting augites, ranging in composition from En_{85} to En_{60}. Phenocryst cores in basic andesites usually are more magnesian than in acid andesites, but there is considerable overlap and variations of 15 mol% En within individual thin sections are not uncommon. Unlike augites, orthopyroxene Fe/Mg ratios are similar regardless of the iron or potassium content of the whole rock. Wo contents usually are 2 to 5 mol%, possibly being lower in high-K than low-K andesites, and in acid than basic andesites.

Groundmass orthopyroxenes are common and can be rimmed with pigeonite or subcalcic augite (Kuno's d $\rightarrow$ c category) as noted above, or can coexist only with augite. The latter assemblage seems typical of andesites which are calcakaline according to Fig. 1.4 (e.g., Kuno 1950a; Kawano et al. 1961), although acid andesites of some suites which are marginally tholeiitic by that criterion also contain orthopyroxene without pigeonite in the groundmass (e.g., Lowder 1970. Heming 1977).

Three aspects of these pyroxene compositions are informative regarding andesite genesis. First, differences in pyroxene assemblages and compositions may reflect differences in equilibration temperatures. Specifically, the frequent difference in groundmass Ca-poor pyroxene between calcalkaline and tholeiitic andesites may indicate that the latter erupt at higher temperatures (Kuno 1950a, 1966). Kuno's argument was based on Hess' (1941) interpretation of pyroxene phase relations in which the low pigeonite structure ($P2_1/c$) was considered a higher temperature polymorph than the orthorhombic structure (Pbca). This has since been confirmed by Ishii (1975) and Ross and Huebner (1979). Characteristic eruption of tholeiitic andesites (pigeonitic rock series) at temperatures higher than calcalkaline andesites (hypersthenic rock series) could indicate different conditions of genesis for the two suites, possibly related to convergence rate and degree of partial fusion (Sect. 7.3), or to degree of crustal contamination.

However, as noted in Section 4.1.1., there is no independent evidence yet that tholeiitic andesites consistently are quenched at higher temperatures than calcalkaline ones, and bulk composition or pressure also may affect which Ca-poor pyroxene precipitates. Groundmass pigeonite is restricted to rocks with

FeO*/MgO > 2, which may be a necessary if insufficient condition for precipitation of pigeonite at low pressure. Also, tholeiitic andesites have higher average *di/hy* ratios than calcalkaline andesites (Sect. 5.2.7). These two factors might cause different Ca-poor pyroxenes to precipitate at the same temperature from magmas of different composition. Finally, pigeonite sometimes is a groundmass mineral in andesites where it is absent as a phenocryst, and occurs as rims on orthopyroxene phenocrysts; the converse is rare. This relationship suggests that the orthopyroxene-pigeonite phase boundary has a positive dT/dP slope (Coombs 1963; Ross and Huebner 1979 suggest 8°C/kb) so that orthopyroxene gives way to pigeonite during isothermal magma ascent, although the relationship can also be explained by isobaric cooling accompanied by iron enrichment, or by a rise in temperature during eruption. Thus, although the difference in groundmass pyroxenes suggests a frequent difference in eruption temperature between tholeiitic and calcalkaline andesites, the issue remains clouded by ambiguity about pyroxene phase relations.

Two other aspects of quadrilateral compositions also reflect equilibration temperatures. The greater difference in Wo contents between augites and orthopyroxenes from high-K versus low-K andesites (Fig. 6.3) could indicate increasing immiscibility due to decreased temperatures as alkali contents increase. However, the differences themselves are uncertain and have alternative explanations, noted earlier. Similarly, temperatures calculated from compositions of coexisting pyroxene phenocryst cores in orogenic andesites using Wood and Banno's (1973) method typically are 970° to 1100°C, regardless of host rock potash content or FeO*/MgO ratio (Table 4.1).

A second important aspect of pyroxene compositions concerns their non-quadrilateral components. In general, concentrations of Al, Na, and Ti reflect the temperature, cooling rate, a_{SiO_2} and coexisting mineral assemblage as well as pressure, which complicates interpretation. However, their typically low concentrations in andesite phenocrysts ($< 3\%$ Al_2O_3, $< 0.5\%$ Na_2O, $< 0.75\%$ TiO_2) at least preclude pressures > 10 kb during their formation, judging from the composition of pyroxenes crystallized from andesite at higher pressure (Green 1972). Al_2O_3 contents $> 5\%$ occur in augites coexisting only with amphibole in some high-K Greecian andesites (e.g., Pe 1973, 1974). However, this may be related to decreased a_{SiO_2} in these suites following suggestions by Le Bas (1962) and Kushiro (1960) for basalts. Subcalcic clinopyroxenes with $> 5\%$ Al_2O_3 occur rarely in andesites and may be high-pressure megacrysts (e.g., Duggan and Wilkinson 1973), but are unknown to me from active volcanoes at convergent plate boundaries.

Al, Na, and Ti contents decrease from augites through subcalcic augites and pigeonites to orthopyroxenes in andesites as in other rocks for crystal chemical reasons, but variations in their concentrations between phenocryst cores and rims and groundmass crystals are inconsistent, and the significance of these variations is not yet known. Similar complexity in lunar basalt pyroxenes may reflect variations in cooling rates (Papike et al. 1976) or f_{O_2} (Kushiro 1973).

Table 6.1. Representative analyses of phenocrysts in orogenic andesites

Host rock	Augite Low-K TH	Med-K TH	Med-K CA	High-K TH	High-K CA	Orthopyroxene Low-K TH	Med-K TH	Med-K CA	High-K TH	High-K CA	Hornblende Med-K CA	High-K CA
SiO_2	53.3	50.8	50.7	51.9	48.8	55.4	54.3	52.3	52.1	53.8	42.4	42.4
TiO_2	0.2	0.5	0.4	–	0.5	0.1	0.2	0.3	0.3	0.2	2.5	3.5
Al_2O_3	2.0	2.6	3.6	2.4	3.8	1.4	0.9	1.6	1.1	1.2	12.1	11.5
FeO^*	8.4	10.5	9.3	11.8	9.6	13.8	17.4	18.4	21.9	19.4	12.8	11.9
MnO	0.2	0.4	0.2	–	–	0.3	0.9	0.4	0.6	–	0.2	–
MgO	17.7	15.8	17.0	14.8	14.0	27.5	23.0	22.8	21.5	25.1	13.8	13.6
CaO	18.2	18.4	18.3	19.3	19.5	2.4	1.8	2.7	1.6	1.0	11.1	11.6
Na_2O	0.2	0.2	0.2	0.5	0.3	0.04	0.03	0.2	0.03	–	2.7	2.4
K_2O	–	–	–	–	0.1	–	–	–	–	–	0.4	0.7
F	–	–	–	–	–	–	–	–	–	–	0.1	0.4
Total	100.2	99.3	99.7	99.6	96.5	100.9	98.3	98.7	99.1	100.7	98.1	98.1
	Atoms per 6 oxygens					Atoms per 6 oxygens					Atoms per 23 oxygens	
Si	1.94	1.91	1.88	1.93	1.89	1.97	2.00	1.95	1.96	1.96	6.24	6.26
Al4	0.06	0.09	0.12	0.07	0.11	0.03	0.00	0.05	0.04	0.04	1.76	1.74
Al6	0.03	0.03	0.04	0.03	0.06	0.03	0.06	0.02	0.01	0.01	0.35	0.26
Ti	0.03	0.01	0.01	–	0.02	0.00	0.00	0.01	0.01	0.01	0.28	0.39
Fe^{2+}	0.39	0.33	0.29	0.37	0.31	0.41	0.54	0.58	0.69	0.59	1.58	1.47
Mg	0.82	0.89	0.94	0.82	0.81	1.46	1.26	1.27	1.21	1.36	3.03	2.99
Ca	0.66	0.74	0.73	0.77	0.81	0.09	0.07	0.11	0.06	0.04	1.75	1.83
Na	0.02	0.02	0.01	0.04	0.02	0.00	0.00	0.01	0.00	–	0.77	0.68
K	–	–	–	–	0.01	–	–	–	–	–	0.08	0.13
%Ca	35	38	37	39	42	5	4	5	3	2	27	29
%Mg	44	45	48	42	42	74	67	65	62	68	48	48
%Fe	21	17	15	19	16	21	29	30	35	30	25	23
References	1	2	3	4	5	1	2	3	6	5	7	8

References: 1. Late, Tonga: Ewart et al. (1973), coexisting phenocrysts. 2. Rabaul, PNG: Heming (1977), coexisting phenocrysts. 3. Tongariro, New Zealand: Ewart (1971), co-existing phenocrysts. 4. San Salvador, El Salvador: Fairbrothers et al. (1978). 5. Giluwe, PNG: Jakeš and White (1972b), co-existing phenocrysts. 6. Shirouma-oike, Japan: Sakuyama (1978). 7. Colima, Mexico: Luhr and Carmichael (1980). 8. Rio Grande, USA: Zimmerman and Kudo (1979)

The final significant feature of pyroxene compositions is their zoning, which can be discontinuous with one pyroxene mantling another as noted previously, or continuous. With respect to Fe/Mg ratios, continuous zoning can be normal with Fe/Mg increasing toward the rim, oscillatory, or reverse. Continuous zoning is more extreme in orthopyroxenes than augites, but rarely exceeds 10 mol%Fs. As with plagioclase, normal and oscillatory zoning are most common and probably reflect isobaric fractionation combined with diffusional supercooling. Oscillatory zoning is limited to < 4 mol% Fs (e.g., Ewart 1976b). Reverse zoning also occurs (e.g., Nakamura and Kushiro 1970b; Nicholls 1971a) and sometimes is accompanied by quartz megacrysts (Smith and Carmichael 1968) or other disequilibrium phenomena (e.g., Sakuyama 1978). Reverse zoning may result from decompression (Ewart et al. 1975), the onset of hornblende (Helz 1973) or magnetite precipitation, or an increase in f_{O_2} due, for example, to a pre-eruptive influx of water (Luhr and Carmichael 1980). However, as with plagioclase, reverse zoning also may indicate magma mixing, especially when accompanied by disequilibrium assemblages.

6.3 Amphibole

Only one amphibole, a hornblende, occurs in orogenic andesites, and for the most part it is restricted to medium and high-K acid andesites. These hornblende-andesites contain an average of 3 to 4 vol.% amphibole as phenocrysts (Ewart 1976a) coexisting with most combinations of augite, pigeonite, orthopyroxene, olivine, and plagioclase. Of these coexisting minerals, plagioclase is most common (Ewart 1976a) and pigeonite least. Rarely amphibole is the sole phenocryst. It is absent from groundmass assemblages.

From experimental work (e.g., Eggler and Burnham 1973) it appears that at least 3 wt.% H_2O in the liquid phase is necessary for amphibole to precipitate within the crystallization interval of orogenic andesite magma (Burnham 1979). However, whether hornblende-andesites differ in nonvolatile composition from other medium or high-K acid andesites is unclear. Larsen et al. (1937a) and Jakeš and White (1972b) concluded they generally do not differ, but hornblende-andesites typically have high Fe_2O_3/FeO ratios (> 0.5), correspondingly high Mg-numbers, and > 3% Na_2O (Jakeš and White 1972b; Cawthorn and O'Hara 1976; Ewart 1976a) which may be distinctive.

Reaction relationships affecting the appearance and disappearance of amphibole in andesites are complex. As discussed by Bowen (1928), Best and Mercy (1967), and others for plutonic rocks, and as implicit in the usual restriction of amphibole to acid andesites, amphibole appears at the expense of pyroxene or olivine due to increased P_{H_2O} and alkali content or decreased temperature or all three. Arguments presented in Section 6.11 suggest that many amphibole phenocrysts are not near-liquidus phases of andesite but, instead, result from magma mixing or continued reaction between liquid and early-formed pyroxene(s) or olivine. Indeed, some single, optically homogeneous amphibole crystals in andesite are intimate intergrowths of monoclinic amphibole and pyroxene (Tomita 1964), and pyroxene or olivine cores in amphibole phenocrysts are not uncommon.

Much more common, indeed almost unfailing, is evidence for amphibole disappearance by resorption where amphibole is replaced by magnetite, plagioclase, and clinopyroxene or orthopyroxene or both. The resorption may yield a corona of these minerals (opacite rim), or destroy the amphibole altogether. Consequently, even pyroxene clots in andesites have been interpreted as resorbed amphiboles (Stewart 1975). Multiple opacite rims can occur in single crystals, indicating that resorption can occur within magma reservoirs. Resorption also occurs at the Earth's surface as indicated by the frequent presence of opacite rims in rocks with crystalline groundmass but their absence in coexisting rocks with glassy groundmass (e.g., Larsen et al. 1937a).

Amphiboles vary from greeen to brown in color, the latter indicating a high Fe^{3+}/Fe^{2+} ratio and called oxyhornblende or basaltic hornblende. Oxyhornblendes occur only in flows or near-surface intrusions, and are most common in oxidized flow interiors (Ujike 1974). Oxidation or resorption or both have been attributed to reduction of P_{H_2O} or increase in temperature or both within magma reservoirs or during eruption.

Amphiboles in andesites are calcic (Ca > 1.34 per 23 O) and most are tschermakites to magnesio-hornblendes according to the nomenclature of Leake (1978); some, especially from high-K suites, have $(Na + K)_A > 0.50$ and thus are pargasitic to edenitic. For compilations of analyses of amphiboles from andesites see Jakeš and White (1972b) and Ujike (1977); two representative analyses are given in Table 6.1. Several aspects of their composition are especially important.

First, amphiboles from andesites are silica-poor, typically containing 39% to 49% SiO_2. Indeed, most amphibole phenocrysts in andesites are *ne*-normative if the mineral's Fe_2O_3/FeO ratio is < 0.5, which was adopted by Allen et al. (1975) and Cawthorn et al. (1973) for amphiboles crystallizing at the NNO f_{O_2} buffer. [Similar values occur in many wet chemical analyses of hornblende phenocrysts (e.g., Leake 1968) or are predicted by the mineral formula calculation procedure of Papike et al. (1974).] Thus, precipitation of amphibole is a highly efficient means by which to increase silica and normative *qz* in magma, as noted by Bowen (1928).

However, there is ~ 10 wt.% variability in SiO_2 contents, with Si usually ranging from 6.2 to 7.0 per 23 O, the more silicic amphiboles being *hy*-normative. Aluminum almost always is sufficient to supplement Si in filling the Z group. Amphiboles with high silica contents (> 6.5 Si per 23 O) are most common in andesites from continental environments such as Colorado, the Cascades, Peru, and Chile; less silicic amphiboles are more characteristic of island arc andesites (Jakeš and White 1972b). If this difference is independent of host rock composition

(Jakeš and White 1972b), the difference probably reflects higher temperature or lower f_{O_2} or both in island arc andesitic magmas, as both phenomena increase Al^{iv} relative to Si at constant P_{H_2O} (Helz 1973). However, the difference also reflects differences in host rock a_{SiO_2} (Ujike and Onuki 1976) and is a braod generalization at best.

Second, compositions of amphibole phenocrysts in orogenic andesites lie within a restricted portion of the basalt tetrahedron and, assuming $Fe_2O_3/FeO = 0.4$ for the amphibole, precipitation of amphibole alone would drive liquids in a vector similar to that of liquids lying along the clinopyroxene + orthopyroxene + plagioclase cotectic discussed in Section 5.2.7 (Cawthorn 1976).

Third, FeO*/MgO ratios of amphibole phenocrysts lie between 0.5 and 2.0, average 1.0, and correlate positively though weakly with FeO*/MgO ratios of their host rocks (e.g., Jakeš and White (1972b). Host rock ratios exceed those of their amphiboles. Similarly, Fe/Mg ratios of amphibole and pyroxene phenocrysts correlate positively, usually being slightly higher in amphiboles than coecisting clinopyroxenes (Ewart 1971; Nicholls 1971a; Jakeš and White 1972b). These observations indicate that fractionation of amphibole will cause Fe-enrichment in a magma, although enrichment, especially relative to silica, will be less than in the case of pyroxene or olivine fractionation.

Fourth, Na + K contents in amphibole phenocrysts correlate positively with Al^{iv} (e.g., Jakeš and White 1972b, Fig. 2), and, therefore, increase with temperature (Helz 1973). Thus, Na + K contents are higher in hornblendes from island arc than continental andesites, even though host rock alkali contents may be lower in island arcs. Amphibole K/Na ratios correlate positively with those of host andesites whose K/Na ratios characteristically are higher than in the amphibole, as they also are in experimental studies (Nicholls 1974; Allen and Boettcher 1978). Indeed, the relationship $(K/Na)_{amphibole\ A\ site} < (K/Na)_{andesite}$ for mol fractions is a necessary though insufficient criterion for equilibrium. This is so because the ratio $(K/Na)_{amphibole\ A\ site}/(K/Na)_{melt}$ is temperature-dependent and, according to Helz (1979), exceeds 1.0 only when temperatures exceed $960°$ to $1037°C$, depending on the ferric iron content of the hornblende. This temperature range is near or above the thermal stability of amphibole in andesite magmas (Sect. 6.9). Equilibration temperatures calculated for most hornblende andesites using Helz' (1979) equation 3 and phenocryst-rock pairs are $850°$ to $1000°C$ and are similar for low-Si and high-Si hornblendes, but neiter groundmass compositions nor unit cell and ferric iron data for the hornblendes are usually available and introduce uncertainties of as much as $200°C$.

Fifth, substitution of F for (OH) in the O3 site enhances the thermal stability of amphibole (Holloway and Ford 1975). The F content of most amphiboles from andesites is unknown. Those summarized by Leake (1968, andesites only), Ishikawa et al. (1980), and Jakeš and White (1972b) average 0.2 wt.% and range from 0.1% to 0.7%. Apparently F contents are no higher in amphiboles from high-K than medium-K andesites within individual volcanic arcs (e.g., New Zealand:

Ewart 1971; Hutton 1944). Thus, there is no reason to expect the distribution of amphibole in andesite to depend heavily on variations in F contents of andesite magma (Sect. 5.3.4).

Finally, the low Al^{vi} contents of amphiboles in andesites (< 0.5 Al^{vi} per 23 O; e.g., Ujike and Onuki 1976) indicate their precipitation at $P_{H_2O} < 9$ kb by comparison with amphiboles crystallized experimentally (Green 1972; Allen et al. 1975; Allen and Boettcher 1978).

Reversed Fe/Mg zoning occurs in hornblendes of andesites and related plutonic rocks. As with pyroxenes, reversals can result from magma mixing (Sakuyama 1978) or increase in f_{O_2} (Mason 1978) perhaps due to an influx of volatiles.

6.4 Olivine

Olivine phenocrysts occur in over 60% of the basic andesites summarized by Ewart (1976a) and in 10% to 40% of the acid andesites, usually in amounts < 1 vol%. Thus, olivine phenocrysts are common in these hy-normative rocks even though some definitions of andesite preclude olivine altogether (Sect. 1.1). Olivine phenocrysts usually give way to pyroxene or amphibole in rocks containing between 50% and 60% SiO_2. The maximum silica content of rocks in which olivine phenocrysts occur varies widely within this range, however. For example olivine is absent at 54% SiO_2 in Tonga (Bauer 1970; Ewart et al. 1973) but is present in andesites containing 60% SiO_2 at Parícutin, Mexico (Fig. 4.7). The reason for this variability is not known in detail, although the proportion of modal olivine correlates positively with whole rock MgO content (Ewart 1976a) suggesting accumulation. Also, olivine stability is enhanced by increased water and alkali contents which depolymerize magma (Kushiro 1975) and decrease a_{SiO_2}. Thus, olivine phenocrysts can persist to higher silica contents in high-K relative to low-K suites, and in calcalkaline relative to tholeiitic ones.

Olivine phenocrysts in andesites can be euhedral and unrimmed or can be embayed and mantled with orthopyroxene, magnetite, augite, pigeonite, or hornblende. Groundmass olivine also occurs, but not in andesites containing phenocrysts of orthopyroxene or hornblende.

The composition of most olivine phenocrysts in andesites lies in the range Fo_{85-65} (Ewart 1976a) which usually represents equilibrium with host rock compositions using the distribution coefficient ($K_D = 0.33$) determined by Roedder and Emslie (1970) for basaltic liquids. For example, the average andesite compositions listed in Table 5.1 would be in equilibrium with Fo_{82-76} olivines if $Fe_2O_3/FeO = 0.3$ in the liquid. Assuming $K_D = 0.33$, only a few examples of olivine phenocrysts which are too Mg-rich for equilibrium with their host andesite have been documented [e.g., sample Kzs-2 of Sakuyama (1978); sample 7 of Bauer (1970); sample AD49 of Marsh (1976a); and sample 4 of Mackenzie (1976).] However, if $K_D = 0.4$ in orogenic andesites (Sect. 5.2.3), then more examples of disequilibrium exist. Moreover, such accumulated or otherwise inherited olivines are relatively common in ocean floor basalts (e.g., Dungan et al. 1978) and probably will be found increasingly in andesites.

Except as occasional rims, Fe-olivines with $< Fo_{55}$ rarely occur in orogenic andesites (e.g., Fairbrothers et al. 1978) although they are often found in icelandites and ferrolatites (e.g., Carmichael 1967b; Baldridge et al. 1973; Leeman et al. 1976). Their absence in orogenic andesites reflects lack of sufficient magmatic Fe-enrichment.

Variation of 20 mol% Fo is common within suites although variation within single samples or zoning within single crystals usually is less than 5 to 10 mol%. Zoning usually is normal in character. Fe/Mg ratios usually are higher in olivines than in coexisting pyroxenes in andesites as in other rocks.

CaO contents in olivine phenocrysts typically lie between 0.15 to 0.3 wt.% and are constant throughout grains as in tholeiitic basalts. This also is the range expected for olivines in equilibrium at 1 atm with melts having the 5 to 10 wt.% CaO contents of orogenic andesites (Watson 1979a). Moreover, these contents are high enough to suggest equilibration at < 10 kb, assuming crystallization at $< 1100°C$ (Finnerty and Boyd 1978). In light of the above, olivine xenocrysts can be identified in some andesites by high Fo contents $(> Fo_{82})$ or low CaO contents $(< 0.1\%)$ or both.

Chromian spinels are frequently included in olivine phenocrysts in andesite. Inclusions of basalt glass can also be present, suggesting magma mixing or crystal accumulation or both (e.g., Anderson 1976).

6.5 Oxides

Three oxides occur as primary minerals in andesite: titanomagnetite (hereafter called magnetite), ilmenite, and chromian spinel. Magnetite is the most ubiquitous and abundant oxide mineral, occurring as a phenocryst in 52% to 65% and 56% to 90% of the basic and acid andesites, respectively, summarized by Ewart (1976a). Because 85% of hornblende-bearing andesites contain magnetite phenocrysts (Ewart 1976a), the high Fe_2O_3/FeO ratio of these rocks probably is a primary feature. Usually 0.5 to 2 vol.% magnetite is present as a phenocryst (0.25 to 1 mm diameter) or microphenocryst often in close proximity to pyroxenes; magnetite is a nearly universal constituent of a crystalline groundmass. Thus, the conclusion that magnetite is absent from andesite, often attributed to Carmichael and Nicholls (1967) or Smith and Carmichael (1968), is incorrect, even in the Cascade province (e.g., Anderson 1974a, 1976). Instead, small amounts of magnetite occur as phenocrysts in basic and acid andesites of all types, including most of the representative suites listed in Table 5.3. However, detailed discussion of the distribution of magnetite phenocrysts in andesites of different types and localities is prevented by insufficient petrographic descriptions of analyzed samples.

In some tholeiitic orogenic basic andesites, FeO*/MgO ratios rise sharply relative to SiO_2 before levelling off (Fig. 5.4). In these suites, as in the basic andesites and icelandites of Thingmuli volcano, Iceland (Carmichael 1964), magnetite phenocrysts occur only in rocks plotting on the relatively level portion of the diagram and the first appearance of magnetite is associated with an abrupt decrease in slope in the diagram (e.g., Oba 1966; Ewart et al. 1973). In some instances this first appearance of magnetite occurs in lavas erupted after prolonged volcano dormancy (Yagi et al. 1972) and also is reflected in a kink on a V vs. FeO*/MgO diagram (Fig. 5.15).

Magnetites sometimes contain lamellae of ilmenite, maghemite, rutile, or pseudobrookite due to subsolidus oxidation; photomicrographs illustrating typical Fe-Ti oxide textures in andesitic rocks are given by Negendank (1972). This oxidation alters the magnetite composition and is associated with reddening of the surrounding host rock; it correlates only weakly with whole rock Fe_2O_3/FeO ratios. Magnetites which occur in opacite rims around hornblendes also have distinctive compositions (Smith and Carmichael 1968). Subsequent comments refer only to primary, homogeneous magnetites.

Ulvospinel ($TiFe_2^{2+} O_4$) contents of magnetite phenocryst cores in andesite usually lie between 20 and 50 mol% and correlate crudely with host rock TiO_2 contents. For example, magnetite phenocrysts and host rocks from Rabaul volcano, Papua New Guinea, contain 15% and 1.0% TiO_2, respectively; in contrast, those from Manam volcano, also in Papua New Guinea, contain 4% and 0.4% TiO_2, respectively (Heming 1974). Al_2O_3 and Cr_2O_3 contents typically are < 3 wt.% and < 500 ppm, respectively, and show no evidence of extensive solid solution toward spinel-hercynite or chromite. These high TiO_2 but low Al_2O_3 and Cr_2O_3 contents may constrain the pressure of last equilibration but calibration results to date are inconclusive (cf. Osborn and Watson 1977, with Osborn et al. 1979). V_2O_3 contents usually lie between 0.75 and 1.25 wt.%, although concentrations as low as 0.2% occur in magnetites of some acid andesites (Marsh 1976a).

Magnetite phenocryst cores are usually poorer in Ti but richter in Fe, Mg, and Al than phenocryst rims or groundmass grains in orogenic andesites (e.g., Lowder 1970; Nicholls 1971a) as in other terrestrial and lunar volcanic rocks (Prevot and Mergoil 1973; Papike et al. 1976). This change in composition occurs during eruption (Ooshima 1970) and can be attributed to reduction in temperature by analogy with behavior in simple systems (Taylor 1963), to reduction of f_{O_2} in magma during eruption (Anderson and Wright 1972), or to change in the composition of coexisting plagioclase and pyroxene (Ewart 1976b).

Ilmenite is much less common than magnetite in orogenic andesites because of low f_{O_2} or a_{TiO_2} or both in the magma. Ilmenite is more common in basalt or dacite than andesite and more common in acid than basic andesite. R_2O_3 contents ($Fe_2O_3 + Al_2O_3 + Cr_2O_3$) of ilmenite phenocrysts in orogenic andesite usually are 15 to 25 mol%.

Finally, chromian spinels (picotites) are present in some basic andesites, but only as inclusions in olivine phenocrysts (e.g., Anderson 1976) and are more common in calcalkaline than tholeiitic rocks of similar silica content (Iida et al. 1961).

The presence of iron oxides has significance for andesite genesis in two ways: as a mechanism for limiting magmatic Fe-enrichment (Sect. 11.3), and as an indicator of temperature and f_{O_2}. Usefulness of the oxides for the latter is restricted by the infrequence of ilmenite in orogenic andesite, by the high R_2O_3 contents of the ilmenites which lie outside the range of compositions calibrated by Buddington and Lindsley (1964) or Taylor (1964), and by the variable amount of Mg and Al in the minerals and alkalies in the magma which were absent in their experiments. However, the experiments of Katsura et al. (1976) and the thermodynamic formulation by Powell and Powell (1977) minimize both of the latter problems. Coexisting magnetite and ilmenite occur in some pyroxene and especially hornblende andesites, and temperatures and f_{O_2} calculated therefrom are given in Table 4.1 and Fig. 5.9, and are discussed in Sections 4.1.1 and 5.3.5.

6.6 Garnet

Garnet phenocrysts or xenocrysts (hereafter called megacrysts) occur infrequently in andesites of continental margins and detached continental fragments including: Miocene acid andesites of New Zealand (Bartrum 1937); Miocene acid andesites of the Inland Sea area, Japan (Fig. 2.5; Morimoto 1953; Tagiri et al. 1975; Ujike and Onuki 1976); Pleistocene acid andesites of Clear Lake, California (Hearn et al. 1975) and Lipari volcano, Eolian arc (Maccarrone 1963); mid-Tertiary basic and acid andesites of the Carpathians in Czechoslovakia (Brousse et al. 1972) and the USSR (Kostuk 1958), of Sikhote Alin, USSR (Boronikhin and Baskina 1975), and acid andesites of Spain (Burri and Parga Pondal 1936; Lopez-Ruiz et al. 1977); Mesozoic basic and acid andesites of the Crimea (Makarov and Suprychev 1964) and Kamchatka, USSR (Sakhno and Lagovskaya 1970); and Ordovician acid andesites of England (Oliver 1956; Fitton 1972). Garnets also are found in Cretaceous dacites and rhyolites of New Zealand (Wood 1974) and Devonian rhyolites of Australia (Green and Ringwood 1968b, 1972; Birch and Greadow 1974), for example. There is no consensus whether all or any of these garnet megacrysts are cognate or accidental.

The volcanologic context of these garnetiferous andesites is not described in detail except for the Clear Lake example where garnets occur in a single flow unit amidst ordinary pyroxene andesites and hornblende dacites. In general, garnet megacrysts are distributed sporadically in rocks of similar composition and otherwise similar mineralogy in the areas cited. The garnetiferous andesites from Japan, the Crimea, Clear Lake, Lipari, Spain, and England are high-K calcakaline acid andesites which are peraluminous (c-normative). (About 25% of high-K acid andesites are peraluminous (Sect. 5.2.7), but only a few contain peraluminous minerals such as garnet.) In contrast, garnet-bearing andesites from Czechoslovakia are medium-K, of variable FeO*/MgO ratio, and *di*-normative. Those from New Zealand, Japan, the Crimea, Kamchatka, Spain, and Clear Lake usually contain phenocrysts of hornblende or biotite or both and quartz, all of which are absent in the English and many of the Czechoslovakian and Liparian examples. Clinopyroxene phenocrysts rarely coexist with garnet. Plagioclase coexists with garnet in all instances. Metapelite schist or amphibolite xenoliths containing garnet and often cordierite occur in the Japanese, New Zealand, Clear Lake, Spanish, and Crimean examples. $^{87}Sr/^{86}Sr$ ratios are > 0.705 in garnetiferous andesites of Czechoslovakia and California (Gill, unpubl. data).

The garnet megacrsts are euhedral to rounded, usually are > 1 mm (up to 2 cm) in diameter, and frequently contain small inclusions of magnetite, apatite, and other possibly magmatic minerals which vary with locality. Some have a corona of plagioclase (Oliver 1956; Lozep-Ruiz et al. 1977) or magnetite and chlorite (Fitton 1972), but most lack coronae altogether; the vermicular cordierite + hypersthene rims noted around garnets in Australian rhyolites (Green and Ringwood 1968b, 1972) are uncommon. Flow lines bend around the megacrysts, indicating the presence of garnet in magma before eruption.

The garnets in andesites are almandine-rich, containing 20 to 30 mol% pyrope and 5 to 20 mol% grossular (Fig. 6.4), < 5 mol% spessartine, and > 0.3 wt.% TiO_2. Garnets in andesites consistently are richer in Mg but poorer in Fe and Mn than garnets in associated dacites and rhyolites (Fig. 6.4; Brousse et al. 1972), reflecting the higher temperatures and higher Mg/Fe ratios of the host andesites. Compositions of garnets occurring as megacrysts and in xenoliths sometimes agree (e.g., Lopez-Ruiz et al. 1977) but were not compared in most of the studies cited.

Many of the megacrysts are zoned in complex fashion, rims usually being enriched in Fe/Mg relative to cores but depleted in Mn. Zoning patterns seem consistent within individual flow units but variable between localities (Brousse et al. 1972). Oscillatory zoning occurs in some dacites (Fitton 1972; Kano and Yashima 1976).

Several of the preceding observations suggest that the garnets are cognate high-pressure phenocrysts. These observations include the locally widespread if irregular occurrence of garnet, their relative euhedrality, and especially their apparent reflection of host rock compositions which frequently are peraluminous. Not only do garnets in andesites differ consistently from those in dacites and rhyolites in major element constituents (Fig. 6.4), but they differ even more strikingly in such trace elements as Y, La, Sc, V, and Cr (Fitton 1972). Coronae, when present, indicate reaction of garnet to an assemblage stable at lower pressure. On the other hand, derivation of the garnets from assimilated garnetiferous metamorphic rocks ("restites") is suggested by the irregularity of garnet distribution, the actual occurrence of such xenoliths in some instances which are cited above, and similarities in composition.

The composition and zoning of the garnets provide ambiguous evidence. Garnets have been crystallized experimentally from three different *di*-normative

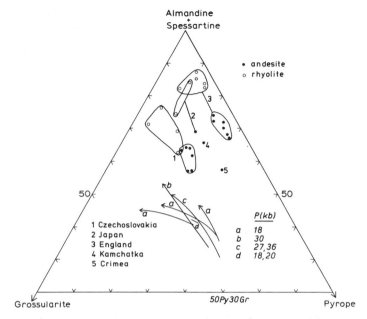

Fig. 6.4. Composition of garnet-megacrysts in calcalkaline lavas, and of garnets precipitated at high pressure from andesitic liquids. Note differences between garnets in andesites *(filled circles)* versus dacites or rhyolites *(open circles)* from the same localities; data sources given in the text. Compositions of experimentally precipitated garnets are from Stern and Wyllie (1978), Green (1972), Green and Ringwood (1968a,b), and Nicholls and Harris (1980); *arrows* point toward decreasing temperatures in the range 1175° to 800°C at the pressures indicated, under conditions ranging from anhydrous to water-saturated. For example, the *three 18 kb lines* show garnets from systems with 2%, 5%, and 10% water, *right to left*

andesite liquids at 9 to 36 kb pressure, from 1400 to 800°C, and with f_{O_2} around the NNO buffer (Sect. 6.9), but the resulting garnets are much richer in Mg and Ca than the naturally occurring megacrysts (Fig. 6.4). The higher the pressure, water content, or temperature during garnet precipitation the worse the discrepany, resulting in increased Ca in the first two instances and in increased Mg in the third. However, garnets of appropriate composition are also uncommin in metamorphic rocks, especially considering the low MnO but high TiO_2 contents of the megacrysts (see Brousse et al. 1972; Kano and Yashima 1976), which imply an igneous origin. Similarly, normal, reverse, and oscillatory zoning could be explained by crystal-liquid equilibria under various conditions, but all three patterns also occur in garnets from metamorphic rocks (e.g., Black 1973).

Thus, although evidence is persuasive that at least some garnets in andesite are cognate, the argument remains open, the appropriate P-T-H_2O conditions remain unknown, and criteria remain absent for distinguishing cognate from accidental garnets. The garnets are much too Ca-poor to be refractory grains of subducted oceanic crust (cf. Gill 1974, Table 1). Even if cognate, the apparent restriction of garnetiferous andesite to continental areas containing pelitic sediments, and the $^{87}Sr/^{86}Sr$ ratios > 0.705 of host andesites, suggest that the peraluminousness leading to garnet resulted from assimilation of pelitic material.

6.7 Other Minerals

Accessory or infrequent but possibly primary minerals in andesite include biotite, silica minerals, apatite, cordierite, anhydrite, and pyrrhotite, in addition to garnet. Biotite is uncommon in rocks with $< 63\%$ SiO_2, usually is restricted to high-K acid andesites with quench temperatures of $800°$ to $950°C$, and almost always coexists with hornblende, occurring in about 20% of the hornblende andesites summarized by Ewart (1976a). It is primarily a phenocryst mineral (e.g., Larsen et al. 1937b; Jakeš and Smith 1970) but can occur in the groundmass (e.g., Wise 1969, p. 982). Biotite in andesite usually has a corona of iron oxide, pyroxene, and plagioclase, and has oxidation features similar to but less well studied than those of amphibole. No systematic study of the distribution or composition of biotite in andesite is available; some analyses are given by Larsen et al. (1937b), Sakuyama (1978), and Dostal et al. (1977c).

Silica minerals in andesite include tridymite or cristobalite or both in the groundmass or cavities, and quartz as a megacryst or in the groundmass. Quartz megacrysts are uncommon even though andesites contain substantial normative quartz (Sect. 5.2.7); however, some samples with a few quartz grains occur in most andesite provinces, with quartz occasionally constituting > 5 vol.% of acid andesite dome lavas. Notable examples are restricted to continental margins or detached continental fragments and include Lassen, USA (Finch and Anderson 1930; Smith and Carmichael 1968); Clear Lake, USA (Hearn et al. 1975); Aegina, Aegean arc (Pe 1973), and Esan, Hokkaido, Japan (Ando 1974). About 20% of the active volcanoes in NE Japan contain rocks with coexisting quartz and olivine (Sakuyama 1978). Quartz may be distributed irregularly in many rock types as at Lassen and Clear Lake, may be restricted to rocks containing hornblende or biotite or both as at Aegina, or may occur in late stage but anhydrous assemblages as at Esan. Almost always quartz megacrysts in andesite are embayed, or are surrounded by a corona a few hundred microns wide mostly of Mg-rich augite, or both (e.g., H. Sato 1975). Where known, $^{87}Sr/^{86}Sr$ ratios of andesites containing quartz megacrysts are no higher than in coexisting quartz-free andesites (Peterman et al. 1970a,b). In general, the presence of quartz could reflect assimilation of sialic rock (e.g., Finch and Anderson 1930), mixing of mafic and acid magmas (e.g., Eichelberger and Gooley 1977), or high pressure crystallization (e.g., Nicholls et al. 1971; Ando 1974).

Apatite is a common accessory mineral in andesites, occurring as inclusions in pyroxenes, amphiboles, and magnetites, and as microphenocrysts in many acid andesites. Whether the difference in P_2O_5 behavior between tholeiitic and calc-alkaline andesites (Sect. 5.2.7) is reflected in different modal apatite distribution is not known, but was suggested by Kuno (1950a). Analyses of apatites from acid andesites are given by Heming and Carmichael (1973).

Primary pyrrhotite is not uncommon in orogenic andesites although only one reported sighting has been published (Heming and Carmichael 1973). Most commonly the pyrrhotites are included in olivine or magnetite phenocrysts and contain at most a few percent Cu or Ni.

Cordierite megacrysts occur in andesites from Spain (Burri and Parga-Pondal 1936), Lipari volcano, Eolian arc (Maccarrone 1963), Kasyo-to volcano, Taiwan (Miyashiro 1957), and Pelée dome, Martinique, Antilles (Lacroix 1904). Cordierite also occurs within xenoliths in andesite (e.g., Aramaki 1961; Tagiri et al. 1975). In both parageneses, the cordierites are Fe-rich, highly disordered, high-temperature polymorphs approaching indialite. The host andesites usually are peraluminous. One Liparian host has an $^{87}Sr/^{86}Sr$ ratio of 0.705, at the high end of ratios for noncordierite-bearing rocks from that volcano (Barberi et al. 1974; Klerkx et al. 1974). As with garnet, the cordierites could either be refractory residues after assimilation of metapelites or cognate precipitates from peraluminous andesite. If the latter, precipitation probably occurred at pressures < 7 kb (Hensen and Green 1973; Green 1977).

Anhydrite occurs as separate irregular grains in some andesites (Taylor 1958) and in xenoliths in others (Nicholls 1971b; Yagi et al. 1972).

6.8 Inclusions in Orogenic Andesites

Magmas erupted at convergent plate boundaries, including basalts and rhyolites as well as andesites, also differ from basalts erupted at mid-ocean ridges or ocean islands in containing more numerous inclusions of rocks other than peridotite nodules. These inclusions have been classified in several ways (e.g., Lacroix 1893, 1901) and include glomero-porphyritic clots, ultramafic to mafic igneous rocks of various textures, and metamorphic rocks. They may represent cognate cumulates, mixed magmas, accidental inclusions (xenoliths) of wall rocks, or refractory residues (restites) of source rocks. Thus, inclusions can provide evidence of pre-eruption crystallization history, of magma mixing, of assimilation, or of source rocks. No estimates of the frequency or relative volume of inclusions in andesites are available except for their occurrence in several historic flow units of which they can constitute up to 5 vol.% (e.g., Lacroix 1904; Steiner 1958).

The most common inclusions in andesite are coarse-grained glomero-porphyritic clots containing 1 to 10 mm sized crystals, often enclosing glass, of the same minerals which occur elsewhere in the sample as phenocrysts, and which usually are attributed to aggregation of phenocrysts. However, clots of plagioclase + augite + orthopyroxene + magnetite were interpreted as decompression breakdown products of amphibole by Stewart (1975) who contended that the average modal proportion of minerals in clots from acid andesites at Crater Lake, USA, is relatively constant but different from that of phenocrysts, and leads to a bulk composition of the clots which is similar to that of amphibole. Because of the inconsistent proportions, oscillatory zoning, and coarse grain size of minerals in most clots, because the eruption temperature of clot-containing pyroxene andesites exceeds the thermal stability of amphibole, and because similar clots occur in anorogenic andesites unlikely to have precipitated hornblende (Sect. 11.9), this suggestion is unlikely to apply to most glomero-porphyritic clots.

Ultramafic nodules occur in ejecta from several volcanoes at convergent plate boundaries [e.g., spinel- and garnet-pyroxenites at Salina volcano, Eolian arc (Keller 1974); lherzolites at Grenada volcano, Antilles (Arculus and Wills 1980); and harzburgites at Klyuchevskoy volcano, Kamchatka (Vlodavetz and Piip 1959)]. However, spinel lherzolites have been found in andesitic ejecta only in the maars of Itinomegata volcano, Honshu, Japan. There the inclusions are similar in Sr and O isotopic composition to their host and contain interstitial glass of andesitic composition (Katsui et al. 1979). This is not, however, unambiguous evidence for the existence of andesitic magma within the upper mantle because alkali basalt magma also has erupted from Itinomegata (hence magma mixing could contribute inclusions) and because andesitic glass also accompanies pargasite breakdown in lherzolites from alkali basalt hosts (e.g., Francis 1976), so could have formed during decompression.

Other inclusions of mafic rocks include nonphenocryst minerals or compositions, or have cumulate or metamorphic or quench textures, and therefore cannot be explained as phenocryst aggregates. The most common of these inclusions have gabbroic to ultramafic mineralogy and cumulate textures, containing calcic plagioclase, olivine, amphibole, augite, orthopyroxene, magnetite, ilmenite, and biotite in various combinations. The assemblages platioclase + olivine ± pyroxene and plagioclase + amphibole ± pyroxene are particularly common, even in hosts lacking olivine or amphibole phenocrysts (e.g., MacGregor 1938; Wager 1962; Yamazaki et al. 1966; Baker 1968; Takeshita and Oji 1968; Aoki and Kuno 1972; Stern 1979; Arculus and Wills 1980). Most of these cumulate inclusions contain > 40% plagioclase; plagioclase-free or magnetite-free inclusions are rare. Minerals in gabbroic inclusions differ from phenocrysts in their relative proportions, with plagioclase and orthopyroxene being less abundant but hornblende and magnetite being more abundant in inclusions. Also, minerals in inclusions are usually less zoned than phenocrysts. Spinel-forming reaction relationships between plagioclase and olivine or orthopyroxene indicative of high pressure are missing. Adcumulus to heteradcumulus textures predominate.

Although similar in kind to phenocrysts, minerals in inclusions can differ in composition with plagioclase being > An90, pyroxenes containing > 5% Al_2O_3, olivines containing $< 0.1\%$ CaO, and magnetites containing $< 6\%$ Al_2O_3 and $> 4\%$ MgO (e.g., Lewis 1973). Indeed, the co-existence of $\geqslant$ An90 plagioclase with Fo80 to 60 olivine is characteristic of gabbroic inclusions (but not phenocrysts) in arc lavas but is rare elsewhere (except for the lunar highlands). These distinctive characteristics suggest equilibration at temperatures or fluid pressures higher than those of phenocryst equlibration. Mineral compostions differ between different mineral assemblages in inclusions, as shown in Fig. 6.5. Basalt or basic andesite glass, more mafic in composition than the host rock and often vesiculated, occurs in interstices and along grain boundaries in many inclusions; analyses and photomicrographs are given by Lewis (1973).

Three interpretations of these gabbroic inclusions can be defended. First, they may be crystal cumulates from more mafic magma, possibly similar in composition to that of the included glass (Wager 1962; Nicholls 1971a; Lewis 1973). If so, the ubiquity of plagioclase and magnetite, the abundance of amphibole which can be adcumulate or poikilitic in texture, and the differences from pheno-

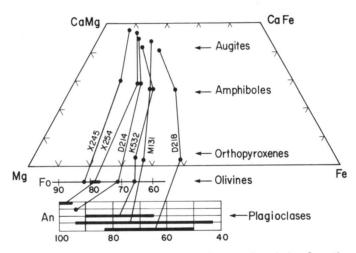

Fig. 6.5. Compositions of minerals in some mafic inclusions in orogenic andesites from the Lesser Antilles. *Bars* show compositional range within single inclusions. Data are from Arculus and Wills (1980) and were selected to show the range in Mg/(Mg + Fe) ratios observed. *Sample numbers* indicate their geographic source: *X* Grenada, *D* Dominica, *K* St. Kitts, and *M* Martinique. Note several features. (1) Mineral compositions vary between inclusions and tielines rarely cross; consequently, equilibrium usually was closely approached but under quite variable P-T-X conditions. (2) Olivine is absent from inclusions whose plagioclase compositions extend to $<$ An65 or which lack cores $>$ An90. (3) Orthopyroxene is absent from inclusions whose olivines are $>$ Fo70. (4) Amphibole coexists with all mineral compositions, roughly decreasing in modal percent as Mg/(Mg + Fe) decreases (i.e., amphibole constitutes over half of samples X254 and D214 but $<$ 15% of K532, M131, and D218). (5) Equilibration temperatures calculated from coexisting Fe-Ti oxides are 850° to 720°C for K532 and 890° to 820°C for M131

crysts in modal proportion and composition of minerals, all have significance for crystal-liquid fractionation models of andesite genesis. Alternatively, the inclusions may be refractory residues left after partial fusion of a mafic source, as argued for mafic inclusions in plutonic rocks (e.g., White and Chappell 1977). Finally, inclusions, especially those with finer-grained rims than cores or acicular mineral morphologies, may be quenched basalt magma mixed with more silica-rich magma (Eichelberger and Gooley 1977).

Because cumulate textures are more common than acicular ones and because mineral compositions when known vary between mineral assemblages in single ejecta units (Fig. 6.5), the cumulate interpretation seems most generally applicable. However, each of these interpretations is compatible with features of some inclusions, and the extent to which each possibility is realized is unknown.

Estimates of equilibration conditions based on the composition of co-existing minerals within cumulates are 700° to 950°C (using oxide and augite-amphibole pairs), f_{O_2} one to two log units above the NNO buffer, and 4 to 10 kb with $a_{H_2O}^{fluid}$ of 0.1 to 0.5 (Powell 1978; Arculus and Wills 1980).

Finally, inclusions of thermally metamorphosed sialic rocks occur in some andesites of continental margins and detached continental fragments, although the original identity of these inclusions rarely is discernable. Some are upper crustal sediments which have been metamorphosed at $> 800°C$, sometimes in the presence of high f_{O_2} and f_{S_2} (e.g., Wilcox 1954; Aramaki 1961; Nicholls 1971b; Yagi et al. 1972; Tagiri et al. 1975). Some of these inclusions have been partially or wholly fused (e.g., Steiner 1958; Nicholls 1971b), thereby providing opportunity for modification of the host magma composition. Also, the garnets and cordierites in andesites described in Sections 6.6 and 6.7 may be relics from inclusions of pelitic rocks, or cognate precipitates from magma enriched in aluminum by assimilation of such rocks. However, clear textural or isotopic evidence of such modification by assimilation is sparse. For example, some inclusion-bearing andesites sometimes have marginally higher $^{18}O/^{16}O$ ratios than associated inclusion-free rocks (Matsuhisa et al. 1973). In contrast, the inclusion-bearing acid andesites from Parícutin volcano described by Wilcox (1954) are no different in $^{87}Sr/^{86}Sr$ ratios than associated inclusion-free basic andesites (Tilley et al. 1967), even though partial melts of such inclusions can have $^{87}Sr/^{86}Sr$ ratios higher than the bulk inclusions themselves (Pushkar and Stoeser 1975). Thus, although there is clear evidence that andesite magma has thermally metamorphosed inclusions of upper crustal rocks, there is little evidence from individual samples that the inclusions have significantly modified the composition of their hosts.

6.9 Mineral Stabilities in Andesite Magma

Experimental studies of mineral stabilities applicable to andesite petrogenesis are legion. Results of some studies which involve simplified analogs are shown in

Table 6.2. Melt compositions and experimental conditions of selected high-pressure synthesis experiments on orogenic andesites

References	1	2	3	4	5	6	7	8	9	10	11	12	13
SiO_2	53.8	53.7	54.6	56.4	57.0	58.6	58.8	59.1	59.1	60.1	60.3	61.1	62.2
TiO_2	0.6	0.8	1.1	1.4	0.7	1.0	0.5	0.9	0.8	0.8	0.7	0.9	1.1
Al_2O_3	14.3	17.4	18.2	16.6	15.4	16.3	19.2	17.8	18.2	17.3	17.0	14.4	17.3
Fe_2O_3	1.7	8.8	6.8	3.0	–	2.9	1.4	1.8	2.3	1.4	0.9	4.2	0.3
FeO	4.5	–	–	5.7	6.0	6.4	3.3	4.8	3.6	4.4	5.4	5.2	5.9
MgO	9.6	4.2	4.2	4.3	9.0	3.0	2.6	3.0	2.5	3.7	3.1	3.0	2.4
CaO	6.9	9.6	7.9	8.5	7.0	7.4	7.6	6.8	5.9	6.2	7.2	5.9	5.2
Na_2O	2.6	3.9	4.2	3.0	3.1	3.4	4.1	4.3	3.8	4.0	3.9	3.7	3.3
K_2O	2.0	1.1	1.1	1.0	1.8	0.5	1.6	1.1	2.2	1.7	1.3	0.6	2.3
P_2O_5	0.16	–	0.26	–	–	0.12	0.18	0.22	0.30	0.28	0.20	0.17	–
P (kb)	4–15	5	2	9–36	13–18	0.5–1.5	1–12	1–23	1–30	0.5–10	9–36	0.3–1.0	18–36
wt.% H_2O	2–4	E [a]	3–6	0	7–16	E	1–5, E	2–5, E [b]	5–20, E	2–5, E	0–10	E	0
fO_2	–	NNO	NNO-HM	–	–	NNO	NNO	MW-MH	NNO	FMQ	–	NNO	–

a E = excess (water-saturated)
b Includes vapor-saturated where X_{water}^{vapor} = 0.25 to 1.0 at 10–22 kb

References:
1. Nicholls and Lorenz (1973). 2. Cawthorn et al. (1973). 3. Ritchey and Eggler (1978). 4. Green and Ringwood (1968a). 5. Kushiro and Sato (1978). 6. Sekine et al. (1979). 7. Maksimov et al. (1978). 8. Eggler and Burnham (1973); Allen et al. (1975); Allen and Boettcher (1978). 9. Piwinskii (1968); Lambert and Wyllie (1974); Stern et al. (1975). 10. Eggler (1972a). 11. Green (1972). 12. Sekine et al. (1979). 13. Green and Ringwood (1968a)

Fig. 5.7; a few others have been mentioned in Sections 6.1 to 6.4. The studies discussed in this section are of mineral stabilites in natural andesite magmas as functions of temperature, pressure, oxygen fugacity, and water content. Representative results for several acid andesites are summarized in Fig. 6.6 and 6.7; compositions and conditions used in published experiments are given in Table 6.2. The effect of pressure and water content on andesite liquidus temperatures are shown in Fig. 4.1.

Plagioclase is the liquidus phase at atmospheric pressure in all andesites studied and remains so until replaced by clinopyroxene above 15 kb in the absence of water. Water suppresses the stability of plagioclase relative to ferromagnesian minerals such that plagioclase is no longer the liquidus mineral of acid orogenic andesite at 150 to 700 bar when water-saturated (Sekine et al. 1979), at 6 kb in a Parícutin andesite containing 2 wt.% H_2O (Fig. 6.6c), or at 10 kb in a Cascade andesite containing 5 wt.% H_2O (Eggler and Burnham 1973). Therefore, the mere ubiquity of plagioclase phenocrysts in andesite (Sect. 6.1) indicates that pre-eruption water contents of andesite magmas are less than about 2 to 5 wt.% (point 7, Sect. 5.3.1). The lower limit is more likely because the Cascade andesite (composition 8, Table 6.2) has 64% normative plagioclase which is anomalously high and therefore has a higher than usual thermal stability of plagioclase. A similar conclusion is implied if the glass inclusions in plagioclase reflect incipient melting due to decompression during magma ascent (Sect. 6.1) because the dP/dT slope for plagioclase stability in andesite magma is positive only when andesite magma contains less than 0.7 to 3 wt.% H_2O (Burnham 1979, and Fig. 6.6.c).

Both augite and orthopyroxene precipitate simultaneously from some andesites over a range of pressures (Rithey and Eggler 1978; Sekine et al. 1979) whereas orthopyroxene precedes augite in others (Fig. 6.6.c) and sometimes is absent altogether (Fig. 6.6.a). Factors controlling these pyroxene relationships are poorly constrained but augite stability in andesite is enhanced relative to orthopyroxene by increased load pressure or increased water contents (Fig. 6.6; Maksimov et al. 1978) or by f_{O_2} above the NNO buffer (cf. Allen et al. 1975, with Allen and Boettcher 1978). Consequently, magma ascent could cause orthopyroxene overgrowths on augite. Pigeonite has not been identified in experiments. Pyroxenes have reaction relationships with amphibole when they coexist.

Unlike the plagioclase case, pyroxene-andesite melt reactions continue to have positive dP/dT slopes in the presence of water. Consequently, because a pyroxene is a common near-liquidus phase in andesite magma, as implied by the modal ubiquity of pyroxene phenocrysts (Sect. 6.2) and by the apparent saturation of most andesites with two pyroxenes (Sect. 5.2.7), the pre-eruption water content of andesite magma must be sufficient to suppress plagioclase enough for pyroxene to be on or near the liquidus. Approximately 2 wt.% H_2O is suggested by Fig. 6.6c.

The low, positive dP/dT slope of the amphibole stability curve at low pressure (Fig. 6.6b,c) indicates that amphibole will be resorbed at shallow depths

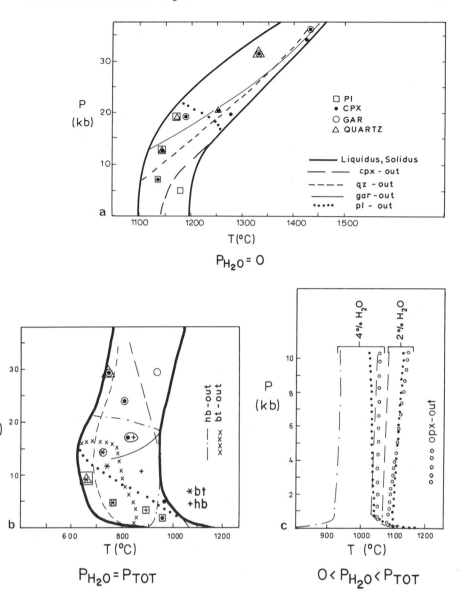

Fig. 6.6 a–c. Experimentally determined phase relationships in orogenic andesite melts at high pressures. **a** Anhydrous conditions; composition 13, Table 6.2. **b** Water-saturated conditions; composition 8, Table 6.2, for near-liquidus temperatures at P < 25 kb, and composition 9 for other conditions. **c** Vapor-saturated but water-undersaturated conditions; composition 10, Table 6.2. Note in **b** and **c** that the maximum thermal stabilites of minerals rise sharply at low pressure. Consequently, near-liquidus melt containing few or no crystals at, e.g., 2 kb *must* precipitate phenocrysts during eruption unless supercooling is achieved due to rapid ascent, lack of vesiculation, or other kinetic factors

during magma ascent, thus explaining the common opacite rims observed and the absence of groundmass amphibole (Sect. 6.3). The maximum thermal stability of halogen-free amphibole in acid andesite melt is 950 ± 20°C depending on the degree of water saturation (Eggler 1972b); it may be stable up to 1000°C in basic andesites (Cawthorn et al. 1973a; Ritchey and Eggler 1978). Although incorporation of F increases the thermal stability of amphibole (Holloway and Ford 1975), the low F contents of andesites (Sect. 5.3.4) and their amphiboles (Sect. 6.3), and the high FeO*/MgO ratio ($\sim$ 1.0) of amphiboles in andesites, suggest that this 930° to 1000°C limit is applicable to natural andesites. As a result, crystallization of amphibole from andesite magma requires low liquidus temperatures or substantial subliquidus cooling. The maximum pressure stability of amphibole in andesite is 18 to 22 kb depending on f_{O_2} and water content (Allen et al. 1975; Allen and Boettcher 1978).

Pressure > 8 kb and water contents > 6 to 10 wt.% are necessary for precipitation of amphibole as a liquidus phase from acid andesite (Fig. 6.6c; Green 1972; Eggler and Burnham 1973); lower limits are still higher for more mafic rocks. Water contents > 3 wt.% in the melt are necessary for amphibole to precipitate at all. Despite the occurrence of plagioclase in over 95% of hornblende andesites (Ewart 1976a), and the frequent occurrence of hornblende gabbro inclusions (Sect. 6.8), amphibole coexists with plagioclase experimentally only under subliquidus conditions, except in fortuitous circumstances in the Cascade andesite unusually rich in normative plagioclase (Eggler and Burnham 1973, Fig. 1). Thus, the phase relations of amphibole, like the geologic context of its occurrence as a phenocryst (Sect. 4.4), suggest that amphibole is a product of low-pressure, late-stage cooling and sub-liquidus water-enrichment, and is not a liquidus phase with which most andesites were saturated during most of their evolution.

Note that olivine precipitates from none of the bulk compositions shown in Fig. 6.6. However, some andesites with high Mg-numbers precipitate olivine under anhydrous conditions, and olivine can become a liquidus phase of others when they have high water contents, much as discussed above for pyroxene (Eggler 1972a; Nicholls 1974; Kushiro and Sato 1978). The maximum pressure stability of olivine increases with increasing water and alkali contents (Nicholls and Ringwood 1973; Kushiro 1975) and with decreasing silica content of the magma (Nicholls 1974). Under water-saturated conditions, olivine is a liquidus phase to 17 kb in some basic andesites with Mg-numbers > 70, and to 10 kb in acid andesites with similar Mg-numbers (Nicholls 1974; Kushiro and Sato 1978). Olivine in acid andesite disappears quickly via reaction relationships to pyroxene or amphibole upon subliquidus cooling (Eggler 1972a). Thus, occurrence of cognate olivine phenocrysts in andesite (Sect. 6.4) is favored by low FeO*/MgO ratios or water contents > 2 wt.% (Fig. 6.6c), and by restricted subliquidus crystallization.

Variation of oxygen fugacity has little effect on the maximum thermal stability of amphibole (Allen et al. 1975) or other silicate minerals, but has a dramatic effect on the stability of magnetite (Fig. 6.7). Neither magnetite nor ilmenite has

Fig. 6.7. Experimentally determined phase relationships on an orogenic andesite melt (column 6, Table 6.2) at atmospheric pressure, after Sekine et al. 1979). This melt composition was chosen because it has the highest FeO*/MgO ratio of those for which results are available. Analogous data for melt compositions 3 and 8 of Table 6.2 are given in the references cited, and for an anorogenic andesite by Thompson (1975). The *stippled field* covers 1.5 log units above the NNO buffer and is the likely level of f_{O_2} in orogenic andesite magmas based on Fig. 5.9

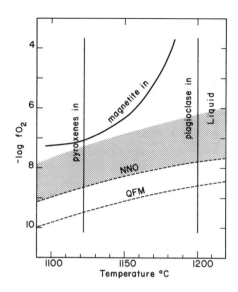

been observed experimentally as a liquidus phase in basic or acid andesites at atmosphere pressure when f_{O_2} is at or below the NNO buffer. However, liquidus magnetite was precipitated by Nicholls (1974) from a Mt. St. Helens (USA) calc-alkaline basic andesite at 1 atm pressure when $f_{O_2} \geqslant 10^{-7}$ at 1200°C, about 1 log unit above NNO. Subliquidus magnetite co-existed with andesite melts in experiments conducted at NNO by Holloway and Burnham (1973). Increased load pressure further reduces the maximum thermal stability of magnetite in hydrous andesite when f_{O_2} is between FMQ and NNO (Eggler and Burnham 1973; Maksimov 1978), apparently by lowering the activity of Fe^{3+} in the melt (Burnham 1979). This instability of magnetite during experiments conflicts with its widespread occurrence as an accessory phenocryst mineral in andesites (Sect. 6.5) and as a constituent of gabbroic inclusions therein (Sect. 6.8) and is not understood. Probably available high pressure experiments have been run under too reducing conditions; other possible explanations are offered in Section 11.3.

At pressures above 15 to 20 kb, garnet replaces plagioclase as the principal aluminous phase in equilibrium with andesite magma (Fig. 6.6a,b), being a liquidus mineral at 30 kb at all water contents (Stern and Wyllie 1978) and becoming one by 20 kb when the magma contains > 10 wt.% H_2O (Green 1972; Allen et al. 1975). The unique pressure at which garnet and clinopyroxene coexist on the liquidus decreases with increasing water content of the magma (Green 1972). Just as the garnets crystallized during these experimental studies differ in composition from garnet megacrysts in andesites (Sect. 6.6), so also the phase relationships differ. Garnet and plagioclase never coexist as near-liquidus phases experimentally despite the universal occurrence of plagioclase in garnetiferous andesites.

Quartz is never a liquidus phase in experimental crystallization of andesite magmas, despite its theoretical stability (Nicholls et al. 1971). Quartz occurs within about 40°C of the liquidus above 30 kb but only in anhydrous acid andesites (Green and Ringwood 1968a; Green 1972); water greatly increases the solubility of quartz relative to ferrromagnesian minerals (Fig. 6.6a,b). However, the dP/dT slope of the maximum thermal stability of quartz remains positive (Green 1972) such that near-adiabatic ascent of magma could cause dissolution of quartz once it formed, regardless of water content. This offers one explanation for the resorption of quartz megacrysts in andesite.

6.10 Trace Element Equilibria Between Minerals and Melt

Concentrations of minor elements in minerals were discussed in some of the preceding sections but are extended here to include the behavior of the trace elements whose concentrations in orogenic andesites were summarized in Section 5.4. At trace level concentrations, elements are distributed between phases coexisting in equilibrium such that

$$D_i^{\alpha/\beta} = c_i^{\alpha}/c_i^{\beta} = \exp\left[(A/T) + a\right] \tag{6.1}$$

where D is a partition coefficient for element i between phases α and β, c is the concentration by weight of i in α or β, A is a function of ΔH of element i in α versus β, T is temperature in K, and a is an integration constant. The bulk distribution coefficient $\overline{D}_i$, is the sum of all $D_i^{\alpha/\beta}$ values times the weight fraction of appropriate phases α in equilibrium with a common phase β, e.g., times the weight fraction of all minerals coexisting with a liquid. At a given temperature, D is constant for specific compositions of α and β as a long as both phases are dilute with respect to i. However, one cannot predict the dependence of D on temperature or on the bulk composition of α or β, nor can one predict the concentration of i in α or β above which D is not constant; experimental determination of these parameters has only begun. Results for natural andesitic melts are limited to values for REE, Sr, Ba, and Y in plagioclase at atmospheric pressure (Drake and Weill 1975) and for REE in garnet at 20 to 30 kb, in augite at 15 to 30 kb, and in amphibole at 10 kb (Nicholls and Harris 1980). Results for Fe-free haploandesitic melts include values for REE in garnet and augite at 20 kb (Mysen 1978) and Ni in olivine at atmospheric pressure (Hart and Davis 1978). Principal conclusions are that the D's increase by half an order of magnitude from basalt to acid andesite due principally to the combined effects of increasing polymerization of the liquid and decreasing temperature.

Approximate partition coefficients applicable to andesitic melts can be estimated from phenocryst/matrix pairs from andesitic whole rocks or andesitic matrices. Table 6.3 summarizes the available data. Note that most D's exceed values for basalts, as predicted above. A few qualitative observations based on these D's follow.

K, Rb, Ba, and Sr all are concentrated in andesite liquid relative to coexisting solids, except for the enrichment of Sr in plagioclase which increases with decreasing temperature. Hornblende precipitation will cause less enrichment of these elements in the liquid than will precipitation of anhydrous minerals. Equilibration with plagioclase or hornblende decreases the K/Rb ratio of liquids; all other minerals and especially biotite have the opposite effect.

Trivalent REE are also concentrated in andesite liquid relative to coexisting olivine, orthopyroxene, biotite, and plagioclase; light REE are preferentially enriched in the liquid relative to heavy REE except by plagioclase whose preferential exclusion of heavy REE and Y increases as temperature decreases (Drake and Weill 1975). Augite and hornblende differ from the above in two important aspects. First, D's for both minerals and especially hornblende change from < 1 to > 1 within the range of compositions and temperatures of orogenic andesite such that equilibration with these minerals may cause initial enrichment but subsequent depletion of REE and Y in liquids. Second, both minerals preferentially accept REE of intermediate ionic radius, with maximum D occurring at about Er (0.88 Å) for augite and at about Tb (0.92 Å) for hornblende. Thus, equilibration with these minerals leads to preferential enrichment of light versus heavy REE in the liquid, but also to slightly concave-up REE patterns. Garnet and apatite each have large and distinctive REE partition coefficients. Although below the empirical value of Table 6.3, experimentally determined D_{Yb}^{gar} values are high, 10 to 30, especially at 20 to 30 kb and water-rich conditions where the garnets are Ca-rich (Fig. 6.4) and therefore more able to accomodate REE.

Eu, with seven 4f electrons, is more susceptible to reduction to divalency than other REE. When $\log f_{O_2} \sim -8$, as in orogenic andesites (Sect. 5.3.5), some Eu may be divalent and therefore have an ionic radius of 1.17 Å which is more like Sr than trivalent REE, making Eu anomalous (cf. Mysen et al. 1978b). With the exception of one hornblende andesite, plagioclases (An 97–77) which have been separated from andesites are enriched in Eu relative to adjacent REE (a positive anomaly) by two to nine times; four fold enrichment was predicted by Weill and Drake (1973) for plagioclase-liquid equilibration under conditions typical for orogenic andesite (T = 1100°C; $\log f_{O_2} = -8.0$). For the same reason, negative Eu anomalies occur in orthopyroxenes, augites, hornblendes, garnet, magnetites, and apatites separated from andesites and related rocks. Nevertheless, because the size of these Eu anomalies is inversely proportional to the absolute partition coefficients, the effect of the anomalous behavior of Eu is more likely to be significant in partial melting than fractional crystallization situations relevant to andesite genesis.

Partition coefficients for Ni, Co, Cr, and Sc are > 1 and generally similar for orthopyroxene, augite, and hornblende; D_{Ni} is higher for olivine than for these minerals. D_V is especially high for magnetite and hornblende. Usually in augite,

Table 6.3. Phenocryst/matrix partition coefficients from orogenic andesite whole rocks

	Plag	Aug	Opx	Ol	Hb	Mt	Gar
K	[a] 0.02–0.20 (2,3,4,12) [c] 0.11 [b]	0.01–0.06 (7,12) 0.02	0.01–0.02 (2,12) 0.01	0.01 (14)	0.33 (2)	0.01	0.01
Rb	0.02–0.19 (2) 0.07	0.01–0.04 (2) 0.02	0.01–0.03 (2) 0.02	0.01 (14)	0.05 (2)	0.01	
Sr	1.3–3.2 (2,3,4,12) 1.8	0.06–0.21 (2,3,12) 0.08	0.01–0.10 (2,3,12) 0.03	0.01 (14)	0.19–0.26 (2,3) 0.23	0.01	0.02 (14)
Ba	0.05–0.36 (2,3,4,11,12) 0.16	0.01–0.15 (2,3,11,12) 0.02	0.01–0.23 (2,3,11,12) 0.02	0.01 (14)	0.08–0.22 (2,3,11) 0.09	0.01	0.02 (14)
Ce	0.06–0.30 (1,10,11,12) 0.20	0.04–0.51 (1,11,12) 0.25	0.03–0.33 (1,11,12) 0.05	0.01 (14)	0.09–0.37 (1,11) 0.25	0.06–0.82 (11,12,13) 0.20	ND
Sm	0.03–0.20 (1,10,11,12) 0.11	0.09–1.4 (1,11,12) 0.75	0.05–0.43 (1,11,12) 0.10	0.01 (14)	0.34–1.7 (1,11) 1.0	0.07–1.4 (11,12,13) 0.30	1.3 (9)
Eu	0.06–1.3 (1,10,11,12) 0.31	0.09–1.2 (1,11,12) 0.80	0.06–0.42 (1,11,12) 0.12	0.01 (14)	0.36–1.9 (1,11) 1.1	0.06–0.66 (11,12,13) 0.25	1.6 (9)
Yb	0.01–0.30 (1,11,12) 0.05	0.09–1.5 (1,11,12) 0.90	0.24–0.67 (1,11,12) 0.46	0.01 (14)	0.46–1.6 (1,11) 1.0	0.11–1.0 (11,12,13) 0.25	56 (9)
Zr, Hf	0.01–0.02 (11,12) 0.01	0.18–0.34 (11,12) 0.25	0.02–0.22 (11,12) 0.10	0.01 (16)	0.34–0.52 (11) 0.40	0.14–1.7 (11,12,13) 0.40	0.5 (16)
Nb, Ta	0.025 (16)	0.30 (16)	0.35 (16)	0.01 (16)	1.3 (16)	1.0 (16)	ND
Th	<0.01 (11,12) 0.01	<0.04 (8,11,12) 0.01	0.04–0.22 (11,12) 0.05	0.01	0.13–0.25 (11) 0.15	0.05–1.2 (11,13) 0.10	ND
Ni	0.01	3.5–9 (3,4,11) 6	5–24 (3,4,11) 8	58 (11)	7–16 (3,5,11) 10	4–19 (4,7) 10	0.6 (15)

Co	0.01	2–9 (3,4,12) 3	3–15 (3,4,12) 6	ND	7–19 (3,5) 13	4–25 (4,6,12,13) 8	2 (9)
Cr	0.01	10–245 (3,4,11,12) 30	7–143 (3,4,11,12) 13	34 (11)	23–90 (3,5,11) 30	1–166 (4,7,12,13) 32	22 (15)
V	0.01	0.9–18 (3,4,11) 1.1	0.5–7.2 (3,4,12) 1.1	0.08 (11)	6–45 (3,5,11) 32	24–63 (4,6,7) 30	8 (15)
Sc	0.01	2.5–17 (3,4,11,12) 3	1.4–7.5 (3,4,11,12) 3	0.30 (11)	8–13 (3,11) 10	1–3 (4,12,13) 2	4 (9)

a Range of reported values
b Suggested value (a weighted mean)
c Data sources:
1. Schnetzler and Philpotts (1970). 2. Philpotts and Schnetzler (1970). 3. Ewart and Taylor (1969). 4. Ewart et al. (1973). 5. Andriambololona et al. (1975). 6. Iwasaki et al. (1962). 7. Duncan and Taylor (1969). 8. Nagasawa and Wakita (1968). 9. Irving and Frey (1978). 10. Ewart (1976b). 11. Luhr and Carmichael (1980). 12. Okamoto (1979). 13. Schock (1979). 14. Arth (1967), recommended value. 15. Gill (1974), recommended value. 16. Pearce and Norry (1979), recommended value

orthopyroxene, and magnetite, $D_{Ni} > D_{Co}$ and $D_{Cr} > D_V$, as predicted by crystal field theory (Burns 1970). Apparent reversal of this trend occurs in some horn-blende and biotites from acid andesites and dacites (Andriambololona et al. 1975; Gill and Till 1978), which is not understood. If true, liquids in equilibrium with these minerals will develop distinctive increasing Ni/Co and Cr/V ratios with increasing fractionation. Garnet also has distinctive partition coefficients for these elements.

As mentioned previously, partition coefficients may vary consistently between low and high-K or tholeiitic and calcalkaline andesites as functions of temperature or liquid composition, but available data are inconclusive. For example, Ni contents are higher in olivines from calcalkaline than tholeiitic basalts and andesites from Hakone volcano, Japan (Iida et al. 1961), but so also are Ni contents in basic members of the respective rock series (Nockolds and Allen 1956; Sect. 5.4.5). A few other ambiguous examples are cited in Section 4.1.1.

6.11 Conclusions

Phenocryst-free orogenic andesites are atypical as well as aphyric. While pheno-crysts cause ambiguity about whether or not a rock was once entirely liquid, they are a natural consequence of the decompression of hydrous magma and provide useful information about the magmas from which, and the P-T conditions at which, they precipitated. Typical phenocryst assemblages are plagioclase + orthopyroxene + clinopyroxene + magnetite ± olivine, and plagioclase + hornblende + magnetite ± clinopyroxene, orthopyroxene; the former are pyroxene andesites, the latter are hornblende andesites. Pyroxene andesites include all varieties of andesite composi-tions defined in Chapter 1; hornblende andesites are mostly restricted to medium and high-K acid andesites. Typically, the ratio of orthopyroxene to clinopyroxene phenocrysts increases with increasing silica contents within suites.

Because of the high yield strength and viscosity of orogenic andesite magmas, the high phenocryst contents of the rocks reflect crystal accumulation as well as cooling history. Plagioclase with > An90, olivine with > Fo82 or < 0.1% CaO, and amphiboles in which $(K/Na)_{amphibole\ A\ site}/(K/Na)_{rock} < 1$, all are unlikely to precipitate from typical orogenic andesites, and therefore probably are accumulated xenocrysts. Rocks with > 10% positive Eu anomalies probably have accumulated > 30% plagioclase. Otherwise, it is ambiguous whether rocks have the composi-tions observed because of the phenocrysts, or vice versa.

Phenocrysts reflect whole rock compositions, e.g., anomalously *di*-poor ande-sites lack clinopyroxene phenocrysts. However, no distinctive phenocryst assem-blage characterizes all orogenic andesites or any of its subdivisions defined in Chapter 1. Hornblende and biotite are mostly restricted to acid andesites of medium and high-K suites, although some high-K suites lack hydrous minerals altogether [e.g., Tanna, New Hebrides (Gorton 1977); San Salvador, El Salvador (Fairbrothers et al. 1978); Irazu, Costa Rica (Krushensky 1972)]. Similarly, Or and Ab contents of plagioclase are higher in high-K than low-K andesites, and Wo contents often are higher in augite phenocrysts but lower in orthopyroxene phenocrysts from high-K relative to low-K andesites. Modal plagioclase/pyroxene ratios are higher in low-K than high-K andesites, as were the corresponding normative ratios (Sect. 5.2.7). Finally, the stability of orthopyroxene is restricted to a narrower range of host rock silica contents as K-enrichment increases. This restriction reflects the increased stability of olivine, augite, hornblende, and biotite relative to orthopyroxene in magmas which are more alkaline or hydrous or both. Both the more calcic nature of augites and the delayed appearance of orthopyroxene in more K-rich andesites is due in part to their reduced a_{SiO_2}, a reduction reflected in their higher normative hy/qz ratios (Sect. 5.2.7).

Tholeiitic and calcalkaline orogenic andesites, as defined in Chapter 1, differ little in phenocryst mineralogy, although the difference in absolute Fe-enrichment is reflected in a more restricted range of pyroxene, especially clinopyroxene, compositions in calcalkaline andesites. However, most andesites containing groundmass pigeonite are tholeiitic, and most andesites containing groundmass hypersthene are calcalkaline according to the definition of Fig. 1.4. This difference in groundmass pyroxene may reflect a difference in temperature of eruption, although alternative explanations are viable.

Orogenic andesites which are erupted on continental margins and detached continental fragments differ mineralogically from those erupted in island arcs only in two minor respects. First, amphiboles from andesites of continental margins or interiors have > 6.5 Si per 23 O, whereas those from island arcs often have < 6.5 Si; amphiboles in dacites and rhyolites from island arcs have also high Si. This contrast in hornblende andesites implies higher temperature or lower f_{O_2} or both for the andesites of island arcs. Secondly, garnet and possibly cordierite seem to be cognate phenocrysts in some peraluminous andesites which are restricted to continental margins or fragments. This restriction indicates that the peraluminous condition sometimes is a result of assimilating pelitic material, rather than analytical inaccuracy (cf. Sect. 5.2.7).

Phenocryst assemblages and compositions constrain the temperature, water contents, and pressure which prevailed during crystal precipitation. Equilibration temperatures calculated from several mineral assemblages in andesites are tabulated in Table 4.1. Pyroxene andesites yield temperatures of 1000° to 1100°C; hornblende andesites of 900° to 1000°C. Although inaccuracies and imprecision are too great to identify consistent and significant differences in equilibration temperatures between andesites of different types or locales, there is a suggestion that calcalkaline and continental andesites erupt at lower temperatures than tholeiitic or island arc andesites, respectively.

Water contents and partial pressures likewise can be constrained mineralogically. The temperature estimates above imply that orogenic andesites contain

1 to 5 wt.% H_2O (Fig. 4.1) as also does the ubiquitous saturation of pyroxene andesites with both plagioclase and pyroxene (Fig. 6.6), the frequent occurrence of plagioclases with sieve textures and with resorbed cores overlain by reversely zoned spikes (Sect. 6.1 and 6.9), and the compositions of co-existing hornblende + pyroxene(s) + magnetite ± plagioclase and olivine (Sect. 5.3.1). The composition of these minerals also implies $P_{H_2O} < P_{TOTAL}$.

Reliable quantitative estimates of the total pressure under which phenocryst precipitation occurs in orogenic andesites are not yet available. Qualitatively, the pressure must be low, certainly < 8 kb and possibly only a few kb. Low pressures are implied by the low Na, Al, and Ti contents in pyroxenes, the low Al^{VI} content of amphiboles, and the relatively high Ca content of olivines. Recognized exceptions are rare. Some of these characteristics differ in the minerals of mafic inclusions in andesites, but the differences may reflect high water rather than load pressures for the inclusions. However, plagioclase-olivine reaction relationships are absent from the gabbroic inclusions, thus restricting pressure to less than about 7 kb. Nevertheless, many phenocrysts in andesite have been affected by decompression which causes breakdown of hornblende and biotite, normal or reverse zoning of plagioclase depending on the water content of the magma, and various kinds of reaction rims around garnet. Decompression also may cause reverse zoning of pyroxenes, pigeonite rims on orthopyroxene, and dissolution of quartz, although these causal effects are less well demonstrated.

Quantitative estimates of load pressure during phenocryst precipitation require knowledge of equilibration temperature, $a_{SiO_2}^{melt}$ or $a_{Al_2O_3}^{melt}$, and their pressure dependence (Nicholls et al. 1971; Carmichael et al. 1977). Available results yield total pressures of 7 to 10 kb and 1 kb for phenocryst assemblages in andesites from Rabaul and Colima volcanoes, respectively (Heming 1977; Ghiorso and Carmichael 1980) and 3 to 10 kb for mafic inclusions in Antillean andesites (Powell 1978; Arculus and Wills 1980).

Two minerals of key importance in explaining why orogenic andesites differ from other intermediate rocks are magnetite and hornblende. Data cited in this chapter imply that magnetite is a frequent though minor phenocryst in most orogenic andesites, but that hornblende is restricted to silica-, soda-, and water-enriched magmas in the upper portions of temporarily closed magma reservoirs, and is not a widespread liquidus phase.

Early precipitation of magnetite from orogenic andesite is suggested by the restricted Mg/Fe ratios of their pyroxenes and olivines, the high An content at which FeO contents of plagioclase go through a maximum, and the ubiquity of magnetite in gabbroic inclusions, as well as the occurrence of magnetite phenocrysts or microphenocrysts in 50% to 90% of orogenic andesites. Portions of andesitic suites characterized by rapid rise in FeO*/MgO ratios relative to SiO_2 (see Fig. 5.4) also lack phenocrystal magnetite whose appearance is associated with reduction of this rise (i.e., suppression of Fe-enrichment). Sometimes this appearance occurs during periods of volcanic dormancy. However, magnetite is absent as a phenocryst from some andesites characterized by lack of Fe-enrichment and is not a liquidus phase of calcalkaline acid andesites under experimental conditions when f_{O_2} is near the NNO buffer.

Amphibole has a maximum thermal stability of 930° to 1000°C in andesite magmas, which is less than the likely liquidus temperature of many pyroxene andesites. Moreover, even acid andesite magmas become saturated with amphibole only when magma water contents exeed 10 wt.% and load pressures exceed 8 kb. These andesites could not have been derived by amphibole fractionation under conditions in which they are not saturated with amphibole. Water contents > 10% are unlikely, as noted above, and at pressures > 8 kb the amphiboles contain more Al^{vi} than is found in hornblende phenocrysts or in the hornblendes of gabbroic inclusions. Also, pyroxenes but not hornblendes are found as inclusions in plagioclase phenocrysts of pyroxene andesites, indicating that a pyroxene, not hornblende, was the mafic phase throughout the crystallization interval of these rocks. Finally, hornblende usually is restricted to topographically and stratigraphically high levels of most andesite stratovolcanoes and, in some cases, to high levels within magma chambers (e.g., Mt. St. Helens, Bezyminanny, Arenal volcanoes). These features collectively suggest that hornblende usually is a sub-liquidus phase produced by reaction of the liquid with early-formed minerals and is of greatest importance under plutonic conditions or during the crystallization of acid andesites and dacites; the features are incompatible with equilibration between most orogenic andesite magmas and amphibole.

Finally, phenocryst mineralogy provides some evidence of crustal assimilation and the most convincing evidence of magma mixing. Assimilation is demonstrated by the occurrence of thermally metamorphosed inclusions of crustal rocks, but mineralogic evidence for resulting modification of the host magma is restricted to the scattered occurrence of garnet and cordierite megacrysts which seem to be neither relics of metamorphic rocks nor high-pressure phenocrysts of typical orogenic andesites. Magma mixing may be indicated by reversely zoned plagioclase and pyroxene phenocrysts, embayed quartz or olivine often with pyroxene reaction rims, mafic inclusions with chilled margins or acicular mineral morphologies, and "outrange" glass inclusions within minerals, i.e., glasses whose composition lies outside the range represented by the bulk rock to groundmass.

Chapter 7 Spatial and Temporal Variations in the Composition of Orogenic Andesites

Comparing orogenic andesites within and between volcanic arcs or over time requires (1) choosing a reference point such as a specified wt.% SiO_2 or FeO^*/ MgO ratio to normalize effects of within-suite differentiation, (2) averaging or generalizing data for individual volcanoes or entire arcs, and (3) comparing data obtained from different laboratories or methods. All introduce error. I have tried to minimize these errors by summarizing analyses of Quaternary rocks by volcano (Appendix) and then averaging these by arc segment (Table 7.1). Some consistent patterns emerge from the comparisons and provide additional constraints on genetic theories.

7.1 Variations in Magma Composition Across Volcanic Arcs

In 17 of the 28 convergent plate boundary segments summarized in Table 2.1, one or more Quaternary volcanoes occurs $\geqslant$ 50 km behind the volcanic front. Reasons why volcanic arc of modest width occurs in some places while merely a volcanic front occurs in others are unclear. For example, displacement of Egmont westward from the volcanic front in New Zealand is attributed by Maori legend to Egmont being banished after courting Tongariro's wife. Geologic legends about arcs of modest width are similarly conjectural though less conjugal, the width being attributed to thick crust, shallow subduction, absence of backarc spreading, or downdip migration of melting or dehydration foci within underthrust lithosphere over time.

A striking characteristic of orogenic andesites and associated rocks within volcanic arcs of modest width is the consistent increase of their incompatible element concentrations, most notoriously their K_2O contents, away from the plate boundary. This pattern parallels the lateral variations in geophysical phenomena summarized in Chapter 3, and was first identified by Tomita (1935) who noticed differences in mineralogy and total alkali contents of basalts across Japan. These differences were later shown also to apply elsewhere by Tomkeiff (1949), Rittman (1953), Kuno (1959), Sugimura (1960, 1968, 1973), and Gorshkov (1962, 1970), among others. Usually the difference in alkalinity is due more to changes in K_2O than Na_2O contents (Dickinson and Hatherton 1967; Hatherton 1969), although exceptions as usual exist (e.g., Johnson 1976b). This is one reason why K_2O was used to define and subdivide orogenic andesites in Section 1.1.

Table 7.1. Geochemical summary of average characteristics of volcanic arcs (from the Appendix)

Arc	1 N	2 % mafic volcanoes	3 % andesitic volcanoes	4 % calcalkaline volcanoes	5 $\overline{X}$ $K_{57.5}$	$\overline{X}$ $FeO^*/Mg_{57.5}$
1. New Zealand	7	14	71	100	1.5	1.7
2. Kermadec	2	100	0	0	0.4	3.0
3. Tonga	5	60	60	0	0.6	3.1
4. Vanuatu	9	100	11	56	2.2	2.9
5. Solomon	4	50	50	67	2.3	2.2
6. Bismarck						
a) East	11	45	64	40	0.9	2.3
b) West	2	100	50	0	0.9	2.7
7. Papua New Guinea	3	33	67	67	2.1	1.6
8. Sunda						
a) Sumatra	5	0	80	80	1.7	2.1
b) Java	16	68	81	56	1.8	2.4
c) Bali-Banda	13	54	77	42	1.8	2.2
9. Halmahera	3	0	33	0	1.9	2.8
10. Sulawesi	1	100	0	100	1.5	2.0
11. SE Philippines	1	100	100	100	1.2	2.2
12. NW Philippines- Taiwan	5	80	40	75	1.7	2.2
13. Ryuku-W. Japan	9	11	67	78	1.5	2.1
14. Mariana-Izu	11	91	45	14	1.0	4.2
15. East Japan						
a) Honshu	27	37	74	67	1.0	2.2
b) Hokkaido	12	50	75	28	1.1	2.6
16. Kuriles	22	32	77	40	1.2	2.2
17. Kamchatka	24	54	58	60	1.3	2.1
18. a) Aleutians	10	60	80	43	1.6	2.5
b) Alaska	5	20	60	100	1.2	1.8
19. Cascades	15	13	53	100	1.3	1.7
20. Mexico	8	38	75	100	1.5	1.4
21. Central America	26	77	46	56	1.5	2.3
22. Columbia-Ecuador	5	40	60	100	1.5	2.0
23. Peru-Chile						
a) $11°–28°S$	17	12	70	100	2.2	1.6
b) $33°–47°S$	8	38	38	33	1.4	2.5
c) $>47°S$	1	0	0	100	<0.7	<1.8
24. Antilles	9	56	89	50	0.9	2.2
25. South Sandwich	–	–	–	–	0.7	4.1
26. Eolian	7	57	43	33	2.4	2.3
27. Aegean	7	14	57	100	2.4	1.7
28. Turkey-Iran	9	11	67	71	2.3	2.2

1. Number of volcanoes per arc. All data from the Appendix. Arcs are keyed to Table 2.1 and Fig. 2.2
2. Percent of the *volcanoes* at which half or more of analyzed ejecta contain $<57\%$ SiO_2
3. Percent of the *volcanoes* at which half or more of analyzed ejecta contain 53 to 63% SiO_2
4. Percent of the *volcanoes* at which FeO^*/MgO is $\leqslant 2.25$ at 57.5% SiO_2 (see Fig. 1.4)
5. Average of the K_2O contents and FeO^*/MgO ratios at 57.5% SiO_2 of all volcanoes within the arc

Consequently, most of the chemical and mineralogic differences between low, medium, and high-K orogenic andesites described in Chapters 5 and 6 also characterize lateral variations across volcanic arcs. Specifically, concentrations of many incompatible elements in addition to potash vary across arcs. Such trace element variations have been summarized by Jakeš and White (1972a) and Gill and Gorton (1973) among others. Mineralogy also varies. Modal and normative pyroxene/ plagioclase ratios, the silica range of rocks containing olivine, hornblende, and biotite phenocrysts, and the likelihood of groundmass orthopyroxene, all increase away from the volcanic front. These differences reflect increasing water and alkali contents and, therefore, lower quench temperatures and a_{SiO_2} across arcs (Sect. 6.11; Sakuyama 1977). Also, sometimes there are across-arc variations in the isotopic composition of magmas (see below), and in the nature of associated ore deposits, with the time-transgressive sequence Fe-(Cu, Mo, Au)-(Cu, Pb, Zn, Ag)- (Sn, W, Ag, Bi) occurring with increasing distance from the plate boundary (Sillitoe 1976).

An up-dated summary of LIL-element variations across nine individual arcs is given in Table 7.2. In so far as data are available for the arcs listed, K, Rb, La, Th and U contents and La/Yb and Rb/Sr ratios *always* increase relative to silica away from the plate boundary; P, F, Sr, Ba, and Zr (or Hf) contents and Th/U, Ba/Sr, and U/Pb ratios either increase or do not change significantly; and K/Rb, Ba/La, and Zr/Nb ratios usually decrease.

In addition, across-arc variations in Fe-enrichment patterns sometimes occur. There is a decrease in maximum FeO* contents and maximum FeO*/MgO ratios across northern Honshu and Hokkaido, Japan, and the northern Kuriles, but not in other active arcs (Appendix). Also, rapid Fe-enrichment relative to silica (tholeiitic behavior in the sense of Wager and Deer; Sect. 1.2), $FeO*/MgO_{57.5}$ ratios > 2.5, and groundmass pigeonite occur only at volcanic fronts, if at all, in volcanic arcs. These two observations led Jakeš and Gill (1970), Jakeš and White (1972a), and Gill and Gorton (1973) to infer a characteristic across-arc transition from tholeiitic to calcalkaline series as well as from low to high-K rock types. However, this inference was misleading both because tholeiitic behavior as defined in Fig. 1.4 is not confined to volcanic fronts, but instead occurs in roughly equal proportions across some volcanic arcs (Miyashiro 1974, 1975), and because rapid Fe-enrichment often is infrequent or absent at volcanic fronts. Consequently, LIL rather than transition elements most uniformly vary in concentration relative to silica across arcs.

Other active arcs which may conform to the pattern of Table 7.2 but for which data are incomplete include: the Sulawesi-Sangihe arc, Indonesia (Jezek et al. 1979); central Honshu, Japan, from Oshima to Dainichi-yama volcanoes (see Kurasawa and Michino 1976; Ui and Aramaki 1978); Kamchatka (Kepezhinskas 1970; Markhinin and Stratula 1973; Leonova 1979); and the South Sandwich arc where Leskov volcano lies 50 km west of the volcanic front and may have higher K, Rb and Sr (Gledhill and Baker 1973). The pattern even prevails in the continental

Table 7.2. Transverse geochemical variations across volcanic arcs. (+ means increase away from plate boundary; − means decrease; = means change absent or hidden in data scatter; na means not available)

Arc	K	Rb	Ba	Sr	K/Rb	La	Yb	La/Yb	Ba/La	Th	U	Th/U	^{87}Sr/^{86}Sr	Zr, Hf	P
1. New Zealand	+	+	na	+	=	na	na	na	na	+	+	=	−	na	+
2. New Britain	+	+	+	=	=	+	=	+	−	+	na	+	=	+	+
3. Java	+	+	+	+	−	+	=	+	=	+	+	+	+	=	+
4. Japan															
a) N. Honshu	+	+	+	na	−	+	=	+	−	+	+	=	−	+	=
b) Hokkaido	+	na	+	na	na	+	=	+	−	+	+	=	−	na	=
5. N. Kuriles	+	+	+	=	−	na	na	na	na	+	+	=	na	na	na
6. Aleutians	+	+	+	+	−	+	=	+	+	na	na	na	−	=	=
7. Cascades	+	+	+	na	=	+	+	na	=	na	na	na	=	na	=
8. South America															
a) S. Peru	+	+	=	=	−	+	=	+	na	na	+	na	na	na	=
b) N. Chile	+	+	+	+	−	+	−	+	−	+	+	na	=	+	=
9. Antarctic	+	+	=	=	−	+	na	+	=	+	na	na	na	=	=

References:
1. New Zealand. Egmont occurs behind the volcanic front but also at the southern end of the tectonically ambiguous volcanic arc along western North Island which may not have a genetic relationship to the present Benioff Zone. Hatherton (1969); Stipp (1968); Cole (1978)
2. New Britain. The Witu Islands are omitted because some of their rocks seem like those of Niua'fou in the Lau Basin. Johnson (1976b); Johnson and Chappell (1979); Blake and Ewart (1974); Arth (1974) Peterman et al. (1970a); Page and Johnson (1974); DePaolo and Johnson (1979)
3. Java. Whitford (1975); Whitford and Nicholls (1976); Whitford et al. (1979a)
4. Japan.
   a) Kawano et al. (1961); Ui and Aramaki (1978); Hedge and Knight (1969); Masuda (1979). F and H_2O also increase (Sakuyama 1977; Ishikawa et al. 1980)
   b) Katsui et al. (1978); Masuda et al. (1975)
5. N. Kuriles. Comparison between Main and Western Zones. Gorshkov (1970); Leonova and Udal'tzova (1970); Leonova (1979)
6. Aleutians. Based on comparison of 1796 Bogoslof andesite and Amak andesites with rocks from volcanic front. Arculus et al. (1977); Kay (1977); Marsh (1976a); Morris and Hart (1980); Marsh and Leitz (1979)
7. Cascades. Adams, Newberry, and Medicine Lake volcanoes are behind the volcanic front. Table entry based on comparison between Shasta and Medicine Lake. Condie and Swenson (1973); Condie and Hayslip (1975); Mertzman (1977)
8. South America.
   a) Peru (zones A_1 and A_2 only; inclusion of shoshonites changes trends for Sr, Ba, K/Rb, U and P). Dupuy and Lefèvre (1974); Dupuy et al. (1976); Dostal et al. (1977a)
   b) N. Chile only (not Bolivia and Argentina because, again, inclusion of shoshonites changes trends for Sr and U). Roobol et al. (1976); Zentilli and Dostal (1977); Klerkx et al. (1977); Déruelle (1978); Dostal et al. (1977b);
9. Antarctic. Saunders et al. (1980)

collision-related Caucasus region (Adamia et al. 1977). Older examples of the pattern include western North America before ridge-trench collision (Christiansen and Lipman 1972; Cameron et al. 1980), Fiji in the Miocene (Gill and Gorton 1973), and Britain in the Ordovician (Fitton and Hughes 1970).

Three factors complicate identification of these patterns. One is the presence of along-strike as well as across-strike variations (Sect. 7.2); another is ambiguity in distinguishing volcanic arc from backarc environs (Sect. 2.1). Specifically, in preparing Table 7.2 I may have unwittingly included "edge effects" (see Sect. 7.2) in the cases of New Zealand, New Britain, and the Cascades. Also, in order to retain focus on orogenic andesites, I have included data for volcanic arcs only, and not for backarc basalts which lack the trace element or isotopic characteristics of subduction-related volcanics which were discussed in Sect. 5.7.

The third complication is the need for an arbitrary decision concerning the silica concentration or range at which to make one's comparisons. That across-arc geochemical variations are more evident in andesites than basalts was presaged in Section 5.2.2 where increased K_2O-SiO_2 slopes were correlated positively with increasing $K_{57.5}$ values. Consequently, across-arc differences in differentiation mechanisms (e.g., in at least the composition and perhaps the kind of fractionating minerals) as well as source compositions are implied. In preparing Table 7.2 I relied on acid andesite compositions whenever possible.

Reversals of the spatial pattern do exist but are uncommon and reflect local tectonic complexities. One exception is the Banks Islands of Vanuatu where basalts from Gaua volcano at the volcanic front contain more K and P relative to silica than do basalts from Mere Lava volcano located 60 km farther east (Mallick 1973). A second pair of exceptions occurs in the Eolian arc and the Flores area of Indonesia where $K_{57.5}$ values scatter randomly and widely despite narrow arc widths. Similarly, despite a factor of three variation in K values along the Central American volcanic front, they do not correlate significantly with depth to the seismic zone (Carr et al. 1979). However, even here the monogenetic basalt cones behind the volcanic front have K values higher than those of over half of the polygenetic stratovolcanoes at the corresponding front. McBirney (1976) delighted in yet another exception where basalts across central Oregon become less rich in K and Zr eastwards. Note that the nonconforming lavas in the Vanuatu, Flores, and Oregon cases are basalts alone. Ad hoc explanations for each of these exceptions can be advanced based on local tectonic complexities involving edge effects or backarc rifting.

Some elements have been cited incorrectly as varying across volcanic arcs. Boettcher (1973) and Best (1975) claimed that Ti increases with K, but usually this is true only if high-Ti alkali basalts in backarcs are included. In general there is no correlation between Ti contents in orogenic andesites and position across volcanic arcs (e.g., Sugimura 1973) or with K contents in lowest-Si andesites (Sect. 5.2.4). Southern Peru is an exception (Frangipane-Gysel 1977). Likewise, Marsh and Carmichael (1974) claim a positive correlation between SiO_2 and depth to the dipping seismic zone. Although such correlation sometimes occurs (e.g., again in the Andes: Déruelle 1978), the opposite is more often true (e.g., Markhinin and Stratula 1973; Sugimura 1973) if, indeed, there is any discernable across-arc variation in silica mode.

Whether there are consistent across-arc variations in the isotopic composition of Sr, Pb, or Nd remains ambiguous, due largely to paucity and insufficient precision of data. If present, such variations are the clearest indication that the source region or the extent of contamination differs across arcs.

Decreasing $^{87}Sr/^{86}Sr$ ratios, despite increasing Rb/Sr ratios with distance from the plate boundary, have been found in northern Honshu, Japan (Hedge and Knight 1969), New Zealand (Stipp 1968), Sulawesi, Indonesia (Jezek et al. 1979), the eastern Aleutians (Morris and Hart 1980), and Fiji (Gill and Compston 1973). However, lateral differences in crustal thickness or age, or in mantle composition, may be more significant than distance from the plate boundary. For example, in Hokkaido, as in northern Honshu, the Quaternary volcanics farthest from the trench and richest in LIL-elements surprisingly have the lowest $^{87}Sr/^{86}Sr$ ratios ($\sim$ 0.7030), but such low ratios also occur where the volcanic front extends into the Sea of Okhotsk (Katsui et al. 1978). This suggests that low ratios occur along coasts and therefore in regions of thinner crust whatever their distance from the plate boundary. Also, south of 40°N latitude, volcanics from islands and seamounts in the Sea of Japan, as well as from southwest Japan, have high initial ratios (0.706 to 0.71; Shuto 1974), implying that isotopic differences amongst Japanese igneous rocks reflect variations in mantle composition rather than crustal thickness. In either case, the across-arc isotopic compositions may be real but fortuitous.

Apparently, no change in $^{87}Sr/^{86}Sr$ ratios occurs across the New Britain (Page and Johnson 1974), Java (Whitford 1975), central Aleutian (Kay et al. 1978), Cascade (Church and Tilton 1973), or South Sandwich arcs (Gledhill and Baker 1973). In contrast, an increase in initial $^{87}Sr/^{86}Sr$ ratios (as well as in silica-normalized K, Rb, Sr, and La contents) with distance from the coast occurs in Mesozoic plutonic rocks of western North America (Early and Silver 1973; Kistler and Peterman 1973; Gromet and Silver 1977). Currently these variations are interpreted as indicating incorporation of more Precambrian sediment into plutons farther from the plate boundary. A similar eastward increase in $^{87}Sr/^{86}Sr$ ratios across the Andes was reported by McNutt et al. (1975), but it is time-transgressive and is lost in data scatter when Quaternary samples alone are considered (Klerkx et al. 1977).

Fewer data are available for Pb and Nd. There is no difference in $^{143}Nd/^{144}Nd$ ratios for rocks collected from, and up to 250 km behind, the volcanic front in New Britain, Papua New Guinea, despite substantial K-enrichment over the same distance (DePaolo and Johnson 1979). Pb results generally are similar to those for Sr, with a decrease in $^{206}Pb/^{204}Pb$ ratios westward across northern Honshu both in young volcanics (Hedge and Knight 1969) and in Miocene ores (K. Sato 1975), and a similar decrease westward across New Zealand (Armstrong and Cooper 1971). No consistent correlation with distance from the plate boundary was found in a 220 km-wide traverse across the volcanic arc in central Honshu, Japan (Tatsumoto and Knight, 1969), nor is there one across the Peru (McNutt et al. 1979), Aleutian (Kay et al. 1978), or Cascade arcs (Church and Tilton 1973).

Thus, across-arc isotopic variations are inconsistent. Where they occur, both ^{87}Sr and ^{206}Pb seem to decrease toward coastlines and may thus reflect crustal thickness or related mantle heterogeneity rather than distance to the plate boundary.

Several attempts have been made to relate some of the geochemical parameters discussed above, principally K_2O, not only to distance from a plate boundary within individual arcs, but also to an absolute depth to the dipping seismic zone (Dickinson and Hatherton 1967; Dickinson 1968, 1975; Hatherton and Dickinson 1969; Hart et al. 1970; Ninkovich and Hays 1972). These attempts have the advantage of pooling data from many arcs, thereby increasing the data base for any given generalization, of predicting subduction geometries or rock compositions given one of the two, and of imputing remarkable regularity to the processes of magma genesis. However, the attempts are weak statistically because there are different element-depth correlations for different arcs (Nielson and Stoiber 1973).

At least four factors contribute to this weakness and they are illustrated vividly by widely descrepant K-h correlations for the Sunda arc (see Whitford and Nicholls 1976, Fig. 5). First, depths to the dipping seismic zone become better known with time and are better known in some arcs than others, leading to systematic errors upon pooling. For example, depths beneath northern Honshu volcanoes now are thought to be about 25% shallower than depths used by Nielson and Stoiber (1973) and Dickinson (1975) due to recent deployment of a local seismic network and use of an improved velocity structure model (Hasegawa et al. 1978). Second, multiple $K_{57.5}$ values can occur at individual volcanoes (Sect. 5.2.2), with the higher values being arguably related to secondary phenomena such as contamination or arc rifting (Whitford and Nicholls 1976; Gorton 1977). Third, inconstancy of slope in $K_2O\text{-}SiO_2$ diagrams (Sect. 5.2.2) and of the SiO_2 content of parental magmas leads to different K_2O-depth correlations depending on the SiO_2 reference point chosen (Cawthorn 1977). Fourth, the thickness or age of crust and the composition of the underlying mantle varies between arcs, resulting in higher $K_{57.5}$ values in continental margin than intra-oceanic arcs relative to a common depth to the dipping seismic zone (Sect. 7.2). Note, however, that variations away from plate boundaries across individual arcs are not necessarily accompanied by increases in crustal thickness; characteristics of crust and depth to the dipping seismic zone exert separate influences.

Thus, within individual arcs there are consistent across-arc increases in K, Rb, La, Th, U and H_2O contents and La/Yb and Rb/Sr ratios in orogenic andesites, decreases in K/Rb and Zr/Nb ratios, corresponding changes in mineralogy, and a change in the character of associated ore deposits. These changes correlate, with local exceptions, to distance from the plate boundary. However, details of the correlations vary between arcs, minimizing their predictive value for new areas. In some cases $^{87}Sr/^{86}Sr$ and $^{206}Pb/^{204}$ ratios decrease away from the plate boundary, but in other cases these ratios plus $^{143}Nd/^{144}Nd$ seem constant or vary inconsistently across the arc.

Used with caution, spatial variations in the geochemical parameters listed above *in contemporaneous rocks* are strong evidence of subduction polarity when geologic criteria indicate that the variations are across-strike rather than along-strike within an inactive volcanic arc. However, the absolute element concentrations provide poor estimates of depth to paleo-seismic zones. Estimates of paleo-polarities which use metamorphosed volcanic rocks should rely on La/Yb and Zr/Nb ratios which increase and decrease, respectively, with distance from the plate boundary, because these ratios are the polarity-sensitive criteria least affected by metamorphism.

7.2 Variations in Magma Composition Along Volcanic Arcs

Variations in the composition or mineralogy or both of orogenic andesites along strike, usually along volcanic fronts, occur as one of four types: random; related to the thickness of crust traversed; regular but apparently unrelated to crustal differences; and edge effects. By random variations I mean the "noise level" of K values (e.g., $K_{57.5}$) along the volcanic front which is about ± 0.25 for most arcs (see Appendix). Greater but irregular variations occur along the front in about a third of the arc segments of Fig. 2.2, such as the Eolian, Flores, and Central American examples cited as exceptions to regular across-arc variations in the previous section. This randomness has no clear genetic significance other than to underscore heterogeneity during arc magma genesis.

In order to assess crustal influences I will use the crustal thicknesses given in Table 2.1 despite their variable reliability, despite ambiguity whether the 7.5 to 8.0 km/s velocities are crustal or mantle or whether there even is a sharp crust-mantle boundary beneath arcs, and despite variations in crustal thickness within arc segments. Likewise I will adopt the classification of crustal types in Fig. 3.3 despite differences within each type in velocity structure and composition or age of basement rocks. In general, island arcs have crust < 25 km thick; continental fragments, peninsulas, and most margins have crust 30 to 40 km thick.

The most obvious and most frequently cited influence of an increased thickness of crust through which magma ascends is to increase the silica mode of ejecta which are erupted. Figure 7.1 shows relationships between crustal thickness and type and the percent of volcanoes at which basalts and basic andesites are the most commonly analyzed rocks in each arc segment. Island arcs usually are regions of mostly basalt and basic andesite volcanoes (the Banda, Kurile, Ryuku,

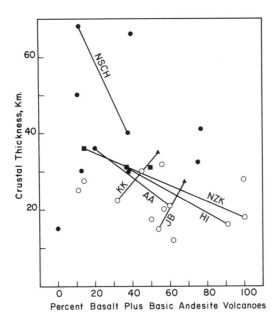

Fig. 7.1. Percent of volcanoes where mafic rocks predominate versus crustal thickness for active volcanic arcs. Data are from Tables 2.1 and 7.1. Percentages refer to relative numbers of volcanoes per arc at which basalts or basic andesites are half or more of the rocks for which chemical analyses are published. They are not relative volumes of ejecta per arc (see Sect. 4.9). *Symbols* indicate crustal type: *dots* mainland continental margins; *triangles* peninsular or insular continental margins; *squares* detached continental fragments; *open circles* island arcs. *Lines* connect portions of the same plate boundary which transect crust of variable thickness: *NZK* New Zealand-Kermadec; *JB* Java-Banda; *HI* Honshu-Izu; *KK* Kamchatka-Kuriles; *AA* Alaska-Aleutians; *NSCH* northern vs. southern Chile

and Aegean arcs are apparent exceptions); continental margins and interiors, peninsulas, and fragments contain mostly more silicic rocks (the New Guinea Highlands and Central America are apparent exceptions). This contrast is most striking between the Kermadec islands and New Zealand, the Mariana-Izu islands and Honshu, Japan, and the Aleutian islands and Alaska. Likewise, the higher andesite/basalt ratio in volcanoes of the northern relative to central Cascades, of the Guatemalan highlands relative to Nicaragua, and of northern relative to central or southern Chile, has been attributed to the presence of older and thicker crust beneath the Washington, Guatemalan, and northern Chilean volcanoes (McBirney 1969; Katsui and Gonzalez 1968; White and McBirney 1979). Large volumes ($> 10^5$ km^3) of rhyolite ignimbrites erupt only on continental crust and not in regions with < 30 km of crust.

Secondly, an overall difference in alkalinity between andesites erupted in continents versus island arcs has been recognized for some time (e.g., McBirney 1969, Table 2; Dickinson 1968, 1975), but the difference is inconsistent and often subtle. For example, Gorshkov (1962, 1970) argued that crustal thickness had no effect on the alkalinity of andesite erupted at northwestern Pacific plate boundaries. Also, reconnaissance studies have found no clear differences other than silica mode in elemental (Forbes et al. 1969) or isotopic composition (Kay et al.1978) between andesites in the Aleutian and Alaskan arcs, despite a doubling of crustal thickness between the two regions.

In contrast, differences both in isotope ratios and LIL-element concentrations (as well as silica mode) occur between volcanic rocks of the New Zealand and Tonga-Kermadec arcs (Ewart et al. 1977), and of central and northern Chile (Klerkx et al. 1977; Lopez-Excobar et al. 1977). Both also are instances of volcanic arcs traversing crust whose thickness approximately doubles along strike; however, depth to the dipping seismic zone also varies along strike in these cases, being shallower under thicker crust in New Zealand but the opposite in Chile (Table 2.1). In both cases, LIL-element contents (K, Rb, Cs, Ba, Th, U, Pb, La, Zr, Nb) are higher relative to silica, and $^{87}Sr/^{86}Sr$ ratios are higher, in andesites erupted through thicker crust.

Less thoroughly documented than the above are along-strike changes in the Sunda and Aegean arcs. $^{87}Sr/^{86}Sr$ ratios in andesites (but not K_{55} values relative to depth to the dipping seismic zone) decrease from Sumatra to Bali (Whitford 1975; Leo et al. 1980) as does crustal thickness (Curray et al. 1977). Both $^{87}Sr/^{86}Sr$ ratios and K_{55} values increase northward from Santorini to the Thebes area in the Aegean arc (Pe and Gledhill 1975) without apparent change in depth to the poorly defined seismic zone but with northward increase in crustal thickness (Makris 1978).

Elsewhere effects are more subtle. Despite Groshkov's claims to the contrary, there are differences in K_{55} and K_{60} values relative to depth to the dipping seismic zone between different segments of northwestern Pacific arcs (Zolotarev and Sobolev 1971; Nielson and Stoiber 1973). These differences correlate with crustal thickness and, therefore, with the magnitude of long-wavelength gravity anomalies (Ui and Aramaki 1978). The differences occur not only between the Izu arc and Honshu, or the Kuriles and Kamchatka, but also longitudinally along the four arc segments themselves. Correspondingly, there is about a 1‰ increase in $^{18}O/^{16}O$ ratios between andesites of Honshu versus those of the Izu or Ryuku islands (Sect. 5.5.6). Similarly subtle correlations of K_2O and Rb with crustal thickness occur in the Cascades (White and McBirney 1979).

Thirdly, Fe-enrichment in andesites is most likely where crust is thin. Fig. 7.2 shows that most volcanoes are tholeiitic according to Fig. 1.4 primarily in arc segments where crust is < 25 km thick. However, the value of this observation is marred by ambiguities in the distinction between the tholeiitic and calcalkaline series which were discussed in Section 1.3. That is, low calcalkalinity in Fig. 7.2

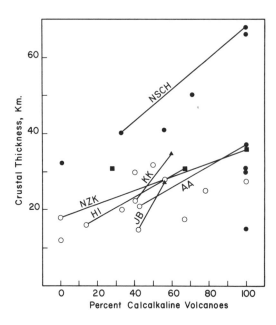

Fig. 7.2. Percent of volcanoes where rocks are calcalkaline (according to Fig. 1.4) for active volcanic arcs. Data are from Tables 2.1 and 7.1. Percentages refer to relative numbers of volcanoes per arc whose rocks have FeO*/MgO ratios <2.25 at 57.5% SiO_2. *Symbols* are as in Fig. 7.1

could indicate frequent occurrence within the arc of rapid Fe-enrichment relative to silica, or regionally high low Mg-numbers, or both. Of the three exceptional cases in which calcalkalinity is low despite crustal thicknesses > 30 km, only in Hokkaido are there examples of rapid Fe-enrichment relative to silica.

Thus, the percentage of calcalkaline series rocks, especially when narrowly defined, increases as crust thickens. This also implies a crude inverse correlation between average Feo*/MgO ratios and crustal thickness (see also Miyashiro 1974, Fig. 15).

To summarize, while continental crust is neither necessary nor sufficient to produce orogenic andesite, the thickness of crust underlying andesite volcanoes crudely affects the composition of magmas erupted. The silica mode, concentrations of LIL-elements relative to silica content, and $^{87}Sr/^{86}Sr$ and $^{18}O/^{16}O$ ratios, all often increase as crust thickens while the likelihood of rapid Fe-enrichment or tholeiitic FeO*/MgO ratios decreases. These differences could reflect assimilation of crustal rocks or fluids derived therefrom, differences in extent or depth of differentiation, or differences between sub-continental versus sub-arc mantle.

In addition to the examples cited above there are along-strike variations in composition which apparently are not accompanied by differences in the crust traversed. One instance is the Antilles arc where the two least K-rich volcanoes (St. Kitts, Statia) lie near (but not at) the northern end of the volcanic front, the four most K-rich volcanoes (Grenada and others) lie at the southern end, and the remainder are intermediate both in composition and latitude (Brown et al. 1977).

$K_{57.5}$ values vary nonuniformly from 0.6 to 1.4 north to south along the volcanic front, although there is no accompanying change in depth to the dipping seismic zone or in crustal thickness. Rb, Ba, and LREE contents behave like K_2O whereas Sr and Zr contents are higher in the south but relatively constant elsewhere. $^{87}Sr/$ ^{86}Sr ratios rise and become more variable southward, passing 0.7040 between Guadeloupe and Dominica, and ϵ_{Nd} values are consistently MORB-like only in the north, being lower elsewhere (Section 5.5). South is of course the direction toward subcontinental versus suboceanic mantle, and toward thicker terrigenous sediment at the plate boundary. [Note that the southward transition from tholeiitic to calcalkaline suites reported by Brown et al. (1977) is a transition in LIL-element concentrations and not Fe-enrichment characteristics; most of their calcalkaline group is tholeiitic according to Fig. 1.4, although no members of either group display rapid Fe-enrichment or groundmass pigeonite.]

A similar example accompanied by even more dramatic isotopic changes occurs in the Banda arc, Indonesia (Jezek and Hutchison 1978; Magaritz et al. 1978; Whitford and Jezek 1979). The northern-most volcano, Banda Api, has erupted low-K tholeiitic dacites with groundmass pigeonite; volcanoes southward along the volcanic front have erupted medium-K (Manuk, Serua) to high-K (Nila, Tuen, Damar) calcalkaline andesites with groundmass hypsersthene. Rb, Cs, Ba, and Sr contents increase southward with K_2O. Isotopic compositions of Sr and O vary considerably along strike but without clear correlation either to each other (if Ambon is omitted) or to elemental concentrations. No significant difference along strike in either depth to the dipping seismic zone or thickness of crust is known.

A final example occurs along the western Bismarck arc, Papua New Guinea, where concentrations of alkalies, Ti, and P at a given silica content reach maxima at one point along the volcanic front (e.g., K_{55} ranges from 0.4 for Vokeo volcano in the west to 1.8 for Long volcano farther east); also, the volume of material erupted apparently increases as LIL-element concentrations increase (Johnson 1976a). Again, no difference in depth to the dipping seismic zone or in crustal thickness is known, although the seismic zone is poorly developed, dips vertically, and is absent altogether in the western half of the arc.

No satisfying explanation of these phenomena is available although they have been attributed tentatively to mantle heterogeneities (Brown et al. 1977), variations in subducted materials (Whitford and Jezek 1979), or varying rates of subduction (Johnson 1976a).

The final type of along-strike variations concerns "edge effects" which occur where convergent plate boundaries terminate at tear faults, orthogonal plate boundaries, or triple junctions (Sect. 2.1 and Fig. 2.4, points B' and G), or where arcs are segmented. At these terminations or segment boundaries cold lithosphere lies adjacent to, as well as under, unsubducted mantle and this situation seems reflected in volcanoes at the surface which are often alkalic (DeLong et al. 1975) and sometimes offset behind the volcanic front. Examples include Egmont, New Zealand; Samoa; the Willaumez Peninsula to Witu Islands, Papua New Guinea;

Sumbawa, Indonesia; Lokon-Empung, Sulawesi, Indonesia; Iki and Oki Dogo islands, southwestern Japan; the Central Depression, Kamchatka; the Anahim Belt of southern British Columbia; Mt. Lassen, USA; Tuxtla, Mexico; Irazu, Costa Rica; and Grenada, Antilles. Each of these volcanoes has erupted magmas richer in LIL elements (sometimes *ne*-normative alkali basalts) than those farther from the edge. In most cases these volcanoes also lie over deeper earthquake foci, thereby clouding the issue of why they are more alkalic, but in others (Sumbawa, Irazu, Grenada) this does not appear so. As noted in Section 2.1, abundant orogenic andesites are found only in those cases where volcanoes which overlie edges also occur on the overthrust plate (i.e., not in Samoa, Oki Dogo, or British Columbia). However, abundant orogenic andesites are not always associated with alkali basalts on the overthrust plate (e.g., Sumbawa, Tuxtla) nor are magmas erupted at edges always alkaline (e.g., Tafahi, Tonga; Matthew and Hunter, Vanuatu; Banda Api, Indonesia; Thule, S. Sandwich arc). Nevertheless, edge effects apparently exist and may account for some along-strike changes in alkalinity in volcanic arcs.

In summary, along-strike variations in composition occur and can be as large in magnitude as across-strike variations, but are far less consistent. Some can be attributed to variations in the thickness of crust encountered by ascending magma and others may result from lying over edges of subducted lithosphere. Some, however, are not understood and may reflect a heterogeneity of source materials or melting conditions which, in turn, may result from either local tectonic complexity or the past geochemical history of the source.

7.3 Effects of Plate Convergence Rate on Magma Composition

The location of andesitic stratovolcanoes with respect to the plate boundary is unrelated to plate convergence rates. There is no significant correlation between convergence rates and depths to the seismic zone beneath volcanic fronts (Sect. 3.5), nor do the data in Table 2.1 show a significant correlation between convergence rates and the dip of the seismic zone (also see Tovish and Schubert 1978). This conclusion differs from that of Luyendyk (1970), whose data set was appreciably smaller.

Nevertheless, convergence rates affect the thermal regime of both the slab and the mantle wedge, and the amount of water released per unit time by slab dehydration. Perhaps as a result, some aspects of andesite composition correlate with convergence rate. Sugisaki (1972, 1976) argued that the relative abundance and tholeiitic behavior of andesites increase while potash contents decrease with increasing convergence rates. Sugimura and Uyeda (1973, Table X) and Miyashiro (1974) also noted that tholeiitic series rocks increase in relative abundance with increasing convergence rates. Note that for Sugimura and Sugisaki, tholeiitic behavior refers to low alkalinity whereas for Miyashiro it means high FeO*/MgO ratios; none use it as did Wager and Deer, Osborn, or Kuno (Sect. 1.3).

Each of the above authors used averages or relative percentages of all analyses available to them for given arc segments. In Fig. 7.3 I have tested Sugisaki's proposed correlations with newer, more complete data. Note the lack of correlations in the now-available data. Although this negative result could be a consequence of how data are aggregated (see caveats in Sect. 4.9 concerning estimates of relative volumes), Fig. 7.3 provides little motivation for pursuing the issue further.

However, while Sugisaki's argument for linear correlation is weak, Miyashiro's (1974) grouping of arcs on the basis of convergence rate is significant and is given in modified form in Table 7.3. Most Group I suites are dominated by basalts or basic andesites, have low K-, Th-, REE-, and Ti-group trace element contents, experience rapid Fe-enrichment, have high K/Rb and low Th/U ratios, and have flat or light-depleted REE patterns; i.e., they are the island arc tholeiitic series of Jakes and Gill (1970). There are exceptions in the Marianas arc where one volcano, Sarigan, is calcalkaline, yet only it has the LIL-depleted characteristics described above. (Note that the Scotia arc convergence rate would be only 2 to 3 cm/yr if

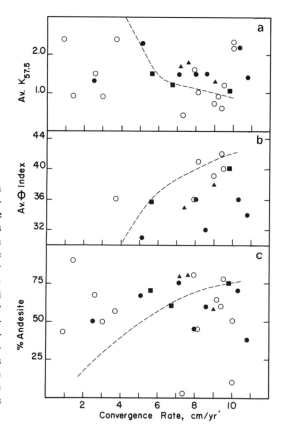

Fig. 7.3 a–c. Lack of correlation between aspects of andesite composition and plate convergence rate in active volcanic arcs. θ stands for tholeiitic. Data are from Tables 2.1 (average convergence rate per arc) and 7.1 except for θ indices and the *dashed correlation lines* which are from Sugisaki (1976, Figs. 2 and 3). *Symbols* are as in Fig. 7.1. **a** and **c** represent the average of values for individual volcanoes per arc; **b** represents the average of all analyses per arc. Sugisaki's lines were based on different convergence rates, as well as different data sets for **a** and **c**

Table 7.3. Volcanic arc classification by convergence rate, crustal thickness, and percent calc-alkalinity

Group I	Convergence rates >7 cm/yr and crust <20 km thick: $>80\%$ of volcanoes tholeiitic		
	Kermadec	Mariana	South Sandwich
	Tonga	Izu	
Group II	Convergence rates >7 cm/yr and crust 30–40 km thick: 33–70% of volcanoes are tholeiitic		
	New Britain	East Japan	Central America
	Vanuatu	Kuriles	Chile, $33°$–$47°$S
	Java	Kamchatka	
Group III	Convergence rates <7 cm/yr or crust >40 km thick: $<50\%$ of volcanoes are tholeiitic		
	New Zealand	Cascades	Chile, $>47°$S
	Sumatra	Mexico	Antilles
	Ryukus	Columbia-Ecuador	Aegean
	Alaska	Peru-Chile, $11°$–$28°$S	Turkey-Iran

ANT/SAM poles alone were used, but actually it is about 9 cm/yr due to the relative eastward movement of the small South Sandwich and Scotia plates; Forsyth 1975). The dipping seismic zone beneath the Scotia arc extends only to 200 km which shows that deep underthrusting is unnecessary for predominance of Group I suites, but all Group I arcs have actively spreading backarc basins behind them.

Group II subduction sites have similarly high convergence rates but include more chemically diverse magmas. Some volcanoes at the volcanic front of New Britain, Honshu, and Hokkaido erupt magmas similar in composition to those of Group I suites, but the similarity is to the "Northeast Japan type" (less depleted type) of island arc tholeiitic series (Masuda and Aoki 1978). This also seems true for the Kuriles and Kamchatka though confirming trace element data are absent. In each case except the central and southern Kuriles, greater chemical diversity at Group II subduction sites reflects both eruption of magmas behind the volcanic front and the eruption of magmas at the volcanic front which differ in composition from Group I suites. Higher potash contents or less iron-enrichment in some Group II magmas apparently reflects increased crustal thicknesses (Sect. 7.2).

Island arc tholeiitic series magmas mostly are absent from Group III subduction sites which usually are dominated by medium-K calcalkaline or slightly tholeiitic series. Slow convergence rates (<4 cm/year) are not accompanied by distinctive magma compositions.

Thus, generation of the island arc theoleiitic series requires convergence rates >7 cm/year, whereas intermediate and high-K tholeiitic or calcalkaline series can be generated at any convergence rate. Apparently a threshold sensitive only to rate of underthrusting must be reached to generate low-K tholeiitic andesites. The most obvious interpretation is that island arc tholeiitic series result from a relatively large degree of partial melting which is possible only when frictional heat at or near the slab wedge boundary is high. This interpretation is consistent with suggestions in Section 6.2 that tholeiitic andesites erupt at higher temperatures than calcalkaline andesites, but the suggestion remains unconfirmed due to inadequate mineralogic data and to different definitions of tholeiitic behavior. Group I and II subduction sites do not differ consistently in such characteristics as upper mantle seismic properties, depth of maximum earthquake foci, presence or absence of backarc, basins or low-Q regions, or heat flow patterns, according to available data.

Three arcs do not fit the Table 7.3 scheme. The Solomon and Aleutian arcs both meet the geologic requirements for Group I but 40% or more of their volcanoes are calcalkaline (according to Table 7.1). The Eolian arc should belong to Group III due to its low convergence rate, but most of its volcanoes are tholeiitic (according to Table 7.1). Both the scheme and these arcs need additional study.

7.4 Relationships Between Compositions of Orogenic Andesites and Adjacent Oceanic Crust

If orogenic andesites are derived entirely or partially from subducted crust, or derived from overlying mantle which has been metasomatized due to dehydration of subducted crust, then differences in the composition of crust subducted beneath various arcs should be reflected in the composition of their andesites. There are at least three variations on the theme of oceanic crust whose "normal" velocity structure is shown in Fig. 3.3, but only a major change in the age of subducted crust or in the influx of terrigenous sediment exert identifiable influences on the composition of magmas in adjacent arcs.

The first variable is the composition of the basaltic portion of oceanic crust (layers 2 and 3). Its pristine composition varies most in LIL-element concentrations, FeO*/MgO ratios and $^{87}Sr/^{86}Sr$ ratios (e.g., Melson et al. 1976; Wood et al. 1979b). Basalts at topographic highs on the seafloor, including but not restricted to ocean islands and seamounts, generally are enriched in LIL-elements, ^{87}Sr, and ^{206}Pb (i.e., are E-type MORB; Sect. 5.4) relative to normal (N-type) ocean floor basalts. As much as 1/4 of pristine MORBs may be of the E-type (Kay and Hubbard 1978), but not that much may get subducted because ocean islands or even topographic highs may instead be accreted (Sect. 2.4).

More significantly, ocean floor basalts vary in their degree of hydrothermal and low-temperature (halmyrolitic) alteration. Both processes increase the water content and $^{87}Sr/^{86}Sr$ ratio of basalt but other effects apparently depend on the temperature, pH, and f_{O_2} of the alteration. Hydrothermal activity, which is estimated to be currently affecting more than 1/3 of the seafloor including crust as old as 55 m.y. (Anderson et al. 1979), and which may sometime have affected as much as 60% of existing ocean crust (Anderson et al. 1980), generally redistributes elements within the crust, leaching here, precipitating there. Low-temperature alteration, on the other hand, results in net increases in the concentration of B, alkalies, Sr, Ba, the light REE except Ce, and U, and increased $^{18}O/^{16}O$ ratios. There is no consensus concerning the overall effect of these two processes on the bulk composition of oceanic crust, but one estimate thereof is provided in Table 8.1. That estimate, plus data from Hart and Nalwalk (1970), Dasch et al. (1973), Hart (1973, 1976), Spooner (1976), and Friedricksen and Hoernes (1978), suggests that ocean floor basalt altered sufficiently to contain 2 to 5 wt.% H_2O also would have $K_2O > 0.5\%$, LIL-element concentrations increased severalfold over values in pristine N- or E-type MORB, $^{87}Sr/^{86}Sr$ ratios around 0.7039, and $^{18}O/^{16}O$ ratios of +8 to 10‰. The isotopic composition of Pb and Nd is unaffected by this alteration so that altered ocean floor basalt remains on the PUM-line of Fig. 5.18 but is displaced to the right of the SNUM-line of Fig. 5.19. In such highly altered rocks, Ba/La ratios are 10 to 30 (Frey et al. 1974; Hart 1976), negative Ce anomalies up to 50% occur (Masuda and Nagasawa 1975), and $^{238}U/^{204}Pb$ ratios are > 8 (Tatsumoto 1978). These are crude empirical generalizations at best because changes in chemical composition due to alteration depend on the variables cited above plus water/rock ratios, and no systematic pattern has been identified.

While the two preceding paragraphs indicate the type and range of basalts which are being subducted, compositional differences between basaltic crust beneath different arc segments cannot be predicted because neither the regional extent of different MORB types nor the regional or geochemical extent of alteration are known. However, because low-temperature alteration increases with time, the chemical characteristics cited above should be most pronounced in old oceanic crust. The crust approaching convergent plate boundaries varies in age from Pliocene (off the Solomons, Cascades, Mexico, and southern Chile) to Jurassic (off the Mariana and central Sunda arcs). The median age of crust currently being subducted is 60 m.y. (Berger and Winterer 1974).

There are two tests of the effects of these age differences. First, the age of incoming crust can be compared with elemental or isotopic characteristics of erupted magmas for all arcs. I found no significant correlations of this kind except the follwoing. Arcs beneath which young crust of Pliocene age is being subducted have unusually high and constant Mg/Fe ratios (i.e., are strongly calcalkaline). (This might account for the anomalous calcalkalinity of the Solomons noted in the previous section.) Calcalkalinity arguably indicates smaller degrees of fusion (Sect. 12.3) which, in turn, may result from the underthrusting of hotter, less altered crust (Sect. 2.6). However, both effect and cause are conjectural at this point. The second test looks for analogous variations along single plate boundaries. The Chilean example, marked by an abrupt change at 47°S, may be the clearest example of magmatic differences related to age (or rate) of underthrust crust, but too few data have been published to judge.

Second, there are significant differences in the thickness and character of ocean floor sediments (layer 1) reaching convergent plate boundaries; see compilations by Scholl and Marlow (1974) and Lisitzin (1972). Surface sediments vary latitudinally. Pelagic carbonates predominate in surface sediments only on ocean crust approaching Chile. Siliceous sediments form mostly at equatorial and high latitudes but only dominate sediments approaching the South Sandwich arc. Pelagic clays constitute most of the surface sediments approaching most arcs and vary somewhat in composition as a function of provenance. For example, illite is the chief constituent of northern Pacific clays but kaolinite or montmorillonite predominate in clays reaching southern hemisphere arcs. Also, there is an eastward increase in the $^{206}Pb/^{204}Pb$ ratio of the northern Pacific surface clays (Sect. 5.5.2). Limestone (bottom), chert, pelagic clay, and hemipelagic-terrigenous mud (top) often underlie one another sequentially due to subsidence of the seafloor and its migration to a subduction site. Either the clay or limestone-chert mixtures make up most of layer 1 approaching most arcs; both usually occur in abundance. However, large volumes of terrigenous turbidites lie in or fill trenches adjacent to New Zealand, Sumatra, the Molucca Sea, southwest Japan, the Kamchatka corner, Alaska, the Cascades, central America, southern Chile, and the southern Antilles. Pelagic carbonates are uncommon or missing from pre-Cretaceous seafloor, and about 300 to 500 m of evaporites occur in sedimentary sequences approaching the Eolian and Aegean arcs (Hsü et al. 1977).

Whether these sediments are scraped off or consumed at subduction sites is much discussed but unresolved (cf. Karig and Sharman 1975, with Scholl et al. 1977). Accretionary forearc sediment prisms attributable to off-scraping are best developed where thick sections of terrigenous sediment occur at the subduction site (see the location listing above). However, mass balance calculations are insufficiently precise to show conclusively whether the influx of sediment can be accounted for in these accretionary prisms or whether, instead, sediments are partially consumed. Moreover, "erosional" forearcs also exist where old crust is exposed near the trench, including at least the Mexico, Chile, Mariana, and Tonga arcs, suggesting consumption rather than accretion of sediments plus cannibalism of these arcs.

Pelagic sediments are stratigraphically lowest on oceanic crust, physically the strongest, and the least represented in accretionary prisms (Moore 1975). They also are trapped within seafloor basalts (Garrison 1974; DSDP holes 332, 422, 442) or beneath later basalt sills (DSDP hole 462). Consequently, they are the sediments most likely to be subducted. Terrigenous sediments are stratigraphically higher, weaker, and dominate accretionary prisms. Consequently, they are less likely to be subducted.

Two aspects of andesite geochemistry correlate with variations in the sediment influx. First, the change in Pb isotopic composition of northern Pacific pelagic sediments cited above reappears in arc magmas although not as endpoints of observed mixing lines (Sect. 5.5.7). Second, magmas in the eastern Indonesian and

Mediterranean arcs have isotopic characteristics which indicate contamination by old, upper sialic crust without having ascended through such crust. Instead, subducted terrigenous sediments are the most proximal contaminants due to impending ARC-ATL collisions in those regions. More subtlely, recycling of terrigenous sediment also explains the Pb isotope arrays of the Aleutian and Cascades arcs, and may explain isotopic compositions in the southernmost Antilles (Sect. 5.5.7).

In general, however, correlations between arc magma compositions and sediment influx face two dilemmas. First, on geochemical grounds, terrigenous sediment is the type most likely to be subducted, whereas on geologic grounds it is least likely. Second and relatedly, some of the "erosional" arcs beneath which pelagic sediments demonstrably are being subducted (e.g., Tonga, Mariana) have magmas which show the least isotopic evidence of a sediment component. See Section 8.4 for a summary of arguments concerning the role of sediment subduction in andesite genesis.

The final variability in the composition of subducted oceanic crust concerns the overall or at least localized concentrations and mineralogy of transition metals which nonuniformly occur as oxide/hydroxide encrustations and as massive sulfide ores on the seafloor or within the ocean crust (Rona 1978). Metallic ores at convergent plate boundaries, such as porphyry copper deposits, may be inherited from such seafloor enrichments (e.g., Sillitoe 1972). While the source of metals in these arc deposits may never be known, in most cases the ores contain a [207]Pb-enriched component (Doe and Zartman 1979) as do many andesites, and sometimes (e.g., Java) this radiogenic Pb is more likely to have come from subducted than crustal-level rocks.

7.5 Changes in the Composition of Orogenic Andesites During Earth History

Secular changes within and between eruptions of andesite, on the one hand, and during the overall evolution of individual volcanoes on the other, were summarized in Sections 4.3 and 4.4; that distinction is arbitrary, for analytic purposes only. Similarly, a distinction between adjacent, often overlapping volcanoes of different age is also arbitrary, although at least it introduces comparisons between lavas erupted from different vents and, therefore, through different conduit systems. Such longer-term secular variations in the composition of orogenic andesites can be seen through three time windows: between successive volcanoes at one site over $< 10^7$ years; during the evolution of individual arc segments over 10^7 to 10^8 years; and during the evolution of the Earth since the Archean.

Differences in LIL-element contents, isotope ratios, and Fe-enrichment trends occur between andesites erupted from successive volcanoes at one place. Often LIL-element contents relative to silica increase with time between successive volcanoes (e.g., Aramaki 1963; Cole 1978) while the degree of Fe-enrichment decreases.

However, there are many exceptions. Unambiguous instances of decreased LIL-element contents relative to silica with time have occurred at Mts. Hood, Jefferson, and Shasta in the Cascades (Wise 1969; White and McBirney 1979; Christiansen et al. 1977). A similar change occurred between the Morne Jacob and Pelée cones, Martinique, Antilles (Gunn et al. 1974). A change from calcalkaline to tholeiitic behavior with time occurred between (not within) Kanaton and the younger, now-active Kanaga volcano, Aleutians (Coats 1952), and at Santorini, Aegean arc (Nicholls 1971a). Finally, the Myoko volcano group, central Honshu, Japan, illustrates a decrease with time by about 0.001 in $^{87}Sr/^{86}Sr$ ratios between cones of Pleistocene age (Ishizaka et al. 1977). Thus, as noted in Section 4.4., the secular changes in composition within and between rocks of overlapping andesite stratovolcanoes are sufficient that pooling of samples acquired in a random or even reconnaissance fashion often is unlikely to yield a consanguineous collection. Moreover, there is no consistent pattern of secular change between overlapping cones; changes, when present, seem random.

Whether similar randomness continues throughout a period of uninterrupted subduction at a convergent plate boundary (i.e., over 10^7 to 10^8 years) is even more difficult to establish. A consistent secular change in volcanic arcs from tholeiitic and low-K rocks to calcalkaline and medium or high-K and finally to alkalic or shoshonitic ones with decreasing age has been claimed (Jakeš and White 1969, 1972a; Gill 1970; Jakeš and Gill 1970) and subsequently often cited but rarely verified in entirety. The tholeiitic or calcalkaline to alkaline or shoshonitic transition has occurred in many Tertiary arcs (e.g., Fiji, New Guinea Highlands, Kamchatka, Greater Antilles, Eolian arc, western USA, Mexico, Eastern Europe, Turkey) but, in most cases, it is a response to cessation of subduction and, therefore, to transition from compressional to extensional stress orientation (Sect. 2.3).

Assessment of whether Fe-enrichment characteristically decreases with time or whether LIL-element concentrations increase within tholeiitic and calcalkaline suites is handicapped by at least three problems. First, the duration of uninterrupted subduction ("congruent development" according to Dickinson 1973) rarely is clear from the geologic record. Estimates for most currently active arcs are given in Table 2.1 but involve interpretations which are often complicated by volcanic episodicity which may reflect varying subduction rates, ridge-trench collisions, or changes of a plate boundary from convergent to transform in type. About 1/3 of active convergent boundaries have maintained their current geometry of subduction since the Mesozoic; others only since the Pliocene.

The second problem, a sub-set of the first, is ambiguity about the provenance of tholeiitic basalt-dominated but andesite-containing basement in areas such as Papua (Milsom and Smith 1975) and Central America (Goossens et al. 1977). These basements tentatively have been interpreted as relict ocean floor, or islands, but they have some arc-like characteristics such as moderate LIL-element concentrations and a moderate percentage of intermediate rocks.

The third problem results from migration of volcanic arcs with time, usually away from the plate boundary (James 1971; Dickinson 1973). That is, active volcanoes lie farther from the plate boundary than do volcanic rocks which apparently were erupted earlier during the same period of congruent development in the Tonga, Mariana, Aleutian, Cascades, Chilean, and Antilles arcs. Likewise, belts of plutonic rocks decrease in age away from the coast in the Alaska, British Columbia, and Sierra Nevada batholiths. Typical rates of migration are 1 to 2 km/m.y. rearward, although the process appears episodic. Oscillatory migration appears to have occurred in the anomalously wide western USA arc during the Tertiary (Coney and Reynolds 1977) and may have begun in South America since the Pliocene (Vergara and Munizaga 1974; Kussmaul et al. 1977). All of these migrations result in time-transgressive changes in composition within volcanic arcs which are attributable to across-arc differences in magma composition (Sect. 7.1) rather than to time-dependent processes at one place.

Despite these caveats, secular changes in the composition of orogenic andesites have been documented in Fiji (Gill 1970), the Mariana-Bonin arc (Shiraki et al. 1978), the Greater Antilles (Donnelly et al. 1971, 1980), and the central Cascades (Loeschke 1979; White and McBirney 1979). In the first three instances, the lowest stratigraphic units belong principally to the island arc tholeiitic series (have low LIL-element concentrations plus rapid Fe-enrichment; i.e., conform to the definition of Jakeš and Gill 1970), whereas overlying units are medium-K and calcalkaline. The same situation also may characterize the Aleutian arc where island arc tholeiitic series rocks occur in at least the Vega Bay Fm. basement of Kiska island (H. Bowman, unpubl. data). Each of these four instances occurs in ensimatic arcs probably constructed on oceanic crust.

There are, of course, exceptions even in the arcs cited above. In the central Cascades, which may also be ensimatic, the Eocene to Oligocene basement is tholeiitic and there are upward increases in Sr and Zr contents plus K/Rb ratios, but island arc tholeiitic series rocks are absent and alkali plus REE contents decrease upwards. In the Marianas, boninites (i.e., LIL-element poor calcakaline rocks) are interbedded with the tholeiites (Meijer 1980).

Unless the tholeiitic rocks of Papua or Central America cited above are of arc versus ocean floor origin, I found no other examples of consistent secular change within active volcanic arcs which could not be attributed to rearward migration of volcanism. Indeed, the up-section changes within and between overlapping cones dicussed above and in Section 4.4 clearly demonstrate some randomness of evolution. Thus, I conclude that consistently increasing LIL-element concentrations plus decreasing Fe-enrichment with time occur only during initial thickening of crust and then only if the convergence rate is sufficient for initial development of an island arc tholeiitic series (Sect. 7.3).

This conclusion gains significance from the repeated discoveries of consistent up-section increases of LIL-element contents relative to silica and changes from tholeiitic to calcalkaline behavior, in Archean volcanic sequences (see below).

Apparently, therefore, some recently active volcanic arcs have re-enacted events especially common during early development of the Earth's crust. Also, to the extent that some andesite stratovolcanoes evolve from tholeiitic to calcalkaline behavior (Sect. 4.4), ontogeny recapitulates phylogeny even amongst andesites.

Orogenic andesites have been erupted on Earth for at least 2.8 billion years. Numerous examples occur throughout the Phanerozoic, usually attributable to association with a convergent plate boundary (Chap. 2). Fewer, though still many, examples of Proterozoic andesites are known and in some areas (e.g., the ~ 1800 m.y. Amisk Group, Churchill Province, Canada or the coeval Yavapai Series, Arizona, USA) they constitute up to 80% of sections 4 to 9 km thick which are associated with regional metamorphism and deformation, and plutonism (Moore 1977; Anderson and Silver 1976). Some little-metamorphosed orogenic andesites also occur in Proterozoic basalt plateaux but they are volumetrically minor (e.g., Annells 1974).

Because volumetric predominance of orogenic andesite today almost always accompanies subduction, much attention has focused on Archean andesites in order to evaluate whether or not subduction occurred then also. Orogenic andesites, approximately as defined in Section 1.1., constitute from 1 to 75 vol.% of Archean greenstone belts (Condie 1976), including about 33% of the Abitibi Belt of Canada, the largest known (Goodwin 1977). Typically, the relative volume of andesite increases up-section, as does the ratio of calcalkaline to tholeiitic series members; orogenic andesite usually is the most common rock type in the calc-alkaline upper portions (Goodwin 1977, Table II).

Few mineralogic details of Archean andesites have been published and often both mineralogy and composition have been modified by metamorphism. Average analyses of andesite from some 2.7 b.y. old Canadian, Australian, and African Archean suites are given in Table 7.4, including data from the Abitibi and Bula-wayan Groups,, each claimed to be the least metamorphosed of any Archean suite yet described. In general, low to high-K, calcalkaline and tholeiitic types are all represented (also see summary by Condie 1976). Ni and Cr contents tend to be higher in Archean than modern andesites but are within the scatter of data in Fig. 5.14. S contents also are higher in Archean andesites (Naldrett et al. 1978). Rb, Zr, and light REE concentrations usually are higher, relative to K_2O, in Archean andesites, whereas heavy REE contents sometimes are uncommonly low (Table 7.4). Consequently, K/Rb ratios typically are lower but La/Yb ratios are higher in Archean andesites. The spread in initial $^{87}Sr/^{86}Sr$ ratios is similar to that in modern andesites, some ratios being consistent with Archean upper mantle values (~ 0.7010 to 0.7015) while others, including those of high-K suites enriched in LIL—elements, are higher (e.g., Hallberg et al. 1976). Initial Nd isotopes in Bulawayan basic andesites were similar to those in the more voluminous under-lying basalts, and are consistent with derivation from a source with a chondritic Sm/Nd ratio (Hamilton et al. 1977).

Table 7.4. Chemical composition of Precambrian andesites (anhydrous basis)

	1	2	3	4	5
N	582	15	–	15	–
SiO_2	58.3	57.1	58.9	62.2	62.1
TiO_2	0.95	0.95	0.60	0.79	0.61
Al_2O_3	16.3	16.4	15.7	15.0	16.8
FeO*	8.0	9.3	7.3	7.0	5.0
MnO	0.17	0.11	–	0.11	–
MgO	4.2	4.2	6.4	3.6	2.8
CaO	6.1	7.0	7.2	5.6	4.4
Na_2O	3.8	4.2	3.0	3.7	4.7
K_2O	1.16	0.83	0.88	1.7	1.4
P_2O_5	0.22	–	–	0.23	–
Rb	30	–	25	49	27
Sr	242	–	250	325	360
Ba	358	–	237	755	–
Zr	145	–	78	204	140
Y	17	–	16	22	26
La	26	–	10	34	21
Yb	4.0	–	1.2	1.6	1.2
Ni	61	115	52	68	–
Cr	111	70	77	93	50
$(^{87}Sr/^{86}Sr)_i$	–	0.701	0.7010	0.7029	0.702

References:
1. Average for entire Superior Province, Canada, 2750 m.y. (Goodwin 1977, Table IV; Condie and Harrison 1976)
2. Average calcalkaline andesite, Abitibi Belt, Canada, 2750 m.y. (Jolly 1977)
3. Average andesite, Maliyami Fm., Bulawayan Group, Zimbabwe, 2720 m.y. (Condie and Harrison 1976; Hawkesworth and O'Nions 1977)
4. Average andesite, Marda Complex, Yilgarn Block, Australia, 2635 m.y. (Hallberg et al. 1976; Taylor and Hallberg 1978)
5. Average Proterozoic andesite, Grenville Province, Canada, 1300 m.y. (Condie and Moore 1977)

  Two generalizations can be drawn from the above observations. First, orogenic andesites have formed throughout at least the last 2/3 of Earth history and have not changed dramatically in mineralogy, elemental or isotopic composition or range of variability thereof, or relative volume during that time. That is, there is as much variation among coeval rocks as between andesites of different ages. However, there may be subtle differences in composition between Archean and modern examples which are related to secular changes of Earth differentiation, although both metamorphism and sampling statistics leave the issue unresolved. That is, many Archean andesites have slightly higher concentrations of S, Rb, Zr, light REE, Ni, and Cr, and some Archean andesites originated from a source not yet depleted in light REE. If true, then the andesite source is likely to have been the upper mantle, which was not by then as differentiated.

Second, orogenic andesites have formed in at least three types of rock associations through time (adapted from Condie 1976; Hallberg et al. 1976). The first includes predominantly bimodal basalt-rhyolite suites in which andesites occur but are uncommon. Archean examples include most of the oldest rocks known (Barker and Peterman 1974); Proterozoic and younger examples include continental flood basalt provinces. The second environment has yielded multi-modal basalt to rhyolite sequences which sometimes become more felsic and less Fe-enriched up-section with calcalkaline andesite being the most common rock type in upper units. Initial $^{87}Sr/^{86}Sr$ ratios are consistent with upper mantle derivation but insensitive to small amounts of sialic recycling. Younger Archean examples include several in the Canadian Superior Province and the Maliyami Fm. (Bulawayan Group), Zimbabwe; Proterozoic and younger terranes may be analogs of modern island arcs where a similar up-section sequence sometimes occurs (see above). The third environment has yielded LIL-element enriched andesites whose intitial $^{87}Sr/^{86}Sr$ ratios exceed the most common upper mantle values and which, therefore, may be derived from atypical mantle or may have incorporated pre-existing sialic crust. Archean examples include the Felsic Fm. (Bulawayan Group), Zimbabwe, and the Marda Complex, Australia. Modern analogs are convergent plate boundaries along old or thick continental margins or fragments.

Chapter 8   The Role of Subducted Ocean Crust in the Genesis of Orogenic Andesite

Derivation of orogenic andesites by partial fusion of ocean crust subducted beneath volcanic arcs was first suggested by Coats (1962). Today the idea commonly is asserted in introductory geology texts and assumed by many, including some of those constructing geophysical models of convergent plate boundaries. The rapid acceptance of this oversimplified belief was due in part to the coincidence of four events in the mid-1960's: (1) the 1968 "Andesite Conference" in Oregon, USA, which emphasized evidence that orogenic andesites originated neither by fractional crystallization nor crustal contamination of basalt; (2) discovery that clinopyroxene and garnet are co-liquidus phases of anhydrous basic and acid andesite at high pressure so that these magmas could be partial melts of basaltic eclogite (Green and Ringwood 1968a); (3) unambiguous demonstration that lithosphere, including ocean crust, is subducted at convergent plate boundaries, thereby providing a constantly replenished supply of basaltic eclogite beneath volcanic arcs (Isacks et al. 1968); and (4) recognition of correlation between magma composition and depth to the dipping seismic zone (Dickinson 1968). Because derivation of orogenic andesites from subducted crust would be their most dramatic link with plate tectonics, this genetic hypothesis is examined first. It is a rather feline hypothesis, having many lives despite repeated death threats from critics.

8.1 Characteristics of Subducted Ocean Crust Beneath Volcanic Arcs

Oceanic crust consists of mafic rocks formed at spreading centers (layers 2 and 3), sediments deposited thereon during transit to a subduction site (layer 1), and any volcanics acquired in off-ridge environments during transit. A typical crustal section is given in Fig. 3.3. There are significant variations in the chemical composition of all three layers which depend mostly on age or latitude; these are summarized in Section 7.4.

   Most of oceanic crust is basaltic in composition. Its suitability as a parent for orogenic andesites depends largely on its extent of alteration. Although estimates of the average composition of layers 2 and 3 are of questionable value, I have included in Table 8.1 the average of analyses of all basaltic rocks from the first 29 legs of the Deep Sea Drilling Project. Note that these basaltic rocks contain 3.7 ± 2.2 wt.% H_2O and recall Anderson et al.'s (1980) estimate that some 60% of layers 2 and 3 were altered to this extent. Consequently, the elemental and

Table 8.1. Chemical compositions of some experimentally produced melts, and of altered ocean floor basalt

	Eclogite melts		Peridotite melts			Altered MORB		
	1	2	3	4	5	6		
SiO_2	62.3	48.8	58.4	64.4	46.6	47.8	Rb	13
TiO_2	–	–	1.0	0.1	0.7	1.7	Sr	264
Al_2O_3	22.7	25.0	20.4	20.6	13.4	15.3	Ba	126
Fe_2O_3	–	–	–	–	–	5.0	Th	1.4
FeO	2.3	8.4	4.5	0.8	7.4	5.7	U	0.4
MgO	0.2	2.9	2.4	3.2	15.0	6.6	La	4.9
CaO	7.9	11.7	9.8	9.4	13.2	9.6	Yb	3.0
Na_2O	3.6	2.5	3.3	1.4	0.9	2.8	Y	38
K_2O	0.7	0.3	0.1	0.1	0.1	0.87	Zr	141
P_2O_5	–	–	–	–	–	0.27	Nb	14
H_2O	–	–	–	–	–	3.7	Hf	2.4
							Ni	107
							Co	50
							Cr	224
							V	296
							Sc	47
							Ba/La	26
							La/Th	3.5
							La/Nb	0.4

References:
1. Stern and Wyllie (1978, Table 8). Calculated liquid at $900°C$, 30 kb from 20% melt of basalt + 5% H_2O
2. As above, 40% melt at $1000°C$
3. Mysen and Boettcher (1975, Table 11). Probed glass formed at $1100°C$, 15 kb when $X^{Vapor}_{H_2O} \sim 0.6$ (peridotite + 14% H_2O). Glass coexisted with ol, opx, cpx, and sp (region IIIP, Fig. 9.1)
4. Mysen and Boettcher (1975, Table 9). Probed glass formed at $1080°C$, 17 kb when $X^{Vapor}_{H_2O} = 1.0$ (peridotite + 17.5% H_2O). Glass coexisted with ol, opx, cpx, amph, and sp (region IP, Fig. 9.1).
5. Scarfe et al. (1979). Same peridotite source as for col. 4. Probed glass formed at $1425°C$, 20 kb from 5% fusion of dry peridotite. Glass coexisted with ol, opx, cpx, and sp (region IVP, Fig. 9.1)
6. Hart (1976). Average of all DSDP basement samples, Legs 1 to 29

isotopic characteristics of extensively altered ocean floor basalts which were summarized in Section 7.4 and Table 8.1 provide current best estimates of the "typical" composition of basaltic rocks being subducted. Ocean crust is, of course, more mafic than this basaltic average since the crust includes gabbroic cumulates. Overall the crust may approach a picritic composition.

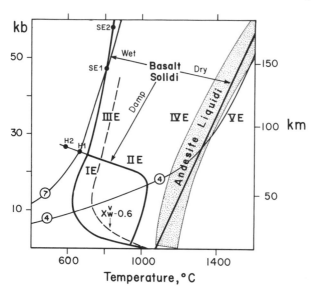

Fig. 8.1. P-T diagram for subducted oceanic crust. The *heavy lines* show representative solidi. Results for water-saturated *(wet)* and dry conditions are from Lambert and Wyllie (1972) and Green and Ringwood (1968a), respectively, for quartz-tholeiites; results for olivine-tholeiites lie at higher temperatures. Because experimentally determined amphibolite solidi in the absence of vapor are unavailable, the vapor-absent *(damp)* solidus follows Burnham's (1979) calculations. The *dashed line* solidus for melting in the presence of a mixed vapor where $X_{H_2O}^{Vapor} \sim 0.6$ follows Holloway and Burnham (1972) and Burnham (1979). Regions *IE* to *VE* refer to melting conditions discussed in the text. The *stippled field* contains liquidus temperatures from Fig. 4.1 for andesites with 0 to 2 wt.% H_2O. *Lines 4* and *7* are possible thermal gradients within subducted ocean crust from Fig. 3.7. *Lettered points* are keyed to Fig. 8.2

The mineralogy of ocean crust varies during subduction as a function of pressure, temperature, and vapor phase composition, moving from zeolite, through blueschist or amphibolite, to eclogite facies assemblages. This prograde metamorphism is accompanied by dehydration which, upon completion, releases the 2 to 5 wt.% H_2O initially present in altered ocean crust plus water from underlying serpentinized peridotite. Dehydration of the crust is expected to occur principally between 80 and 125 km depths over a wide temperature range, although biotite and 14 Å-chlorite may persist deeper (Anderson et al. 1976, 1978; Delany and Helgeson 1978). Water contained in underlying ultramafic rocks is released when first serpentine and then talc dehydrate at temperatures above about 600° and 700°C, respectively (Fig. 9.1). Dense hydrated magnesian silicates such as clinohumite apparently are stable to depths up to 450 km but require temperatures < 700° at > 200 km depths in order to form (Ringwood 1975; Yamamoto and Akimoto 1977; Delany and Helgeson 1978). Consequently, geothermal gradients as low as line 5 in Fig. 3.7, or deep initial serpentinization of sub-oceanic mantle, are necessary for these phases to play a significant role.

The load pressure within subducted crust beneath volcanic arcs can be estimated more confidently than the temperature. Ocean crust apparently is 110 to 135 km vertically beneath volcanic fronts on average (Sect. 3.4) and therefore at pressure of 33 to 38 kb.

Estimates of temperature within ocean crust beneath volcanic arcs vary more (Fig. 3.7), ranging from 150°C (Bird 1977) to 1350°C (Griggs 1972) at 30 kb, with the principal uncertainties being the magnitude of shear stress, the amount of heat absorbed by endothermic dehydration reactions, and the effects of convection in the wedge (Sect. 3.4). As a result of the variations in estimated geothermal gradients, one cannot predict confidently what will melt or where melting occur during subduction. In principle, any one of the following four possibilities can occur beneath volcanic arcs: (1) both slab and wedge melt (see geotherms 1 to 4, Fig. 3.7; as a result the mantle wedge is expected to be configured as an Fig. 3.5a or b); (2) neither slab nor wedge melts (geotherm 5, Fig. 3.7); both asthenosphere and andesitic volcanism are therefore absent as in parts of Peru and Chile, Sect. 2.1); (3) the slab dehydrates but the wedge alone melts (geotherm 6, Fig. 3.7; wedge configured as in Fig. 3.5a or b); or (4) the slab alone melts (wedge configured as in Fig. 3.5c).

Although one cannot a priori rule out possibilities 1, 3, or 4 beneath volcanic arcs, in principle one can distinguish between 1 and 3 on the one hand versus 4 on the other by mapping Q structures (Sect. 3.3), and distinguish between all three by geochemical arguments. (In practice, however, Q structures are difficult to determine and the geochemical characteristics of slab-derived fluids in possibilities 3 and 4 merge). One can also specify conditions which must be met for each possibility to be realized. Fig. 8.2 together with similar ones by Wyllie (1979) will help to clarify these specifications. Fig. 8.2 shows the spatial distribution of the dehydration and melting reactions which are shown in Figs. 8.1 and 9.1. It assumes a low geothermal gradient of 2.4°C/km below 60 km depth along the slab/wedge boundary (line 7, Fig. 3.7), following Anderson et al. (1978). Choice of higher gradients rotates lines around P, causes lines SE1–SE2 and SG1–SG2 to overlap, and moves SE1–SE2 toward or even above H1–H2 (see Wyllie 1979, Figs. 16A, 17A). Note that during slab dehydration fluids are released due to amphibole breakdown (at H1–H2) primarily beneath the forearc, not volcanic arc, regardless of geothermal gradient.

Possibility 1 above (slab + wedge fusion) occurs under the widest variety of conditions, ranging from high geothermal gradients within nearly-dry eclogite (region IVE, Fig. 8.1) to relatively low gradients within water-saturated eclogite (Wyllie 1979, Fig. 17B; region IIIE, Fig. 8.1). Note, however, in Fig. 8.2 that this possibility does not occur *beneath the volcanic arc* because temperatures there are too low even for fusion of water-saturated eclogite. Obviously, low gradients also preclude fusion of nearly-dry eclogite beneath volcanic arcs.

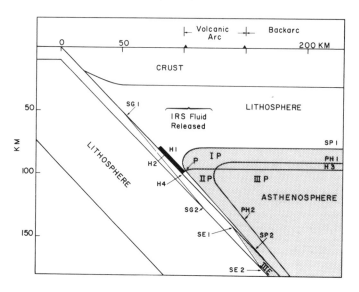

Fig. 8.2. Schematic cross-section of a convergent plate boundary showing melting regions. Compare with other similar cross-sections in Figs. 2.3, 3.5, and 3.6. Also compare with figures in Wyllie (1979) plus P-T diagrams, Figs. 8.1 and 9.1, to which the labeled points are keyed. The geotherm implied here is line 7 on Fig. 3.7. *Lines SG1–SG2* and *H1–H2* represent the serpentine and amphibole dehydration reactions in subducted peridotite and ocean crust, respectively, Aqueous solutions (*IRS* fluids) are released by these reactions 80 to 125 km from the plate boundary for the 45° dip geometry shown. The *heavy bar* represents the depth region beneath Honshu of high seismic b-values which arguably indicate the principal location of these solutions (Anderson et al. 1980). *Line SP1–SP2* is the water-saturated peridotite solidus, and *lines H3–H4* and *PH1–PH2* are the stability limits of amphibole and phlogopite in peridotite. The *stippled region* shows the distribution of partial melt in the presence of pure water vapor. *Regions IIIE* and *IP* to *IIIP* refer to melting conditions discussed in the text

Possibility 3 (wedge-only fusion) is impossible assuming most published geothermal gradients (Wyllie 1979) and requires a configuration such as illustrated in Fig. 8.2, but even so is problematical. If geothermal gradients are low enough to suppress slab fusion, then overlying peridotite will melt beneath the volcanic front only in the presence of a relatively CO_2-free vapor, and only if convection within the wedge causes very high thermal gradients *adjacent to* (not along) the slab-wedge boundary. For example, the configuration in Fig. 8.2 requires about a 100°C/km gradient between points P and H4! Further discussion of this possibility is deferred to Section 9.3.

Possibility 4 (slab-only fusion) requires high geothermal gradients within hydrous but not necessarily water-saturated eclogite (Wyllie 1979, Figs. 16B, 17A; regions I to IIIE, Fig. 8.1). However, the 1000° to 1100°C temperatures of pyroxene basic andesite magmas cannot be reached within the slab unless the overlying wedge also melts. Consequently, slab-only fusion is unlikely to yield melts with the composition of orogenic andesites.

The consistency with which volcanic fronts overlie ocean crust 110 to 135 km deep, and the lack of correlation between this depth and convergence rates, suggest that melting results more from depth-dependent reactions than temperature-dependent phenomena such as strain-heating (Sect. 3.5). That is, if subducted ocean floor basalt were vapor-saturated, then melting could occur at any depth along the wet (vapor-saturated) solidi in Fig. 8.1, depending on the geothermal gradient. Since thermal gradients increase as convergence rates increase, there is no mechanism for keeping the depth of melting consistent. Instead, this depth must reflect a pressure which is specified for one of two reasons. First, it could be a pressure at which there is a rapid drop in solidus temperatures for either water-deficient basalt or peridotite. Indeed, such a drop occurs between 25 and 35 kb for both due to the restricted stability ranges of amphibole and carbonate minerals (see "damp solidi" in Figs. 8.1 and 9.1). Alternatively, it could be a pressure at which sufficient water consistently is added to cause vapor-saturation and consequent sharp reduction of solidus temperature, or to cause the percent of interstitial melt to exceed the holding capacity of the rock fabric. Indeed, slab dehydration and therefore water release occurs in the 80 to 125 km depth interval (Figs. 8.1 and 9.1).

Consequently, both explanations of the consistent volcanic front locations above the dipping seismic zone are viable. The first is necessary if slab dehydration followed by wedge-only fusion proves untenable. It does not, however, demonstrate whether arc magmas are solely slab-derived or, instead, high-pressure mixes of melts from both slab and wedge. The second explanation is preferred if wedge-only fusion works, i.e., if dehydration chills the slab as much as is shown by geotherms 5 to 7 in Fig. 3.7.

8.2 Circumstantial Evidence of Slab Recycling in Arc Volcanism

Many observations in the preceding chapters circumstantially imply that subducted oceanic crust is recycled in orogenic andesites. The fourth possibility raised in the preceding section, namely that orogenic andesites are primary melts of subducted oceanic crust, is the strongest form of the recycling argument. Alternatively, slab recycling also could mean that slab-derived basaltic magma differentiates to andesite during ascent, that slab-derived siliceous magma mixes with the low-melting fraction of the wedge to yield andesite, or that an aqueous fluid carrying various elements in solution is exhaled from the slab during dehydration.

The strongest argument for some kine of recycling is the presence, during or not long before volcanism, of ocean crust 70 km or more beneath almost all areas in which orogenic andesites predominate (Chap. 2). Where subducted lithosphere has been absent or too shallow, or when subduction is prevented by a collision, orogenic andesites are absent or subordinate. After subduction ceases, predominance of orogenic andesite also ceases within 1 to 5 m.y. Moreover, andesites predominate only within a portion of the region at convergent plate boundaries which is characterized by volcanism, high heat flow and conductivity, low P and S wave velocities, and high attenuation of seismic waves; i.e., there is (by definition) a distinction between volcanic arc and backarc regions in terms of relative volumes of magma types. Specifically, predominantly andesitic volcanism occurs only above that portion of the dipping seismic zone which is about 80 to 250 km deep.

Several lines of evidence, all of them ambiguous to interpret, indicate that the slab melts or dehydrates as well as unbends within this depth range. First, the angle of dip appears to increase beneath some volcanic arcs whereas the number of earthquakes per unit area and time, and the b-values, appear to decrease (Sect. 3.4). Second, a sharp decrease in seismic velocities (about 6% within a few km) has been inferred to occur just above the slab/wedge boundary beneath some volcanic arcs and backarcs (Sect. 3.4). Both observations suggest the presence of a fluid (magma or vapor) above the boundary. The observations do not yet distinguish whether such fluid underlies forearcs, volcanic arcs, backarcs, or all three. Third, the along-strike spacing of volcanoes within volcanic arcs crudely approximates depth to the dipping seismic zone, possibly suggesting that this also is the depth from which magma begins ascent (Sect. 4.6). Fourth, earthquakes which arguably are precursors to eruptions of andesite originate at 100 to 200 km depths within the dipping seismic zone and ascend at km/day rates, implying the ascent of something (if only stress) from the slab even though the physical significance of these ascent rates remains unknown (Sect. 3.5).

Slab recycling is also suggested by andesite geochemistry in three ways. First and least ambiguous are the arc magmas which appear from their isotopic compositions to have been contaminated by old upper sialic crust without having ascended through it, but beneath which such crust has been subducted in the form of terrigenous sediment (Sect. 7.4). This is most obvious prior to ARC-ATL collisions, but also occurs where island arcs terminate near continents.

Second and more universal is the geochemical distinctiveness of volcanism at convergent plate boundaries. That is, the volatile contents and ratios, the relative enrichment of K-group and Th-group elements but depletion of Ti-group elements with respect to the REE, and the frequent enrichments in $^{87}Sr/^{86}Sr$ relative to $^{143}Nd/^{144}Nd$ plus $^{207}Pb/^{204}Pb$ or $^{208}Pb/^{204}Pb$ relative to $^{206}Pb/^{204}Pb$ ratios of magmas at convergent plate boundaries are, in varying degrees, distinguishable from characteristics of magmas in other tectonic environments (Sect. 5.7).

Whether this distinctiveness reflects slab recycling is less clear. Qualitatively, the increased concentrations of H_2O, Cl, K-group and Th-group trace elements, and increased $^{87}Sr/^{86}Sr$ ratios suggest inheritance from altered oceanic crust (see Table 8.1) or subducted sediment; the increased $^{207}Pb/^{204}Pb$ ratios suggest subducted terrigenous sediment from old sial. Decreased relative concentrations of the Ti-group elements may reflect either their greater compatibility (higher partition coefficients) in ordinary silicate mineral/melt or mineral/vapor equilibria, or the existence of a refractory titaniferous phase such as rutile in either the slab or wedge or both. However, the strength of this argument for recycling depends on its quantitative persuasiveness: can mass balance calculations simultaneously explain all the distinctive characteristics identified? Despite several recent attempts (e.g., Doe and Zartman 1979; Armstrong 1980; Kay 1980), solutions remain nonunique, weakly constrained and geochemically incomplete. However, most of these characteristics are common to orogenic magmas erupted through crusts which differ widely in thickness, composition, and age, but are not shared with anorogenic andesites erupted through continental crust. Therefore they are not fundamentally due to crustal contamination during ascent, although this process may have similar results which cannot be distinguished yet from slab recycling in continental settings.

Finally, there would be an even more powerful third argument if the consistent across-arc changes in andesite compositions (Sect. 7.1) were controlled by equilibria with refractory minerals within subducted crust. However, this hypothesis is contentious at best. Consider two examples. First, the increase in K_2O contents and decrease in K/Rb ratios away from convergent plate boundaries were attributed by Jakeš and White (1970) and Fitton (1971) to a change with increasing pressure from amphibole to phlogopite as the K-bearing phase in subducted crust, and by Beswick (1976) to successive fusions of phlogopite eclogite. However, neither explanation is valid. Assuming the initial K/Rb ratio of the material being fused remains constant, liquids in equilibrium with refractory phlogopite (corresponding to magmas erupted above deeper earthquake foci) will have K/Rb ratios higher than those in equilibrium with refractory amphibole, using the distribution coefficients of Table 6.3; this prediction is opposite to the pattern observed (Table 7.2). For phlogopite equilibrium alone to explain the observations, phlogopite must remain refractory during successive partial melting episodes each sufficient to yield andesite and, indeed, must constitute a larger percentage of the residue at large degrees of fusion than at small ones (Beswick 1976, Table 1). Both are unlikely, especially if degrees of fusion are 20% to 40%.

As a second example, the increase of K_2O in andesites away from convergent plate boundaries was attributed by Marsh and Carmichael (1974) to buffering of K_2O contents by a melting reaction involving sanidine; for this reaction, an increase in pressure results in an increase of K_2O in the magma. However, the reaction predicts that SiO_2 like K_2O increases across volcanic arcs which is incorrect (Sect. 7.1), and also requires sanidine to be a refractory phase in eclogite after 20% to 40% fusion whereas sanidine is uncommon even in subsolidus eclogite assemblages (see Eggler and McCallum 1974, for the only known example). Consequently, while slab equilibria may influence across-arc geochemical variations, no internally consistent model of such influence has been developed as yet and the variations cannot therefore be used as evidence of slab-derivation of andesites.

Finally, remember that all of the circumstantial evidence cited above implies only that subducted lithosphere underlies the site of volcanism *which is dominated by* orogenic andesite and, even then, underlies that site *either before or during* the time of volcanism. Andesitic volcanism per se has occurred without the benefit of underlying subducted lithosphere in several places (Sect. 2.8 and 5.6) and even andesite-dominated volcanism arguably has occurred near subduction sites but when not underlain by subducted lithosphere (Sect. 2.9). Although both reminders indicate that orogenic andesites are not (always) primary melts of subducted lithosphere, they do not undercut arguments for recycling in their weakened forms.

8.3 Are Orogenic Andesites Primary Melts of Subducted Ocean Floor Basalt? No

If shear heating is sufficient to cause melting in addition to dehydration of subducted oceanic crust (geotherms 7 and 1 to 4, Fig. 3.7), then orogenic andesites could be primary (undifferentiated) partial melts of basalt, or differentiates thereof. This fusion could occur in any of the five P-T regions identified in Fig. 8.1. All such primary melts must have temperatures, chemical compositions, and liquidus phases consistent with equilibrium with subducted crust. The following discussion shows that few if any orogenic andesites or basalts meet these requirements, thereby indicating that slab recycling is indirect.

Fusion at pressures $<$ 25 kb is least likely to be applicable because subducted crust beneath volcanic arcs usually is under load pressures $\geqslant$ 33 kb (Fig. 3.7). Nevertheless, melts could be in equilibrium with amphibole beneath volcanic arcs (region IE, Fig. 8.1) if *all* of the following conditions are met: subducted crust is (anomalously) as shallow as 85 km or less beneath volcanic arcs; amphiboles within metamorphosed oceanic crust are sufficiently rich in Mg or F (Holloway and Ford 1975; Boettcher 1977) or there is a vapor phase sufficiently rich in CO_2 (Holloway 1973) for amphibole to remain stable at pressures $>$ 25 kb; and average thermal gradients exceed 8°C/km.

Amphibole coexists with small volumes of melt for over 100° of the melting interval above the water-saturated solidus at $<$ 25 kb (region IE, Fig. 8.1), and for a smaller but unknown temperature interval above the vapor-absent solidus (region IIE). Melts within region IE are lower in temperature and more siliceous than andesite (Wyllie 1977). Also they are water-saturated near the solidus (about 22 wt.% H_2O at 25 kb), and therefore unable to ascend without fractionating or freezing, and unlikely to precipitate plagioclase as a liquidus phase at low pressure. Clearly these melts are not the orogenic andesites we seek.

Characteristics of melts near the solidus in region IIE, where beginning of melting in the absence of vapor is associated with the breakdown of amphibole, are less predictable but at least could have temperatures up to 1050°C and be water-undersaturated (5 to 22 wt.%). They might be similar to some orogenic andesites (Boettcher 1973, 1977; Burnham 1979). Burnham (1979) guessed that melts in region IIE range from trondhjemitic compositions for 5% melts near 700°C to andesitic compositions for 40% melts near 1050°C. However, the water contents and temperatures of region IIE melts apply only to hornblende acid andesites (Sect. 6.11), and even the latter lack amphibole as a liquidus phase in water-undersaturated conditions when $X_{H_2O}^{Vapor} < 0.75$ (Allen and Boettcher 1978), thereby precluding their formation at the region IIE solidus. (Melts in region IIE which are far enough above the solidus not to coexist with amphibole will be analogous to melts of regions IVE and VE, discussed below.)

The presence of refractory amphibole would, however, make partial melts in region IIE more similar in composition to orogenic andesites than in its absence. Melting in which the residuum is only amphibole ± pyroxene will not yield trace element, especially REE, concentrations similar to those of most orogenic andesites (Ewart and Bryan 1973; Lopez-Escobar et al. 1976), but such models are unrealistic because at pressures above about 15 kb garnet will be a refractory phase. In contrast, melting of garnet amphibolite in which garnet is about 10% of the residuum will produce liquids without either convave-up REE patterns (due to equilibrium with amphibole) or excessive heavy REE depletion (due to equilibrium with garnet), and will better match the La/Sm ratios and Sc contents of orogenic andesites, depending on distribution coefficients. However, equilibrium with amphibole will exacerbate the Ba/La ratio problem, and does nothing to resolve the other trace element and isotopic dilemmas enumerated below.

Consequently, it remains for experimental petrologists to show that near-solidus melts in region IIE have the bulk composition (including water contents) of some orogenic andesites; the odds seem slim. And, although melts in region IIE should resemble orogenic andesites in REE contents more clsely than do melts formed at higher pressure, significant mismatch of trace element concentrations and isotope ratios nevertheless occurs. Thus, it is not likely that orogenic andesites are primary melts from region IIE even if the unlikely physical conditions stipulated above are met.

The pressures of regions III, IV, and VE are more likely to be applicable to andesite genesis; consequently, fusion and crystallization of basalt and andesite under these conditions have been studied experimentally by Green and Ringwood (1968a), Green (1972), and Stern and Wyllie (1978). If any hydrous vapor is present, melting will begin in region IIIE with the solidus temperature increasing as $X_{H_2O}^{Vapor}$ decreases, as shown schematically in Fig. 8.1 for $X_{H_2O}^{Vapor} \sim 0.6$. Near the solidus, these melts, like those in region IE, are too siliceous, hydrous, and low in temperature to be orogenic andesites. However, as temperatures and melt fractions increase through the melting interval, liquids become more andesitic and eventually basaltic as the liquidus is approached. Calculations indicate that melts having andesitic compositions (see Table 8.1) are produced by 20% to 40% fusion between 900° and 1000°C at 30 kb if the subsolidus assemblage contained 5 wt.%

water (Gill 1974; Stern and Wyllie 1978). These andesitic melts coexist with garnet + clinopyroxene ± kyanite in proportions estimated to be 1:1.5:0 to 1:2:0.2. Although the melt and residual mineral proportions are approximations only, and although the particular melts of Stern and Wyllie (1978) also are too low in temperature to be typical orogenic andesites, the calculations illustrate that andesite genesis by eclogite fusion is possible, provided that garnet constitutes 25% to 40% of the residue. Furthermore, basic and acid andesites with 0 to 4 wt.% water all have both garnet and clinopyroxene as co-liquidus phases, as required (Stern and Wyllie 1978).

Note, however, that under vapor-present conditions there is nothing special about andesite; it is merely one of a wide range of compositions produced within the melting interval. Moreover, the Ca/(Mg + Fe) ratios of the melts in Stern and Wyllie's experiments, especially the melts containing > 53% SiO_2, differ significantly and consistently from the Ca/(Mg + Fe) ratios (and concentrations of these elements) in magmas erupted in volcanic arcs (Fig. 8.3). Qualitatively, sphene would help eliminate this discrepancy if refractory (Green 1972, 1980); however, over 25% of the residuum would have to be sphene to overcome Stern and Wyllie's (1978) objection (based on data for 20% fusion of basalt in their Table 8).

In contrast, it has been claimed that acid andesite is the eutectic-like melt of dry basaltic eclogite in region VE (Green and Ringwood 1968a; Green 1972; Marsh and Carmichael 1974), in a way analogous to granite being the minimum-temperature melt of pelitic sediments within the crust. The claim is based on acid andesite having a liquidus temperature ($\sim$ 1400°C) which is lower than for basalt or rhyolite at 30 kb in the absence of water, and similar to the dry solidus temperature of quartz tholeiite (Fig. 4.1). Partial melting of dry quartz eclogite through approximately a 30° interval is expected to yield acid andesite (point j' in Fig. 8.3) until coesite disappears; subsequent equilibrium melting (but not fractional melting) through the melting interval is expected to yield acid andesite to basalt with increasing temperature.

However, three observations preclude this dry melting hypothesis. First, for quartz eclogite having the bulk composition of ocean floor basalt, the eutectic-like melt at 30 kb has a Ca/(Mg + Fe) ratio too high for orogenic andesites (j' in Fig. 8.3). Second, although clinopyroxene is a liquidus phase of dry basic to acid andesite above 25 kb, the composition of liquidus clinopyroxene in acid andesite differs significantly from the composition of clinopyroxenes within the melting interval of dry basalt (Green and Ringwood 1968a; Stern and Wyllie 1978), thus precluding equilibrium between andesite and the basalt residue. Third, temperatures around 1400°C beneath volcanic arcs require unrealistically high shear stresses (> 10 kb; Anderson et al. 1978).

Arguments based on thermodynamic properties of melts and residua are inconclusive. Marsh and Carmichael (1974) demonstrated that a basic andesite could be in equilibrium with eclogite at 1100° to 1600°C beneath volcanic arcs. However, their calculations assumed refractory sanidine, and the andesite chosen

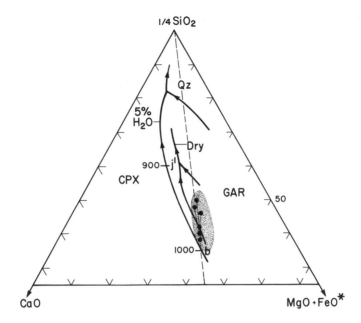

Fig. 8.3. Liquidus phase relations at high pressure in the pseudoternary $1/4SiO_2$ + CaO + (MgO + FeO*), after Stern and Wyllie (1978, Fig. 15). Hydrous results are for 30 kb, anhydrous results for 27 kb. *Arrows* point down-temperature. The *numbered tickmarks* along the garnet-pyroxene "cotectic" are temperature estimates which bracket likely andesitic liquids, based on Stern and Wyllie's Table 8. The *dashed line* is their "average calc-akaline trend". The *joined dots* are the average low, medium, and high-K orogenic andesites of Table 5.1 and the *stippled field* encloses the representative suites of Fig. 5.1 to 5.3. Note the constant CaO/(MgO + FeO*) ratios of orogenic andesites, in contrast to the decrease in this ratio with increasing temperature for liquids in equilibrium with garnet + clinopyroxene. Point *j'* is the projection of the three-phase cotectic, i.e., the putative dry minimum melt composition of andesitic composition (Green and Ringwood 1969a). Point *b* is composition 2 of Table 8.1

was one of the quartz basalts with a high Mg-number noted in Section 5.2.3. Calculations using a similar quartz basalt from Utah gave analogous results (Hausel and Nash 1977), but calculations with more typical orogenic andesites have precluded equilibrium with eclogite (Heming 1977).

It altered ocean floor basalt contains 0.9% K_2O (see Table 8.1), and < 0.4% H_2O after being metamorphosed to 30 kb, then it could contain up to 0.8% mica without water vapor; 14 Å-chlorite also may persist to these depths. Fusion of such mica-or chlorite-eclogite in the absence of vapor (region IVE), is analogous to fusion of vapor-free amphibolite at lower pressure (region IIE). Although neither temperatures nor compositions of melts in region IVE have been determined experimentally, both could be appropriate for orogenic andesites.

Melts in regions III to VE and above the thermal stability of amphibole in region IIE will co-exist primarily with garnet and clinopyroxene ± kyanite, coesite, and mica (or sanidine), and may co-exist with accessory phases such as rutile, sphene, apatite, sulfides, or minerals of the humite or perovskite groups. [Experimental reports of accessories during eclogite fusion are given by Green (1972, 1980), Merrill and Wyllie (1975), and Stern and Wyllie (1978).] Several studies have shown that melts of ocean floor basalt coexisting solely with a garnet plus clinopyroxene residuum do not have trace element concentrations similar to those of orogenic andesites (DeLong 1974; Gill 1974, 1978; Lopez-Escobar et al. 1976, 1977; Thorpe et al. 1976; Dostal et al. 1977b; Kay 1977). Many discrepancies exist which are merely summarized below; details are given in the papers just cited.

First, high-K orogenic andesites have K, Rb, Ba, and Th contents too high for 20% to 40% melts of even altered ocean floor basalt with a composition such as given in Table 8.1 (see Fig. 5.17b,c). For low- and medium-K suites these discrepancies are not as severe as maintained by Armstrong (1971) and others, provided that a large enough percentage of oceanic crust is altered significantly before subduction. However, such extensive alteration would also produce O, He, and Th isotopic compositions which differ significantly from those observed in orogenic andesites, assuming that the crust being subducted is 60 m.y. old on average (Sect. 5.5.7 and 7.4). Second, Sc contents of most orogenic andesites are too high (> 10 ppm) for the andesites to be melts of eclogite unless $D_{Sc}^{gar} > 8$. [Ni contents in orogenic andesites have been considered inexplicably low if due to equilibrium with eclogite (e.g., Gill 1974; DeLong 1974), but pyroxene and garnet distribution coefficients are high enough for refractory eclogite sometimes to be an adequate Ni sink (Gill 1978; Leeman and Lindstrom 1978), thus removing this generalized objection; however, see Fig. 5.17b,c for an example where it still applies.] Relatedly, Cr/Ni ratios are < 1 in eclogite melts but > 1 in orogenic andesites (Fig. 5.17d). Third, Sr concentrations are too constant, whereas Cr, V, and sometimes Ni concentrations within andesite suites are too variable, to be explained by garnet + clinopyroxene residua; see Fig. 5.17b,c for an example.

Fourth, light REE contents are too low in medium and low-K orogenic andesites (< 50 times chondritic abundances), whereas heavy REE and Y contents are too high (> 5 times chondritic) in almost all orogenic andesites (Fig. 5.17a). Melts in equilibrium with eclogite could have heavy REE contents around 10 times chondritic if the weight fraction of subsolidus garnet is < 0.3, the percent fusion is > 0.3, and D_{HREE}^{gar} values are 5 to 10. While not impossible, such a fortuitous combination seems unlikely. Consider the D_{HREE}^{gar} values, for example. While experimentally determined coefficients are as low as 6 to 9 for garnets coexisting with basic andesite liquids in region IVE, values appropriate to acid andesites in the same region are about 20 to 40 (Nicholls and Harris 1980). Moreover, eclogite fusion produces a negative correlation between Yb or Y and SiO_2 in the absence of refractory quartz and kyanite, especially if D_{HREE}^{gar} values decrease during fusion along the "cotectic" in Fig. 8.3. Such negative correlations are uncommon in suites containing andesite.

Fifth, Ti, Zr, Hf, and Nb concentrations are lower in orogenic andesites than ocean floor basalts although neither D^{cpx} nor D^{gar} are expected to be above 1 for these elements (Table 6.3). Finally, eclogite equilibria provide no mechanism by which to explain the across-arc geochemical variations summarized in Section 7.1.

Qualitatively, refractory accessory minerals, especially rutile, biotite, and sphene, might explain some of these discrepancies (Gill 1974; Marsh 1976a; Green 1980), although there is little experimental evidence that these minerals coexist with andesitic melt in regions III to VE. Indeed, sphene has been shown to be unstable there, and rutile stability also may be limited to regions I and IIE (Hellman and Green 1979). The quantitative influence of accessories on trace element concentrations cannot be judged yet due to lack of information about their modal abundance within the melting inverval and their distribution coefficients, although the inability of biotite (phlogopite) to account for across-arc variations in K/Rb ratios was noted in Section 8.2. If they are refractory, some of these minerals provide attractive explanations for the depletion of Ti-group elements (e.g., the high La/Nb ratios) in arc magmas.

The isotopic composition of Sr and Pb in orogenic andesites generally is permissive but inconclusive regarding whether or not subducted ocean crust is a viable source, although these isotopes usually indicate that such crust was altered or contained a sedimentary component, or both. However, the Sr isotopic similarity between orogenic andesites and silica-undersaturated basalts in some arcs implies that the wedge, not subducted oceanic crust, is the source of all magmas in at least those arcs (Sect. 9.4). Nd and especially Th isotopic compositions of orogenic andesites are incompatible with derivation from either fresh or altered ocean floor basalts (Sect. 5.5.7) and, as noted above, the isotopic compositions of O and He are incompatible with their derivation from highly altered ocean floor basalts.

While the conclusion of the preceding discussion is that most magmas erupted in volcanic arcs are not primary melts of altered and then subducted ocean floor basalts, some magmas may be such melts or differentiates thereof. Such primary partial melts are likely to be low- to medium-K in character with a La/Yb ratio > 20, < 1 ppm Yb, < 15 ppm Y, < 10 ppm Sc, possibly a negative Ce anomaly, and the following isotopic characteristics: $^{238}U/^{204}Pb > 8$, $^{18}O/^{16}O > 8$, $^{3}He/^{4}He < 7$ times atmospheric, $\epsilon_{Nd} \geqslant +9$ but $^{87}Sr/^{86}Sr$ displaced to the right of the SNUM line in Fig. 5.19, and initial $^{230}Th/^{232}Th$ activity ratios > 2. Primary acid andesites should have CaO > (MgO + FeO*). A few rocks meet some of these requirements (Condie and Swenson 1973; Lopez-Escobar et al. 1977; Kay 1978) but none meet most. Ironically, these few candidates also have high Mg-numbers and Ni contents and therefore are more likely to have been derived from (or at least to have interacted with) peridotite than from basaltic eclogite.

Alternatively, subducted oceanic crust, especially if picritic, may be the source of basalts which differentiate to andesite upon ascent. Composition b in Fig. 8.3 is an example of such a basalt produced by 40% fusion. However, of the requirements cited above, only the Yb and Y content ones will be removed low-pressure crystal fractionation so that, again, few if any examples of this possible process are known.

8.4 The Sediment Solution

The chemical composition of orogenic andesites is more closely matched by partial melts of subducted oceanic crust containing a few percent of sediment than by partial melts of sediment-free basalt. This closer match (the "sediment solution") occurs both because illite-rich sediments have high concentrations of LIL-elements, notably Ba and Pb, and because terrigenous sediments have the isotopic characteristics of their provenance. Consequently, partial melts of basalt sediment mixtures have higher Ba/La and lower U/Pb ratios, higher $^{87}Sr/^{86}Sr$ and $^{207}Pb/^{204}Pb$ ratios, and lower $^{143}Nd/^{144}Nd$ ratios than do melts from altered ocean floor basalt alone. Indeed, earlier in this chapter I used the capability of sediment involvement to account for the geochemical distinctiveness of arc magmas as a circumstantial argument for slab recycling in arc magma genesis.

Evidence for the sediment solution inherently is difficult to distinguish from evidence for crustal contamination of magmas during their ascent. Consequently, the examples of arc magmas which have some of these distinctive characteristics but have not risen through continental crust provide the least ambiguous evidence of sediment incorporation during magma genesis (Sect. 7.4). The Banda arc, Indonesia, is the best-documented example (Magaritz et al. 1978; Whitford et al. 1979b) but similar arguments apply to Pb isotopic compositions in the Aleutian and Antilles arcs, Sr compositions in the Eolian, Aegean, and southernmost Antilles arcs, and Nd compositions in all oceanic arcs exept Scotia (Sect. 5.5.7).

Geochemical evidence for a sediment solution also comes from models which offer explanations of global geochemical budgets (e.g., Fyfe 1978; Armstrong 1980; Kay 1980). These models assume that sediment involvement is indirect, i.e., via wedge metasomatism or mixing as discussed in Section 9.4.

Quantitative arguments for sediment involvement in the genesis of specific rocks are usually geochemically incomplete, highly unconstrained, or unsuccessful. For example, high Sr/Nd, like high Ba/La ratios are among the distinctive geochemical features of arc magmas (Sect. 5.7). Basalts and andesites from the Mariana and New Britain arcs share these characteristics and yet have Sr and Nd isotopic compositions which show little evidence of a sediment input. The same will be true if Irazu, Costa Rica, andesites have anomalously high Th/La ratios despite their mantle-like Th activity ratios. Similar quantitative problems also arise in budgets of Pb, U, and Th contents as well as Pb isotopic compositions (following section).

Indeed, only a few of the trace element and isotopic arguments against slab-only derivation of orogenic andesites presented in the preceding section (i.e., only

Ba/La and U/Pb ratios and Nd, Pb, and Th isotopic compositions) are resolved by invoking the sediment solution.

Fusion of silicate sediments will occur at or just below the region I to IIIE solidus of Fig. 8.1 if they are water-saturated, or in regions II to IVE if undersaturated. Subducted carbonates, even in the presence of cherts, are refractory except in region VE (Wyllie 1977). CO_2 produced by decarbonation reactions may form carbonate scapolites, thus maintaining high $X_{H_2O}^{vapor}$ (Burnham 1979).

For all of the above reasons, orogenic andesites seem little more likely to be derived from partial melts of altered ocean floor basalt plus sediment than of the basalt alone. Consequently, the sediment solution is indirect, e.g., via recycling in the wedge (Sect. 9.4).

Sediments constitute at most a few percent of the mass of the source rock for orogenic andesites or their parent magmas, although sediments may contribute a larger percentage of some elements (e.g., Ba, Th, Pb). The limiting percentage depends on the compostion of the sediment, whose LIL-element concentrations, negative Eu anomaly, and $^{18}O/^{16}O$ ratio provide the chief constraints. Terrigenous sediments are more likely than pelagic ones to be involved for three reasons. First, the sediment effect is observed only in arcs with high terrigenous/pelagic sediment influx ratios. Second, Pb isotope arrays from specific volcanoes or arcs have terrigenous, not pelagic, sediments as a mixing component. Third, apart from regional variations in $^{206}Pb/^{204}Pb$ ratios, none of the spatial or temporal differences in the pelagic sediment influx to arcs is reflected in the compostion of their magmas. Consequently, even though terrigenous sediments are less likely to be subducted than pelagics, they are the ones incorporated into the mantle wedge by thrusting or fluid transport.

Possible exceptions involving more direct incorporation of subducted sediment in magma genesis are the examples of forearc volcanism within accretionary prisms cited in Section 2.6. These rocks have the appropriate isotopic and peraluminous features, but are rarely andesitic.

8.5 IRS Fluids and Maxwell's Demons

Two conclusions of this chapter are that altered ocean floor basalts and sometimes sediments are recycled in orogenic andesite magmas but that most, perhaps all, of these magmas are neither primary partial melts of basalt ± sediment nor differentiates thereof. Consequently, recycling is indirect, involving mixing within the mantle wedge. If both slab and wedge melt (possibility 1, Sect. 8.1), then melts with the characteristics summarized in the two previous sections will be mixing components as discussed in Section 9.4. But what if wedge-only fusion occurs beneath volcanic arcs, as illustrated in Fig. 8.2?

Slab dehydration across lines H1–H2 and SG1–SG2 in Fig. 8.1, 8.2, and 9.1 can release copious volumes of water, For example, dehydration of crust alone will release 3 to 9 X 10^9 gm of H_2O per year gobally if 50 X 10^{-6} km^3/yr of crust with a density of 3 gm/cm^3 is subducted and releases 2 to 6 wt.% H_2O.

Assuming that some carbonate minerals and pore fluids are present until these dehydration fronts are reached, the composition of the fluid which is released will lie within the H_2O-CO_2-Cl system. At high pressures and temperatures, such fluids are effective solvents. Some 20 to 40 wt.% SiO_2 dissolves in water in equilibrium with orthopyroxene ± olivine at 15 kb, although less dissolves in mixed H_2O-CO_2 fluids (Nakamura and Kushiro 1974). Elements with high volatility or water-solubility such as K_2O and other alkalies, Pb, and U also will be dissolved (e.g., Stern and Wyllie 1978, re. potash). Even elements as refractory as the REE partition strongly into the fluid relative to eclogite solids (Mysen 1979), an effect which is enhanced by increased Cl activity in the fluid (Flynn and Burnham 1978).

Consequently, the fluid released during slab dehydration will be rich in *incompatible* trace elements, in *radiogenic* Sr and He and possibly Nd and Pb, and in *silica*. I therefore call this an IRS fluid, which is a peculiarly American acronym because it denotes an agency which "extracts under pressure" or "recycles to achieve a more even distribution" depending on one's political perspective.

The properties of IRS fluid are conveniently unknown, making tests of its influence or even existence difficult to construct. Hence the Maxwell's demon metaphor is invoked, after Bowen (1928). For example, if IRS fluid is sufficiently oxidizing to explain the relatively high f_{O_2} of arc magmas, then perhaps it should dissolve more U than Pb, and especially more U than Th, as would be true at crustal pressures. Indeed, (^{230}Th/^{238}U) ratios $<$ 1 in arc rocks may indicate source enrichment in U shortly before magma genesis (Sect. 5.5.5). This can be tested by comparing U/Th and U/Pb ratios in orogenic andesites with the time-averaged ratios of their source regions which are implied by their Pb isotopic compositions. While there are differences between observed and implied ratios, no type of orogenic andesite has higher U/Th and higher U/Pb ratios than did its time-averaged source (Sect. 5.4.3 and 5.5.2).

Similarly, decreasing Ba/La and ^{87}Sr/^{86}Sr ratios away from the volcanic front in some volcanic arcs (Sect. 7.1) may result from release of IRS fluid principally beneath the volcanic front, but in the previous section I noted the enigma of anomalously high Sr/Nd ratios in arc basalts which have Sr and Nd isotopic compositions on the SNUM line. Obviously, to be persuasive, these sorts of geochemical arguments must be more quantitative than is now possible.

Finally, the extent of mobility of IRS fluid is disputed because the mechanism of migration is unknown. Movement by diffusion through solids is slow, whereas movement by hydrofracturing or infiltration along grain boundaries is rapid (cf. Marsh 1976b, with Fyfe and McBirney 1975, or Mysen et al. 1978a).

8.6 Conclusions

Subducted ocean crust is involved in the genesis of orogenic andesites at convergent plate boundaries, but the nature and time scale of involvement remain unclear after a decade of active research. Indeed, the geochemical distinctiveness of magmas erupted within volcanic arcs seems to be the result of slab recycling, but it is difficult to distinguish slab from crustal-level contributions where both are possible.

Recycling could occur through dehydration or fusion of subducted crust. Dehydration occurs beneath forearcs and volcanic fronts; fusion could occur anywhere from forearc to backarc or not at all, depending on the geothermal gradient. However, melts with the temperature of basic pyroxene andesites cannot be produced within the slab without the overlying wedge also melting.

The composition of anatectic melt from altered, then subducted, ocean floor basalt varies geographically as a function of the age (hence alteration) and proportion of N-type versus E-type MORB being subducted. It also varies between regions I and VE of Fig. 8.1., but in none of these regions are the melts likely to resemble the composition of orogenic andesites in detail. This conclusion is weakened only slightly by considering partial melts of basalt plus sediment. Therefore, slab recycling occurs during mixing in the mantle wedge.

Differences between anatectic melts on the one hand, and IRS fluids (i.e., fluids enriched in incompatible trace elements, radiogenic isotopes, and silica) released during dehydration on the other, are much less at pressures beneath volcanic arcs than at crustal pressure. For example, silicate melts at the slab/wedge boundary beneath volcanic arcs can contain 20 to 30 wt.% H_2O, whereas IRS fluids there can contain 20 to 40wt.% SiO_2. This makes the distinction between whether the slab melts or dehydrates less important for arc magma genesis as well as more difficult to determine.

While orogenic andesites are not primary melts of subducted oceanic crust, the question of whether the weight fraction of slab-derived fluid in arc magmas is dominant or trivial remains open. The two most viable ideas involving large weight fractions are (1) mixing of slab-derived rhyolite to andesite with wedge-derived basalt, and (2) fusion of subducted picrite (MORB + cumulates) plus terrigenous sediment to yield basalt, followed by minor mixing of basalts in the wedge and differentiation during ascent.

Chapter 9 The Role of the Mantle Wedge

The upper mantle is considered parental to most terrestrial magmas with basalt being the typical offspring. However, Poldevaart (1955) and subsequently others proposed that the addition of water, such as that lost from subducted ocean crust during dehydration, would cause the progeny to become, instead, orogenic andesite. This proposal, like that discussed in the preceding chapter, provides a clear link between plate tectonics and andesite genesis, yet is oversimplified at best.

9.1 Characteristics of the Mantle Wedge

The mantle wedge is laterally and vertically heterogeneous in physical properties at individual convergent plate boundaries (Fig. 2.3) and the heterogeneity may differ from one plate boundary to another (Sect. 3.3). The most dramatic lateral change apparently occurs beneath volcanic fronts (Fig. 3.5a) such that mantle underlying both volcanic arcs and backarcs is characterized by high heat flow and conductivity, by low P and S wave velocities and Q factors, and by low density. These features probably indicate the presence of a few percent of interconnected fluid within the asthenosphere beneath both volcanic arcs and backarcs. However, this interpretation is equivocal (Goetze 1977) and, even if correct, cannot distinguish partial melt from hydrous fluid. Although widespread pyroxenite beneath the volcanic arc would also satisfy observed P wave velocities (Matsushima and Akeni 1977), hot peridotite ± fluid satisfactorily explains all the geophysical characteristics of asthenosphere within the mantle wedge.

The thermal structure of the wedge was discussed in Section 8.1 and illustrated in Figs. 2.3, 3.6, and 8.2. The discussion concluded that the mere existence of melts at 1100°C in the volcanic arc essentially requires fusion of the wedge ± slab. However, for wedge-*only* melting to occur beneath volcanic arcs (e.g., Fig. 8.2), IRS fluid must migrate kilometers into the wedge which, in turn, must convect across the vapor-saturated peridotite solidus (SP1−P−SP2 in Figs. 8.2 and 9.1) before melting occurs. This is so because the mantle into which the fluid rises is too cold to melt and must be moved to a hotter environment before melting can occur.

This question of whether or not the wedge convects is very important. Convection is necessary for wedge-only fusion to occur, as noted above. However, it also is required if the wedge is even the principal source of arc magmas, as argued later. This requirement follows from the $\sim$ 10 km^3/m.y. per km of plate boundary eruption rate in arcs (Sect. 4.8) which would deplete an entire subarc wedge (e.g., Fig. 2.3) in a few ten's of million years.

Ascending convection limbs beneath actively spreading backarc basin axes are generally accepted (Karig 1971; Andrews and Sleep 1974; Hsui and Toksoz 1979). However, the fluid dynamic situation in the subvolcanic arc region is ambiguous, especially in an evolving convergent zone. I will assume that subvolcanic arc mantle is replenished by convection throughout the shaded region of Fig. 8.2., and that depleted mantle is mixed back into the larger convecting mantle reservoir. Justification includes arguments for subbackarc convection in the articles cited above, plus the estimates of low wedge viscosity (Sect. 3.3). Note that my assumption requires magma ascent beneath volcanic arcs in a direction *opposite* to that of mantle convection. If true, this would contribute to slower magma ascent rates.

Some samples of the mantle wege before, during, or after subduction are available as ultramafic xenoliths in volcanic rocks. Few such xenoliths occur in andesites or other rocks erupted within volcanic arcs (Sect. 6.8). More commonly, the xenoliths are samples of the wedge beneath backarcs (e.g., Japan, Aoki 1968; New Zealand, Rodgers et al. 1975; Aleutians, Francis 1976), or samples of mantle beneath areas which previously had been a volcanic arc until subduction ceased (Sect. 2.3). Although no systematic comparison has been made of xenoliths from active backarcs or previously active volcanic arcs with xenoliths from elsewhere, three observations are nevertheless pertinent.

First, the mantle beneath volcanic arcs may be more hydrous and less hot than the mantle beneath backarcs or elsewhere. For example, inclusions in Quaternary orogenic andesites and alkali basalts of Itinomegata volcano (see Sect. 6.8) which is located near the volcanic arc-backarc boundary of northern Honshu, Japan, characteristically contain minor pargasitic hornblende and equilibrated at temperatures 100° to 300° lower than did the characteristically amphibole-free but otherwise similar inclusions in alkali basalts of the south-western Japan backarc (Takahashi 1978).

Second, xenoliths from some backarcs and former volcanic arcs typically contain evidence of metasomatic modfication by fluid which could have been derived from subducted lithosphere. Such evidence includes pyroxenite dikes (Wilshire and Pike 1975), interstitial amphibole (Francis 1976), high LIL-element contents in otherwise refractory assemblages (Frey and Prinz, 1978), and the isotopic systematics of Sr and Nd (Menzies and Murthy 1980). However, third and probably most important, ultramafic xenoliths from presently or formerly active volcanic arcs or backarcs seem similar in character, diversity, and relative proportions of rock types, to xenoliths from eleswhere. The low equilibration temperatures, amphiboles, and anomalous LIL-element concentrations noted above, are found elsewhere as well. Elsewhere, metasomatism is attributed to reaction with partial melt ascending from asthenosphere, not from subducted lithosphere.

Thus, studies of xenolithic samples of the mantle wedge have not demonstrated that the wedge differs significantly, consistently, or permanently from the upper mantle elsewhere. This conclusion is reinforced by the eruption of high-Ti alkali basalts in former volcanic arcs after the cessation of subduction (Sect. 2.3 and 2.4), which implies that a significant portion of the wedge was not permanently modified to a composition that yields andesite and related magma upon partial melting, However, during subduction, the mantle beneath volcanic arcs apparently differs consistently from the mantle beneath backarcs in both trace element and isotopic characteristics. This conclusion stems from the contrast summarized in Table 5.4 and applies either to cases where backarc spreading is present or absent. Consequently, the distinctive geochemistry of arc magmas (Sect. 5.7) apparently is inherited from a wedge whose distinctiveness is restricted both temporally and spatially.

In addition to the spatial and temporal heterogeneities just mentioned, the mantle wedge probably varies beneath different volcanic arcs, and especially beneath island arcs and continental interiors. Upper mantle beneath continents differs from that beneath ocean basins by being richer in LIL-elements (including Rb, U, and light REE) such that $^{87}Sr/^{86}Sr$ ratios may be $\geqslant 0.7047$, ϵ_{Nd} approach 0, and Pb isotopic compositions lie above the PUM line of Fig. 5.18 (Sect. 5.5.7). This may be true even if, perhaps ironically because, sub-continental mantle has been more depleted in its basaltic component (Jordan 1978). LIL-element-rich sub-continental mantle has been inferred to occur beneath the South American and New Zealand volcanic arcs (Brooks et al. 1976a), and may be expected beneath other continental margins, peninsulas, and fragments as well. Upper mantle beneath island arcs may include *both* the LIL-element-depleted type which feeds mid-ocean ridges, and the less depleted type which feeds ocean islands (see Gill 1976c; Masuda and Aoki 1978), but arc isotopic traits (i.e., $\epsilon_{Nd} < 9$ but $^{206}Pb^{204}Pb < 19$) are typical of neither mantle source.

9.2 Circumstantial Evidence that Arc Magmas Originate Within the Mantle Wedge

In addition to the thermal arguments presented in Section 8.1, two other observations suggest that partial melting in the mantle wedge is more important during andesite genesis than is fusion of subducted lithosphere. First, even if lithosphere is subducted to depths > 70 km, active volcanism is absent unless the mantle wedge contains asthenosphere (Sect. 2.1). This is anomalous if conduits of magma ascend from subducted lithosphere. Second, although orogenic andesites usually are accompanied by subduction, this is not necessary (Sects. 2.1, 2.4, and 2.8). Differences in composition between andesites at subduction sites and elsewhere can be explained by differences in the composition or mineralogy of the mantle source or the differentiation path followed (Sects. 5.6 and 11.9).

Although some conduit systems which feed stratovolcanoes within volcanic arcs clearly extend into the mantle wedge (Sect. 3.1 and 3.6), the composition of magma within these conduits is unknown. Historic eruptions of all volcanoes for which mantle-depth conduits have been inferred have yielded basalt or basic andesites (Sect. 3.6). And, although interstitial acid andesite glass has been found in ultramafic xenoliths from volcanic arcs and elsewhere, this glass is not clear evidence of acid andesite at mantle depths (Sect. 6.8). Consequently, circumstantial evidence suggests that partial melting occurs principally within the mantle wedge and produces basalt to basic andesite magma.

9.3 Are Orogenic Andesites Primary Melts of Only the Mantle Wedge? Rarely

Recent reviews of the mineralogy, composition, and partial melting behavior of the upper mantle have been provided by Wyllie (1971, 1979), Ringwood (1975), and Yoder (1976). A key feature is that the upper mantle typically consists of minerals whose $Mg/(Mg + Fe^{+2})$ ratios are 0.88 ± 0.02. Representative solidi for one such peridotite are given in Fig. 9.1; results for different peridotites were compared by Milhollen et al. (1974). Melts in region IP of Fig. 9.1 could coexist with both amphibole and phlogopite depending on water and potash contents; those in IIP could coexist with phlogopite; those in III or IVP coexist with neither.

Partial melts of the wedge in the presence of IRS fluid will be more siliceous than partial melts of mantle elsewhere. This conclusion is based on melting experiments in the relatively simple systems enstatite-water, forsterite-diopside-silica-water, and forsterite-plagioclase-silica-water, between 1965 and 1975, which demonstrated that the incongruent melting behavior of orthopyroxene persists to high pressure in the presence of water and allows silica-saturated liquids to coexist with forsterite as well as with pyroxenes (Kushiro 1972, 1974). Thus, whereas increasing load pressure causes near-solidus melts of peridotite to become increasingly silica-undersaturated, increasing water pressure has the opposite effect. These results suggested that a wide range of magmas (e.g., basalts to rhyolites; Yoder 1973) could be generated from the mantle wedge under varying degrees of water-saturation.

Analyses in Table 8.1 illustrate this effect for a natural complex system. Columns 4 and 5 are compositions ,as determined by direct microprobe analysis, of glasses quenched from melting experiments which used the same peridotite under water-saturated and anhydrous conditions, respectively, at similar pressures. Column 5 represents near-solidus melting in region IVP of Fig. 9.1; column 4 represents an unknown percentage of melt in region IP.

The interpretation of such data and, consequently, the extent of silica-enrichment which is possible for a liquid derived from peridotite, became matters of sharp controversy amongst experimentalists (Mysen et al. 1974). Mysen and

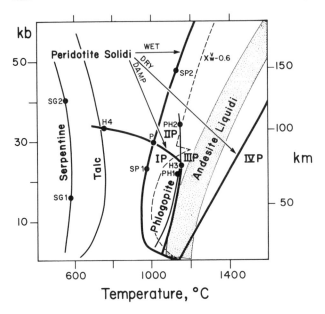

Fig. 9.1. P-T diagram for the mantle wedge. The *heavy lines* show representative solidi from Green (1973). The *dashed-line solidus* is for melting in the presence of mixed vapor where $X_{H_2O}^{vapor}$ ~0.6 (Wyllie 1979). Stability limits for serpentine and talc are from Wyllie (1979) and Delany and Helgeson (1978); the limit for phlogopite (which includes points PH1 and PH2) is from Wendlandt and Eggler (1980), assuming water-saturated conditions. The *line H3–P–H4* is the amphibole stability limit. Regions *IP* to *IVP* refer to melting conditions discussed in the text. Andesite liquidi are from Figs. 4.1 and 8.1. *Lettered points* are keyed to Fig. 8.2

Boettcher (1975) reported that compositions such as in Table 8.1, columns 3 and 4, were characteristic of melts formed at 15 to 25 kb under water-saturated conditions and at temperatures up to 200°C above the solidus in regions I and IIIP (and by inference in region IIP also). However, similar glasses formed during experiments by others were attributed to crystallization during quench and, therefore, as not being in equilibrium with a peridotite residue (Green 1973, 1976).

Whether or not the experimentally produced glasses were equilibrium fusion products, very few orogenic andesites are likely to be primary partial melts of the mantle wedge for two reasons: the compositions of orogenic andesites are dissimilar to those of the experimental glasses, and are not those of liquids in equilibrium with peridotite. First, while compositions of the experimental glasses meet my minimal definition of orogenic andesite, their feldspathic components (CaO and Al_2O_3) are higher, their Fe contents are much lower, and their incompatible element contents (TiO_2 and K_2O) are lower than in most orogenic andesites (cf. Table 8.1, cols. 3 and 4, with Fig. 1.1). As with eclogitic parents, experimental melts of peridotite have $CaO > (MgO + FeO^*)$ whereas orogenic andesites have the opposite. Moreover, the inability of ordinary upper mantle to supply the concentrations of incompatible elements found in medium or high-K orogenic andesites is even more dramatic when trace elements are considered (e.g., Lopez-Escobar et al. 1977; Fig. 5.17a). Consequently, pre-enrichment of these elements is necessary by mechanisms such as discussed in Section 9.4.

Second, few orogenic andesites could be in equilibrium with peridotite under upper mantle conditions. This conclusion is based on a consideration of the water and Ni contents, Mg-numbers, a_{SiO_2}, and liquidus mineralogy of andesites.

The experimental glasses produced by Kushiro (1974) and Mysen and Boettcher (1975) needed to be vapor-saturated and to contain > 14 wt.% H_2O to be andesitic in composition. However, most orogenic andesites contain only 1 to 5 wt.% H_2O and are water-saturated only at very shallow depths (Sect. 5.3.1). Moreover, water-saturated magmas must crystallize upon decompression (see Fig. 6.8b) and, therefore, are very unlikely to reach the surface. Thus, orogenic andesites cannot be products of fusion of water-saturated peridotite.

Similarly, about 95% of orogenic andesites have Mg-numbers too low for liquids in equilibrium with mantle olivine (Fo88 ± 2) which is the dominant refractory phase of peridotite (Sect. 5.2.3). This conclusion can be confirmed graphically by plotting orogenic andesite compositions on Fig. 2 of Hanson and Langmuir (1978); most lie below the 0% melt field, precluding primary partial melt status and indicating that fractionation has occurred. Also, most orogenic andesites have < 40 ppm Ni (Sect. 5.4.5), which is too low for liquids in equilibrium with peridotite containing about 2000 ppm Ni (Figs. 5.14 and 5.17d) unless D_{Ni}^{ol} is > 50. Such a coefficient is unlikely, even for temperatures around $1100°C$ and for liquids with andesitic MgO contents (Gill 1978). Indeed, because compositions of orogenic andesites plot below the partial melt curve in Fig. 5.14, fractionation during andesite genesis again is indicated.

The possibility of equilibration can also be evaluated thermodynamically by calculating the pressure and temperature at which the activity of components in andesite liquid equals the activity of those components in refractory peridotite. For example, acid andesites with $\log a_{SiO_2} \gtrsim -0.3$ at 1 bar and $1000°C$ (which are typical values given by coexisting groundmass phases) cannot have equilibrated with spinel or garnet lherzolite in the mantle wedge (Nicholls and Carmichael 1972). Results of similar calculations for basic orogenic andesites are ambiguous, sometimes yielding results consistent with derivation from spinel peridotite (e.g., Heming 1977), but other times not (Nicholls and Carmichael 1972), with differences in conclusions resulting more from the components and activities adopted or the other thermodynamic data used than from differences between rocks being evaluated.

Finally, magmas in equilibrium with typical peridotite should have En88 orthopyroxene and Fo88 olivine as liquidus phases at high pressure; i.e., the melts should precipitate such crystals or at least not dissolve them when added even though olivine may be in reaction relationship with the liquid. The thermodynamic considerations noted above indicate that as a_{SiO_2} of lavas increases, the maximum depth at which equilibrium with peridotite is possible decreases. Even under water-saturated conditions, most quartz-normative andesites precipitate (or fail to dissolve) Fo88 olivine ± orthopyroxene only to 10 kb or less; water-saturated, olivine-normative calcalkaline andesites with FeO*/MgO ratios < 1 meet this requirement only up to slightly higher pressure (Green 1973; Nicholls and Lorenz 1973; Nicholls and Ringwood 1973; Nicholls 1974). One exception is a quartz-normative boninite with FeO*/MgO = 0.67 (Table 6.2, col. 5) which is olivine + orthopyroxene-saturated at 17 kb when water-saturated, and at 14 kb when the charge contained 7 wt.% H_2O (Kushiro and Sato 1978). The maximum pressure for olivine ± orthopyroxene-saturation decreases rapidly with decreasing water content of the melt. Although the pressure range for this double saturation of most orogenic andesites containing 1 to 5 wt.% H_2O is unknown (indeed, one may not exist), at best it is probably appropriate only to crustal depths and, therefore, is inapplicable to partial melting within the mantle wedge.

Some parts of the preceding discussion refer only to the peridotite plus water system. Because orogenic andesites may contain as much CO_2 as H_2O (Sect. 5.2.2), the effects of CO_2 on peridotite melting must also be considered. However, andesites are even less likely to be primary partial melts of peridotite which contains mixed CO_2-H_2O vapor.

In the presence of CO_2 plus H_2O, andesitic glasses have been found experimentally only in peridotites which were partially melted under vapor-saturated conditions at pressures < 27 kb where $X_{H_2O}^{vapor} \geqslant 0.6$ (Eggler 1975; Mysen and Boettcher 1975). That is, andesitic liquids lie between the solid and dashed vapor-saturated solidi of region IP, Fig. 9.1 at low pressure; a representative "liquid" composition of this type is given in Table 8.1, col. 3. At higher pressure a carbonate mineral becomes a refractory phase and liquids become carbonatitic (Wyllie 1979). At lower $X_{H_2O}^{vapor}$, liquids become olivine or nepheline-normative. Andesitic glasses formed under these vapor-saturated conditions differ from orogenic andesites in all the ways discussed above, including having > 14 wt.% water. Because CO_2 expands the stability field of orthopyroxene at the expense of olivine, the presence of CO_2 decreases the maximum pressure at which andesite and olivine are in equilibrium, and thereby reduces further the possibility of obtaining orogenic andesites from mantle depths.

Under more realistic conditions in which peridotite contains hydrous or carbonate minerals but relatively little or no vapor, vapor-undersaturated liquid may form. However, this liquid will be quartz-normative only if CO_2 is missing altogether. Melting in the presence of small amounts of a mixed vapor probably results in nephelinitic liquids in the presence of amphibole, or in melilitic to kimberlitic liquids at higher pressure in the presence of refractory carbonate (Eggler and Holloway 1977).

A silicate liquid not in equilibrium with carbonate minerals will have a higher H_2O/CO_2 ratio than will a coexisting mixed vapor because H_2O is more soluble than CO_2 in the liquid (Sect. 5.3). Consequently, orogenic andesite with molar $CO_2/(CO_2 + H_2O)$ ratios around 0.5 (Sect. 5.3.2) would have to be produced from peridotite in which $X_{H_2O}^{vapor}$ is < 0.5, and melts produced under these conditions are olivine- or nepheline-normative basalts rather than andesites. The CO_2-rich volatiles must, therefore, have been inherited from a basaltic parent by differentiation.

As a result of the preceding discussions, criteria can be set for orogenic andesites which possibly are primary partial melts of typical peridotite, as was also done for partial melts of subducted ocean crust in the preceding chapter. Peridotite-derived andesites will have FeO*/MgO ratios < 1, Mg-numbers > 67, Ni contents > 100 ppm and near the partial melt curve in Fig. 5.14, high water contents, and little or no CO_2. Oxide activities should be consistent with equilibrium with peridotite and olivine plus orthopyroxene should be liquidus phases at mantle pressures. Plagioclase phenocrysts probably will be absent and olivine present due to the high water contents and Mg-numbers, but hornblende or pyroxene could be the sole phenocrysts due to a reaction relationship between olivine and the liquid. LIL-element concentrations will be unusually low unless pre-enrichment of the mantle wedge has occurred. Possible examples of peridotite-derived orogenic andesites include some of those with high Mg-numbers noted in Section 5.2.3. In all cases pre-enrichment with LIL-elements is implied.

While less than 5% of orogenic andesites meet any of these criteria and few, therefore, can be considered primary, the maximum silica content of magma which can be derived at pressures of 10 to 30 kb from typical peridotite containing, say, 0.2 to 1 wt.% H_2O is unknown. Linear extrapolation between data for the system Fo-Di-Qz-H_2O suggests there is less than 0.5 wt.% SiO_2 difference between the anhydrous composition of partial melts produced at 20 kb from dry peridotite versus peridotite containing 1% H_2O (Presnall et al. 1978). This slight difference includes effects due to incongruent melting of orthopyroxene but omits similar effects due to melting of hydrous minerals. It also omits the effect of the alkalies present in IRS fluid (Kushiro 1975). Estimates of melt compositions resulting from fusion of natural peridotites in the presence of small amounts of hydrous minerals and mixed volatiles are only educated guesses. In one attempt at synthesis, Wyllie (1979) suggested that at pressures above 10 kb, qz-normative liquids are produced near the solidus only under vapor-saturated conditions, when $X_{H_2O}^{vapor} > 0.5$, and to a maximum pressure of 20 kb (i.e., region IP, Fig. 9.1). Liquid compositions were not otherwise described and, of course, even ol-normative liquids can contain 55% SiO_2 or more. Nevertheless, a reasonable conclusion is that primary magmas from the mantle wedge will be slightly closer to andesite in composition than are primary magmas elsewhere (e.g., more siliceous, lower compatible trace element contents), due to the effect of water and alkalies. How much closer is the issue.

However, water and alkalies may not be the sole agents responsible for production of relatively siliceous magma from peridotite. As noted in Sections 2.8

and 5.7, tholeiitic basalts to basic andesites with 50% to 54% SiO_2 and < 6% MgO occur and sometimes dominate in flood basalt provinces where they are remarkably uniform in composition, voluminous in scale, and were erupted at rates equal to or greater than those of magmas anywhere. These three features collectively though circumstantially argue against the origin of these basalts to basic andesites by differentiation or contamination even though their Mg-numbers (< 50) and Ni contents (< 25 ppm) are below values cited above as primary. Consequently, fertile mantle which is atypically Fe-rich or olivine-poor (pyroxenite) may exist on regional scales, especially beneath continents (Wilkinson and Binns 1977; cf. Jordan 1978). This caveat can be ignored if these flood "basalts" are shown to be crustally contaminated.

9.4 Fluid Mixing, Metasomatism, and Demonology in the Mantle Wedge

If arc magmas are geochemically distinctive and this distinctiveness can be attributed neither to direct slab derivation nor to crustal contamination (Sects. 5.7 and 8.5), then it must result from slab recycling via mixing processes within the mantle wedge. This argument is strengthened by the occurrence in Grenada, Antilles, and in Sumbawa, eastern Indonesia, of basanites which must be derived from peridotite yet which have these same distinctive trace element and isotopic characteristics (Whitford et al. 1978; Hawkesworth et al. 1970c). Also, global geochemical budgets require such mixing (Sect. 8.5).

Other arguments for recycling are more ambiguous. Whether LIL-element *concentrations* in potentially primitive arc magmas require pre-enrichment via mixing within the wedge (as would be necessary if orogenic andesites themselves were primitive: Sect. 9.3) is not yet known and awaits better criteria for primitiveness plus independent estimates of the percent fusion involved. Similarly, although the discussion of thermal structures in Sections 8.1 and 9.1 favors mixing due either to wedge + slab fusion or fusion of IRS-contaminated mantle, wedge-only fusion *without a contribution from the slab* is possible (e.g., in region IIP of Figs. 8.2 and 9.1).

Moreover, evidence for recycling sometimes is contradictory or at least enigmatic. The presence of high Sr/Nd and Ba/La ratios in Mariana and New Britain arc rocks whose Sr and Nd isotopes (plus Pb, O, and He isotopes in the Mariana example) require no slab contribution was cited in Section 8.4. Similarly, calculations by Mysen (1979) and Kay (1980) show that the low La/Yb ratios of island arc tholeiitic series rocks also preclude a slab contribution, whereas such suites from Tonga, Japan, and the South Sandwich arcs have the highest Ba/La ratios known. Finally, the occurrence in South Sandwich Sea (backarc) basalts of Sr whose isotopic composition lies to the right of the SNUM line in Fig. 5.19 (Hawkesworth et al. 1977) is anomalous because a subducting slab does not yet extend beneath that active spreading center.

If recycling does occur, then abstractly one can distinguish between two agents of mass transfer and between two mixing processes. The agents could be IRS fluid or anatectic melt from the slab; the processes could be subsolidus metasomatism or veining on the one hand, or fluid mixing on the other. Neither distinction is easily made in practice because the effects are so similar. Consequently, neither distinction may be important to petrogenetic theory in the long run.

Fluid mixing provides a single-stage process in which peridotite fuses in the presence of IRS fluid, thereby dissolving the fluid, or in which peridotite-derived basalt plus slab-derived andesite to rhyolite mix at high pressure. First consider fluxing by IRS fluids. The greatest effect of the fluid is to promote melting. Addition of water may cause vapor-saturation where none had been, thereby reducing solidus temperatures by up to $200°C$ and moving melting conditions in Fig. 9.1 from region IIIP to IP. Where vapor-saturation already exists, addition of water increases the percent of melt present at any temperature. Either phenomenon may increase melt contents beyond the storage capacity of the peridotite parent, resulting in porous segregation and magma ascent (Sect. 3.5). The IRS fluid dissolves in melts which are produced if they are vapor-undersaturated, as expected. If the fluid composition is, say, 50 wt.% H_2O + 40 wt.% SiO_2 + 10 wt.% others, including IR-components (Sect. 8.5), then a maximum of about 1 wt.% SiO_2 can be gained by water-poor melts parental to orogenic andesites.

A simple example will illustrate that sufficient IRS fluid is available to supply the water in arc magmas. Suppose that 50 km^3 of magma are derived from a 250 km^3 zone of magma genesis which stretches 50 km along the strike of an arc, and that half of this is vented to form one volcano in 25,000 years (Fig. 3.7; Sects. 4.7 and 4.8). This yields 20 km^3 of ejecta/m.y. per km of plate boundary, which lies within the range of estimates in Table 4.3. If following 25% crystal fractionation the magma was andesite containing 2.7 wt.% H_2O (Fig. 12.5) and its density was 2.7 gm/cm^3 (Fig. 4.3), then initially the 50 km^3 of magma would have contained 2.7×10^{15} gm H_2O. Subduction at 10 cm/yr beneath this zone of magma genesis would exhale five times that much water in 25,000 years even if only one km of crust contained the 4 wt.% H_2O suggested for altered ocean floor basalt in Table 8.1, assuming all the water was lost.

Consequently, fluxing seems a viable combination of agent and process by which to extract quartz-normative basaltic liquids containing a few percent of the IRS fluid which is released during slab dehydration. Recall from Sections 8.1 and 9.1 that this combination requires low geothermal gradients *along* the slab/wedge boundary, high gradients *normal* to the boundary, and convection within the wedge. Some implications of this hypothesis are explored in Figs. 12.1 to 12.4 and accompanying text.

Alternatively, slab-derived siliceous melts may rise into the hotter mantle wedge, becoming superheated before they mix with peridotite-derived basalt or simply react with the peridotite. Mixing dilutes the hydrous, siliceous melt with less hydrous basalt and in principle can yield a wide range of intermediate compositions with appropriate temperatures and water contents (Kay 1978; Burnham 1979).

In both fluid mixing cases, either andesites or their parent magmas could be "primary" in the sense of ascending without differentiation without being "primary" in the sense of being in equilibrium with a reasonable, specifiable residuum.

Because both IRS fluids and slab-derived melts are rich in silica, LIL-elements, and water, few criteria distinguish between their effects during mixing. If the slab remains cold, IRS fluid must be the transporting agent. However, two geochemical arguments favor the melt. First, the lack in andesites of U/Th and U/Pb ratios which are higher than implied for their time-averaged source by their Pb isotopic compositions seems less consistent with transport of elements in an oxidizing aqueous medium than in a melt (Sect. 8.5). Second and more importantly, relatively low water contents of orogenic andesites favor mixing with water-undersaturated melt instead of IRS fluid. Especially if they are water-undersaturated, slab-derived melts will have lower water/silica ratios than will IRS fluids. Consequently, if sufficiently silicic basalts cannot be produced in the presence of a few tenths wt.% H_2O (Sects. 9.3, 12.2), then high-pressure magma mixing is the only way to obtain water-poor yet silica-rich parental basaltic melts for andesites.

Next, consider metasomatism. Composite ultramafic nodules, secondary hydrous minerals in peridotite nodules, high LIL-element concentrations in refractory peridotite minerals, and Sm/Nd isotopic systematics of basalts, all indicate that mantle metasomatism or veining precedes and may cause basaltic volcanism, and results in considerable mantle heterogeneity, in many tectonic environments. Arcs differ only in involving siliceous melts plus IRS fluids as well as basaltic melts plus H_2O-CO_2 mixed vapors, and possibly differ in the scale envisioned. Either IRS fluid or silicate melts are potential agents of extensive metasomatism or veining in the mantle wedge. For example, the same parameters of slab dehydration discussed above would add as much as 1.8 wt.% H_2O and SiO_2 to peridotite during the 25,000 year lifespan of the hypothetical volcano if metasomatism were confined to the 250 km^3 region of eventual magma genesis.

Subsolidus reaction between lherzolite and siliceous melt would lead to pyroxenite or eclogite veins or both (Ringwood 1975). It is unclear whether such a metasomatized wedge could subsequently yield melts which lack the requirements summarized in Section 9.3 for equilibrium with peridotite. If these requirements need not apply, then by combining less than one percent of slab-derived melt having the characteristics summarized in Section 8.4 with ordinary upper mantle one can, in ad hoc fashion, produce a source which upon a few percent fusion yields melts similar in composition to arc basalts (Kay 1980). However, as long as the same minerals remain refractory after melting, their relative percentages are unimportant for the composition of eutectic-like liquids. Hence, only if major changes in mineralogy are produced can the requirements of Section 9.3 be waived. While neither the geophysical characteristics of, nor ultramafic inclusions from, the mantle wedge provide much evidence for such major changes (Sect. 9.1), the issue is open, and the possibility that continental flood "basalts" (i.e., basic andesites) are primary (Sect. 9.3) gives one additional pause.

Few criteria distinguish between the effects of metasomatism versus fluid mixing. One such criterion involves the spatial and temporal limits of distinctive arc geochemistry, because fluid mixing involves buoyant materials over short time intervals whereas metasomatism is a permanent feature of the wedge. Consequently, the presence of distinctive magmas within volcanic arcs but not backarcs (Sect. 5.7) and the replacement of orogenic andesites by high-Ti alkali basalts following cessation of subduction (Sect. 2.3) both favor fluid mixing. Alternatively, the presence of a recycled component in Antillean and Indonesia silica-undersaturated basanites (see above) is more consistent with fusion of metasomatised peridotite than with fluid mixing involving a silica-rich liquid.

In summary, both fluid mixing and metasomatism in the mantle wedge are likely but currently untestable processes, i.e., they are like Maxwell's demons. Arguments that either process occurs are circumstantial yet need to be detailed and quantitative to be persuasive because some arguments apparently are contradictory. Criteria for distinguishing between processes or agents of mixing are inadequate and the distinction may prove unimportant. However, it is clear that orogenic andesites are not likely to be undifferentiated melts from the mantle wedge unless both slab and wedge fuse and the two resulting melts mix within the wedge. Otherwise primary melts are basaltic and the principal question is how silica-rich and magnesia-poor these basalts can be.

Note that if the weight fraction of a wedge-derived basalt component in arc magmas is high, then the subvolcanic arc wedge becomes more important than mid-ocean environments as a site of permanent mantle differentiation. Convection of mantle through this corner would therefore play a critical role in the evolution of both crust and upper mantle.

Chapter 10   The Role of the Crust

10.1 Circumstantial Evidence for Crustal Involvement in Orogenic Andesites

Because average sialic crust is similar in composition to orogenic andesite its total fusion can yield andesite, and its assimilation into basaltic magma can change the composition of the mixture towards andesite. However, crustal involvement cannot be necessary for either the genesis or geochemical distinctiveness of orogenic andesite because the distinctive andesites predominate even when sialic crust is thin or absent, because of arguments presented in Chapters 8 and 9 that most andesites or their parent magmas originate below the crust, and because many andesites are isotopically incompatible with crustal derivation. Qualitative geochemical arguments for crustal-level assimilation and for slab recycling are often similar because the contaminants are similar. Thus, circumstantial distinctions between the two processes which rely on geologic situations where one is inapplicable (e.g., ARC-ATL collisions) or where one is constant but the other not (e.g., along-strike variations in crustal thickness) are useful.

There is strong circumstantial evidence that sialic crust affects fom andesites. First, as a general rule, volcanic arcs situated on thicker crust are characterized by more silicic, more differentiated magmas which, relative to given silica content, are richer in LIL-elements (especially K, Rb, Cs, Ra, U, and REE), have lower K/Rb ratios, and have higher as well as more variable $^{87}Sr/^{86}Sr$ and $^{18}O/^{16}O$ ratios but lower $^{206}Pb/^{204}Pb$ and $^{143}Nd/^{144}Nd$ ratios than are magmas erupted through thinner crust (Sects. 5.8 and 7.2). The higher silica mode could reflect reduced density contrast between magma and wall rocks where crust is thick (Fig. 4.3). This reduced contrast retards ascent, promoting differentiation. Second, half a dozen qualitative isotopic arguments which are summarized in Section 5.5.7 also imply crustal-level recycling in andesites from several arcs. Third, some andesites contain thermally metamorphosed and even partially fused inclusions of sialic rocks, or garnet or cordierite megacrysts, or both. With the possible exceptions of Pelée and Lipari, the volcanoes from which such andesites are erupted overlie sialic crust > 25 km thick. Similarly, there are disequilibrium mineral assemblages in some andesites which may include refractory residues after crustal fusion or assimilation (Chap. 6). Finally, secular decline in K_{55} values within some volcanic arcs (Sect. 7.5) might result from gradual assimilation and depletion of LIL-elements from the crust beneath sites of volcanism (White and McBirney 1979).

Thus, the important quesions become the weight fraction of crustal involvement within specific suites, the relative abundance and geographic distribution of
affected suites, and the physical processes of involvement. At least three such processes are possible: partial to total fusion; interaction with (e.g., assimilation of)
silicic crustal rocks; and mixing between basalt and crustally derived silicic magmas. The first two are considered below; the third is included within the larger
discussion of magma mixing in Section 11.7.

Definitive criteria by which to answer the questions posed above are unavailable due to inadequate knowledge of the composition and mineralogy of the lower
crust and of subcontinental upper mantle, due to inadequate knowledge of crustal
temperatures beneath volcanic arcs, and due to nonunique interpretations of the
chemical and mineralogic data cited above as evidence for crustal involvement.
In this regard, recall from Sections 3.1 and 3.2 that at least the upper crust in most
active volcanic arcs is dominated by intermediate igneous rocks of Mesozoic or
Tertiary age, whereas the lower crust is not dominated by amphibolite or eclogite,
and that temperatures at the base of the crust beneath volcanic arcs locally may
be 1000° to even 1200°C. Also, recall from Sections 5.5.7 and 9.1 that some
uppermost mantle, especially beneath continents, is enriched in LIL-elements
(especially $U > Pb$, $Rb > Sr$, and light $>$ heavy REE) relative to subocenaic
mantle and perhaps relative to planetary values. Consequently, partial fusion of
such mantle might yield magma with $^{87}Sr/^{86}Sr$ ratios > 0.7047, with $^{207}Pb/^{206}Pb$
systematics similar to those of orogenic andesites, and with Sr-Nd isotopic compositions displaced toward the origin or to the right of the SNUM line of Fig. 5.19.

10.2 Crustal Anatexis

Origin of andesite by fusion of crustal rocks was proposed by Holmes (1932) and
continues to be defended cogently (e.g., Pichler and Zeil 1969, 1972). The viability
of this mechanism is demonstrated clearly by the Manicouagan impact melt sheet
which was formed on the Canadian Shield 214 m.y.B.P. when impact melting of
an approximately 60:40 mafic:felsic rock mixture produced hundreds of km^3 of
chemically homogeneous high-K calcalkaline acid andesite (Table 5.5, col. 12).
Relative to the high-K andesites of Tables 5.1 and 5.2, the Manicouagan rocks
have higher Co, Ba and light REE contents, and a high initial $^{87}Sr/^{86}Sr$ ratio of
0.71, but are otherwise similar in composition though, of course, not texture
(Floran et al. 1978).

Three kinds of arguments are allied against routine derivation of andesite
from crustal sources. First, LIL-element concentrations in orogenic andesites,
especially in low and medium-K varieties, are poorly matched by the same linear
combination of crustal rocks which match their major element concentrations
(Taylor 1969), and the isotopic composition of their Sr, Pb, Nd, He, and O is not
that of "average sialic crust", such as that which fused to form the Manicouagan

sheet, for example. However, pre-Mesozoic sialic crust is missing from many volcanic arcs or may previously have been depleted in LIL-elements such as Rb and U by dehydration or anatexis. For example, assuming $0.1 = $ Rb/Sr (melt) $\geqslant$ Rb/Sr (source), it would take $\geqslant 370$ m.y. (since the Devonian) for source rocks with an initial $^{87}Sr/^{86}Sr$ ratio of 0.7035 to reach even 0.7050. Consequently, lower crust capable of generating at least high-K andesites with $^{87}Sr/^{86}Sr$ of 0.705 to 0.709 usually cannot be ruled out a priori. Indeed, the isotopic composition of Pb in andesites from Wyoming, USA, and Peru is more similar to that thought characteristic of the lower crust than either upper crust or mantle in those regions (Sect. 5.5.7). Such crust should, however, have Sr-Nd isotopic compositions to the left of the SNUM line of Fig. 5.19; andesites with such compositions are unknown.

Second, derivation of andesite *liquid,* or even a crystal-liquid mush with $< 50\%$ crystals, requires temperatures $> 1000°C$ at 10 kb if the liquid has < 5 wt.% H_2O (Wyllie 1977, Fig. 11). Because this temperature exceeds that recorded in metamorphic rocks or that normally expected within the crust (apart from meteorite impacts!), formation of andesite liquid usually has been discounted. However, appropriately high crustal temperatures apparently can be reached beneath volcanic arcs when crust is > 25 km thick, if surface heat flow is > 2 HFU yet heat generation within the crust is only ~ 1 HGU (Uyeda and Horai 1964); these three criteria are met in some island arcs and continental regions (Sect. 3.1 and 3.2). Consequently, conditions appropriate to andesite genesis, such as 30% to 60% fusion of rocks with basaltic composition under water-saturated conditions at 5 kb (Helz 1976) or near-total fusion of intermediate gneisses, can be achieved locally within volcanic arcs due to intrusion there of mantle-derived magma. Hornblende, pyroxene, or complexly zoned and resorbed plagioclase "phenocrysts", garnet or cordierite megacrysts, and gabbro or amphibolite inclusions *could* be refractory residua (Pichler and Zeil 1972; White and Chappel 1977).

Third, arguments that the sialic crust, hydrosphere, and atmosphere have resulted from andesite volcanism throughout Earth history (e.g., Taylor 1967; Anderson 1974b), and that the volume of sialic crust increases with time within specific volcanic arcs (e.g., James et al. 1976) preclude andesite genesis solely through crustal recycling. However, these objections are only to the mechanism as an exclusive cause and even so are negated if crustal evolution is a multi-stage phenomenon involving growth by addition of mantle-derived basalt followed by differentiation through anatexis.

Thus, some orogenic andesites may result from partial to total crustal fusion. Such andesites will occur within suites in which relative volumes of rocks decline with decreasing silica contents (i.e., in which basalts and basic andesites are uncommon or genetically unrelated to the silicic rocks or both). Their elemental and isotopic composition must, of course, reflect the composition and age of the lower crust in that region: i.e., andesites probably will be high-K calcalkaline varieties with $^{87}Sr/^{86}Sr$ ratios > 0.705, with $^{207}Pb/^{204}Pb$ above the PUM line in Fig. 5.18, with $\epsilon_{Nd} < 0$, with $^{3}He/^{4}He$ less than atmospheric, and with $^{18}O/^{16}O > 8\%o$.

Their compositions are unlikely to conform to regular across-arc variation patterns such as those described in Section 7.1. These conditions are most likely to be met when crust is thick or during collision events.

Despite this possibility of crustally derived andesites, I have neither found, nor been able to develop, convincing quantitative arguments for the mechanism in any specific case. Data for modern orogenic andesites from southern Peru, northern Chile, Turkey, and Iran, or for older examples from the USA interior and Sardinia all are suggestive but ambiguous. At most, internally consistent arguments can be made for derivation of single samples from plausible but ad hoc parents; suite members are related by fractional crystallization rather than by partial fusion of a common source following arguments of Section 5.4.7. Moreover, the areas cited above may be underlain by (therefore generated from) LIL-element-enriched upper mantle, and there are across-arc geochemical variations in each of the modern volcanic arcs cited which are fortuitous if crustal anatexis predominates in these regions.

For example, Savalan volcano, northwestern Iran (volcano number 28.6 in the Appendix) has erupted rocks with 59% to 68% SiO_2 which are characterized by relatively constant FeO^*/MgO ratios of 1.9, Na_2O/K_2O ratios of 1.8, and K/Rb ratios of 274, by a crude inverse correlation between REE and SiO_2, and relatively high Ni contents (Dostal and Zerbi 1978). There is no evidence of subducted lithosphere beneath Savalan. Ad hoc schemes for derivation of magmas with these characteristics from the crust are possible but problematical. For example, the observed Ce/Yb ratios are too high for melts derived from typical sialic crust in the amphibolite facies if the distribution coefficients of Table 6.3 are applicable (see Fig. 5.17a). Moreover, the two- to sevenfold within-suite variation in Ni, Cr, Co, and Sc contents indicates that suite members are related more by fractional crystallization than differential partial melting (Sect. 5.4.7). Also, the greater alkalinity of rocks from Savalan than Ararat, located closer to the plate boundary, is just fortuitous if this explanation is correct.

10.3 Crustal Assimilation

The second family of mechanisms for crustal involvement includes contamination of mantle-derived magma by assimilation of or interaction with upper or lower crustal rocks; these ideas derive from Daly (1933), Wager and Deer (1939), Kuno (1950a), Tilley (1950), and Wilcox (1954). Evidence that assimilation of sialic rocks does occur includes most of the mineral and chemical features discussed in the preceding section, but in relaxed forms. Obviously, if geothermal gradients and sources can be compatible with andesite genesis through crustal anatexis, then crustal contamination also can occur in the same environments. Specifically, occurrence of noncumulate xenoliths or minerals arguably derived therefrom by disaggregation, and heavy element isotopic heterogeneities in otherwise consanguinous suites, provide the strongest prima facie indications of contamination. However, unambiguous evidence that assimilated rock disaggregates or dissolves on a significant scale is indirect at best and usually absent altogether (Sect. 6.8). Petrographic observations appealed to as such evidence include: corroded plagioclase and quartz crystals, especially when containing inclusions of silicic glass; garnet and cordierite megacrysts; and gradational contacts between xenoliths and host. See Kuno (1950a) and Steiner (1958) for examples, but recall from Chapter 6 that interpretation of such evidence is ambiguous.

Tests of crustal-level contamination during andesite genesis are somewhat easier to construct than are the analogous tests of contamination within the mantle wedge which were discussed in Section 9.4. Qualitatively, some andesite suites are consistent with a mixing history. Two such are in New Zealand (Steiner 1958; Ewart and Stipp 1968) and northern Chile (Klerkx et al. 1977). For example, consider data in Figs. 10.1 and 10.2. If these rocks result from binary mixing of basalt magma and sialic crust, then data plotted in a ratio-ratio diagram should define a hyperbola whose asymptotes constrain the ratios of the end members (Langmuir et al. 1978). The New Zealand data do crudely define such a mixing line on an $^{87}Sr/^{86}Sr$ vs. K/Rb plot (Fig. 10.1). Data for the same samples should then define a straight line on a "companion plot" such as an $^{87}Sr/^{86}Sr$ vs. $^{87}Rb/^{86}Sr$ diagram (Fig. 10.2) and individual data points should retain the same relationship to each other as in the original diagram since that relationship is determined by the mixing process. Again the New Zealand data define a mixing line, thereby yielding a pseudoisochron whose 110 m.y. age is that of the Rangitata Orogeny which affected most of New Zealand including, presumably, crustal rocks beneath the modern volcanic arc. The validity of Ewart and Stipp's (1968) conclusion that about a 1:1 mixture of basalt and sediment would yield andesite with appropriate $^{87}Sr/^{86}Sr$ and Rb/Sr ratios is illustrated in Fig. 10.2.

In detail, however, binary mixing can explain neither these data nor those for any other suite of orogenic andesites for three reasons. First, element-element diagrams for orogenic andesites rarely have the characteristics required for mixing (see Sect. 5.4.7; Fig. 5.17) although data scatter often obscures relationships. Although not shown, C_H-C_T and C_T-C_T diagrams for the New Zealand andesites are most consistent with a fractional crystallization relationship. Relatedly, different mixing diagrams indicate inconsistent relative proportions of the end members which form the same samples. For example, the four data points numbered in Figs. 10.1 and 10.2 do not retain consistent mutual relationships. Second, the predicted end members are unlikely and do not match possible candidates. The individual andesites in Figs. 10.1 and 10.2 do not lie on a mixing curve between typical New Zealand basalt and the sediment mixture(s) which, on major element grounds (Cole 1978, Table 11) can form average New Zealand andesite when mixed with the basalt. Third, although porphyritic, the andesites contain insufficient crystals to be about 50% contaminated, as required. That is, because at least as much crystallization as dissolution must occur during assimilation (Bowen

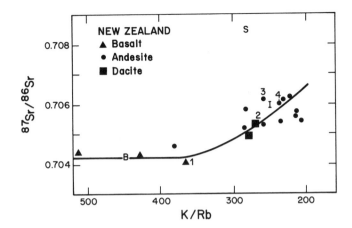

Fig. 10.1. Mixing diagram for New Zealand volcanics using $^{87}Sr/^{86}Sr$ versus K/Rb ratios. Data from Ewart and Stipp (1968). One basalt with coordinates 0.7042, 650 is missing. Letters *B*, *S*, and *I* are average basalt, sediment, and ignimbrite values. Numbered points *1* to *4* have decreasing K/Rb ratios, requiring an increased wt. fraction of the low K/Rb mixing component. Mixing curve was fit by eye

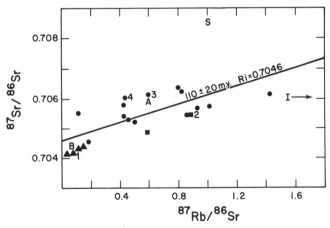

Fig. 10.2. Companion plot to Fig. 10.1., using ^{86}Sr as a denominator common to both axes. Data define a straight line (pseudoisochron) as shown. *Symbols* and *letters* as in Fig. 10.1 except for *A* which is the average for andesites. The average ignimbrite *(I)* lies off scale with coordinates 0.7060, 2.3. Note that *numbered points* have different Rb/Sr than K/Rb relative ratios

1928), the similarity in phenocryst content between the New Zealand basalts and andesites is anomalous.

Indeed, the third point above is the tip of the iceberg which explains why the first two tests of mixing hypotheses consistently yield negative results: i.e., assimilation of solids by nonsuperheated liquids *must* be accompanied by crystallization. Hence, the results of assimilation plus fractional crystallization are *necessarily* additive. Even though heats of solution of dissolving phases (quartz, alkali feldspar) are smaller than the heats of crystallization of precipitating phases (plagioclase, pyroxenes), the general driving forces of solidification will cause the weight fraction of precipitates to exceed that of assimilated rock, possibly by large amounts. Consequently, crystallization effects will dominate major and compatible trace element behavior, whereas mixing effects are most evident in ratios of incompatible trace elements or isotopes. For example, fractional crystallization has limited effect in Fig. 10.1 so that the variations observed must reflect mixing of some kind if all samples are genetically related. Fig. 10.2 can be interpreted similarly (Briqueu and Lancelot 1979).

Thus, at least bulk assimilation plus fractional crystallization must be involved. However, there are no quantitative demonstrations to date that this combined process can account for the geochemistry of any well-analyzed suite of andesites. For example. although bulk assimilation alone of 10% to 20% Canadian gneiss into Ecuadorian andesite can yield mixtures with the Sr and Nd isotopic characteristics of northern Chilean andesites (Thorpe and Francis 1979), the detailed Sr systematics of similar though less international rocks in southern Peru require fractional crystallization as well as assimilation (Briqueu and Lancelot 1979). Moreover, if both Sr and O isotopes of the Peruvian rocks are considered, then even a joint fractionation plus assimilation process is inconsistent with the data (Magaritz et al. 1978; Taylor 1980). Instead, other processes such as assimilation of metamorphic waters or anatectic melts (this time within the crust, not mantle wedge), diffusion, isotopic exchange (Pankhurst 1969), or zone refining (Harris 1957) may permit selective contamination by incompatible elements and radiogenic isotopes while escaping the mass balance requirements of bulk assimilation plus fractionation. Untestability again is an asset. Consequently, these alternative contamination processes remain possible explanations of the circumstantial evidence for crustal involvement despite the quantitative failure of even realistic mixing models.

In summary, the role of the crust appears threefold. First, sialic crust is a density filter, promoting differentiation by retarding ascent. Second and popular opinion to the contrary, in principle sialic crust could be a source for certain orogenic andesites. However, in practice, no crustally derived andesites have been clearly identified. Third, for several circumstantial and qualitative geochemical reasons, selective crustal-level contamination seems to have occurred in some orogenic andesites but neither a comprehensive geochemical budget nor a contamination mechanism have been established for any specific example.

It is this furtive, subtle manner by which crust affects the andesites passing through it which, geochemically, most distinguishes orogenic andesites from the batholiths which are often thought to be their subsidiary reservoirs. While relationships between andesites and salic plutons are another story, the data presented here suggest to me that andesites and granodiorites are often unrelated genetically, and that andesites recycle much less crust than do the salic plutons.

Chapter 11   The Role of Basalt Differentiation

That orogenic andesites and their plutonic equivalents result from crystal fractionation of basalt has been a commonplace in petrology since the idea was
developed by Bowen (1928). Indeed, the alternatives discussed in the three preceding chapters arose only in response to perceived inadequacies of Bowen's proposal and especially to its lack of ready explanation for the close association
between orogenic andesites and convergent plate boundaries discussed in Chapter 2. However, because each attempt to explain orogenic andesites as primary
partial melts or mixtures thereof, or as crustally contaminated primary melts, has
been found inadequate in most instances, we should now re-examine differentiation mechanisms carefully.

11.1 General Arguments for and Against Differentiation

The oldest and most straightforward argument that andesites originate by differentiation of basalt is that they often are related spatially, temporally, and mineralogically to basalts, dacites, or rhyolites whose chemical compositions collectively
define pleasingly smooth scatter in element-element diagrams which, therefore,
can be interpreted as representing liquid lines of descent. However, the argument
is suggestive only. Data scatter typically precludes graphical distinction between
differentiation, anatexis, and mixing processes. There are discontinuities within
some data sets. Phenocrysts preclude confidence that lines are of liquid descent,
and relative volumes of rock types often are not known reliably or are at odds
with differentiation mechanisms.

Second, andesites, including orogenic andesites, sometimes occur elsewhere
than convergent plate boundaries (Sect. 2.8). Usually these andesites in anorogenic locations differ more in relative volume than in chemical composition from
their kin at subduction zones, although compositional differences do exist
(Sects. 5.6 and 5.7). The anorogenic andesites indicate that differentiation of
basalt can be a sufficient mechanism for producing andesite either within or outside volcanic arcs, an argument developed more fully in Section 11.9.

Third, evidence for crustal-level magma reservoirs beneath active volcanoes
from which acid andesites are erupted (Sect. 3.6), and evidence from ejecta of
individual eruptions for compositional or mineralogic gradients within these
reservoirs (Sect. 4.3) both suggest that differentiation occurs within the reservoir

system beneath andesite stratovolcanoes during periods of repose. Moreover, stratigraphic evidence that rocks typically become more differentiated, or change from tholeiitic to calcalkaline, up-section within volcanoes during one or more cycles (Sect. 4.4), and that the extent of differentiation increases as the length of respose increases, is difficult to explain by any process other than in situ differentiation.

Fourth, the occurrence in orogenic andesites of arguably cognate mafic xenoliths and xenocrysts (Sect. 6.8), especially those indicating more mafic compositions than likely for the host, indicate crystal fractionation or mixing or both. Fifth, trace element systematics, such as seen in compatible element versus incompatible element diagrams, sometimes unequivocally demonstrate that crystal fractionation rather than anatexis or mixing has occurred (Sect. 5.4.7). Sixth, the isotopic similarity between andesites and associated basalts which is necessary for magmatic differentiation alone to be a viable explanation, usually though not always exists (Sect. 5.5). Finally, the low Mg-numbers, Ni-contents, and (possibly) H_2O/CO_2 ratios of orogenic andesites require differentiation if melts are derived mostly from peridotite within the mantle wedge (Sect. 9.3).

Objections to differentiation seize on exceptions to each of the above. Chemical variation diagrams contain scatter in excess of analytical imprecision (e.g., Figs. 5.1 to 5.3 and 5.17). Even variations in the composition of ejecta from single or successive eruptions sometimes cannot be explained quantitatively by crystal fractionation (Sects. 4.3 and 4.4). Xenoliths can be interpreted as coagulated phenocrysts precipitated at low pressure subsequent to development of diverse liquid compositions at greater depth, and xenoliths are not universal anyway. Isotopic uniformity has alternative explanations and is violated in at least a third of the active volcanic arcs.

However, the principal objections are even more fundamental. If differentiation of basalt is the cause, why is orogenic andesite so restricted to convergent plate boundaries? Why are the presumptive parents and the necessary crystal cumulates so often missing or volumetrically minor? Why do orogenic andesites frequently differ in composition, especially in Fe-enrichment behavior, from other differentiates of basalt? These obvious reservations are long-standing because answers are subtle. I shall return to them in Section 12.2.

Other objections are less obvious. Minor and trace element contents of orogenic andesites rarely can be accounted for quantitatively by crystal fractionation schemes (e.g., Taylor et al. 1969b; Gill 1978), and will be discussed separately below. Examples of mineralogic disequilibrium, such as summarized in Section 6.11, and especially instances of incompatible mineral assemblages or multi-mineralic reverse zoning in excess of that attributable to decompression, are inexplicable by crystal fractionation alone and require its augmentation by mixing or assimilation processes as well.

Also, what little is known about the physics of magma ascent and storage is unfavorable to simple crystal fractionation models. Crystal fractionation clearly is prevented if magma ascends with superheat due either to adiabatic decompression of nearly anhydrous liquid, or along the more realistic decompression paths calculated by Marsh (1978), or at the km/day rates discussed in Section 3.5. This objection does not apply to isobaric crystallization within magma storage areas, but viscosities and volumes in these areas probably are such that turbulent convection occurs which, together with the appreciable yield strength of andesite liquid, will retard or preclude crystal-liquid separation except by flow differentiation or filter pressing (Sect. 4.1.4). Nevertheless, until the fluid dynamic and thermal behavior of magma is better known, these potentially powerful objections can be subordinated to other better-developed arguments.

Consequently, circumstantial arguments for andesite genesis via magmatic differentiation are strong but not without serious objections. Specific differentiation mechanisms are considered separately below.

11.2 Roles of Plagioclase, Pyroxenes, and Olivine

The simplest differentiation mechanism is crystal fractionation involving phenocryst minerals. Such a mechanism seems plausible for many orogenic andesites. Plagioclase, augite, and orthopyroxene are the three most ubiquitous and abundant phenocrysts in orogenic andesites (Sects. 6.1 and 6.2) whose whole rock compositions define an apparent plagioclase + augite + orthopyroxene cotectic trend within the basalt tetrahedron (Sect. 5.2.7). Experiments show that plagioclase together with a pyroxene are the liquidus or near-liquidus phases of orogenic andesites at low load and water pressures (Sect. 6.9). Appropriately, the compositions of these phenocrysts, or of these minerals in mafic xenoliths, are consistent thermodynamically with equilibrium between mineral and host at crustal depths and realistic temperatures (Sect. 6.11). Consequently, many orogenic andesites seem to be multiply saturated with all three phases which should, therefore, figure prominently in differentiation schemes. Such crystal fractionation would have to take place at crustal level pressures or at the crust-mantle interface to account for the significant role of plagioclase, as discussed in Section 6.9, and to account for the phenocryst compositions whose low-pressure traits were summarized in Section 6.11.

Indeed, these three minerals are quite successful in accounting for the iron-free composition of some orogenic andesites. Specifically, graphical, normative, and least squares calculations of solid residua which must be removed to generate one andesite from another or from basalt within possibly consanguineous suites usually require that 50 to 70 wt.% of the solids be plagioclase with the orthopyroxene/augite ratio increasing as the suites become more silicic (Table 11.1). Clearly, water contents cannot be high enough to suppress plagioclase crystallization dramatically. These inferred relative mineral percentages rarely match those of phenocrysts, with plagioclase being still more abundant amonst phenocrysts.

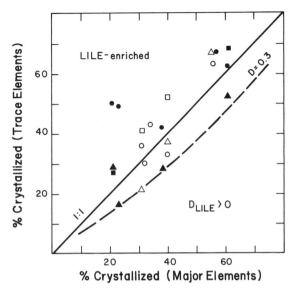

Fig. 11.1. Comparison of percents of crystallization (f_s) involved during andesite evolution. Estimates using major elements are from Table 11.1. Estimates using trace elements assume that the bulk distribution coefficient $\bar{D}_i = 0$ so that $f_s = (C_i^P/C_i^D) + 1$, where C_i^P and C_i^D are concentrations of i in parent and daughter, respectively. Symbols: *circles* i = Rb or Cs; *squares* i = Th; *triangles* i = Zr or Hf. *Open symbols* are from island arcs; *filled symbols* from continental margins. Agreements between estimates lie on the *solid line* and are expected if the crystal fractionation models based on major element considerations in Table 11.1 are correct, if trace element data are accurate, and if $\bar{D}_i = 0$. Data *above the line* indicate one or more condition is unmet, such as LIL-element enrichment due to magma mixing, crustal contamination, or vapor fractionation as well as crystal fractionation. The continental margin examples here are not consistently more LIL-element-enriched than island arc samples. Data *below the solid line* may indicate that $\bar{D}_i \neq 0$. The *dashed line* is for $\bar{D}_i = 0.3$ and fits most of the Zr data

However, the inferred percentages are broadly similar to those in pyroxene gabbro xenoliths (cf. Lewis 1973; Stern 1979) and are appropriate to the cotectic trend within the basalt tetrahedron described above. Thus, the excess plagioclase phenocrysts probably accumulated due to their low density contrast with liquid.

Fractionation of olivine instead of, or in addition to, orthopyroxene is necessary to account for compositional variations in some of the suites summarized in Table 11.1, and for derivation of basic andesites from potentially primary basalt. This is a realistic supplement for several reasons. As noted in Section 6.4, olivine crystals occur in a majority of basic andesites, but begin a reaction relationship with liquids containing between about 52% and 60% SiO_2 to produce orthopyroxene or hornblende, thus limiting olivine's role in andesite genesis. The maximum silica content of olivine-phyric andesites is greater in high-K versus low-K and in calcalkaline versus tholeiitic varieties, presumably indicating where a_{SiO_2} is least and where the role of olivine fractionation is correspondingly greatest.

Table 11.1. Crystal fractionation models of andesite genesis [a]

	Low-K		Medium-K	
	Tonga		Medicine Lake	
Match [b]				
SiO_2	54.1 → 57.6	57.6 → 64.8	54.2 → 55.7	56.1 → 62.2
FeO*/MgO	2.0 → 3.6	3.6 → 5.1	1.7 → 2.3	2.4 → 4.0
K_2O	0.49 → 0.70	0.70 → 1.11	0.85 → 1.07	1.15 → 1.81
Disagreements	none	none	none	none
Conditions [c]				
% crystallized	40	31	23	38
Xpl	0.63	0.38	0.74	0.53
Xaug	0.24	0.42	0.11	0.23
Xopx	0.10	0.03	–	0.10
Xol	–	–	0.15	0.03
Xmt	0.03	0.17	–	0.11
Xhb	–	–	–	–
Requirements [d]				
$\bar{D}_{Ni}$	3.7	4.0	5.0	–
$\bar{D}_{V}$	0.5	4.8	0.9	3.6
$\bar{D}_{Cr}$	5.3	2.1	–	–
$\bar{D}_{Yb\ or\ Lu}$	–	–	–	–
References	1	1	2	2

a   All based on least squares solutions to linear mixing equations wherein phenocryst minerals are subtracted from parent liquids in proportions which result in the closest approximation of derivative liquid compositions

b   Characteristics of observed parent-daughter relationships used. Disagreements are oxides where observed and calculated concentrations differ by 0.1 wt.% or more. (+) means that concentrations are greater in the observed derivative liquid than in the calculated one. If complete results are unavailable in the reference, the sum of squares of residuals (R^2) is given instead. Arrows connect observed compostions of parent and daughter

c   Results of calculations. Removal of the percent shown of crystals, having the weight fraction (Xpl, etc.) given, accounts for the "Match" specified in b

d   Bulk trace element distribution coefficients ($\bar{D}$) over the crystallization interval listed in "Conditions" (c) which are required by the observed concentrations in parent and derivative rocks. Parentheses indicate values which are incompatible with the range of coefficients in Table 6.3

Colima	Atitlan	Agrigan	Grenada	High-K San Salvador	Filicudi
57.6 →60.4	50.9 →58.3	53.3 →57.4	50.5 →56.8	51.8 →60.7	54.4 →59.5
1.5 → 1.7	1.5 → 2.1	2.4 → 4.1	1.3 → 2.0	2.2 → 5.5	1.8 → 1.9
1.16→ 1.40	0.66→ 1.57	1.20→ 1.80	1.14→ 1.60	1.23→ 2.71	2.02→ 2.78
none	none	$R^2 = 0.06$	K_2O (+)	SiO_2 (−)	$R^2 = 0.05$
			TiO_2 (−)	Na_2O (+)	
21	61	34	56	57	32
0.50	0.63	0.49	0.44	0.55	0.50
0.07	0.14	0.28	0.12	0.32	0.30
0.06	−	0.06	−	−	0.11
−	0.15	0.09	−	0.06	−
0.02	0.07	0.07	0.06	0.06	0.09
0.34	−	−	0.39	−	−
3.8	2.2	−	(1.3)	1.9	2.8
(0.6)	1.7	−	(1.4)	2.7	−
4.2	1.9	−	(1.1)	(1.2)	2.3
0.6	0.7	−	−	−	0.6
3	4	5	6	7	8

References:

1. Ewart and Bryan (1973); Ewart et al. (1973)
2. Mertzman (1977)
3. Luhr and Carmichael (1980)
4. Woodruff et al. (1979)
5. Stern (1979)
6. Arculus (1976); Brown et al. (1977), (based on averaged analyses)
7. Fairbrothers et al. (1978); Mayfield et al. (1980)
8. Villari and Nathan (1978)

Some olivine crystals are more magnesian that would precipitate in equilibrium from their hosts, suggesting accumulation of these xenocrysts from a more mafic parent. Also, olivine occurs in mafic xenoliths even when absent as phenocrysts (Sect. 6.8), and is a liquidus phase of some orogenic andesites at relatively high water contents (Sect. 6.9).

The extent of olivine fractionation depends on the parental liquid. If parents are picritic basalts, then extensive olivine fractionation is necessary. If, instead, quartz-normative basalts or basic andesites with Mg-numbers $\geqslant 66$ but MgO and Ni contents of 8% to 10% and 100 to 150 ppm, respectively, can be produced in equilibrium with the mantle wedge under hydrous conditions, then relatively little olivine fractionation need occur. Specifically, addition of 5 to 20 wt.% olivine to basic andesites or associated basalts in several volcanic arcs is sufficient to produce a potentially primary liquid composition, i.e., one whose Mg-numbers and Ni contents are at least compatible with a peridotitic residuum (e.g., Jakeš and Gill 1970; Nicholls and Ringwood 1973; Nicholls and Whitford 1976; Lopez-Escobar et al. 1977).

Thus, fractionation of plagioclase and pyroxene(s) ± olivine can quantitatively account for the iron-free major element variations within some suites of orogenic andesite, and account for the evolution of such suites from potentially primary basaltic precursors. However, because pyroxenes and olivine have FeO*/MgO ratios lower than those of the magmas from which they precipitate, and because fractionation of these minerals is an inefficient means of increasing the silica content of the residual liquid, this proposal is able to explain neither the relative lack of Fe-enrichment nor the unusual abundance of andesites in orogenic suites. Consequently, it is necessary to invoke precipitation of magnetite, hornblende, or garnet, all of which have high FeO*/MgO ratios but low silica contents.

11.3 Role of Magnetite and the Plagioclase-Orthopyroxene/Olivine-Augite-Magnetite (POAM) Model

Precipitation of magnetite as a means of promoting silica-enrichment while preventing iron-enrichment was proposed by Kennedy (1955) and Osborn (1959) and amplified in a series of subsequent papers (Osborn 1962, 1969, 1976, 1979; Hamilton et al. 1964; Presnall 1966; Roedder and Osborn 1966; Hamilton and Anderson 1968; Eggler 1974). Briefly, Osborn and his colleagues argued from experimental determinations of phase boundaries in increasingly complex systems that crystallization of basalt or andesite leads to absolute iron-enrichment in the liquid until and sometimes even after magnetite saturation occurs. The more constant f_{O_2} remains, and the higher f_{O_2} is in the primitive magma relative to temperature, the less silicate fractionation occurs before magnetite-saturation (see Figs. 11.2 and 12.2) and the less relative iron-enrichment occurs after magnetite-saturation. Precipitation of magnetite forces liquid compositions toward SiO_2 and

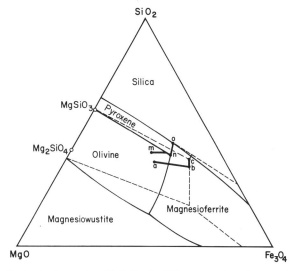

Fig. 11.2. Phase relationships at 1 atm in the system $MgO\text{-}Fe_3O_4\text{-}SiO_2$ at constant f_{O_2} equal to that of air *(solid lines)* and at constant $CO_2{:}CO$ ratio of 132 *(dashed lines)*, after Osborn (1976). *Lettered points* are keyed schematically to Figs. 12.1 to 12.4, although the order of pyroxene and magnetite appearance differs between this figure and those. The line *abc* represents a tholeiitic fractionation trend under relatively reducing conditions. Olivine alone precipiates along segment *a-b*. Magnetite (magnesioferrite) appears at *b*, preventing further increase in Fe_3O_4/MgO ratio and causing silica enrichment along segment *b-c*. Under more oxidizing conditions *(solid lines)*, magnetite appears at *n* after less silicate crystallization. Phase relationships here imply unrealistically constant FeO/Fe_2O_3 ratios, but similar relationships exist in more complex systems (see references in text)

away from FeO*, resulting in a kink on an FeO*/MgO versus SiO_2 diagram among other things. Thus, crystal fractionation of plagioclase, pyroxene(s), and magnetite ± olivine is probably the most frequently invoked differentiation mechanism by which to explain andesite genesis. Its apparent successfulness is clear in Table 11.1, whose first two column pairs also illustrate differences before and after magnetite saturation occurs.

Although the ability of magnetite fractionation to cause silica-enrichment is obvious, the similar role of calcic plagioclase fractionation in the POAM model is at least as great. Inclusion of 2 wt.% magnetite within pyroxene gabbro removed from basaltic or andesitic liquid reduces the silica content of the gabbro by about 1 wt.%, thereby increasing its effectiveness for silica-enrichment. If the gabbro contains at least 50 wt.% plagioclase (Table 11.1), then an increase of 8 mol.% An in the plagioclase has the same effect. Consequently, the influence of water on the shape of the plagioclase melting loop which increases the An content of crystals in equilibrium with any given hydrous versus anhydrous melt (Johannes 1978; Fig. 6.1) is as much responsible for silica-enrichment in orogenic andesitic liquids as is the earlier precipitation of magnetite therein.

Following Carmichael and Nicholls' (1967) and Smith and Carmichael's (1968) studies of Cascade andesites it has been common to discredit the POAM model by asserting the absence of magnetite phenocrysts in orogenic andesites and associated basalts. Also, magnetite is a phenocryst in some basalts whose segregation veins (i.e., whose differentiates) are Fe-rich, whereas magnetite phenocrysts are absent from other basalts or basic andesites whose segregation veins are similar or even depleted in FeO* contents relative to their hosts (Fig. 5.4b); this is opposite to Osborn's prediction. However, magnetite occurs as microphenocrysts (0.1 to 0.25 mm), groundmass microlites, or inclusions in silicate phenocrysts in virtually all orogenic andesites, and is reported as a "phenocryst" in 50% to 90% of them, (Ewart 1976a), including many from the Cascades. Thus the issue of modal occurrence usually is one of grain size and relative amount rather than existence.

Although relevant kinetic experiments are unavailable, it is possible that the smallness of magnetite crystals results from magnetite having a slower growth rate than silicates (Árakai 1968) or easier nucleation (Carmichael 1964). Magnetite clearly differs greatly in crystal structure and composition from silicates and has a much higher entropy of fusion (ΔS_m = 17.3 entropy units; Henriquez and Martin 1978) which affects crystallization behavior. However, conventional wisdom predicts that crystallization rate is directly proportional to ΔS_m and therefore predicts that magnetite crystals should exceed silicate crystals in size if all are near-liquidus phases.

Nevertheless, several geologic, mineralogic, and geochemical arguments for the early precipitation of magnetite strongly suggest that magnetite is a near-liquidus phase in those portions of many suites of orogenic andesite having at most a mild increase in FeO*/MgO ratios relative to silica. In addition to the modal data cited above, these arguments include the restricted Mg/Fe ratios of silicates in such andesites (Chap. 6), the early (high An content) maximum of FeO* contents in plagioclase (Sect. 6.1), small magnetite inclusions in silicate phenocrsts and the ubiquity of coarse magnetite in mafic inclusions (Sect. 6.8), and the coincidence between the initial appearance of magnetite phenocrysts in ejecta following dormancy during stratovolcano evolution and an irreversible change from tholeiitic to calcalkaline iron-enrichment patterns (Sect. 4.3). The exotic occurrence of gas-charged magnetite lava flows interbedded with Plio-Pleistocene calcalkaline andesite lavas at El Laco, near Lascar volcano, northern Chile (Henriquez and Martin 1978), attests to locally significant concentrations of magnetite, as precipitates or an immiscible liquid, at such volcanoes.

The geochemical arguments for magnetite precipitation are twofold. The similarity between Fe^{2+} and Ti, Zr, Hf, and possibly Nb and Ta behavior in orogenic andesites (Sects. 5.2.4 and 5.4.4) strongly suggests magnetite involvement, at least in calcalkaline suites, because magnetite alone amongst possible anhydrous liquidus minerals of andesite has distribution coefficients greater than 1.0 for these cations and, therefore, can account for their negative correlations with silica (Table 6.3). (Magnetite fractionation also will increase Th/U and δO^{18} ratios in liquid but these effects, if present, are too small to be identified.)

Second, although similarity between V contents of average andesites and basalts has been cited as evidence against magnetite fractionation (Taylor et al. 1969b), within specific suites such similarity is present only while rapid Fe-enrichment occurs. More frequently, V contents decrease steadily while silica increases and FeO*/MgO ratios remain relatively constant (Sect. 5.4.5). Indeed, a bulk distribution coefficient of 2 to 5 for V is necessary to explain V behavior quantitatively in suites lacking rapid Fe-enrichment (Table 11.1 and Sect. 5.4.7) and this usually requires magnetite participation (Table 6.3). Thus, V behavior provides evidence for, not against, magnetite fractionation during andesite genesis. (The similar ability of hornblende to explain both these geochemical observations is noted in the following section.)

There remain, however, five other significant objections to the POAM fractionation model. Obviously in those instances where modal magnetite is absent and indirect arguments for its presence are missing or ambiguous the model is inapplicable or contrived. Enhanced magnetite stability at high pressure has been invoked to explain its efficacy despite absence as a phenocryst, an absence which could, therefore, be explained by resorption upon decompression (Osborn 1969, 1979). However, the beneficial effect of pressure on magnetite stability is debatable at best (Eggler and Burnham 1973; Eggler 1974).

Second, the amount of magnetite needed to account for changes in major element or V contents within suites usually is 6 to 11 wt.% of the solid phases (e.g., Table 11.1). Even when magnetite phenocrysts are present, this amount typically exceeds their modal proportion amongst phenocrysts by factors of two or more. Note, however, that the wt.% of magnetite usually is about 1.5 times its vol.%, that magnetite would settle more rapidly than silicates once liquid yield strength is exceeded, that magnetite crystals are included in most silicates, and that filter pressing or chamber-wall plating could explain liquid-magnetite separation in much the same way that they have been invoked to explain the role of augite during evolution of MORB (Hodges and Papike 1976; Dungan and Rhodes 1978). Although admittedly ad hoc, all four observations may explain the apparent under-representation of magnetite amongst phenocrysts. Significantly, magnetite commonly constitutes 5 to 15 modal % of cumulate-textured xenoliths found in hosts containing only 1 to 5 modal % magnetite phenocrysts (e.g., Arculus and Wills 1980).

Third, magnetite apparently is not a liquidus phase of either orogenic andesite or basalt in experimental charges when f_{O_2} is near or below the NNO buffer (Sect. 6.9), which has been thought maximum for island arc magmas (Eggler and Burnham 1973). However, the extent of prima facie evidence for magnetite precipitation which was cited above suggests that this experimental discrediting of magnetite fractionation is wrong for one or more of the following reasons. Oxygen fugacity in natural orogenic andesite magmas apparently is up to one log unit above NNO (Sect. 5.3.3) and therefore natural conditions are more oxidizing than in the experiments. Magnetite stability is highly sensitive to f_{O_2} and magnetite becomes a near-liquidus phase of hydrous andesite at f_{O_2} between the NNO and HM buffers (Fig. 6.7). Second, Fe probably was lost to crucibles during some of the experiments leading to underestimates of magnetite stability. Third, demonstrations that magnetite is a sub-liquidus phase at 1 atm pressure ignore the suppression of the liquidus caused by a few wt.% water. Fourth, magnetite stability is a function of Fe^{+3}/Fe^{+2} ratios which increase as the activity and radius of cations which modify the polymerized silicate liquid network increase (Paul and Douglas 1965; Lauer and Morris 1977). Thus, magnetite has higher thermal stability in high-K relative to low-K orogenic andesites, at constant f_{O_2}. Finally, the crude correlation betwen P_2O_5 and FeO* behavior in andesites (Sect. 5.2.6) and demonstration that apatite affects magnetite stability (Philpotts 1967), suggests that $a_{P_2O_5}$ influences magnetite solubility in andesites and differs between experiments and natural magmas.

The fourth objection to the POAM model is that it is not always able to account for variations in trace element contents within orogenic andesite suites. This can be demonstrated in two ways. First, one can use results of least squares calculations such as summarized in Table 11.1 to determine the percent solidification and weight fraction of minerals necessary to explain major element variations within suites of arguably consanguinous rocks. These values, together with distribution coefficients (Table 6.3), allow one to calculate likely changes in trace element concentrations due to Rayleigh-type fractionation. Typical examples are available in the references cited in Table 5.3. Frequently there is a more rapid increase in LIL-element concentrations within orogenic andesite suites than is predicted by models of low-pressure anhydrous crystal fractionation, a less rapid increase in REE, especially heavy REE, less of negative Eu anomaly, and a less precipitous drop in transition metals, especially Ni and Cr, than precipitated by calculations.

Alternatively, trace element data alone can be used to determine bulk distribution coefficients, possible fractionating minerals, and degrees of crystallization (Allègre et al. 1977; Minster et al. 1977). This approach has not yet been successfully applied to suites of orogenic andesites and, even when successful for basalts, it breaks down when over 50% crystallization occurs and intermediate differentiates are encountered.

For illustration, a combination of the two approaches just described is given for the ten parent-daughter andesite-containing pairs in Table 11.1 and Fig. 11.1. Instead of comparing calculated and observed trace element contents in daughters, I calculated representative compatible element bulk distribution coefficients which must apply if the fractionation scheme predicted from major element considerations is applicable. Most necessary coefficients lie within the range of possibilities represented by data compiled in Table 6.3. However, as mentioned above, necessary heavy REE coefficients are about 0.6 which requires D> 1 for pyroxenes and magnetite, or fractionation of apatite (see below), or both. I also assessed observed versus predicted incompatible trace element enrichments by comparing degrees of crystallization determined by assuming $\overline{D} = 0$ for certain elements with the degree of crystallization determined using major elements (Fig. 11.1). Note that the POAM fractionation model rarely accounts rigorously for incompatible trace element behavior even when K_2O contents are explicable, that alkali and especially Th often are enriched beyond expectation, and that Ti-group elements are not H-type trace elements in the sense of Section 5.4.7 during differentiation to andesites, i.e., they have bulk distribution coefficients > 0.1. The infrequence and small magnitude of Eu anomalies in orogenic andesites, like the stability of magnetite, is attributed to the factors listed earlier which cause high M^{3+}/M^{2+} ratios in andesitic melts.

The examples of POAM fractionation collected in Table 11.1 are among the most convincing available. Many are worse, resulting in more disagreements and higher R^2 values for major elements, and more inexplicable trace element behavior. Nevertheless, predicted trace element changes usually are appropriate in sign and discrepant only in magnitude. Consequently, failures of the POAM model can be assigned to lack or consanguinity of the selected samples, the effect of accumulated phenocrysts, or the incorrectness of distribution coefficients, as well as to inapplicability or incompleteness of the model (e.g., to disequilibrium or to the simultaneous occurrence of other differentiation processes; see Fig. 5.17b,c,d, for an example where the POAM model is generally successful but must be supplemented by magma mixing in some cases, assuming all compositions are accurate and those of liquids.)

In some cases, however, trace element data clearly disallow the model even when circumstantial arguments favor it. One example is the discovery that, although formation of some calcalkaline from tholeiitic magmas along the volcanic front of northeastern Japan seems to accompany precipitation of magnetite during repose late in the history of cone construction (Sect. 4.4), other calcalkaline andesites there have Cr and LIL-element contents too high to be derived from their tholeiitic predecessors by crystal fractionation (Masuda and Aoki 1979). Nevertheless, trace element data seem to me less at odds with low-pressure crystal fractionation of plagioclase, pyroxene(s), and magnetite ± olivine than with other proposed mechanisms (Gill 1978).

The final objection is that POAM fractionation at shallow depths cannot explain suites such as in southwestern Japan (Aoki and Oji 1966) or Grenada, Antilles (Arculus 1976) where rock compositions cross the Fo-Di-Ab thermal divide of the basalt tetrahedron. However, such crossings are uncommon and anhydrous crystal fractionation at medium pressures can produce the observed result anyway, involving fractionation of olivine, plagioclase, and augite at 3 to 5 kb or aluminous augite and spinel plus either plagioclase or olivine at 5 to 20 kb (Presnall et al. 1978). Hornblende fractionation can achieve the same end but is not necessary to do so.

In summary, although potentially insuperable problems exist, crystal fraction of plagioclase-orthopyroxene/olivine-augite-magnetite (POAM) seems the most viable though by now commonplace explanation of the geology, mineralogy, and chemical composition of many (probably most) orogenic andesites. Judging from Table 11.1 and arguments summarized in Section 11.2, approximately 5 to 40% of POAM crystallization from a primary basalt is necessary to produce basic andesites, depending on criteria for primaryness and the relative proportions of minerals crystallizing. Amongst andesites themselves, approximately 5% to 15% of POAM crystallization is needed for each 1 wt.% increase in SiO_2 content of derived liquid, the figure decreasing as differentiation increases. These overall degrees of crystallization during andesite evolution are less than commonly believed (e.g., Kuno 1968b), in part due to use of all major elements in calculations instead of using only incompatible elements as constraints (see Fig. 11.1).

11.4 Role of Amphibole

One of the most obviously distinctive characteristics of magmas at convergent plate boundaries is that they are hydrated (Sect. 5.3.1). However, POAM-fractionation is equally applicable to anhydrous or hydrous magmas and shows only reluctant concessions to the presence of water, viz. by increasing the An content of plagioclase while suppressing its melting point to near that of pyroxenes and olivine, by increasing f_{O_2} and therefore magnetite stability, and by decreasing liquid density. Although subtle, these effects do contribute to differences in fractionation trends between orogenic and anorogenic subalkaline magmas (Sect. 12.2), including the need for modest P_{H_2O} to cause liquidus temperatures of the orogenic variety to decrease with increasing FeO^*/MgO ratios (Sect. 4.1.1).

In contrast, a more dramatic consequence of water in magma is to stabilize a hydrous mineral such as amphibole as a liquidus phase. This consequence is implicit in Bowen's reaction series and was invoked by him to explain the origin of Alaskan andesites and related rocks (Bowen 1928). The role of amphibole fractionation in andesite genesis has been re-emphasized more recently by Yamazaki et al. (1966), Lewis (1973), Boettcher (1973, 1977), Allen and Boettcher (1975, 1978), and Cawthorn and O'Hara (1976).

The strongest evidence for a significant role of amphibole fractionation is its effectiveness in explaining some geochemical phenomena. Essentially, amphibole fractionation can explain the same phenomena as can magnetite fractionation, and others also. Salient aspects of the composition of amphiboles precipitating from orogenic basalts or andesites are their low silica contents (most are *ne*-normative), relatively high FeO^*/MgO ratios (higher than other silicates but lower than host liquid), relatively high K/Na ratios (but again lower than host liquid), position within basalt tetrahedron, and high distribution coefficients for V, heavy rare earth, and Ti-group trace elements (Sects. 6.3, 5.2.7, and 6.10). Consequently, amphibole fractionation could explain enrichment of liquid in silica but not iron thereby maintaining high Mg/Fe ratios of other minerals during differentiation, could also explain the apparent cotectic trend of suites containing orogenic andesites within the basalt tetrahedron including penetration of the Fo-Di-Ab low pressure thermal divide, and explain the apparent compatibility of some trace elements, especially heavy REE, in orogenic andesites including crossovers of REE patterns and negative correlations between heavy REE (or Y) and silica. Fractionation of amphibole accompanied by little or no plagioclase also could explain development or normative corundum in some orogenic andesites as well as of suites which have nearly constant normative *ab/or* or Na_2O/K_2O ratios.

Experimentally, conditions necessary for amphibole saturation have been thought more realistic than those necessary for magnetite saturation of orogenic basalts and andesites so that, in principle, amphibole fractionation has seemed likely as well as effective, and preferable to the POAM model. Moreover, hornblende and plagioclase are mutually exclusive as liquidus phases of andesite so that hornblende-only fractionation is more likely than hornblende + plagioclase fractionation from an experimental point of view.

However, discovery of the water contents, temperatures, and f_{O_2} prevailing during andesite genesis have reversed the situation described above. Magnetite saturation now seems likely despite experimental evidence to the contrary whereas, as discussed in Section 6.11, conditions necessary for amphibole to occur *on the liquidus of orogenic andesite* (i.e., pressures > 8 kb, water contents > 10 wt.%, and temperatures $< 1000°C$) all are unlikely or at best of restricted applicability (e.g., to high-Mg andesites which contain only hornblende phenocrysts).

Physical evidence for amphibole fractionation in orogenic andesite suites, like that for magnetite, is contentious because hornblende is an unambiguous phenocryst in relatively few andesites. Tschermakitic hornblende inclusions or phenocrysts are found in some medium-K basalts and basic andesites (Lacroix 1904; Jakeš and White 1972b; Arculus 1978; Bultitude et al. 1978; Mertzman 1978), and andesite literature frequently asserts that hornblende "ghosts" haunt most andesites, an assertion usually attributed to Yoder (1969). However, only about 18% of Ewart's (1967a) data set of volcanic arc rocks contained even identifiable pseudomorphs of hornblende, and only 9% of the orogenic andesites in Chayes' data file (see Sects. 1.1 and 5.1) originally were called hornblende andesites. Some of these hornblendes result from magma mixing (Sect. 11.7) or sub-liquidus reactions (Sect. 6.11). Even so, hornblende phenocrysts are largely restricted to medium and high-K andesites, to topographically and stratigraphically high levels of stratovolcanoes, and to high levels within magma chambers (Sect. 4.4), although I have no statistical measures of "largely". Thus, the available systematic data cited above indicates that these megacrysts and ghosts are far from ubiquitous and, indeed, uncommon except in the special circumstances listed.

However, amphibole is especially susceptible to breakdown during decompression and therefore to loss of credit for its role at high pressure in andesite genesis.

The shallow positive dP/dT slope of amphibole stability causes this breakdown to anhydrous products during decompression (Sect. 6.9), with degree of breakdown depending on temperature and rate of ascent. The results are opacite rims on, and eventually ghosts of, hornblende. However, as summarized above, both have limited and quite specific occurrences in andesites. Also, those hornblendes which do exist have low Al^{VI} contents and therefore formed at low pressure (Sect. 6.3).

Much more ubiquitous are crystal clot mafic inclusions of plagioclase + pyroxene(s) + magnetite (Sect. 6.8) whose mineralogy is that to which hornblende breaks down. These inclusions, therefore, have been interpreted as relict hornblendes (Stewart 1975). I disbelieve this interpretation in most instances because the minerals and mineral proportions in clots are variable and often do not sum to hornblende equivalence, because clot minerals are invariably similar in kind and often in composition to phenocrysts, because the clot plagioclase usually is zoned oscillatorily, and because the grain size of clots is much coarser than in opacite rims (Sect. 6.3). Also, hornblende is rarely included in other minerals in either the inclusions or the phenocrysts of clot-containing pyroxene andesites, and therefore was not a co-precipitate. Finally, similar clots occur in anorogenic andesites where hornblende is an even less likely liquidus phase (Sect. 11.9).

The most persuasive physical evidence of amphibole fractionation comes from mafic inclusions containing cumulus-textured hornblende which, although less common than the pyroxene gabbro clots discussed above, occur in many volcanic arcs, and especially those studied in England (Sect. 6.8). Mineral compositions in the inclusions (Fig. 6.5) often suggest precipitation from more mafic liquids or at higher P_{H_2O} than for phenocryst minerals. However, in these xenoliths, as in plutonic rocks in general, there is no assurance that the hornblende precipitated near the liquidus rather than due to lower temperature reaction relationships between liquid and olivine or pyroxene. This is especially true when hornblende is poikilitic in texture, as in most inclusions. Similarly, there is no assurance that the liquids from which the inclusions precipitated were not atypically hydrous, thereby explaining their crystallization at depth rather than eruption.

Arguments based on elemental behavior are ambiguous and largely qualitative. Quantitative arguments such as discussed for the POAM model in the preceding section are hampered by one's inability to specify amphibole compositions with sufficient rigor to choose between arithmetically satisfactory solutions. In some instances, however, fractionation of amphibole, alone or together with other minerals, can be shown to account for major element variations within suites. Nevertheless, note in the examples chosen for Table 11.1 (columns 5 and 8) that 30 to 40 wt.%.hornblende fractionation is necessary but inconsistent with the behavior of one or more compatible trace elements, and that 40 to 50 wt.% of the solid assemblage is plagioclase, whereas plagioclase and hornblende are mutually exclusive as near-liquidus phases of andesitic liquids. One or the other problem is common to all hornblende fractionation models. Furthermore, if $D_V^{hb} > D_{Cr}^{hb}$ and

$D_{Co}^{hb} > D_{Ni}^{hb}$ (Sect. 6.10), then the behavior of ratios of these elements in orogenic andesites is inconsistent with hornblende-dominated fractionation.

Ironically, both hornblende and magnetite fractionation are appealing ideas which suffer from similar lacks of prima facie petrographic evidence and experimental confirmation. However, the absence of modal hornblende is more widespread than that of magnetite, the reasons for absence of magnetite from experiments are more obvious, and the indirect geologic and geochemical evidence for magnetite fractionation is more persuasive to me. Some, possibly most, hornblende in andesite forms at temperatures *below the liquidus* (Sect. 6.11). Thus, I conclude that hornblende fractionation is relatively uncommon and usually restricted to medium or high-K acid calcalkaline andesite magmas within the upper regions of crustal-level magma reservoirs.

In summary, the kind of hornblende-only fractionation during andesite genesis proposed by Boettcher (1973, 1977) and Cawthorn and O'Hara (1976) is uncommon if applicable at all. Liquids generated thereby can cross the Fo-Di-Ab plane within the basalt tetrahedron and will be identifiable by concave-up and crossing REE patterns, development of normative corundum, unusually slight increases in *ab/or* or Na/K ratios, slight Fe-enrichment, and possibly by increasing Ni/Co and Cr/V ratios. Alternatively, plagioclase + magnetite + hornblende ± pyroxene ± olivine fractionation may occur once water contents in the liquid phase of magma exceed 3 wt.%, although suitable P-T conditions for this are as yet unknown. Possible well-studied examples include the Grenada and Colima suites represented in Table 11.1, and Ararat, Turkey (Lambert et al. 1974). Apart from petrographic evidence of hornblende, the best clues to its involvement during multimineralic fractionation are low quench temperatures and greater apparent compatibility of Ti-group and heavy rare earth elements, but the latter criterion awaits improvement in resolving the distribution coefficients of augite and hornblende.

11.5 Role of Garnet

Crystal fractionation dominated by plagioclase is applicable to differentiation within the crust or at its base in island arcs, whereas fractionation dominated by amphibole is applicable mostly to depths between 25 and 65 km. At still higher pressure, appropriate to depths within or immediately above subducted lithosphere, garnet joins clinopyroxene as a near-liquidus phase of orogenic basalts and andesites. Consequently, eclogite fractionation provides another possible mechanism of andesite genesis (Green and Ringwood 1968a; Green 1972). According to this proposal, like the preceding one, differentiation of basalt occurs at high pressure but mineralogic evidence of the process is lost due to crystal resorption during ascent into shallow reservoirs where the observed phenocrysts would precipitate from already differentiated liquids.

As with amphibole fractionation, the strongest argument for eclogite involvement is that it also enriches liquids in silica but not iron due to, in this case, the high FeO*/MgO ratios of garnets in equilibrium with orogenic andesite, especially under hydrous conditions, at least at temperatures $< 950°C$ (cf. Green 1977; and Stern and Wyllie 1978). Eclogite fractionation also would explain instances of steep REE patterns, negative correlations between heavy REE and silica, crossovers in REE patterns, and relatively constant Ni contents (Sect. 6.10). Moreover, it should do so discreetly at great depth so that absence of parental magma or cumulates or both could be disregarded more easily, especially because the density contrast between minerals and liquid is higher in this than in other models, promoting efficient separation. Absence of crustal-level magma chambers (Sect. 3.6) is accounted for.

As required by the model, orogenic basalts and andesites become saturated with both garnet and clinopyroxene between 20 and 30 kb in the presence of < 5 wt.% H_2O (Sect. 6.9); at higher water contents andesites may have only garnet as a liquidus phase. Indeed, some orogenic andesites contain garnet megacrysts which arguably are cognate and sometimes are rimmed by lower pressure mineral assemblages, suggesting reaction during decompression (Sect. 6.6).

However, many things conspire to preclude this suggestion. First, what few garnet megacrysts there are in andesites occur in uncommon peraluminous varieties which probably assimilated pelitic sediments and precipitated garnet as a penalty. This interpretation is consistent with the high $^{87}Sr/^{86}Sr$ ratios of garnetiferous andesites and with their restriction to continental fragments and margins. The assimilation-precipitation process, at least when accompanied by cordierite megacrysts, probably takes place at depths < 20 km, although the specific conditions under which the Mn-poor almandines of orogenic andesites precipitate from their hosts is unknown. At least these garnets are not relict high-pressure phases (Sects. 6.6, 6.7, and 6.9).

Second, the success of eclogite fractionation in accounting for the major element composition of orogenic andesites is illusory. In detail it causes liquid compositions to diverge significantly from the trend of orogenic andesite suites (tholeiitic or especially calcalkaline, low-K or high-K) either in Ca/(Mg + Fe) ratios or in an AFM diagram (Stern and Wyllie 1978; and Fig. 8.3). Third, although fractionation involving high clinopyroxene/garnet proportions could account for the major element behavior of some tholeiitic suites, the mechanism cannot explain the abrupt change from tholeiitic to calcalkaline differentiation trends (e.g., the kinks in FeO*/MgO versus SiO_2 or versus V diagrams: Figs. 5.4 and 5.15), or the associated change in Ti-group element behavior. Fourth, garnet-pyroxene phase boundaries within the basalt tetrahedron at high pressure (Yoder and Tilley 1962) cut the orogenic andesite trend at a high angle and therefore cannot explain it. Finally, eclogite fractionation dramatically fails to explain the trace element features, most notably Sr, Sc, and heavy REE contents, cited in Section 8.3 as evidence against derivation of orogenic andesite from subducted eclogite. Although many models fail trace-element testing. apparently there are no loopholes in tests of eclogite fractionation (Gill 1978).

11.6 Role of Accessory Minerals: Apatite, Chromite, Sulfides, Biotite

Because element distributions in magma are affected by distribution coefficients multiplied by the weight fraction of the appropriate phase, accessory minerals (with very low weight fractions) must have high distribution coefficients to be influential during fractionation. Four possible candidates are considered.

Apatite is the most ubiquitous accessory mineral in igneous rocks, occurring as small euhedral inclusions in silicates or magnetite, and as microphenocrysts in many calcalkaline acid andesites (Sect. 6.7). Indeed, apatite fractionation in amounts of 0.5 to 1.0 wt.% of solids is necessary to prevent phosphorous enrichment in liquid. Because such enrichment occurs only in low-K, tholeiitic basic andesites (Sect. 5.2.6), minor apatite fractionation is necessary to explain P_2O_5 behavior everywhere else. This contrasting behavior, similar to that of magnetite, suggests that increased oxygen or water fugacity increases the thermal stability of apatite as well as magnetite. Although no apatite/andesite distribution coefficients are available, REE (especially Nd to Dy) favor apatite by factors of 10 in basalt to 30 in dacite (Nagasawa 1970; Irving 1978). Consequently, minor apatite fractionation will increase the apparent "compatibility" of REE and F, somewhat alleviating two trace-element objections to POAM crystal fractionation (Sect. 11.4).

Cr-spinels occur only as inclusions in olivine crystals, at least some of which are xenocrystal, in andesites. Because $D_{Cr}^{sp} \geqslant 100$ (Irving 1978), spinel fractionation will dramatically reduce Cr contents in liquid, probably contributing along with pyroxene fractionation to low Cr contents in orogenic basalts and andesites generally (Jakeš and Gill 1970).

Due to their relatively low Fe contents and temperature but high f_{O_2}, orogenic andesites are likely to be S-saturated, especially if derived from basalts by crystal fractionation (Sect. 5.3.3). Distribution coefficients between andesite and sulfide liquids favor concentration of Ni and Cu in the sulfide by factors of 460 and 243, respectively (Rajamani and Naldrett 1978), and the density of the sulfide liquid is high. Consequently, silicate-sulfide fractionation could contribute to the low Ni content of orogenic basalts and andesites, and even to the curtailment of Fe-enrichment. Although the small weight fraction of modal sulfide minerals (Sect. 6.7) and the lack of correlation between Ni and Cu (Sect. 5.4.6) indicate that effects of sulfide fractionation during andesite genesis are small, it is an attractive way to produce liquids with anomalously low Ni/MgO ratios (e.g., region A of Fig. 5.14).

Biotite is a phenocryst in some high-K orogenic andesites and occasionally is present in their mafic inclusions (Sect. 6.7 and 6.8). Biotite fractionation would impede or reverse enrichment of K, Rb, and Ba in liquids while accelerating depletion of Co and V (Table 6.3), and for this reason was invoked by Jakeš and Smith (1970) and Dostal et al. (1977c) to explain suites of biotite-phyric andesites. In addition, because FeO*/MgO ratios of biotites approximate or exceed those of coexisting hornblendes, biotite fractionation will also restrict magmatic Fe-enrichment. No quantitative tests of biotite's role are available, but its fractionation would help explain instances of negative K_2O-SiO_2 correlations (Sect. 5.2.2).

Thus, each accessory phase may have a role and fractionation of apatite and pyrrhotite seem likely in most andesites. The effects on liquid compositions probably are minor but are not well known.

11.7 Role of Magma Mixing

Mixing of fluids during andesite genesis has been invoked twice already in this book, once as a mechanism of slab recycling at high pressure within the mantle wedge (Sect. 9.4), and once as a mechanism of selective assimilation within the crust (Sect. 10.3). In both cases, relative fluid proportions are unknown, and one of the fluids may be impure steam rather than magma. In this section I shall focus instead on mixing of two magmas within the crust as a process of creating or modifying orogenic andesite. Conceptually, the claimed significance of such mixing ranges from major (i.e., as an alternative to crystal-liquid fractionation: Bunsen 1851; Eichelberger 1975) to minor (i.e., back-mixing of liquids which are mutually related via crystal-liquid fractionation: McBirney 1980). The claimed frequency of magma mixing ranges from ubiquitous (Anderson 1976) to rare.

Magmas of diverse composition demonstrably coexist within reservoir systems, and even erupt simultaneously (MacDonald and Katsura 1965; Yoder 1973; Anderson 1976). Several prominent eruptions of andesite within volcanic arcs have vented coexisting magmas: e.g., Fuji, Japan, in 1707 when dacite (SiO_2 = 68%) and andesite (SiO_2 = 53%) erupted simultaneously from adjacent vents (Tsuya 1955); Lassen, USA, in 1915 when dacitic (SiO_2 = 65%) and andesitic (SiO_2 = 60%) glasses were intermingled in pumice but homogenized in lava (MacDonald and Katsura 1965); Soufrière, St. Vincent, Antilles, in 1902 when dacite (SiO_2 = 65%) and andesite (SiO_2 = 57%) were ejected in about a 1:2 ratio during the initial stages of eruption (Carey and Sigurdsson 1978); and the catastrophic eruption of Santorini, Aegean, which may have terminated Minoan civilization around 3400 years B.P. when rhyolite (SiO_2 = 70%) and andesite (SiO_2 = 54%) were extruded (Sparks et al. 1977). The 1970 eruption of Hekla, Iceland, provides an example of co-eruption of dacite (SiO_2 = 67%) and andesite (SiO_2 = 54%) from a different tectonic environment (Thorarinsson and Sigvaldason 1972; Sigvaldason 1974).

As is clear from Chapter 6, most orogenic andesites are not obviously banded or variegated as in some of the examples just cited; if mixtures, most andesites were well homogenized before eruption. There is, nevertheless, ambiguous evidence that such homogenization occurs. The most obvious and widespread feature of orogenic andesites possibly indicative of mixing is disequilibrium amongst phenocrysts: i.e., polymodal phenocryst compositions, minerals or their compositions which are inappropriate to their bulk host or its groundmass, reverse zoning, resorption textures, and coexistence of phases such as olivine and quartz. Good examples of such features in rocks with andesitic bulk compositions are given by Larsen et al. (1938), Pe (1974), Eichelberger (1975), Eichelberger and Gooley (1977), Anderson (1976), Sakuyama (1978), and Johnston (1978).

These mineralogic features and the ambiguity of their significance were discussed in Chapter 6. If no process other than magma mixing produced disequilibrium among phenocrysts, then most andesites are mixed. Indeed, when the proportion of crystals is less than the proportion of necessary contaminant, features such as reversals in the composition of plagioclase phenocrysts by > 10 mol.%. An, and the coexistence of olivine and quartz crystals each with pyroxene overgrowths, can be explained only by mixing. However, the tendency of andesite liquid to accumulate crystals as mementoes of their past history, the slow diffusion rate in crystals, the dependence of mineral compositions on P_{H_2O}, f_{O_2}, and coexisting phases as well as on the rest of the liquid composition, and the possibility of crystal settling within a chemically zoned magma chamber, all preclude confidence that all petrographic examples of disequilibrium have magma mixing as their common cause.

The mineralogic case for mixing is strongest when disequilibrium features are multiple and consistent throughout simultaneous ejecta. For example, late-stage hornblende andesites and dacites of Shirouma-oike volcano, central Honshu, Japan, contain fine-scale banding (0.5 to 10 cm wide) and, even within single optically homogeneous samples (e.g., Kzs-2), there are two distinct mineral assemblages (Sakuyama 1978). One assemblage includes olivine and normally zoned clinopyroxene which equilibrated with each other at about 1000°C; the other includes hornblende, biotite, quartz, ilmenite, and reversely zoned orthopyroxene ($\Delta 2$ to 6 mol.% En) which equilibrated at about 800°C. The olivine is too Mg-rich to be in equilibrium with the host (Sect. 6.4), although neither olivine nor quartz have pyroxene rims. Some plagioclase phenocrysts are reversely zoned, i.e., have An-rich sheaths ($\Delta 30$ mol.% An) overlying a frittered region and andesine core. Taken together these observations strongly indicate mixing of mafic and silicic magmas.

My experience looking at orogenic andesites petrographically and reading about them in the literature is that such substantially diverse mineral assemblages are uncommon, although so also is the extensive study by electronmicroprobe which produces convincing evidence for mixing. For example, only about 20% of the volcanoes in the East Japan volcanic arc contain any, much less many, rocks with coexisting olivine and quartz (Sakuyama 1978). Also, as noted for plagioclase in Section 6.1, minerals which are included in phenocrysts in orogenic andesites usually coexist as phenocrysts. Thus, mixing usually seems restricted to magmas with the same mineralogy.

A second strong argument for magma mixing during andesite genesis is based on the composition of glasses included in phenocrysts. (The volatile contents of these glasses were discussed in Sect. 5.3.) Although glass compositions sometimes lie between that of the bulk host and its groundmass (e.g., Heming 1977), i.e., lie within the range of compositions from which the phenocryst could have crystallized if it is cognate and not accumulative, other times they do not. Indeed, Anderson (1976) reported that such "outrange glasses" occur in all but one of the andesite samples he had studied. The simplest interpretation of the "outrange glasses" is that they represent magmas from which the host phenocrysts precipitated and which subsequently were mixed and homogenized before eruption. For example, single andesite specimens can contain olivine and bytownite crystals with basalt glass inclusions, but orthopyroxene, sodic plagioclase, or quartz crystals with dacite or rhyolite glass inclusions (Anderson 1976; Eichelberger and Gooley 1977). Typically the glass compositions are quite variable, possibly indicating a complex mixing history, but possibly also reflecting analytical difficulties, post-inclusion crystallization of the host crystal, or failure of included glass to represent the bulk liquid composition. Although neither the frequency of "outrange glasses" nor the significance of their diversity are known, the glasses suggest mixing even when mineralogic evidence is moot.

A third argument for magma mixing interprets mafic inclusions as basalt pillows in silicic magmas (Eichelberger and Gooley 1977). As discussed in Section 6.8, this argument is strong only when inclusions have fine grained or glassy rims or have acicular mineral morphologies (both characteristic of quench); such inclusions are uncommon in orogenic andesites in my experience.

A fourth argument for mixing is the convergence of rock compositions toward andesite with time during evolution of some stratovolcanoes (Sect. 4.4). Persistent use of a common conduit beneath polygenetic volcanoes, and the existence of crustal-level subsidiary reservoirs, both provide ample opportunity for mixing of new parental liquid with older residual differentiates, followed or accompanied by further crystal fractionation. However, even the data of Yanagi and Ishizaka (1978) cited in Section 4.4 are inconclusive and more suggestive than proof of mixing. That is, few of the earliest rocks in each cycle at Myoko volcano have disequilibrium mineral assemblages (Hayatsu 1976, 1977) and the heterogeneous $^{87}Sr/^{86}Sr$ ratios found there by Ishizaka et al. (1977) are not explicable by cyclic mixing.

A final argument for mixing is its ability to explain incompatible and some compatible trace element concentrations in excess of those predicted by crystal fractionation (Hart and Davis 1978; Rhodes et al. 1979). Using Fig. 11.1 as an example I commented in Section 11.3 that LIL-element concentrations in orogenic andesites sometimes increase at rates exceeding those predicted by POAM fractionation. Note in Fig. 5.17b,c that for a given intermediate composition, mixing (point a′) yields higher concentrations of incompatible and highly compatible trace elements than does crystal fractionation (point a). Consequently, the excess LIL-element enrichments of Fig. 11.1 may reflect magma mixing instead of crustal assimilation or inapplicability of the POAM model. This idea is illustrated in Fig. 5.17b,c by data for the Hargy-Galloseulo volcano, New Britain,

Papua New Guinea. While most samples define a line which indicates consanguinity via crystal fractionation rather than partial melting or mixing, several samples have higher trace element contents, consistent with back-mixing of andesite (a) plus dacite (D) previously derived therefrom.

I have not cited straight line trends of orogenic andesite suites on element-element diagrams as evidence for magma mixing because data scatter usually precludes clear distinction between straight and curved line correlations and because I know of no example where such evidence for mixing includes isotopes and incompatible trace elements as well as major elements. For example, mixing lines are clearly inapplicable to the suites shown in Figs. 5.1 to 5.3 and 5.17.

Indeed, a serious problem for magma mixing models of andesite genesis, as for crustal assimilation models, is the *lack* of evidence within andesite-dominated suites for binary mixing behavior. To illustrate, the observations made about New Zealand andesites in Section 10.3 argue equally against their origin by mixing of basalt and rhyolite, which was proposed by Eichelberger (1975). The average composition of New Zealand rhyolite (point I) is shown in Figs. 10.1 and 10.2 to be no more satisfactory than Mesozoic sediment (point S) as an end-member mixing component. Similarly, although the anorogenic andesites erupted in 1970 from Hekla volcano, Iceland, provide reasonably likely examples of magma mixing, their bulk compositions are not internally consistent with binary mixing between realistic and members (see Sigvaldason 1974, Figs. 9, 10 and Table 10, especially for P and trace elements). However, as with crustal assimilation, *magma* mixing (i.e., when crystals are present) doubtless is accompanied by crystallization as well as dissolution so that binary mixing models are unrealistic; the effects of fractionation and mixing will be superimposed. Nevertheless, I found no orogenic andesite (much less andesitic suite) for which a mixing model provided a convincing quantitative explanation of bulk rock compositions, such as was given by Wright and Fiske (1971) for basalts and andesite from Kilauea volcano, Hawaii, for example.

A second problem concerns the unlikelihood of homogenization between large volumes of two liquids having significantly different temperature, densities, and viscosities (McBirney 1980). Such issues may also limit mixing to consanguineous magmas initially similar to each other in mineralogy as well as composition.

The pros and cons of mixing summarized above relate to its frequence. No consensus is in sight, but probably much less than a quarter of orogenic andesites involved mixing of magmas as diverse as basalt and dacite, judging from the infrequence of olivine-quartz assemblages. However, this is the section of the book most certain to be outdated before publication.

More important genetically, however, is the origin of the magmas being mixed. If cogenetic, then mixing is a second-order aspect of differentiation. As such, magma mixing may contribute significantly to difficulties identifying the first-order processes (Sect. 12.1) and may be the immediate cause of some andesite eruptions (Sparks et al. 1977), but it is neither a process whereby crust is recycled in andesites nor the principal mechanism of andesite genesis.

In contrast, if the silicic component is a partial melt of crust, as proposed by Sigvaldason (1974), Eichelberger (1975), and others, then its participation in andesites constitutes an alternative mechanism of crustal involvement to those discussed in Chapter 10. However, the crustal provenance of silicic "outrange glasses" is speculative at best; those reported by Anderson (1976) do not lie near the minimum in the Ab-Or-Qz-H_2O system, for example. Even andesites erupted within volcanic arcs characterized by large volumes of potentially contributing rhyolite are separated from the rhyolites in space (e.g., New Zealand: Cole 1978) or time (e.g., northern Chile: Pichler and Zeil 1972). Also, if mixing usually is restricted to magmas with similar mineralogy (see above), then the silicic component in pyroxene andesites is not likely to be crustally derived rhyolite which usually is hornblende and biotite-bearing. Thus, crustally derived liquids seem to influence andesite genesis more through some kind of open-system selective leaching of certain elements from wall rocks (Sect. 10.3) than by being the principal silicic component involved in mixed magmas having andesite bulk composition.

11.8 Role of Other Differentiation Mechanisms

The roles in andesite genesis of two differentiation mechanisms in addition to crystal-liquid fractionation have already been discussed: selective leaching of, or isotopic exchange with, crustal-level wall rocks (Sect. 10.3), and magma mixing (Sect. 11.7). Both seem likely in a minority of cases although neither's magnitude has yet been demonstrated quantitatively in a specific instance. Other possible mechanisms include liquid immiscibility, diffusion, and vapor phase fractionation. Each has ever been invoked to explain differences between calcalkaline andesites and the differentiates of anorogenic tholeiitic sills such as the Skaergaard which have been attributed mostly to crystal-liquid fractionation.

Liquid immiscibility was appealed to by Fenner (1948) for this reason, but no theoretical, experimental, or textural evidence for immiscible silicate liquids has been reported for orogenic andesite bulk compositions, although such evidence is possible for other rock types (Philpotts 1978; Roedder 1978). Specifically, iron and alkali contents are too low in orogenic andesites. I have, however, noted possible immiscibility between andesite and sulfide liquids (Sects. 5.3.3 and 11.6), and similar immiscibility between andesite and oxide liquids has been suggested to explain the apparently magmatic magnetite + apatite deposits such as the Chilean lava flows mentioned in Section 11.3 (Philpotts 1967). There is no evidence as yet that either process commonly plays a significant role in andesite genesis.

The role of chemical diffusion in magmas has been de-emphasized since Bowen (1928) noted that rates of conductive heat loss greatly exceed those of chemical diffusion, resulting in solidification before significant change in composition. As noted in Section 4.1.4, likely chemical diffusion rates in anhydrous orogenic andesite magmas are about 10^{-8} cm^2 s^{-1} for most elements or for water.

Consequently, if magmas which differentiate to andesite exist for at most 10^6 years (Sect. 4.5), then a static diffusion front whithin the magma will move only 10 m, whereas conductive heat loss in that interval will result in the solidification of kilometers of magma. Although the lower activation energies for transport phenomena in hydrous magmas and their lower viscosities suggest that diffusion will be more effective in orogenic andesite magmas than in dry basalts, these effects are offset by lower temperatures.

Diffusion can be expected to have its greatest effect during andesite genesis in two circumstances: across areas with high thermal gradients such as near apices of magma reservoirs (e.g., Hildreth 1979); and at the margins of convecting magma (e.g., Shaw 1974). For example, if thermally stratified magma overlies uniform, convecting magma within reservoirs (e.g., Hildreth 1979, Fig. 16; McBirney 1980, Fig. 12), and if Soret effects turn out to be significant (Rice 1978), then these effects will augment or locally exceed those of crystal-liquid fractionation, thereby explaining atypical trace element behavior in the initial ejecta of eruptions. Such diffusional effects beneath andesitic stratovolcanoes are thought to be less than beneath large intra-continental calderas due to smaller magma volumes and shorter periods of repose.

The relatively high volatile contents of magmas in volcanic arcs may make gas transfer (vapor fractionation) of elements an especially important process there (Fenner 1926; Stanton 1967). Indeed, exsolution of volatiles from ascending and crystallizing magma is expected (Sect. 5.3.1) and has long been accepted as central to an explanation of explosive volcanism, and of the association and isotopic similarities between many hydrothermal ores and igneous rocks in this environment (e.g., Burnham 1967, 1979; Holland 1972). In orogenic andesites, the composition of this vapor lies in the H-C-O-S-Cl-F system and will be determined as much by relative solubilities as relative concentrations. That is, whereas H_2O is the most abundant volatile constituent in orogenic andesite liquid, that liquid is more likely to become first saturated with CO_2 or a sulfur species than with H_2O (Sect. 5.3). Of course, once the vapor phase forms, H_2O like everything else partitions between it and residual liquid, resulting in some $a_{H_2O}^{vapor}$ which increases as the vapor decompresses (Holloway 1976). The resulting vapor, especially when Cl-rich, is an effective solvent for alkalies and divalent metals as well as silica (Burnham 1967; Holland 1972), an effectiveness which increases with pressure. Consequently, exsolution of the vapor causes differentiation and dessication of residual silicate liquid due to its depletion in the elements which partition strongly into the vapor.

The vapor bubbles ascend buoyantly at 10's to 100's of cm/year, resulting in one or more of the following: failure of the surrounding rock due to increase of vapor pressure above load pressure; loss of vapor to the atmosphere and surrounding rock, resulting in lava fountaining or explosive eruption, and in pegmatites or hydrothermal ores, respectively; or enrichment of elements in near-surface silicate liquids by release of solutes due to reduced solvency of the vapor at lower pressure.

This mechanism of depleting deep, residual liquids while enriching near-surface ones results in vertical differentiation and stratification of a magma reservoir without sinking of crystals, thereby accounting for compositional gradients inferred to exist beneath volcanoes whose successive ejecta (from increasingly lower positions within the reservoir) contain no evidence of crystals inherited from upper, more differentiated levels.

Little is known about element complexing and partitioning or isotopic mass fractionation between silicate liquid and vapor under relevant conditions. The principal cationic constituents in vapor exsolved from andesite would be Si, Na, K, and Fe with the Na/K ratio in vapor exceeding that in the melt, and solutes would constitute at most 2 to 10% wt.% of the vapor at pressures < 10 kb, depending on a_{Cl}^{vapor} and $a_{CO_2}^{vapor}$ (Burnham 1967, 1979; Holloway 1971; Holland 1972). Divalent transition metals such as Mn and Zn have distribution coefficients around 20 favoring the vapor when Cl concentrations in the vapor are about 3 mol/kg (Holland 1972), thus facilitating their extraction from silicate liquid. In contrast, trivalent REE prefer the silicate liquid at low pressure such that vapor loss causes three effects in residual silicate liquid: an overall increase in REE contents; greater enrichment in heavy than light REE; and a negative Eu anomaly (Flynn and Burnham 1978). Mass fractionation of light elements will occur, enriching vapor in light isotopes and causing, for example, dO^{18} and $^3He/^4He$ ratios of the residual silicate liquid to decrease.

Qualitatively, upward transport and subsequent loss to (not from) magma of solutes from vapor provides yet another explanation of why acid andesites forming at apices of magma reservoirs sometimes have concentrations of alkalies and compatible trace elements in excess of that attributable to crystal fractionation alone. However, this effect should be greater for Na than K, and much greater for Mn and Zn than alkalies, neither of which has been recognized, and the mechanism does not alleviate the problems in accounting for REE behavior which I cited in Sect. 11.3.

The quantitative effect of vapor phase fractionation on melt composition depends on four factors in addition to solute concentrations in the vapor: the weight fraction of volatiles initially in magma; the percent of outgassing from residual silicate liquid; percent of solutes added to enriched silicate liquid as a result of decompression of the vapor; and the relative volumes of enriched versus residual silicate liquids. For example, if volatiles constitute 5 wt.% of initial magma, are completely exsolved from residual liquid, lose all their solute to overlying and thereby enriched silicate liquid, and the proportion of residual to enriched liquid is 100:1, then near-surface liquid would be enriched in an element (e.g., Na) by 50% if solutes constitute 10 wt.% of the vapor at the site of exsolution and if solutes contain the same concentration of that element as did the original magma. Indeed, up to 100% enrichment of Na $>$ K was observed experimentally in a 1-cm long capsule when a maximum of 5 wt.% H_2O exsolved from acid andesite between 18 and 10 kb (Sakuyama and Kushiro 1979).

That volatile fractionation does occur beneath some andesite stratovolcanoes is indicated by the common enrichment of S relative to Cl prior to eruption of andesite (Sect. 5.3.3), by disequilibrium enrichments of Ra and Rn in andesite ejecta (Sect. 5.5.5), by an apparent distribution coefficient for U which is higher than attributable to crystal fractionation alone at Irazu volcano, Costa Rica (Allègre and Condomines 1976), and by the emission of more SO_2 from Pacaya volcano, Guatemala, than could have been supplied by erupted material (Stoiber and Jepsen 1973).

However, apart from the Ra and U data cited above, there is little evidence that vapor phase fractionation significantly alters concentrations of nonvolatile elements in subalkaline silicate liquids. Even in near-surface basalt lava lakes where vesiculation unambiguously occurs there is no evidence for alkali enrichment of uppermost silicate liquid (Wright et al. 1976).

More fundamentally, there is little evidence that volatile contents in orogenic andesites or their parents are high enough to cause vapor saturation throughout large volumes of magma prior to eruption (Sects. 4.2 and 5.3). The exceptions are Anderson's (1974b) analyses of glass inclusions in phenocrysts whose S, Cl, and inferred H_2O contents are sufficiently constant despite differences in Si and K that the glass compositions appear to have been buffered by a volatile phase. However, most orogenic andesites contain less than 5 wt.% H_2O and therefore would exsolve water vapor only at depths < 4 km (Sect. 5.3.1), which is too shallow to be the major site of precipitation and crystal fractionation. Thus, I attribute Anderson's data to one of three factors. First, crystals with glass inclusions large enough to analyze may be atypical, forming only in the uppermost portions of magma reservoirs and only during eruption. Second, volatile contents of the inclusions may be atypical of the bulk melt, reflecting phenomena associated with liquid entrapment in crystals (Wilcox and Kuo 1973). Or third, the vapor may be water-poor and CO_2- or SO_2-rich. Such vapor is a less effective solvent which reduces the efficacy of the vapor fractionation mechanism.

Thus, while none of the three mechanisms of differentiation discussed above seems likely to play as great a role as does crystal fractionation in andesite genesis, all of them and especially vapor fractionation are able to play a spoiler role. That is, orogenic andesites must vesiculate somewhere during ascent, affecting an unpredictable volume of magma for an unpredictable length of time. Exsolution causes differentiation of residual silicate liquid, and may also affect overlying liquid as ascending vapor re-equilibrates with that liquid during decompression. Effects may be sufficient in magnitude to explain some of the inconsistencies in trace element behavior noted in Section 11.3, but more quantitative constraints are needed.

11.9 Differentiation Processes Leading to Andesite
in Anorogenic Environments

Occurrences of andesite, sometimes orogenic andesite, at places which clearly or possibly were unrelated to subduction during volcanism were summarized in Section 2.8 where I streassed that such andesites differed from their counterparts at convergent plate boundaries more in relative abundance than composition. In turn, similarities and differences in composition were noted in Section 5.6. If andesites in all environments originate largely by crystal fractionation, as argued above, then differences in composition as well as volume require explanation.

Because andesites in anorogenic environments are relatively uncommon, so also are thorough studies of them. Nevertheless, low pressure crystal fractionation of phenocryst phases is the principal mode of origin inferred in the most complete of such studies, summarized below.

Andesites occasionally occur as oozes in lava lakes or as lavas along rift zones or volcano flanks at intra-plate ocean islands such as Hawaii. For example, the major element composition of rare Kilauea andesites can be explained quantitatively by 80% to 90% crystallization (Wright and Fiske 1971) of a parent similar in composition to one which D. Green (1970) demonstrated to be primary by virtue of having Fo90 and En90 as liquidus phases at 15 to 20 kb. Olivine, minor spinel, and possibly orthopyroxene fractionate until MgO reaches about 8 wt.% after which plagioclase (An60–70), augite (Wo36 Fs20), and pigeonite (Wo9 Fs35) dominate the mineral assemblage, being removed in about 3:3:1 proportions. Magnetite, ilmenite, and apatite begin to precipitate from liquids with about 55% SiO_2, such as the Kamakaia flow. Olivine settling occurs in a near-surface reservoir but multi-mineralic crystal fractionation is thought to accompany conduit flow between reservoir and flanks or to occur during filter pressing. Some andesites, including the most silicic lava known at Kilauea (Table 5.5, col. 5), result from mixing of magmas which themselves can be related by crystal fractionation.

Andesites on ocean floors or on islands at divergent plate boundaries are only slightly more common. Low-K ocean floor andesite similar to those cited in Table 5.5 is explicable by about 40% crystallization of plagioclase (An55), augite (Wo38 Fs25), and magnetite in roughly 2:2:1 proportions, plus minor pigeonite and apatite, from ferrobasalt (52% SiO_2, 16% Fe_2O_3) which itself is the result of at least 75% crystallization (plagioclase > augite > olivine) of a potentially primary parent (Byerly et al. 1976). Mafic inclusions occur in the andesite and may be basalt pillows or cumulates. Trace element data apparently are consistent with the model (F. Frey, written comm. 1977). Reverse zoning of plagioclase and augite phenocrysts in the andesite suggests magma mixing.

The more voluminous basic andesites and icelandites of Hekla and Thingmuli volcanoes, Iceland (Table 5.5, col. 3), also have been interpreted as derived from basalt by over 90% crystal fractionation of plagioclase (An50–60), augite (Wo38–

31 Fs16–30), and lesser olivine, which are joined by magnetite at silica contents above about 49% (Carmichael 1964; Baldridge et al. 1973). However, detailed tests of crystal fractionation models are not published for either volcano and problems might be suspected because phenocryst contents decrease to almost nothing with time during the 1947 and 1970 eruptions of Hekla, indicating that the necessary crystals were absent from the pre-eruption reservoir, and because simultaneous eruption of dacite and basic andesite from Hekla in 1970 indicates the likelihood of magma mixing (Sigvaldason 1974). Also, most Hekla basic andesites lack negative Eu anomalies (Thompson et al. 1974) despite extensive plagioclase fractionation with f_{O_2} at the FMQ buffer. However, it is difficult to quantify these issues because at Hekla, as at many andesite volcanoes in volcanic arcs, basalts are notably absent and this makes choice of a parent composition arbitrary.

The "ferrobasalts and ferrolatites" of the Craters of the Moon area (COM), Snake River Plain, USA, are an alkali-enriched variation on this same theme (Leeman et al. 1976; Table 5.5, col 7). Like the suites already discussed, COM lavas have lower Al_2O_3 contents but higher FeO* contents and FeO*/MgO ratios than any orogenic andesites at convergent plate boundaries, and lack modal orthopyroxene despite high normative hy/di ratios. Nevertheless, COM lavas are calcaklaine in the sense that their analyses point toward the alkali apex of an AFM diagram. Unlike the andesites discussed above, some COM lavas actually are orogenic andesites by my definition, being more fractionated relative to silica than are Icelandic andesites in the sense that Ti, Mg, and Ca contents are lower whereas LIL-element contents are higher in COM rocks.

Andesite genesis by crystal fractionation of olivine (Fo70–60), plagioclase (An50–40), and magnetite in proportions of about 2:1:1 from basalt is consistent with the observed phenocryst mineralogy, with experimentally determined phase equilibria, and with the presence of mafic inclusions (glomeroporphyritic clots), and also can account for changes in major trace element compositions quantitatively (Leeman et al. 1976). Approximately 45% crystallization of COM "basalt" is necassary to produce a liquid containing 52% SiO_2, but the presumptive parental basalt is nonprimary (4.9% MgO, 10 ppm Ni) and itself andesine-normative. In addition, minor spinel and about 10% apatite fractionation are necessary to account for Cr, and P_2O_5 plus REE behavior, respectively. Although magnetite fractionation is implied by its presence as a phenocryst and by elemental behavior, magnetite is not a liquidus mineral in COM lavas until f_{O_2} equals or exceeds that of the NNO buffer (Thompson 1975).

Despite the successfulness of crystal fractionation as a mechanism of differentiation, Sr and Pb isotopes are more radiogenic in differentiated magmas than basaltic ones, resulting in a pseudoisochron (Leeman et al. 1976). Large-scale crustal assimilation is precluded because COM lavas are richer in many LIL-elements than are ordinary crustal rocks, indicating that crust has influenced the andesites selectively, affecting Sr and Pb isotopic compositions more than elemental concentrations.

Note above that the parental "basalt" is itself differentiated and, indeed, andesitic by some definitions (Thompson 1973). Quantitative efforts to relate the trace-element composition of the COM parent to the more voluminous basalts of the Snake River Plain with $> 6\%$ MgO have not been successful (Leeman et al. 1976) and the two lava types may form a bimodal population, suggesting that they have complex genetic ties (Thompson 1972), if any.

A rather similar example of andesites in anorogenic environments consists of the "andesine basalts and trachytes" of the Pleistocene Boina Center, Ethiopia (Barberi et al. 1975). The Center is made of transitional basalts capped by a shield volcano with intermediate to silicic rocks (trachytes and pantellerites) at its top, constituting less than 10% of the volcano's volume. Again at least some of the rocks are orogenic andesites by my definition.

Internally consistent use of trace element data shows that basaltic andesite at Boina (54% SiO_2) can be produced by 70% crystallization of olivine, plagioclase, and minor augite from a presumptive parent which could be primary. Magnetite first precipitates at about 46% SiO_2 (60% crystallization) after which, for the first time, there is rapid increase of silica in the liquid with increasing degrees of solidification. Ni and Cr are low and constant (5 to 10 ppm) within the andesite silica range which represents 70% to 80% crystallization. Cu and Zn correlate positively with silica during rapid Fe-enrichment but negatively with silica in the andesites. Negative Eu anomalies begin to develop only within the andesites. $^{87}Sr/^{86}Sr$ ratios generally increase with differentiation but cannot be interpreted quantitatively as the result of simple crustal assimilation; selective leaching of, or equilibrium with, crustal rocks was invoked by Barberi et al. (1975).

The preceding comparisons clearly indicate that crystal fractionation sometimes results in andesite, even orogenic andesite, elsewhere than at convergent plate boundaries, and that magnetite fractionation in particular is the principal agent which suppresses Fe-enrichment and accelerates silica-enrichment in liquids there. Similarly, the involvement of minor spinel and apatite fractionation in andesite genesis elsewhere heuristically supports their similar roles in volcanic arcs. Occurrence of mafic inclusions elsewhere confirms their interpretation as cumulates rather than decompressed hornblendes.

Problems with crystal fractionation models of andesite genesis are not unique to volcanic arcs. For example, magnetite is not a near-liquidus phase of COM lavas at f_{O_2} between the FMQ and NNO buffers in experimental charges despite clear evidence that magnetite fractionation occurred. Also, negative Eu anomalies are uncommon in nonarc basic andesites despite extensive plagioclase fractionation during their genesis. The more reducing nonarc magmas are, the more of a puzzle this is. Ni and Cr contents, when low, can be relatively constant in nonarc andesites despite extensive pyroxene fractionation. Isotope compositions are heterogeneous in andesites erupted through continental crust even when elemental concentrations can be explained by crystal fractionation alone and when bulk assimilation can be shown not to have occurred. Such problems, then, are no more forceful against crystal fractionation in volcanic arcs where volume relations are equivocal than outside arcs where basaltic parentage is taken for granted.

There are, or course, differences from, as well as similarities to, orogenic andesites in volcanic arcs in the preceding comparisons. Differences in volatile and trace element ratios (Sect. 5.7) reflect source compositions. Larger weight fractions of crystals must be removed before melts in anorogenic environments become andesitic. Lower f_{O_2} in anorogenic melts causes magnetite saturation to occur later in their crystallization histories. Plagioclase constitutes a somewhat larger mass fraction of fractionating phases in volcanic arcs than elsewhere (about 2/3 vs. 1/2). Presumably this difference reflects higher Al_2O_3 contents in volcanic arc basalts rather than differences in P_{H_2O}. Fractionating plagioclase elsewhere also is more sodic, which makes residual liquids less enriched in silica. Orthopyroxene fractionation largely is restricted to volcanic arcs, usually being absent even in tholeiitic suites elsewhere. Volatile contents are expected to be lower elsewhere but H_2O in Hekla basic andesite magma was estimated to be 2.5 to 6 wt.%, based on compositions of coexisting plagioclase and glass (Baldridge et al. 1973).

Two differentiation mechanisms other than crystal fractionation of anhydrous minerals plus apatite were invoked in these explanations of nonarc andesite genesis. They are mixing of magmas which previously had differentiated by crystal fractionation, and selective isotopic modification of magma by sourrounding crust.

Chapter 12   Conclusions

12.1 Andesite Genesis by POAM-Fractionation: the Most Frequent Mechanism

From Chapters 8 to 11 it should be clear that the name "orogenic andesite" denotes rocks sufficiently diverse that they can result from a wide variety of processes acting alone or multiply, a variety which has made andesites a problem to explain by any single mechanism. However, how often and to what extent each possible process actually operates is a different matter. My conclusion is that crystal fractionation of phenocryst phases from basalt, i.e., usually of plagioclase + orthopyroxene/olivine + augite + magnetite (POAM), is by far the most common and extensive process, supplemented to an unknown extent by magma mixing, selective interaction with the crust, and vapor fractionation. A typical andesite event may develop as follows.

Once upon a time beneath a distant kingdom by the sea, oceanic crust was subducted and dehydrated as it was metamorphosed to eclogite facies assemblages. I call the resulting aqueous phase "IRS fluid" because, at high pressure, it carries substantial amounts of incompatible elements, radiogenic nuclides, and silica. Its composition depends on the age, alteration, and sediment component of ocean crust which is subducted (Sect. 8.5). Whether this IRS fluid or, instead, siliceous melt ascends from subducted crust depends on the geothermal gradient and, therefore, on the rate and duration of subduction and the age of subducted crust. Mass transfer from slab to mantle wedge (slab recycling) occurs in either case. This transfer accounts for uplift of volcanic arcs (Sect. 3.1), changes in slab behavior beneath volcanic arcs (Sect. 3.4), and geochemical differences (especially in volatile, incompatible trace element, and isotopic ratios) between magmas in volcanic arcs versus elsewhere, including backarcs (Sect. 5.7).

Addition of either IRS fluid or slab-derived melt to the mantle wedge causes LIL-element enrichment, formation of new accessory minerals such as amphibole, phlogopite, rutile, or dense hydrated magnesium silicates, and possible formation of local pyroxenites. The net effect is to lower the density, viscosity, and sometimes the melting point of the wedge, resulting in melting, melt extraction, and convection (Sects. 3.3 and 9.1). Melting may occur farther from the plate boundary than does slab dehydration depending on the geothermal gradient (Sect. 8.1) except, of course, in unusual cases of forearc volcanism (Sect. 2.6). This need for horizontal displacement can be met by convection of metasomatized mantle across the P-SP1 or P-H3 lines of Figs. 8.2 and 9.1, a process which also returns sub-arc mantle to the larger upper mantle system. Melting occurs near the slab-wedge boundary (Sect. 3.4). However, judging from studies of heat flow, body wave attenuation and velocities, and surface wave velocities, partially molten mantle (i.e., the asthenosphere) usually extends to shallower depths beneath volcanic arcs than beneath intraplate volcanoes due to nearer surface diapiric ascent of material whose density has been reduced by hydration or partial fusion (Sect. 3.3). This sequence of slab dehydration or fusion, mantle modification by IRS fluid or melt, convection, and mantle fusion, collectively explains the low density and seismic velocities, and the high conductivity, seismic attenuation, and heat flow observed or inferred for the mantle wedge beneath volcanic arcs and, sometimes, backarcs (Sects. 3.1 and 3.3), and explains why sub-arc asthenosphere is a necessary condition for volcanic arc formation (Sect. 2.2).

Liquid coalesces and is extracted from the partially molten wedge by buoyancy forces plus periodic compression and relaxation of the wedge (Sect. 3.5) which in a general sense explains the spacing of active volcanoes within volcanic arcs (Sect. 4.6), and correlations between the timing of volcanic and seismic activity there (Sect. 4.7). [Peak episodes of arc volcanism (Sect. 4.7), such as in the Quaternary, may directly reflect planet-wide episodicity of melting in the upper mantle, or reflect it indirectly via more rapid subduction due to more rapid spreading.]

The composition of primitive liquids in the mantle wedge when they segregate from refractory peridotite and begin upward ascent is unknown in most critical respects but probably differs from liquid in asthenosphere elsewhere. Primitive sub-arc melts probably are olivine- or even-quartz normative depending on the pressure and percent fusion, have Mg-numbers > 67, and probably contain 50% to 55% SiO_2 and lower concentrations of Ni and Cr than are typical for liquids of similar Mg-numbers. H_2O/CO_2, Cl/S, Ba/La, and Th/La ratios are atypically high and Nb/La ratios low. As long as sufficient terrigenous sediment is subducted, $^{87}Sr/^{86}Sr$ ratios will be high relative to $^{143}Nd/^{144}Nd$, and $^{207}Pb/^{204}Pb$ ratios will be high relative to $^{206}Pb/^{204}Pb$. H_2O contents probably are 0.5 to 3 wt.% and f_{O_2} is unusually high, possibly above the NNO buffer.

Shallow depths of liquid segregation could explain the quartz-normativeness, but slab recycling is responsible for the other characteristics. The higher water and alkali contents of the mantle wedge lead to more siliceous melts due both to fluid mixing and to the incongruent melting of orthopyroxene (and possibly of accessory phases such as amphibole and phlogopite). The same high water contents lead to lower temperatures of melting per degree of fusion and, therefore, to lower Ni and Cr contents in the liquid because distribution coefficients for olivine and pyroxene will be higher. Finally, unspecified titaniferous refractory phase(s) within the slab or wedge or both could explain the distinctively low concentrations of Ti-group elements. Of course, great variety in the concentration of incompatible elements and in isotope ratios is possible in these primitive melts due, for example, to differences in the composition of subducted crust, whether IRS fluid or melt is released from the slab, the duration of subduction and therefore the extent of metasomatism, and percent fusion before melt extraction from the wedge. Two examples of this diversity are points *a* and *m* in Figs. 12.1 to 12.4, schematically representing 25% and 10% melts, respectively, of a compositionally uniform source.

Assuming that magma ascends by "magma-fracturing" in a region mostly under compression, ascent will be less rapid than in regions under tension (Sects. 3.5 and 4.7). Some fractionation of olivine, pyroxene, spinel, and eventually plagioclase occurs within the mantle, e.g., due to crystal settling or plug flow of a Bingham-type fluid. As differentiation and ascent occur, the viscosity of liquid increases while its density contrast with surroundings decreases, eventually resulting in stagnation and formation of a subsidiary reservoir at the base of or within the crust

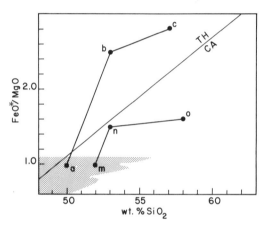

Fig. 12.1. FeO*/MgO vs. SiO_2 diagram showing schematic evolution of calcalkaline and tholeiitic suites of orogenic andesites. *Stippled region* shows possible range of unfractionated liquids from the mantle wedge with Mg-numbers $\geqslant 67$; the upper silica limit is unknown but increases as water contents increase. Melts *a* and *m* represent 25% and 10% melts, respectively. Figures 12.1 through 12.4 assume a common peridotite source for both which contains 0.1 wt.% H_2O, 0.1% K_2O, 1.25 ppm Rb, and 2000 ppm Ni. IRS fluid (water:SiO_2:other in proportions of 5:4:1) is added to this peridotite in 0.3:100 proportions and then is dissolved in the melt, thereby constituting 1.2 wt.% of *a* and 3% of *m* so that *a* has 1.0 wt.% H_2O and *m*, 2.43 wt.% H_2O.

Line *abc* is a tholeiitic fractionation trend in which *a–b* represents 30% crystallization (pyroxene $>$ olivine $>$ plagioclase), and *b–c* represents 20% crystallization (plagioclase $>$ pyroxene $>$ magnetite). Melt *c* therefore has 57% SiO_2, 0.7% K_2O and 1.75% H_2O: a low-K tholeiitic andesite.

Line *mno* is a calcalkaline fractionation trend. Segment *m–n* parallels *a–b* but represents only 10% crystallization. Segment *n–o*, like *b–c*, represents 20% crystallization but is longer and flatter due to a larger weight fraction of magnetite. Melt *o* has 58% SiO_2, 1.4% K_2O, and 3.4% H_2O: a medium-K calcalkaline andesite

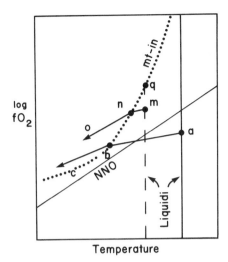

Fig. 12.2. Dimensionless T-f_{O_2} diagram showing schematic evolution of calcalkaline and tholeiitic suites. Temperature increases to the right in reciprocal units so that the NNO buffer is a straight line. See Fig. 12.1 for explanation of melts.

The *solid line* represents the liquidus for melt a which crystallizes 30% silicates over segment $a-b$, becoming slightly oxidized in the process due to precipitation of ferrous oxide-bearing minerals. At b magnetite saturation occurs. Segment $b-c$, therefore, involves 20% crystallization of silicates plus magnetite, resulting in less net oxidization.

The *dashed line* represents the liquidus for melt m. At f_{O_2} higher than point q, magnetite becomes the liquidus phase and the liquidus temperature becomes f_{O_2}-dependent. The liquidus temperature for m is less than for a because m contains 2.5 times more water. The oxygen fugacity of m is assumed to be higher because it contains more water from wedge peridotite, and because it has a higher IRS fluid proportion. Due to both the lower liquidus temperature and higher f_{O_2}, only 10% crystallization of silicates occurs (segment $m-n$) before magnetite saturation at n. Subsequent crystallization along segment $n-o$ is shown as involving more net reduction due to the larger weight fraction of magnetite crystallizing. Note that although tholeiitic basalt a has a higher liquidus temperature than does calcalkaline basalt m, calcalkaline andesites n and o have higher temperatures than do tholeiitic andesites b and c, respectively, which have similar silica contents

(Sect. 3.6). This effect of the crust as a "density filter" contributes to the higher silica mode of rocks within volcanic arcs overlying thicker crust (Sect. 7.2).

Differentiation continues within subsidiary reservoirs. Crystal settling (not flotation, even of plagiocalse, except in very anhydrous tholeiitic basic andesites within the mantle), convective flow differentiation, and filter pressing all may occur during solidification. Plagioclase, olivine or orthopyroxene, augite, and magnetite (POAM) constitute most of the mass fraction of solids removed. Magnetite and calcic plagioclase fractionation is the principal cause of a rapid increase in the silica content of derivative liquids (Sects. 11.2 and 11.3). Magnetite saturation

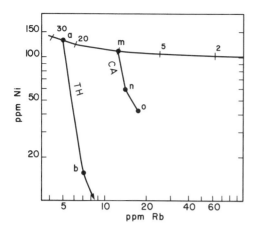

Fig. 12.3. Ni versus Rb diagram showing schematic evolution of calcalkaline and tholeiitic suites. See Fig. 12.1 for explanation of melts. Segments $a-b$ and $m-n$ are parallel and assume $\overline{D}_{Ni}$ = 7 for pyroxene > olivine > plagioclase crystallization. Segment $a-b$ represents 30% crystallization whereas $m-n$ represents only 10%. Slopes flatten at b and n as plagioclase dominates the solids. Numbered *tickmarks* are percent fusion

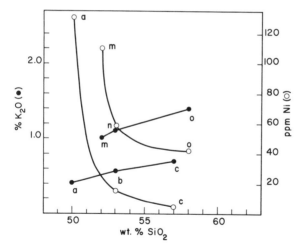

Fig. 12.4. K_2O and Ni versus SiO_2 diagram showing schematic evolution of calcalkaline and tholeiitic suites. See Fig. 12.1 for explanation of melts. At a given silica content, calcalkaline suites are more alkalic because they represent smaller melt fractions and because the potash-silica slope is steeper due to fractionation of correspondingly more sodic plagioclase. At a given silica content calcalkaline suites have higher contents of Ni and other compatible trace elements because the silica content of the parent magma is higher and there is greater change in silica per unit crystallization once magnetite saturation occurs

is shown schematically at points *b* and *n* in Figs. 12.1 to 12.4 after a small weight fraction of crystallization takes place. This saturation occurs earlier in the fractionation history of magmas from arcs than elsewhere due to the relatively high f_{O_2} but low liquidus temperature of the primitive basalt. Accessory apatite and sulfide fractionation contributes slightly to rapid silica-enrichment but primarily affects minor and trace elements. The reservoir can become compositionally zoned with its apices enriched in differentiated liquid and gas. Eventually alkali and water contents suffice for hornblende to precipitate, especially near these apices, although this typically occurs late in the differentiation history when liquid contains, say, 60% SiO_2, except in those medium or high-K suites where water contents are higher to start with. This sequence of events explains changes within single eruptions and within cycles during the stratigraphic history of andesite stratovolcanoes (Sect. 4.3 and 4.4), explains the low pressure features of phenocryst minerals (Chap. 6), and the presence of crustal level magma bodies beneath especially those volcanoes producing acid andesites (Sect. 3.6).

The entire process summarized above, from slab exhalation to magma eruption, may take months (see Blot 1972, 1976; Rose et al. 1978; and Sect. 3.5) to millenia (see Allègre and Condomines 1976, and Sect. 4.4). The constraints are few and poor. However, multiple magma batches continue to use a common conduit or closely adjacent ones during the 10^3 to 10^5 year lifespan of a given volcano, i.e., until one period of closure terminally clogs the conduit or until the energy or fertility of the underlying mantle wedge is spent. There are two consequences of this longevity. First, since there are endless variations on the above theme (e.g., different slab contributions, degrees and depths of melt segregation, ascent paths in P-T space), not all ejecta from the same volcano will have followed a common line of liquid descent from a common primitive parent (Sects. 4.5 and 5.1). Second, magma mixing is likely, resulting in suites which blend linear effects of mixing with nonlinear effects of crystal fractionation, and resulting in suites which converge on andesite with time (Sect. 4.4).

Two additional differentiation processes — selective crustal interaction and vapor fractionation — also may be quite common but neither the magnitude, mechanism, nor frequence of their influence is known yet. Crustal interaction apparently is selective, affecting isotope ratios more than elemental concentrations. In part this difference is the result of the crystal fractionation which inevitably accompanies the assimilation of solids, but even fractionation plus assimilation cannot account for the geochemical budgets of andesites in detail (e.g., Taylor 1980). Nevertheless, some such selective interaction is necessary to explain many circumstantial observations and isotopic characteristics (Sects. 5.5.7 and 10.1), and therefore probably occurs. Whether large volumes of orogenic andesite magma are vapor-saturated for long periods prior to final ascent and eruption is unknown but, if so, the vapor probably is richer in CO_2 than H_2O. However, should an H_2O-Cl-rich vapor be distributed throughout a large volume of reservoir, due to convection for example, vapor fractionation could both enrich overlying magma and deplete underlying magma in many elements (Sect. 11.8). Both these processes probably occur at most volcanoes, but usually have little geochemical influence because they are limited kinetically by the time available for diffusion or bubble separation and re-equilibration.

Obviously the "andesite problem" would not remain if the preceding explanation, which is not original except perhaps in its scope, accounted for all observations in specific instances or had universal applicability. Some inconsistencies are discussed in the following section. Furthermore, in principle some orogenic andesites may be primary melts of subducted ocean crust, the mantle wedge, or continental crust, or high-pressure mixtures of slab + wedge anatectic melts. Criteria for recognizing such unfractionated melts are given in Sections 8.3, 9.3, and 10.2, respectively. Few orogenic andesites qualify. No complete suites qualify, indicating that the two alternate source materials at most yield magmas which differentiate to orogenic andesite via crystal fractionation. Fractionation of hornblende or garent also may occur occasionally and criteria for recognizing these liquid descent lines are given in Sections 11.4 and 11.5, respectively. Again, few orogenic andesites qualify; those which do are mostly high-K types.

12.2 Some Outstanding Problems Requiring Clarification

While the storyline just presented seems to me least at odds with current knowledge, there are serious problems to address before acceptance or revision can occur. Ten of these problems are selected for discussion below and some partly rhetorical questions are asked which future work will need to answer; still other problems are cited in Section 11.3 and elsewhere in earlier chapters.

To answer these questions it will be necessary to thoroughly study rock suites which range from potentially primary to andesitic, and whose stratigraphic relations and absolute ages are compatible with consanguinity. Ideally, a range of such suites is needed, representing a variety of tectonic environments, alkali contents, Fe-enrichment trends, and phenocryst assemblages. However, few if any such suites were available in 1980 (Table 5.3 lists the best known), and little progress will occur until the situation improves.

First and of longest standing is the problem of why orogenic andesites predominate mostly at convergent plate boundaries if they are simply products of basalt differentiation which occurs anywhere. Three replies are pertinent. If unfractionated liquids in the mantle wedge are more siliceous than elsewhere because the water and alkali contents of the system are higher, then even if the extent of solicification prior to eruption were similar in all tectonic environments, the products of, say, 20% to 50% crystallization in volcanic arcs might be andesites by my definition, whereas elsewhere derivative liquids would still be basalts.

However, if orogenic andesites contain only 1 to 5 wt.% H_2O and K_2O (Sects. 5.3.1 and 5.2.2) and result from 20% to 50% crystallization of primary basalt (Sect. 11.3), then we are talking about melting peridotite in the presence of only about 0.02 to 0.2 wt.% H_2O and K_2O if the primary melt results from 5% fusion, or 0.15 to 1.2 wt.% H_2O and K_2O if 30% fusion is more applicable. For example, in Figs. 12.1 to 12.4 I assumed that primary melt a, which represents 25% fusion in the presence of 0.25 wt.% H_2O and 0.1 wt.% K_2O, contains 50 wt.% SiO_2 (including 0.5 wt.% silica which is inherited from IRS fluid that constituted 0.3 wt.% of the total system or 1.2 wt.% of melt a). Similarly, I assumed that primary melt m, which represents 10% fusion of the same source, contains 52 wt.% SiO_2 (including 1.2 wt.% silica which is inherited from the same initial mass of IRS fluid, which constitutes 3 wt.% of melt m). As noted in Section 9.3, the maximum silica contents of melts which can be produced in the presence of such small water and potash contents is unknown, but there is reason to doubt figures as high as I have assumed here.

Thus, *Question 1: How siliceous are peridotite-derived basalts at P, T conditions appropriate to the mantle wedge in the presence of 0.2 to 1 wt.% H_2O and even less K_2O?*

If the silica contents assumed in my example are overestimates, then high pressure magma (not IRS fluid) mixing involving a water-undersaturated, slab-derived silicic melt (Sect. 9.4) will be required.

Secondly, crystal fractionation characteristically may be more extensive in volcanic arcs than elsewhere. This might occur because ascending magmas are more hydrous and less Fe-enriched and therefore have a greater density contrast with crystals, are cooler and therefore more viscous, and because they traverse regions mostly under horizontal compression. These conditions may result in more efficient crystal-liquid separation, in ascent which is interrupted more frequently and therefore results in lower eruption rates of magmas in volcanic arcs than elsewhere (Sect. 4.8), or in ascent through conduits with larger surface area to volume ratios and therefore greater heat loss. This line of reasoning obviously is a deadend if the 0.5 to 1.8 km/day "ascent rates" of earthquake foci preceding eruptions of some orogenic andesites reflect magma ascent (Sect. 3.5), but that issue is unresolved.

Nevertheless, *Question 2: What is the magnitude and distribution of stress within the mantle wedge beneath volcanic arcs and what is their effect on ascent rates and degrees of crystallization of liquids having viscosity of 10^2 to 10^4 poise and a yield strength of 10^3 to 10^5 dyne/cm^2?*

Thirdly, andesites may be more common in volcanic arcs because magnetite precipitates earlier (i.e., at a lower weight fraction of crystallization) due to lower initial temperature or higher initial f_{O_2} of the primitive basalt or both (see Fig. 12.2), and because fractionating plagioclase is unusually calcic. Both result in enhanced silica enrichment, causing derivative liquids to be andesitic rather than basaltic (Sect. 11.3).

The second problem is where the necessary crystal cumulates are. Osborn (1969), for example, suggested that alpine ultramafics are the complement but these are mostly refractory harzburgites which are inappropriate in mineral composition and texture. The answer seems twofold: the cumulates are minor in volume, and part of a gabbroic lower crust. The emphasis above on a more siliceous parent implies that less crystallization is necessary to get andesite than is often assumed. If, for example, only 20% crystallization separates primary liquid from basic andesite and a further 20% crystallization yields "average andesite" (i.e., a total of about 36% crystallization; see Table 11.1 and Figs. 12.1 to 12.4), then an eruption rate of about 10 km^3 of andesite per m.y. per km of plate boundary (Sect. 4.8) requires addition of only 5.6 km^3 of cumulates to the crust per m.y. per km of plate boundary. Even less addition is necessary if some of the crystallization occurs within the mantle or if significant volumes of less fractionated magma contribute to that eruption rate. Assuming volcanism during a total of 10 m.y. since the Eocene (see Kennett et al. 1977) and distribution of cumulates beneath a volcanic arc 50 km wide (Sect. 2.1), this amounts to average crustal thickening by addition of cumulates of only 1.1 km in 30 m.y. In so far as Tertiary island arcs have lower crust which is at least 5 km thicker than that of adjacent ocean basins (Fig. 3.3), this rate of thickening is obviously plausible.

Moreover, most of the cumulates will be gabbroic (Table 11.1 and Sect. 11.2), which is consistent with the most common kind of inclusions found in andesites (Sect. 6.8). The P wave velocity (6.6 to 7.2 km/s) and density (< 2.9 g/cm^3) of the lower crust beneath most volcanic arcs are appropriate to gabbroic mineralogy.

However, *Question 3: What are realistic eruption rates and volumes of magma in volcanic arcs?*

And *Question 4: What is the mineralogy of the lower crust beneath active and formerly active volcanic arcs?*

A third long-standing problem, and its answer, are implicit in the preceding discussions. If basalt is the parental magma, why is it so uncommon in volcanic arcs? Again there are two responses. First, one can follow McBirney (1976) in arguing that basalt under-representation is an illusion created by biased media coverage which is slanted to the violence and beauty of stratovolcanoes. Indeed, basalt is "uncommon" only where crust is thick (Sect. 7.2). However, available relative abundance estimates, though very crude, rarely yield proportions of rock types commensurate with weight fractions of liquids in a differentiation series (Sect. 4.9). Alternatively, one can attribute basalt under-representation to crustal filtering and to magma ascent path. Basalt is under-represented because 20% to 40% crystallization, for example, is typical of ascent paths from asthenosphere, the percentage increasing as crust thickens.

Nevertheless, *Question 5: What are realistic relative proportions of basalt and andesite in volcanic arcs overlying crust of variable thickness?*

The fourth problem concerns the putative absence of magnetite phenocrysts from basalts and andesites in volcanic arcs, and from the liquidus of their compositions in experimental charges. For reasons cited in Section 11.3 I believe that the small magnetite crystals which ubiquitously occur in orogenic andesites are liquidus phases, and that the experiments in question were conducted under unrealistically reducing conditions, or lost Fe to capsules, or both.

However, *Question 6: What is the range of* f_{O_2}, P_{H_2O}, *and load pressure of orogenic basalts and andesites during differentiation, and is magnetite a liquidus phase under these conditions?*

Question 7: What are the nucleation, growth, and settling rates of magnetite in andesitic liquids, and does magnetite exert its influence on liquid descent lines directly as a liquidus phase, or indirectly as a result of filter pressing or the like (possibly similar to the role of clinopyroxene in MORB fractionation)?

The fifth problem concerns the inability of the POAM model to account for trace element compositions of some orogenic andesites quantitatively (Sect. 11.3). I believe that most rock suites available for testing the model lack sufficient geologic and isotopic constraints to judge whether or not they are truly consanguinous. Because multiple lines of liquid descent are represented in single volcanoes, great care in sample selection is necessary, but difficult and uncommon. Even so, such accounting for trace-element behavior is more successful for this model than for any other, and within errors of distribution coefficients, approach to surface equilibrium, likely mineral weight fractions, and degrees of crystallization; the other models are not (Gill 1978). Furthermore, disagreements in magnitude but agreement in kind of behavior may also result from second-order phenomena such as magma mixing, crustal interaction, or vapor fractionation, each of which is likely on some scale.

Consequently, *Question 8: If one can demonstrate that a suite of superposed, variably differentiated basalts to andesites were erupted within the plausible lifespan of an appropriate magma volume, and that these rocks are isotopically homogeneous and lack unequivocal petrographic evidence of magma mixing, then are their trace element concentrations consistent with crystal fractionation of their phenocryst phases?*

The sixth problem concerns the ascent path and fluid dynamics of magmas. For example, the ascent paths calculated by Marsh (1978) predict superheating and therefore preclude crystal fractionation except at shallow crustal levels, and Thompson (1972) suggested that orogenic (and other) andesites separate from their crystal residuum during filter pressing. We have seen that crystal-liquid separation would be retarded or prevented by high liquid yield strengths or by convection or turbulent flow of magma, but also would be enhanced by plug flow of a Bingham fluid. Each is possible for orogenic andesites (Sect. 4.1). Convection also could enhance the effectiveness of diffusion rate-controlled crustal interaction.

Thus *Question 9: What are the rheologic properties of andesite magma, the likely dimensions and thermal boundary conditions of its reservoirs, and the fluid dynamic characteristics of these reservoirs, and which processes of crystal-liquid separation do these constraints allow?*

The seventh problem is how much slab recycling occurs. The elemental and isotopic ratios of incompatible elements in orogenic andesites and related rocks suggest recycling of sialic crust (Sects. 5.5.7, 8.2, 8.5, and 9.4), but criteria are needed for distinguishing between crustal-level assimilation versus mixing near the slab-wedge boundary. Attempts thus far rely on special geologic circumstances such as impending ARC-ATL collisions (Sect. 2.4). However, especially because elemental and isotopic arguments for recycling are not always mutually consistent (e.g., Mariana arc andesites seem isotopically almost unaffected by subduction yet have volatile and trace element ratios which apparently require slab recycling), recycling arguments must be quantified to be persuasive, much less to be useful for global geochemical budgets.

Thus, *Question 10: What is the agent (IRS fluid or melt), the process (fluid mixing or metasomatism), and the geochemical budget of mass transfer from slab to wedge, how do these vary between subduction sites, and how do the agents differ from their crustal-level analogs?*

The final three problems are the ubiquity and magnitude of three second-order phenomena. Considerable evidence for occasional crustal interaction and magma mixing, plus evidence that vapor saturation is likely, were cited in Sections 10.1, 11.7, and 11.8. However, the processes remain dei ex machina. Because none of these processes alone can account for the composition of orogenic andesites in as quantitatively satisfying a way as can crystal-liquid fractionation, I invoke them only to make the inexplicability of some data seem trivial instead of seeming a fundamental flaw in the argument. However, they are extremely important issues and generate the following questions.

Question 11: Can processes other than "selective crustal interaction" account for isotopic compositions within otherwise consanguinous-seeming suites of orogenic andesites which are more heterogeneous and radiogenic than those of intra-oceanic basalts? For example, is mantle heterogeneity a viable alternative?

And *Question 12: What is the "selective crustal interaction" process and its geochemical budget?*

As regards magma mixing, *Question 13: What percent of orogenic andesites contains mineral compositions incompatible with other minerals or their host, outrange glasses, or mafic inclusions with quench textures, and do the bulk compositions of these apparently mixed magmas define internally consistent fractionation + mixing lines for many elements and isotopes?*

As regards vapor fractionation, *Question 14: Do volatile contents exceed solubility limits for appreciable times during andesite genesis, what are appropriate vapor/ silicate liquid distribution coefficients, and does bubble-re-equilibration during ascent release significant amounts of solute to overlying magma?*

Still other problems for the POAM fractionation model concern its applicability to different types of andesite. These are discussed in the following two sections.

12.3 Origin of Tholeiitic Versus Calcalkaline Andesites

Can POAM fractionation account for the contrasts between theoliitic and calcalkaline orogenic andesites which have been noted throughout this book? Before answering yes, I will reiterate some of the differences.

Although most rocks with $> 53\%$ SiO_2 in volcanic arcs are calcakaline in the sense that there is relatively little change in their FeO^*/MgO ratios over a wide silica range, there are substantial differences in the absolute value of this ratio and differences in groundmass Ca-poor pyroxenes such that many orogenic andesites are tholeiitic according to various definitions. For example, 40% of the 2500 surveyed in this study are tholeiitic according to Fig. 1.4. Using this definition, tholeiitic orogenic andesites usually have higher ratios of normative plagioclase to pyroxene, lower concentrations of transition metals and, especially when rocks closely related in space and time are compared, lower concentrations of LIL-elements, than do calcalkaline orogenic andesites with similar silica contents (Sect. 5.8 and Fig. 12.4). Concentrations of Fe, Ti, P, V, and possibly Nb, Cu, and Zn all remain constant or go through a maximum in tholeiitic orogenic andesites but consistently correlate negatively with silica in calcalkaline ones (Sects. 5.2 and 5.4). Low-K andesites which also are tholeiitic in the sense of having a rapid increase in FeO^*/MgO ratios relative to silica and having pigeonite in the groundmass (i.e., members of the "island arc tholeiitic series") are restricted to volcanic fronts of arcs where plates converge at > 7 cm/year (Sect. 7.3) and sometimes precede calcalkaline varieties during the evolution of a single stratovolcano (Sect. 4.4) or of entire provinces (Sect. 7.5). While the above remarks suggest that arc tholeiitic magmas have higher liquidus temperatures, lower initial water contents, and lower initial f_{O_2}, available data are insufficient to confirm these suggestions.

Three explanations of these differences between tholeiitic and calcalkaline orogenic andesites have been discussed implicitly in Chapters 8 to 11. First, both may result from differentiation of the same basaltic parent but differ in f_{O_2} or the extent of crustal interaction (Kuno 1950a, 1968b), or in the depth and therefore mineralogy of crystal fractionation (e.g., Hawkesworth and O'Nions 1977; White and McBirney 1979). That is, a tholeiitic (or high-alumina) basalt may yield tholeiitic andesite differentiates until f_{O_2} or P_{H_2O} increases sufficiently, or unless the load pressure is great enough, to stabilize magnetite, amphibole, or garnet whose subsequent fractionation produces calcalkaline liquids. This is an especially attractive explanation for up-section changes in composition and mineralogy within single stratovolcanoes where the transition between rock series is associated with dormancy and with the first modal appearance of magnetite or amphibole phenocrysts (Sect. 4.4). However, the proposal cannot explain the common enrichments of transition metals such as Ni and Cr in calcalkaline versus tholeiitic andesites at similar silica contents, nor the magnitude of differences in LIL-element contents between the suites.

A second explanation is that the basaltic parents of tholeiitic and calcalkaline andesites originate from different degrees and depths of partial fusion of a common peridotite source, with tholeiitic parents reflecting larger melt fractions (Jakeš and Gill 1970; Miyashiro 1974; Kushiro and Sato 1978; Masuda and Aoki 1979). Figures 12.1 to 12.4 illustrate this interpretation schematically. Assume that parental lherzolite has a Mg-number of 88, 2000 ppm Ni, 0.1% K_2O and H_2O, and 1.25 ppm Rb. Further assume that a primary tholeiitic arc basalt (point a) represents 25% fusion of this lherzolite in the presence of some IRS fluid. The basalt would have a Mg-number of 67, 131 ppm Ni (if $\overline{D}_{Ni} = 20$), 0.4 wt.% K_2O and H_2O, and 5 ppm Rb. If the IRS fluid were 5 parts water, 4 parts silica, and 1 part incompatible, radiogenic "others", and amounted to 1.2 wt.% of the melt which formed (in which it then dissolved), then the resulting unfractionated mixture would have 1.0 wt.% H_2O and, say, 50% SiO_2 with f_{O_2} below the NNO buffer. This liquid then precipitates primarily olivine, pyroxene(s), and later plagioclase during its evolution from point a to b where magnetite saturation occurs. About 44% total crystallization is necessary to reach point c, at 57% SiO_2, where the liquid would have low Ni contents and about 0.7% K_2O and 1.75% H_2O.

In contrast, the calcalkaline parent (point m) represents only 10% fusion despite being fluxed with the same mass of IRS fluid. As a consequence, point m is more alkaline, more hydrous, and has higher f_{O_2}. I have assumed it is also more siliceous. Because the liquidus temperature is lower and the initial f_{O_2} is higher in the calcalkaline case, less silicate fractionation is necessary before magnetite saturation occurs (Fig. 12.2), but this nevertheless happens at the same silica content of the liquid (point n) as in the tholeiitic example. Only 10% crystallization might be necessary to reach point n where the liquid would contain more Ni, K_2O, Rb, and H_2O than did the tholeiitic andesite of comparable silica content.

Although idealized, the foregoing exercise illustrates that the POAM fractionation model presented in Section 12.1 applies to both tholeiitic and calcalkaline suites and qualitatively can explain geochemical differences between them. While I found no adequately analyzed natural examples where both suites were erupted in common space and time, analyses of Quaternary Japanese samples from adjacent volcanoes but an unknown stratigraphic sequence do, with one exception, conform to the patterns shown (Masuda and Aoki 1979).

A third possibility is that tholeiitic and calcalkaline magmas differ in their relative proportion of recycled material from the slab, or in the agent or process of recycling. At one extreme, entirely different sources may yield the two series. For example, some argue that tholeiitic series melts come from subducted oceanic crust whereas calcalkaline melts come from the mantle. Principal evidence supporting this idea includes the higher Mg-numbers and Ni and Cr contents of the calcalkaline varieties, the low water and LIL-element contents in arc tholeiitic magmas (Ewart et al. 1977), the higher Ba/La and $^{87}Sr/^{86}Sr$ ratios of arc tholeiitic magmas, and the occasional predominance of tholeiitic magmas early in arc history when temperatures in subducted crust may be highest (Anderson et al. 1978).

Others have suggested the exact opposite: that the tholeiitic magmas are from the mantle wedge while calcalkaline ones are from subducted crust. Reasons for this view include evidence from Ni contents and REE patterns for olivine-dominance of fractionation from tholeiitic parents which, upon olivine-addition, become viable candidates for primary mantle liquids (Nicholls 1974; Ringwood 1975), and greater heavy REE depletion in calcalkaline than tholeiitic andesites, suggesting equilibration of the former with subducting eclogite (Kay 1978). The occasional predominance of tholeiitic magmas early in arc history would reflect magma genesis from thermally perturbed mantle during initial stages of subduction, prior to the availability of subducted ocean crust during the period of fertility of non-convecting subvolcanic arc mantle (Donnelly and Rogers 1980).

Still others have opted for less extreme positions: that one or the other series involves more of a slab-derived component. For example, although the mass of IRS fluid added to melts a and m in Figs. 12.1 to 12.4 is the same, recycled material constitutes a larger fraction of m, helping to make it more siliceous, hydrous, alkaline, and oxidizing. Hence, calcalkaline series rocks may recycle more slab or at least contain a relatively larger slab-derived mass fraction than do those of the tholeiitic series. Because of both the gradational differences between series and the unlikelihood of obtaining either directly from the slab, this less extreme position is favored.

In practice, all three processes — differences in fractionating mineralogy, in percent fusion, and in the amount of slab recycling — probably operate and overlap. The first two processes merge and fade if eutectic-like melting maintains similar silica contents, temperatures, and f_{O_2} over a range of melt fractions. Similarly, the second and third processes merge if smaller melt fractions consistently contain larger weight fractions of recycled material, as assumed in Figs. 12.1 to 12.4.

Any of the three processes discussed above might explain the long-term transition from tholeiitic to calcalkaline behavior in young island arcs and in Archean greenstone belts which were noted in Section 7.5. Changes in fractionating phases plausibly can be linked with secular thickening of the crust in two ways. Thicker crust slows magma ascent by reducing density contrast with wall rocks. Longer residence time within sialic crust provides greater opportunity to assimilate oxygen, thereby increasing magnetite stability (Miyashiro 1974; Goodwin 1977). Alternatively, thicker crust may cause magma stagnation and fractionation at greater depth which might enhance magnetite stability (Osborn 1979; White and McBirney 1979) or permit garnet fractionation (Hawkesworth and O'Nions 1977). Such mechanisms also have the virtue of explaining the positive correlation between the percent of calcalkaline rocks and crustal thickness in modern volcanic arcs (Sect. 7.2).

Alternatively, if calcalkaline rocks result from smaller percentages of fusion, then their relative increase with time is analogous to the tholeiitic to alkalic transition on ocean islands, presumably attributable to diminishing heat sources or diminishing mantle fertility with time.

Finally, if tholeiitic magmas contain a smaller recycled component, then the transition might reflect increased wedge metasomatism with time (Green 1980).

Other ideas which have been proposed to explain tholeiitic to calcalkaline transitions each require andesite genesis from the slab [tholeiitic andesites in the case of Anderson et al. (1978), calcalkaline andesites in the case of Condie and Moore (1977), or Donnelly and Rogers (1980)] or from the lower crust (Goodwin and Smith 1980). The weight of Chapters 8 and 10 is against these. However, given the lack of constraints, the entire topic remains unresolved.

12.4 Origin of Across-Arc Geochemical Variations

The across-strike differences in composition of orogenic andesites summarized in Section 7.1 are, in my opinion, too regular to be fortuitous. Many explanations of these variations have been proposed, invoking the following sorts of regular differences in magma genesis with increased distance from the plate boundary: increased crustal thickness and resulting contamination (Condie and Potts 1969; Zentilli and Dostal 1977); increased crustal thickness causes crystal fractionation to occur at greater depth and therefore with higher pyroxene/plagioclase ratios (Ui and Aramaki 1978); increased wall-rock reaction or zone refining within the mantle wedge due to longer magma ascent paths (Harris 1957; Green and Ringwood 1967; Ninkovich and Hays 1972; Best 1975); the effect of pressure on distribution coefficients (Dickinson 1968; Marsh and Carmichael 1974); horizontal or vertical heterogeneities in the composition of the mantle wedge which either predate subduction or result from variable degrees of slab recycling (Ringwood 1975; Whitford and Nicholls 1976; Saunders et al. 1980); increased influence of subducted sediment relative to subducted mafic rocks in the source of arc magmas (Tatsumoto 1969); changes in refractory phases during fusion of subducted crust (Jakeš and White 1970; Fitton 1971; Boettcher 1973); or decreasing percent fusion plus increasing depth of liquid segregation in the mantle wedge (Kuno 1959; Sugimura and Uyeda 1973; Miyashiro 1974; Whitford et al. 1979a). Because $^{87}Sr/^{86}Sr$ and $^{207}Pb/^{204}Pb$ ratios are more sensitive to crustal interaction than are elemental concentrations (Sect. 10.3) yet usually remain constant or decrease across volcanic arcs (versus batholiths) despite increasing Rb/Sr and U/Pb ratios (Sect. 7.1), appeal to variable crustal contamination to explain these geochemical trends is unconvincing. Similarly, no known changes in refractory phases account for the patterns summarized in Table 7.2 (Sect. 8.2). In one form or another, however, all the other explanations are viable; definitive choice between them is not yet possible. Thus I will merely show that the scheme presented in the preceding sections is consistent with the spatial variations in composition observed.

Across-arc (and some along-arc) increases in LIL-element concentrations in orogenic andesites can be attributed largely to decreased percent fusion of chemically homogeneous peridotite with increased distance from (or along) the plate boundaries (cf. points c and o in Fig. 12.4). Clear demonstration of this idea's validity requires establishing the kind of regionally consistent pattern within one volcanic arc that Frey et al. (1978), for example, found in the anorogenic basalts of southeastern Australia. Such a test cannot yet be undertaken in volcanic arcs largely because potentially primary basalts are uncommon for the reasons discussed earlier. However, the explanation is at least qualitatively consistent with water contents seeming higher and temperatures lower in more K-rich suites (Sects. 5.8 and 7.1), and with the volume of ejecta decreasing with increased alkalinity across (Sect. 2.1) and sometimes along volcanic arcs (Sect. 7.2).

Despite this internal consistency, there is no satisfying theoretical explanation why percent fusion usually should decrease away from the plate boundary because, at constant pressure, temperatures will, if anything, increase in that direction (Fig. 3.6). However, if percent fusion is governed by the amount of IRS fluid which is added, then melt fractions will be greatest above the principal sites of slab dehydration. These are expected to be near the volcanic front (Fig. 8.2).

Reasons to suspect that differences in percent fusion alone cannot account for the spatial variations observed include the following. First, variations observed in isotope and LIL-element ratios (e.g., Na/K, K/Rb, Rb/Sr, Th/U, La/Sm, Zr/Nb) are incompatible with variable melting of a homogeneous source, leaving only olivine and pyroxenes as refractory phases. However, if amphibole, phlogopite, or other accessory minerals are more refractory during generation of the parent magmas for high-K versus low-K andesites, for example, then some variability of LIL-element ratios will result. Such a suggestion requires quantitative testing not yet possible for volcanic arcs due to the lack of thoroughly analyzed potentially primary rocks, as noted above, although one such effort concluded that the K/Na and La/Yb ratios in the mantle wedge beneath Java must increase with distance from the trench or with depth (Nicholls and Whitford 1976; Whitford et al. 1979a). A heterogeneous mantle wedge also accounts more easily for along-strike variations in composition. Note that the decreases in K/Rb, Ba/La, and $^{87}Sr/^{86}Sr$ ratios across some arcs (Table 7.2) are consistent with a reduced fraction of recycled slab material away from the dehydration front, but the three changes do not always occur within the same arcs (e.g., New Britain) and are at odds with Th-U-Pb systematics and increasing La/Yb ratios (Sect. 8.5).

Secondly, occasional decreases in heavy REE and Y contents across volcanic arcs suggest a larger weight fraction of refractory garnet and, therefore, greater depth of liquid segregation with distance from the plate boundary. Greater depth also would result in decreased silica-saturation, a characteristic of higher-K arc suites, whereas reduction in percent fusion was assumed to have the opposite effect in Figs. 12.1 to 12.4. Negative correlations between percent fusion and depth of liquid segregation are common in basalt petrogenesis arguments and presumably result from ascent of partially fused mantle along a thermal gradient which is steeper than the peridotite solidus.

Thirdly, across-arc differences in the conditions (e.g., T, f_{O_2}) and phases of fractionation will affect liquid compositions, possibly accounting for changes from positive to negative correlations between heavy REE or Y and silica, and the decreased likelihood of negative Eu anomalies.

Even if the scenario sketched in Section 12.1 for andesite genesis is fundamentally correct in many instances, the numerous possible variations on the theme provide ample opportunity to develop the diversity of compositions summarized in Section 5.8. That differences in percent fusion can account for as much of this diversity as has been shown in the last two sections is surprising, though still conjectural. That there *are* consistent spatial and temporal variations in alkalinity and Fe-enrichment (Chap. 7) is more impressive still. While this consistency neither proves nor precludes specific genetic options, I have shown that it is at least consistent with the POAM fractionation model developed in Section 12.1. Criteria for choosing between explanations of the consistency await fuller geochemical constraints for both examples and processes.

12.5 Epilog

A common obstacle to solving problems, such as the "andesite problem", is disagreement over problem definition. Just as three blind men each can repeatedly verify their observations that an elephant has the characteristics of a trunk or tusk or tail, depending on their perspective and range of experience, so also explanations of andesite genesis often satisfy only certain kinds of observations, e.g., certain field, petrographic, analytical, or experimental data. Consequently, I have tried to assemble, organize, and account for as many observations relevant to andesite genesis as possible.

It is premature, may always be premature, to consider the problem "solved" because apparently as many roads lead to andesite as to Rome. However, evidence that the principal means of andesite genesis is crystal fractionation of phenocryst phases from mantle-derived basalt plus recycled slab seems common to most suites. Thus, the early promise that plate tectonics offered an elegant explanation of andesites as primary melts has proven empty. The arguments made fifty years ago by Bowen, thirty years ago by Kuno, and twenty years ago by Osborn seem essentially correct today even though tectonically uninformed. However, as with perceptions of elephants, there is some truth in each vision. There *is* a contribution from the slab. Consequently, undifferentiated melts in the mantle wedge are more siliceous than in asthenosphere elsewhere. Ascent paths *are* affected by the compressional stress regime. Fractionating minerals are *less* silicic due to the effects of more dissolved water and oxygen. All three insights are outgrowths of plate tectonics and constitute essential ammendations of older differentiation hypotheses.

It is valuable now to recall reasons why all the human labor which I have summarized concerning andesitic volcanism was undertaken in the first place. I have emphasized the *origin* of orogenic andesite magmas, a fundamental topic for understanding how the Earth operates. My principal conclusion is that few orogenic andesites are unfractionated primary melts. Most are derivative, the residues of low pressure crystal fractionation. In some cases they also reflect blending of successive magma batches, interaction with the crust during their ascent, and vapor fractionation. The parental melts are hybrid basalts which contain contributions from both the chemically heterogeneous mantle wedge and subducted oceanic crust. This hybridization does not require contemporaneous subduction, limiting the paleosignificance of orogenic andesites.

However, the relative proportion of slab versus wedge contributions to parental melts, the details of the crystal fractionation process, the frequency of other differentiation mechanisms, and the spatial and temporal variations in all of the above, remain speculations. Together they constitute the chief andesite problems for the 1980's.

In a broader sense, the 1980's also will see continued attempts to apply the conclusions summarized above to other problems. For example, if the weight fraction of the mantle wedge contribution to parental melts is high, then orogenic andesite (not MORB) volcanism is the principal mechanism which controls the geochemical evolution of the upper mantle, as well as the crust. Also, the details and differences of genetic histories of andesites should affect civil defense and metals exploration planning, if understood. However, these are matters for other books by other people.

Appendix Chemical Composition of Ejecta from Recently Active Volcanoes at Convergent Plate Boundaries*

Arc[1]	Volcano[2]	No.[3]	<53% SiO_2[4]	53–57	57–63	>63	$K_{57.5}$[5]	FeO*/ $MgO_{57.5}$[6]	Ref.
1. New Zealand (Fig. 1)									
1.1	White Island	16	0	1	11	4	1.1	1.4	1, 126
1.2	Okataina (Tarawera)	25	0	0	0	25	–	–	1, 129
1.3	Taupo	4	0	0	0	4	–	–	1
1.4	Tongariro	66	3	18	45	0	0.8–1.6	1.4	1, 2
1.5	Ngauruhoe	4	0	2	2	0	1.3	1.6	1, 2
1.6	Ruapehu	16	0	3	13	0	1.5	1.6	1, 2
1.7	Egmont	22	4	16	2	0	2.2	2.3	1, 3
2. Kermadec (Fig. 1)									
2.1	Raoul	13	8	3	0	2	0.5	3.0	4, 139
2.2	Macauley*	9	8	0	0	1	0.4	–	5, 139
3. Tonga (Fig. 2)									
3.1	Hunga Ha'apai	2	0	2	0	0	–	–	6
3.2	Tofua	11	0	8	2	1	0.5	2.8	8, 9
3.3	Metis Shoal	2	0	0	0	2	–	–	6, 7
3.4	Late	10	0	9	1	0	0.6	2.7	6
3.5	Fonualei	15	0	0	3	12	(0.8)	(3.8)	6
4. Vanuatu (Fig. 3)									
4.1	Suretamatai	16	12	3	1	0	1.9	(3.4)	138
4.2	Gaua	8	6	2	0	0	>5.0	>3.7	140
4.3	Aoba	42	42	0	0	0	–	–	10
4.4	Ambrym	8	8	0	0	0	–	–	10, 11
4.5	Lopevi	9	5	1	3	0	1.4	(2.0)	139
4.6	E. Epi*	14	11	3	0	0	1.6	–	10
4.7	Tongoa	22	18	0	0	4	(2.0)	–	10
4.8	Erromango	92	77	13	2	0	1.4 + 3.5	2.5	147
4.9	Tanna	9	0	7	2	0	(2.5)	(3.0)	10
5. Solomon Islands (Fig. 4)									
5.1	Savo	4	0	1	1	2	(1.9)	–	13
5.2	New Georgia	17	12	5	0	0	3.0	2.1	14
5.3	Bagana	40	1	34	5	0	1.7	2.5	15
5.4	Balbi	2	0	0	2	0	2.5	2.0	15

* See p. 326 for explanation of footnotes, pp. 327/328 for references, and pp. 329–336 for figures

Arc[1]	Volcano[2]	No.[3]	<53% SiO_2[4]	53–57	57–63	>63	$K_{57.5}$[5]	$FeO^*/$ $MgO_{57.5}$[6]	Ref.
6. Bismarck (Fig. 4)									
6.1	Rabaul	21	5	3	8	5	1.7	(2.7)	16, 17
6.2	Ulawun	22	19	3	0	0	0.5	(1.7)	19
6.3	Bamus	12	0	8	4	0	0.4	2.8	19
6.4	Hargy-Galloseulo	14	0	4	7	3	0.6	2.3	19
6.5	Lolobau	15	1	5	3	6	0.7	2.6	19
6.6	Sulu*	10	0	7	0	3	0.6	1.6	19
6.7	Witori*	12	0	0	3	9	0.5	3.0	18, 19
6.8	Krummel-Hoskins	35	1	14	13	7	0.7	2.0	18, 19
6.9	Talasea	19	1	0	12	6	1.3	2.1	19, 20
6.10	N. Willaumez	15	3	2	5	5	1.3	2.5	19, 20
6.11	Witu I.	29	13	7	4	5	0.7 + 1.7	–	19
6.12	Karkar	14	0	13	1	0	0.9	2.7	21, 22
6.13	Manam	16	11	4	1	0	–	–	21
7. Papua New Guinea (Fig. 4)									
7.1	Highlands	30	15	8	6	1	1.5– 2.5	2.5	23
7.2	Lamington	29	1	10	14	4	2.2	1.3	12, 13
7.3	Victory, Trafalgar	19	0	3	16	0	2.0	1.1	13, 24
8. Sunda (Fig. 5)									
a) Sumatra									
8.1	Sorikmarapi	2	0	0	2	0	(1.6)	(2.0)	25
8.2	Talakmau	3	0	0	3	0	1.6	2.1	25, 26
8.3	Marapi	6	0	3	3	0	2.2	(2)	25–27
8.4	Tandikat	5	0	1	3	1	1.7	2.0	25–27
8.5	Krakatau	34	7	4	5	18	1.4	2.3	25, 26
b) Java									
8.6	Tangkuban	21	3	10	8	0	2.3	2.7	25, 28
8.7	Guntur	3	2	0	1	0	0.8 + 1.4	2.2	25, 28
8.8	Papandajan	12	0	2	8	2	1.2	2.1	25, 28
8.9	Galunggung	4	1	3	0	0	0.8	2.2	25, 28
8.10	Tjerimai	15	0	9	6	0	1.6	1.9	25, 28
8.11	Slamet	22	18	4	0	0	(1.7)	(2.0)	25, 28
8.12	Dieng	12	1	3	8	0	1.8	2.2	25, 28
8.13	Sundoro	14	2	6	6	0	2.0	2.6	28
8.14	Sumbing	16	1	3	12	0	1.8	2.2	25, 28
8.15	Ungaran	19	10	8	1	0	3.1	2.2	25, 28
8.16	Merabu	13	6	2	5	0	2.0	2.2	25, 28
8.17	Merapi	30	8	22	0	0	2.1	2.6	25, 28
8.18	Kelud	6	1	2	3	0	1.0	2.4	25, 28

Arc[1]	Volcano[2]	No.[3]	$<53\%$ SiO_2[4]	53–57	57–63	>63	$K_{57.5}$[5]	$FeO^*/$ $MgO_{57.5}$[6]	Ref.
	8.19 Semeru	9	2	1	6	0	1.4	2.9	25, 28
	8.20 Bromo	7	0	5	2	0	2.9	(2.5)	25, 28
	8.21 Raung	9	4	3	2	0	1.8	(2.8)	25
c)	Bali-Banda								
	8.22 Batur	9	4	0	0	5	(1.7)	–	28
	8.23 Agung	8	2	5	1	0	1.6	(2.3)	28
	8.24 Seraja	4	1	3	0	0	1.7	>2.5	28
	8.25 Paluweh	12	5	2	5	0	2.2	2.5	25
	8.26 Ili Boleng	3	1	1	1	0	2.1	2.7	25, 29
	8.27 Lewotolo	7	2	2	3	0	(3.5)	(2.4)	25, 29
	8.28 Sirung	5	2	1	0	2	1.8	(3.2)	29
	8.29 Damar	5	0	0	5	0	2.5	1.8	29
	8.30 Teun	4	0	0	4	0	(1.8)	(1.8)	29
	8.31 Nila	5	0	4	1	0	2.2	1.8	29
	8.32 Serua	6	0	2	4	0	1.1	1.4	29
	8.33 Manuk	5	0	2	3	0	1.0	1.5	29
	8.34 Banda	8	1	0	0	7	(0.5)	(2.8)	29
9.	Halmahera (Fig. 6)								
	9.1 Dukono	6	1	1	1	3	2.0	(2.7)	25
	9.2 Ternate	6	0	2	3	1	1.8	2.9	30
	9.3 Tidore*	5	2	0	1	2	1.8	–	31
10.	Sulawesi (Fig. 6)								
	10.1 Lokon-Empung	5	2	1	0	2	1.5	–	25
11.	S.E. Philippines (Fig. 7)								
	11.1 Mayon	55	6	47	2	0	1.2	2.2	32, 33, 34
12.	N.W. Philippines Taiwan (Fig. 7)								
	12.1 Taal	4	2	1	1	0	(2.2)	3.0	32
	12.2 Kuei-shan-tao	5	0	1	4	0	1.1	1.8	35
	12.3 Tatun area	16	9	3	3	1	1.8	2.1	35
	12.4 Keelung*	9	1	4	4	0	1.6	1.7	35
	12.5 Agin-court	3	3	0	0	0	–	–	35
13.	Ryuku-West Japan (Fig. 8)								
	13.1 Suwanosezima	13	0	3	10	0	1.3	2.1	36, 37
	13.2 Nakano-zima	4	0	0	2	2	0.9	1.7	36
	13.3 Kutinoerabu-zima	3	0	0	3	0	(<1.5)	(<1.9)	36, 37
	13.4 Tokara-Iwo-zima	12	2	1	0	9	(0.8)	–	36
	13.5 Sakurajima	83	0	0	37	46	1.4	2.1	36, 37
	13.6 Kirisima	21	2	7	11	1	1.9	2.4	36, 37

Arc[1] Volcano[2]	No.[3]	<53% SiO₂[4]	53–57	57–63	>63	K₅₇.₅[5]	FeO*/MgO₅₇.₅[6]	Ref.
13.7 Unzen	7	0	0	3	4	1.6	1.9	36
13.8 Aso	35	9	11	9	4	2.2	2.1	36
13.9 Kuzyo	9	1	3	4	1	1.5	2.7	36
14. Mariana-Izu (Figs. 9, 10)								
14.a Sarigan	4	1	2	1	0	0.7	1.9	150
14.b Pagan	16	14	2	0	0	>1.4	>2.4	36, 38, 150
14.1 Agrigan	29	17	10	2	0	1.8	4.1	151
14.2 Asuncion	6	0	6	0	0	–	>2.8	150
14.3 Uracas	5	2	3	0	0	1.2	3.4	36, 37
14.4 Tori-zima	8	5	2	1	0	0.4	2.3	36
14.5 Aoga-zima	4	2	0	2	0	0.4	3.8	36
14.6 Hatizyo-zima	32	19	4	5	4	0.5	3.0	36, 39
14.7 Miyake-zima	12	4	8	0	0	0.5	5.9	36
14.8 Kozu-zima	5	0	0	0	5	–	–	36
14.9 Osima	7	7	0	0	0	–	–	36
15. East Japan (Fig. 10)								
a) Honshu								
15.1 Omuro-yama	13	5	6	2	0	1.0	1.8	36
15.2 Hakone	15	3	3	4	4	0.6	3.4	36
15.3 Fuji	18	13	1	0	4	–	–	36
15.4 On-take	15	2	1	7	5	1.7	(2.4)	36, 135
15.5 Myoko	55	16	20	14	5	1.5	2.0	36, 148
15.6 Asama	73	0	11	36	26	0.8	2.0	36, 41, 159
15.7 Kusatu-sirane	4	0	0	2	2	(1.3)	–	36
15.8 Akagi	8	0	3	2	2	1.0	2.2	36
15.9 Takahara*	8	3	2	1	2	0.9	–	40
15.10 Nasu	15	4	4	7	0	0.9	1.9	36, 40, 42
15.11 Bandai	17	0	0	17	0	1.0	2–3	142
15.12 Adatara	4	0	1	3	0	–	–	36
15.13 Azuma	8	1	3	4	0	1.2	1.9	36, 40
15.14 Gassan*	4	1	0	3	0	1.9	1.9	40
15.15 Zao	5	2	1	2	0	(1.1)	1.8	36, 40
15.16 Funagata*	7	5	1	1	0	0.8	2.0	40
15.17 Narugo	4	0	0	1	3	(0.5)	(2.1)	36
15.18 Kurikoma	2	0	1	1	0	0.8	1.8	36
15.19 Chokai	15	1	9	5	0	1.8	2.2	36
15.20 Akita-Komagatake	36	8	6	21	1	0.5	3.0	36, 43
15.21 Iwate	13	7	5	1	0	(0.6)	>2.6	36
15.22 Hatimantai	34	8	8	13	5	0.7	–	36, 40
15.23 Yakeyama	4	0	0	3	1	1.0	2.0	36
15.24 Kayo	6	1	0	4	1	0.8	(2.4)	134

Arc[1]	Volcano[2]	No.[3]	>53% SiO_2[4]	53–57	57–63	>63	$K_{57.5}$[5]	$FeO*/MgO_{57.5}$[6]	Ref.
	15.25 Iwaki	13	0	7	5	1	1.0	2.5	36, 40
	15.26 Hakkoda	4	0	0	3	1	0.8	<1.8	40
	15.27 Towada	22	3	3	7	9	0.6	2.8	40, 133
b) Hokkaido									
	15.28 Osima-osima	4	2	1	1	0	>2.2	–	44
	15.29 Esan	7	0	0	7	0	0.5	2.7	132
	15.30 Komaga-dake	6	0	0	6	0	0.7	2.4	36
	15.31 Usu	20	6	8	0	6	(0.7)	–	44, 46, 47
	15.32 Yotei*	4	0	0	3	1	1.0	2.6	44
	15.33 Tarumai-Shikotsu	10	1	0	5	4	0.8	2.2	36, 44
	15.34 Tokati	5	2	1	1	1	(1.8)	–	36, 44
	15.35 Daisetsu	5	0	3	1	1	1.6	2.1	36
	15.36 Rishiri*	8	5	1	0	2	–	–	44
	15.37 Me-akan	6	0	3	3	0	1.1	–	36, 37
	15.38 Atosanupuri	12	1	1	2	8	(1.0)	(2.7)	36, 44
	15.39 Mashu*	38	5	11	13	9	0.3	3.5	44, 45
16. Kuriles (Fig. 11)									
	16.1 Golovnin	9	0	1	4	4	0.5	2.8	48
	16.2 Mendeleev	5	1	0	2	2	0.7	(2.4)	48
	16.3 Tiatia	4	3	1	0	0	–	(>2.6)	48
	16.4 Ivan Groznyi	7	0	1	6	0	1.1	1.7	48
	16.5 Medvezh'ya*	6	1	1	1	3	0.8	2.9	48
	16.6 Urup Is.	47	7	10	21	9	0.8	2.2	48, 49
	16.7 Goryashchaya Sopka	10	0	1	8	1	1.2	1.7	48, 49
	16.8 Zavaritzki	18	1	6	5	6	0.8	2.5	48, 49
	16.9 Milne*	13	0	7	6	0	–	1.9	48, 49
	16.10 Uratman*	14	1	7	2	4	0.9	2.2	48
	16.11 Ushishir Is.	5	0	1	3	1	0.6	(2.0)	48
	16.12 Shiashkotan Is.	11	0	3	8	0	1.0	2.3	48
	16.13 Kharimkotan Is.	6	0	0	5	1	1.1	2.4	48
	16.14 Tao-Rusyr*	10	4	3	1	2	1.3	3.0	48
	16.15 Nemo	9	0	4	3	2	0.9	2.3	48
	16.16 Ebeko	14	3	4	6	1	1.8	2.0	48
	16.17 Bogdanovich*	8	0	3	5	0	2.0	2.2	48
	16.18 Fuss Pk.	3	0	2	1	0	2.4	2.0	48
	16.19 Alaid	6	6	0	0	0	–	–	48
	16.20 Shirinki*	2	0	0	2	0	1.9	–	48
	16.21 Ekarma	3	0	0	3	0	1.2	1.8	48
	16.22 Chirinkotan	6	0	2	4	0	1.9	1.8	48

Arc[1]	Volcano[2]	No.[3]	<53% SiO$_2$[4]	53–57	57–63	>63	K$_{57.5}$[5]	FeO*/MgO$_{57.5}$[6]	Ref.
17. Kamchatka (Fig. 12)									
17.1 Ksudach		9	3	2	3	1	(1.0)	(3.3)	50, 51
17.2 Gorely									
	Khrebet	23	8	3	6	6	2.0	2.0	51
17.3 Avachinsky		20	2	11	6	1	0.7	–	50
17.4 Koriavsky		15	5	4	5	1	–	2.6	51
17.5 Aag*		6	0	0	6	0	1.4	1.6	51
17.6 Arik*		7	0	1	4	2	1.3	1.1	51
17.7 Dzenzursky		6	0	0	3	3	–	–	51
17.8 Karymsky		13	0	1	9	3	1.4	1.8	51
17.9 Zentralny									
	Semiachik	4	0	1	3	0	1.0	1.9	50
17.10 Burliastchy		11	0	2	3	6	1.0	2.1	51
17.11 Uzon		6	3	0	2	1	1.3	3.0	50
17.12 Kronotzky		7	6	0	1	0	(0.7)	(2.5)	51
17.13 Gamchen		8	0	8	0	0	(1.1)	–	51
17.14 Malaya									
	Udina*	13	6	2	5	0	(1.3)	2.5	51
17.15 Plosky									
	Tolbachik	29	18	9	2	0	–	(2.6)	50, 51
17.16 Ostriy									
	Tolbachik*	9	8	1	0	0	–	–	51
17.17 Ziminykh*		14	2	6	3	3	(1.4)	–	51
17.18 Bezymianny		15	0	1	12	2	(0.9)	1.5	50
17.19 Kamen*		8	4	2	2	0	–	(2.2)	51
17.20 Klyuchev-									
	skoy	32	12	20	0	0	(1.6)	–	50
17.21 Sheveluch		27	3	5	18	1	1.3	1.6	50
17.22 Khangar*		8	2	0	1	5	(1.6)	–	51
17.23 Ichinsky		23	3	1	6	13	(1.8)	–	50, 51
17.24 Anaun*		17	10	6	1	0	(1.7)	–	51
18. Aleutians-Alaska (Fig. 13)									
18.1 Kiska		4	0	1	3	0	1.5	2.0	52-R
18.2 Little Sitkin		13	1	3	5	4	1.1	2.0	52-H
18.3 Cerberus (Semisopoch- noi Is.)		15	5	3	6	1	2.0	3.0	52-O
18.4 Kanaga		4	0	3	1	0	1.7	2.4	53
18.5 Moffet (Adak Is.)*		14	9	2	3	0	1.3	2.5	52-C, 55
18.6 Recheschnoi*		4	0	1	2	1	1.4	–	52-L, 54
18.7 Vsevidof		4	0	1	1	2	(1.3)	–	52-L
18.8 Okmok (Umnak Is.)		20	11	2	1	4	1.5	(3.2)	52-L
18.9 Bogoslof		2	1	0	1	0	(2.6)	–	52-L, 56
18.10 Makushan (Ulnalaska Is.)		12	6	4	2	0	(1.5)	2.2	52-S

Arc[1]	Volcano[2]	No.[3]	<53% SiO_2[4]	53–57	57–63	>63	$K_{57.5}$[5]	FeO*/ $MgO_{57.5}$[6]	Ref.
18.11	Trident	2	0	0	2	0	(1.2)	<2.1	57
18.12	Katmai	18	0	4	6	8	0.9	–	58
18.13.	Redoubt	2	0	0	2	0	(1.5)	<2.3	57
18.14	Drum	30	0	1	8	21	1.5	1.1	156
18.15	Edgecumbe	9	4	2	2	1	0.7	1.7	59
19. Cascades (Fig. 14)									
19.1	Garibaldi*	16	3	2	6	5	(1.2)	1.7	60
19.2	Baker	2	0	0	2	0	(1.7)	–	61
19.3	Glacier Pk*	27	4	5	6	12	1.3	1.2	62
19.4	Rainier	10	0	1	7	2	(1.4)	–	61, 63
19.5	St. Helens	5	2	0	1	3	(0.8)	–	64
19.6	Adams	2	0	0	2	0	(2)	–	65
19.7	Hood	43	9	5	26	3	1.0	<2.0	66
19.8	Jefferson*	31	5	13	3	7	1.0	1.5	63, 67, 69
19.9	Three Sisters*	25	4	3	4	14	1.3	2.0	69, 71
19.10	Newberry*	72	20	12	6	34	1.6	(2.0)	70
19.11	Crater Lake*	37	2	10	8	17	1.3	1.5	68, 69 71
19.12	McLoughlin*	35	4	21	10	0	1.2	1.7	71
19.13	Shasta	10	1	0	7	2	–	–	63, 72
19.14	Medicine Lake	31	8	5	5	13	1.4	(1.6)	72, 73
19.15	Lassen	146	7	14	57	68	1.4	1.4	72, 74
20. Mexico (Fig. 15)									
20.1	Ceboruco	9	0	0	6	3	1.9	2.0	130
20.2	Colima	16	0	2	14	0	1.2	1.5	75, 157
20.3	Paricutin	23	0	12	11	0	1.3	1.2	76, 77
20.4	Nevado de Toluca*	7	0	1	3	3	1.6	1.0	137
20.5	Chichináutzin Group*	30	1	12	16	1	(1.5)	1.6	78
20.6	Xitli	14	12	1	0	1	–	–	75, 77
20.7	Popocatepetl	11	0	0	9	2	(1.7)	1.3	75, 77
20.8	Tuxtla*	26	26	0	0	0	–	–	144
21. Central America (Fig. 15)									
21.1	Chicabel*	4	1	1	2	0	1.0	–	79
21.2	Siete Orejas*	18	2	4	10	2	(1.4)	–	79
21.3	Santa Maria	56	20	18	3	15	1.3	(2.1)	80
21.4	Cerro Quemado	21	0	0	6	15	1.5	(2.2)	79
21.5	Atitlán	·9	1	8	0	0	1.4	2.1	79, 154
21.6	Tolimán	6	0	5	1	0	2.0	1.8	79
21.7	Fuego	49	47	2	0	0	–	1.8	81

Arc[1] Volcano[2]	No.[3]	$<53\%$ SiO_2[4]	53–57	57–63	>63	$K_{57.5}$[5]	$FeO*/$ $MgO_{57.5}$[6]	Ref.
21.8 Agua	4	1	1	2	0	1.7	–	79, 152
21.9 Pacaya	19	19	0	0	0	–	–	79
21.10 Moyuta*	6	0	1	4	1	1.0	2.2	79
21.11 Tecuamburro	5	0	1	3	1	0.9	2.2	79
21.12 S.Ana-Izalco	11	5	5	1	0	2.2	2.2	82, 152, 154
21.13 S. Salvador	30	7	12	11	0	2.0	3.7	82, 149
21.14 Ilopango	6	0	0	1	5	(1.1)	2.0	155
21.15 S. Vincent	13	1	5	7	0	1.3	2.3	155
21.16 Tecapa	10	7	2	1	0	1.9	2.4	155
21.17 S. Miguel	10	7	3	0	0	(1.5)	>3	155
21.18 Conchagua	4	3	0	1	0	1.3	(2.5)	155
21.19 Telica	3	3	0	0	0	1.7	–	82, 152
21.20 Cerro Negro	16	16	0	0	0	–	–	75, 82, 83
21.21 Masaya	8	8	0	0	0	–	–	84
21.22 Apoyo*	7	4	0	1	2	(1.4)	–	84
21.23 Mombacho	6	3	1	1	1	(1.7)	2.1	85
21.24 Concepcion	4	1	1	2	0	(1.8)	>2.7	85
21.25 Arenal	8	0	7	1	0	0.7	>2.3	86
21.26 Irazu	4	0	4	0	0	2.6	1.8	82, 87
22.Columbia-Ecuador (Fig. 16)								
22.1 Galeras	3	0	1	2	0	(1.7)	2.0	88
22.2 Azufral de Tuquerres	3	0	0	1	2	(1.6)	–	88
22.3 Sumaco	4	4	0	0	0	–	–	88
22.4 Cotopaxi	41	0	5	19	17	1.4	(2.1)	88, 158
22.5 Tungurahua	3	0	2	1	0	(1.5)	–	88
23.Peru-Chile (Fig. 17)								
a) 11° to 28°S								
23.1 Chachani	12	0	1	10	1	2.2	–	89, 124
23.2 Pichu-Pichu	2	0	1	1	0	(2.0)	–	89
23.3 El Misti	6	0	1	4	1	1.9	–	89, 124
23.4 Ubinas	2	0	0	2	0	(2.4)	–	89
23.5 Ticsani	6	0	0	2	5	–	–	89, 124
23.6 Calientes	8	0	0	6	2	–	–	89
23.7 Guallatiri	8	0	0	5	3	2.4	(1.8)	93
23.8 Aucanquilcha*	12	0	0	3	9	–	–	94
23.9 Oilague	8	0	1	5	2	2.6	–	94
23.10 Chela	6	0	2	4	0	1.9	–	94
23.11 Polapi	9	0	3	4	2	2.2	–	94
23.12 Araral*	9	0	0	4	5	2.4	–	94
23.13 Ascotan*	5	0	1	3	1	2.4	–	94
23.14 Azufre*	8	0	0	3	5	2.0	–	94

Arc[1]	Volcano[2]	No.[3]	<53% SiO_2[4]	53–57	57–63	>63	$K_{57.5}$[5]	FeO*/ $MgO_{57.5}$[6]	Ref.
	23.15 San Pedro-San Pablo	58	0	4	33	21	1.8	(1.4)	94, 95
	23.16 Southwest Bolivia*	20	0	2	4	14	(2.5)	<1.8	91
	23.17 Northwest Argentina*	20	6	9	4	1	(2.5)	1.3	92
b) 33° to 47°S									
	23.18 Tupungatito	5	0	0	5	0	2.0	(1.7)	96
	23.19 Antuco	6	3	1	2	0	1.1	2.9	97
	23.20 Cerro Azul	10	0	1	3	6	1–2	2.2	98
	23.21 Nevados de Chillan	12	0	0	2	10	(1.3)	–	99
	23.22 Carran	6	4	2	0	0	–	(>2.5)	90
	23.23 Puyehue	11	3	2	0	6	(1.3)	(>2.8)	90
	23.24 Tronador	9	5	2	1	1	1–2	2–3	100
	23.25 Hudson*	2	0	0	2	0	(2.5)	(>3)	145
c) >47°S									
	23.26 Burney	5	0	0	2	3	<0.7	<1.8	128
24. Antilles (Fig. 18)									
	24.1 Saba	3	0	0	3	0	1.1	1.4	103
	24.2 St. Kitts	173	27	59	73	14	0.6	2.5	101,102
	24.3 Nevis	6	0	1	5	0	0.9	2.2	103
	24.4 Montserrat	11	2	4	5	0	0.7	(2.2)	104
	24.5 Dominica	280	21	47	118	94	0.9	2.4	101
	24.6 Martinique (Pelée)	22	4	8	5	5	1.0	2.4	131
	24.7 St. Lucia	8	0	2	3	3	1.2	–	103
	24.8 St. Vincent	10	1	9	0	0	0.7	>2.4	103,105
	24.9 Grenada	264	129	49	78	8	1.2	1.7	101,106
25. South Sandwich (Fig. 19)									
	(whole arc)	34	14	8	8	4	0.7	4.1	107
26. Eolian (Fig. 20)									
	26.1 Stromboli	6	6	0	0	0	–	–	108
	26.2 Panarea*	19	0	0	12	7	1.8	–	109
	26.3 Lipari	28	0	11	6	11	2.0 + 2.8	–	109,141
	26.4 Salina*	18	5	6	5	2	1.2	(2.7)	110,111
	26.5 Vulcano*	22	5	2	5	10	5	(2.3)	108,110, 141
	26.6 Filicudi*	40	26	8	6	0	2.2	1.8	112,113
	26.7 Alicudi*	9	8	0	1	0	2.0	–	109,113

Arc[1]	Volcano[2]	No.[3]	<53% SiO₂[4]	53–57	57–63	>63	K₅₇.₅[5]	FeO*/MgO₅₇.₅[6]	Ref.
27. Aegean (Fig. 21)									
	27.1 Thebes area*	15	3	8	4	0	(3.0)	–	114
	27.2 Kromyonia (Susaki)	5	0	0	1	4	<2.4	<1.9	115
	27.3 Saronic Gulf (Methana, Aegina)	23	1	4	15	3	1.9	1.2	115,116
	27.4 Milos	5	0	0	0	5	–	–	115
	27.5 Santorini	77	21	14	4	38	1.5	1.5– 2.0	115,117, 118
	27.6 Nisyros	39	0	12	5	22	(1.8)	2.0	119
	27.7 Kos	4	0	0	4	0	3.9	–	115
28. Turkey, Caucasus, Iran (Fig. 22)									
	28.1 Erciyas	27	2	5	5	15	(1.2)	<1.8	120
	28.2 Sürphan	4	0	1	1	2	(1.9)	(2.8)	125
	28.3 Tendurek	3	1	0	2	0	<3	<3	136
	28.4 Ararat (Agri)*	12	1	0	1	10	(1.3)	(>2.3)	121
	28.5 Ishkhansar*	12	2	5	4	1	3.9	1.9	146
	28.6 Savalan	16	0	0	11	5	(2.2)	<1.8	143
	28.7 Damavand	6	0	1	5	0	3.7	1.5– 2.0	122
	28.8 Yazd-Kerman area*	26	3	0	10	13	(2.5)	–	123
	28.9 Bazman	33	3	10	14	6	1.0	1.9	153

1. Arcs are keyed to Fig. 2.1 and Tables 2.1 and 7.1
2. An incomplete list of Quaternary to Holocene volcanoes for which two or more whole rock chemical analyses of at least one rock type had been published before 1980. Volcanoes are numbered by arc and their locations are shown in accompanying maps. Most volcanoes have been active historically and are taken from the *List of World Active Volcanoes* (Katsui 1971). Volcanoes identified by an asterisk (*) are excluded from that list, but included here because they are young and have been studied geochemically. The list is incomplete both in number of analyses per volcano and number of volcanoes for which analyses exist. However, about 5000 chemical analyses and 60% of the active volcanoes at convergent boundaries are included
3. Total number of chemical analyses of different ejecta. Multiple references to single analyses or multiple analyses of single ejecta units are counted once
4. Subdivisions in wt.% SiO_2 corresponding to basalt, basic andesite, acid andesite, and dacite or rhyolite, respectively
5. K_2O at 57.5% SiO_2 estimated from variation diagrams. Parentheses ( ) indicate the estimate is especially uncertain due to poor K-Si correlation or few data; hyphens (–) indicate no estimate was possible for these reasons; pluses (+) indicate two or more separate K-Si correlations exist; ranges given are maxima and minima
6. FeO*/MgO at 57.5% SiO_2 estimated from variation diagrams. Parentheses and hyphens as for 5

References to Appendix

1. Cole and Nairn (1975)
2. Cole (1978)
3. Gow (1968)
4. Brothers and Searle (1970)
5. Brothers and Martin (1970)
6. Ewart et al. (1973)
7. Melson et al. (1970)
8. Bauer (1970)
9. Baker et al. (1971)
10. Gorton (1977) and pers. comm. 1976
11. Williams and Warden (1964)
12. Taylor (1958)
13. Jakes and White (1969)
14. Stanton and Bell (1969)
15. Taylor et al. (1969a), Bultitude et al. (1978)
16. Heming and Carmichael (1973)
17. Heming (1974)
18. Blake and Ewart (1974)
19. Johnson and Chappell (1979)
20. Lowder and Carmichael (1970)
21. Morgan (1966)
22. McKee et al. (1976)
23. Mackenzie and Chappell (1972), Mackenzie (1976)
24. Jakes and Smith (1970), Smith and Davies (1976)
25. Neumann van Padang (1951)
26. Westerweld (1952)
27. Leo et al. (1980)
28. Whitford and Nicholls (1976)
29. Whitford and Jezek (1979), Jezek and Hutchison (1978)
30. Brouwer (1920), Wichmann (1925)
31. Kuenen (1935)
32. Neumann van Padang (1953)
33. Moore and Melson (1969)
34. Newhall (1979)
35. Wang (1970)
36. Kuno (1962)
37. Ono (1962)
38. Larson et al. (1975)
39. Isshiki (1963)
40. Kawano et al. (1961)
41. Aramaki (1963)
42. Kato (1964)
43. Aramaki (1971)
44. Katsui (1961)
45. Katsui et al. (1975)
46. Katsui et al. (1978)
47. Oba (1966)
48. Gorshkov (1970)
49. Piskunov (1975)
50. Vlodavetz and Piip (1959)
51. Ehrlich (1968)
52. U.S.G.S. Bull. 1028
53. Coats (1952)
54. Byers (1961)
55. Marsh (1976a)
56. Arculus et al. (1977)
57. Forbes et al. (1969)
58. Fenner (1926)
59. Brew et al. (1969)
60. Mathews (1957)
61. Coombs (1936)
62. Tabor and Crowder (1969)
63. Condie and Swenson (1973)
64. Verhoogen (1937)
65. Sheppard (1967)
66. Wise (1969)
67. Greene (1968)
68. Ritchey (1980), Williams (1942)
69. McBirney (1968)
70. Higgins (1973)
71. McBirney (1976, written comm.) and unpublished thesis by Maynard, Univ. of Oregon
72. Smith and Carmichael (1968)
73. Anderson (1941), Condie and Hayslip (1975), Mertzman (1977)
74. Fountain (1979)
75. Mooser et al. (1958)
76. Gunn and Mooser (1970)
77. Wilcox (1954)
78. Bloomfield (1975)
79. Rose (1976, written comm.) based on unpublished theses by Gierzycki (1976) and Johns (1975) at Michigan Tech. Univ. and by Eggers (1971) and Carr (1974) at Dartmouth College
80. Rose (1972a,b), Rose et al. (1977)
81. Rose et al. (1978)
82. Weyl (1961)
83. Di Scala (1969)
84. Ui (1972)
85. McBirney and Williams (1965)
86. Melson and Saenz (1973)
87. Krushensky (1972)

88. Hantke and Parodi (1966)
89. Lefèvre (1973)
90. Roa (1976)
91. Fernandez et al. (1973), Kussmaul et al. (1977)
92. Hormann et al. (1973)
93. Katsui and Gonzalez (1968)
94. Roobol et al. (1976)
95. Francis et al. (1974)
96. Thiele and Katsui (1969)
97. Vergara and Katsui (1969)
98. Oyarzun and Villalobos (1969, Vergara (1969)
99. Deruelle and Deruelle (1974)
100. Larsson (1941)
101. Brown et al. (1977)
102. Baker (1968)
103. Robson and Tomblin (1966)
104. Rea (1974)
105. Aspinall et al. (1973)
106. Arculus (1976)
107. Baker (1978)
108. Imbo (1965)
10 9. Barberi et al. (1973)
110. Pichler (1967)
111. Keller (1974)
112. Villari (1972), Villari and Nathan (1978)
113. Hoppenberger and Kiesl (1975)
114. Ninkovich and Hays (1972)
115. Georgalas (1962)
116. Pe (1973)
117. Nicholls (1971a)
118. Pichler and Kussmaul (1972)
119. Di Paola (1974)
120. Ayranci and Weibel (1973), Innocenti et al. (1975)
121. Lambert et al. (1974)
122. Gansser (1966)
123. Forster et al. (1972)
124. James et al. (1976)
125. Innocenti et al. (1976)
126. Black (1970)
127. Thorpe et al. (1976)
128. Stern et al. (1976)
129. Ewart (1969)
130. Thorpe and Francis (1976)
131. La Croix (1904), Gunn et al. (1974), Westercamp (1976)
132. Ando (1974)
133. Taniguchi (1972)
134. Kohari (1974)
135. Kobayashi et al. (1975)
136. Ota and Dincel (1975)
137. Whitford and Bloomfield (1976)
138. MacFarlane (in press)
139. Ewart et al. (1977)
140. Mallick (1973)
141. Kiesl et al. (1978)
142. Nakamura (1978)
143. Dostal and Zerbi (1978)
144. Thorpe (1977)
145. Ponce (1976)
146. Gushchin et al. (1976)
147. Colley and Ash (1971)
148. Hayatsu (1977)
149. Fairbrothers et al. (1978)
150. Dixon and Batiza (1979)
151. Stern (1979)
152. Carr et al. (1979)
153. Dupuy and Dostal (1978)
154. Woodruff et al. (1979)
155. Mayfield et al. (1980)
156. Miller et al. (1980)
157. Luhr and Carmichael (1980)
158. Paulo et al. (1979)
159. Okamoto (1979)

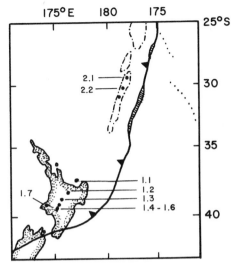

Appendix – Fig. 1

Appendix – Fig. 2

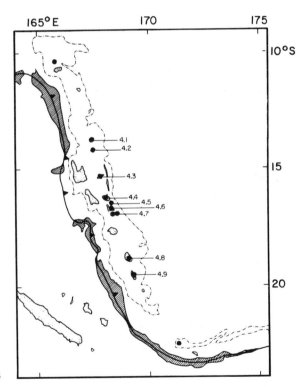

Appendix – Fig. 3

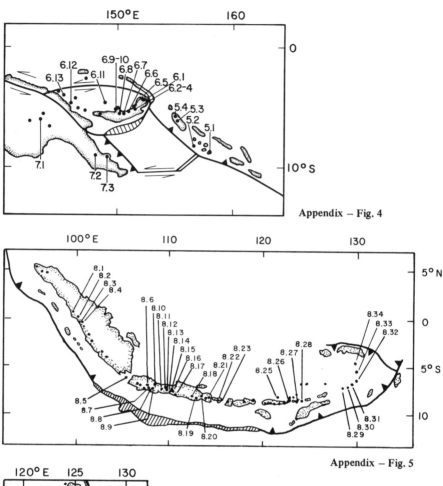

Appendix – Fig. 4

Appendix – Fig. 5

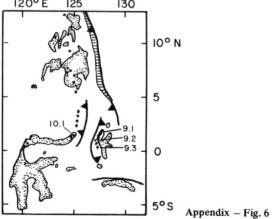

Appendix – Fig. 6

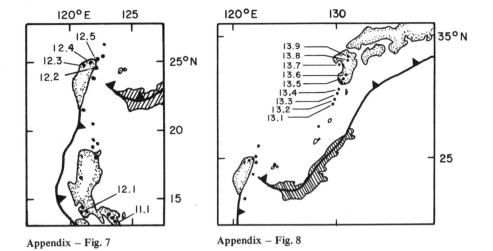

Appendix – Fig. 7

Appendix – Fig. 8

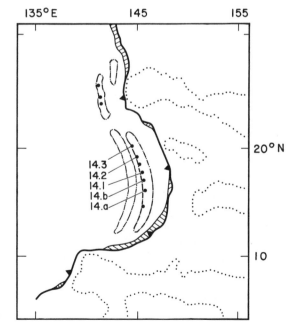

Appendix – Fig. 9

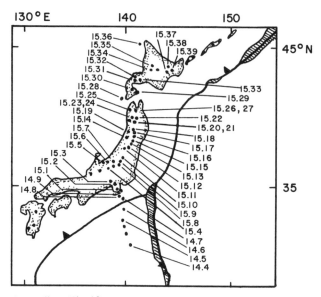

Appendix – Fig. 10

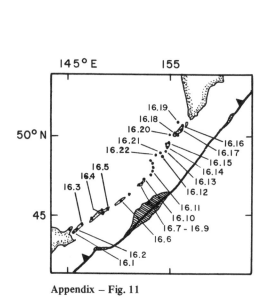

Appendix – Fig. 11

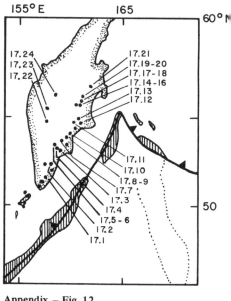

Appendix – Fig. 12

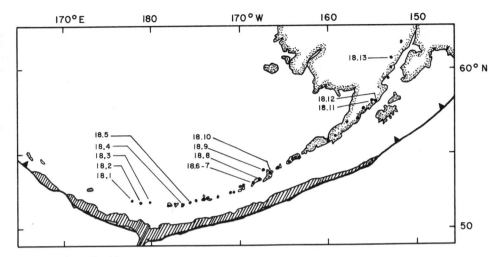

Appendix – Fig. 13

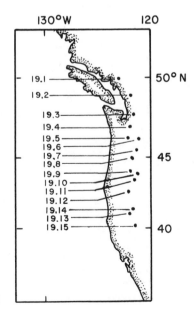

Appendix – Fig. 14

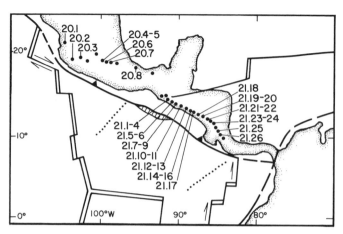

Appendix – Fig. 15

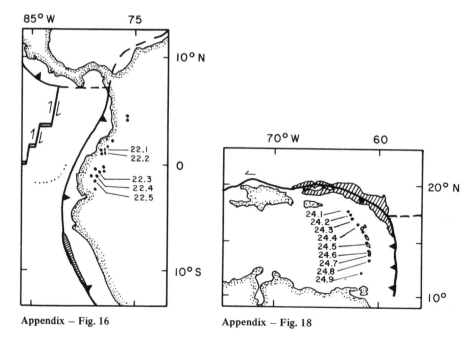

Appendix – Fig. 16

Appendix – Fig. 18

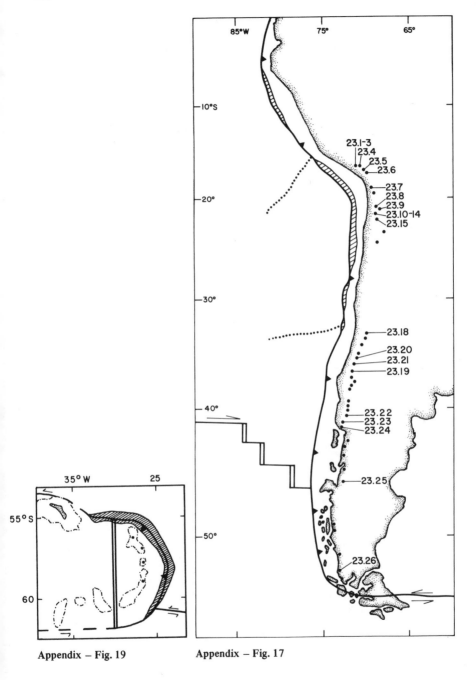

Appendix – Fig. 19   Appendix – Fig. 17

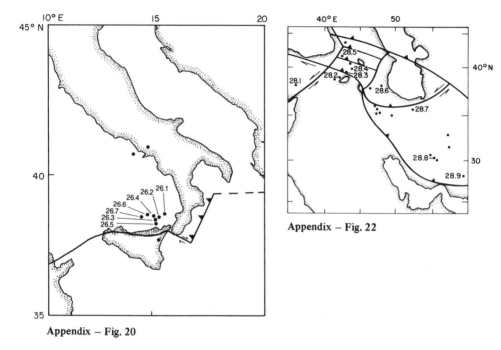

Appendix – Fig. 20

Appendix – Fig. 22

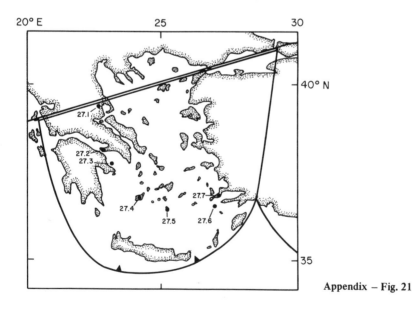

Appendix – Fig. 21

References

Abbey S (1977) Studies in "Standard Samples" for use in the general analysis of silicate rocks and minerals, part 5. 1977 ed of "usable" values. Geol Surv Can Pap 77-34: 31

Acharya HK, Aggarwal YP (1980) Seismicity and tectonics of the Philippine Islands. J Geophys Res 85: 3239–3250

Adamia ShA, Lordkipanidze MB, Zakariadze GS (1977) Evolution of an active continental margin as exemplified by the alpine history of the Caucasus. Tectonophysics 40: 183–189

Adams RD, Ware DE (1977) Subcrustal earthquakes beneath New Zealand; locations determined with a laterally inhomogeneous velocity model. N Z J Geol Geophys 20: 59–84

Aggarwal YP, Barazangi M, Isacks B (1972) P and S traveltimes in the Tonga-Fiji region: a zone of low velocity in the uppermost mantle behind the Tonga island arc. J Geophys Res 77: 6427–6434

Akasheh B (1972) Thickness of the crust in Iran. Bull Fac Sci Tehran Univ 4: 63–69

Allègre CJ, Condomines M (1976) Fine chronology of volcanic processes using $^{238}U-^{230}Th$ systematics. Earth Planet Sci Lett 28: 395–406

Allègre CJ, Minster JF (1978) Quantitative models of trace element behavior in magmatic processes. Earth Planet Sci Lett 38: 1–25

Allègre CJ, Treuil M, Minster JF, Minster B, Albarede F (1977) Systematic use of trace element in igneous process, part I. Fractional crystallization processes in volcanic suites. Contrib Mineral Petrol 60: 57–75

Allen JC, Boettcher A (1978) Amphiboles in andesite and basalt II. Stability as a function of $P-T-fH_2O-fO_2$. Am Mineral 63: 1074–1087

Allen JC, Boettcher AL, Marland G (1975) Amphiboles in andesite and basalt: I. Stability as a function of $P-T-fO_2$. Am Mineral 60: 1068–1085

Anders E (1977) Chemical composition of the moon, earth, and eucrite parent body. Philos Trans R Soc London Ser A 285: 23–40

Anderson AT (1974a) Before eruption H_2O content of some high alumina magmas. Bull Volcanol 37: 530–552

Anderson AT (1974b) Chlorine, sulfur, and water in magmas and oceans. Geol Soc Am Bull 85: 1485–1492

Anderson AT (1975) Some basaltic and andesite gases. Rev Geophys Space Phys 13: 37–55

Anderson AT (1976) Magma mixing: petrological process and volcanological tool. J Volcanol Geotherm Res 1: 3–33

Anderson AT (1979) Water in some hypersthenic magmas. J Geol 87:509–531

Anderson AT, Gottfried D (1971) Contrasting behavior of P, Ti, and Nb in a differentiated high-alumina olivine tholeiite and a calcalkaline andesitic suite. Geol Soc Am Bull 82: 1929–1942

Anderson AT, Wright TL (1972) Phenocrysts and glass inclusions and their bearing on oxidation and mixing of basaltic magmas, Kilauea Volcano, Hawaii. Am Mineral 57: 188–216

Anderson CA (1941) Volcanoes of the Medicine Lake Highland, California. California Univ Publ Geol Sci Bull 25: 347–422

Anderson CA, Silver LT (1976) Yavapai Series – a greenstone belt. Ariz Geol Soc Dig 10: 13–26

Anderson DL, Sammis C (1970) Partial melting in the upper mantle. Phys Earth Planet Int 3: 41–50

Anderson OL, Grew PC (1977) Stress corrosion theory of crack propagation with applications to geophysics. Rev Geophys Space Phys 15: 77–104

Anderson RN, DeLong SF, Schwarz WM (1978) Geophysical and geochemical constraints at converging plate boundaries, part II: a thermal model for subduction with dehydration in the downgoing slab. J Geol 86: 731–739

Anderson RN, Hobart MA, Langseth MG (1979) Geothermal convection through oceanic crust and sediments in the Indian Ocean. Science 204: 828–832

Anderson RN, Uyeda S, Miyashiro A (1976) Geophysical and geochemical constraints at converging plate boundaries, part I: dehydration in the down-going slab. Geophys JR Astron Soc 44: 333–357

Anderson RN, Hasegawa A, Umino N, Takagi A (1980) Phase changes and the frequency-magnitude distribution in the upper plane of the deep seismic zone beneath Tohoku, Japan. J. Geophys Res 85: 1389–1398

Ando S (1974) Geology and petrology of Esan Volcano, Hokkaido, Japan. J Jpn Assoc Mineral Petrol Econ Geol 69: 302–312

Ando S (1975) Minor element geochemistry of the rocks from Mashu Volcano, eastern Hokkaido. J Fac Sci Hokkaido Univ Ser IV 16: 553–566

Andrews DJ, Sleep NH (1974) Numerical modelling of tectonic flow behind island arcs. Geophys JR Astron Soc 38: 237–251

Andriambololona RD (1976) Les éléments de transition dans les suites andesitiques et shoshonitiques Sud du Pérou. Unpubl PhD Thesis, Univ Sci Tech Languedoc

Andriambololona RD, Lefevre C, Dupuy C (1974) Repartition de quelques éléments de transition à l'état de traces dans les roches calco-alcalines du Sud du Pérou. CR Acad Sci Paris Ser D 279: 1971–1974

Andriambololona RD, Lefevre C, Dupuy C (1975) Coefficient de partage des éléments de transition dans les minéraux ferro-magnésians extraits des dacites. CR Acad Sci Paris Ser D 281: 1797–1800

Annells RN (1974) Keweenawen volcanic rocks of Michipicaten Island, Lake Superior, Ontario. An eruptive centre of Proterozoic Age. Geol Surv Can Bull 218

Ansell JH, Smith EGC (1975) Detailed structure of a mantle seismic zone using a homogeneous station method. Nature (London) 253: 518–520

Aoki K (1968) Petrogenesis of ultrabasic and basic inclusions in alkali basalts, Iki Island, Japan. Am Mineral 53: 241–256

Aoki K, Kuno H (1972) Gabbro-quartz diorite inclusions from Izu-Hakone Region, Japan. Bull Volcanol 36: 164–173

Aoki K, Oji Y (1966) Calc-alkaline volcanic rock series derived from alkali-olivine basalt. J Geophys Res 71: 6127–6135

Arakai P (1968) Correlations of rate of cooling, texture, and mineralogical composition in the pyroxene andesite complex of the Cserhat Hills. Acta Geol Acad Sci Hung 12: 11–30

Aramaki S (1961) Sillimanite and cordierite from volcanic xenoliths. Am Mineral 46: 1154–1165

Aramaki S (1963) Geology of Asama Volcano. J Fac Sci Tokyo Univ Sect II 14: 229–443

Aramaki S (1971) Chemical composition of the rocks and rock-forming minerals [of Akita-Komagatake]. Bull Volcano Soc Jpn 16: 184–201

Aramaki S, Katsura T (1973) Petrology and liquidus temperature of the magma of the 1970 eruption of Akita-Komagatake Volcano, northeastern Japan. J Jpn Assoc Mineral Petrol Econ Geol 68: 101–124

Arculus RJ (1976) Geology and geochemistry of the alkali basalt-andesite association of Grenada, Lesser Antilles island arc. Geol Soc Am Bull 87: 612–624

Arculus RJ (1978) Mineralogy and petrology of Grenada, Lesser Antilles Island Arc. Contrib Mineral Petrol 65: 413–424

Arculus RJ, Wills KJA (1980) The petrology of plutonic blocks and inclusions from the Lesser Antilles island arc. J Petrol 21: (in press)

Arculus RJ, DeLong SE, Kay RW, Brooks C, Sun SS (1977) The alkalic rock suite of Bogoslof, Aleutian island arc. J Geol 85: 177–186

Armstrong RL (1971) Isotopic and chemical constraints on models of magma genesis in volcanic arcs. Earth Planet Sci Lett 12: 137–142

Armstrong RL (1975) Episodic volcanism in the central Oregon Cascade Range: confirmation and correlation with the Snake River Plain. Geology 3: 356

Armstrong RL (1980) Radiogenic isotopes: the case for crustal recycling on a near-steady-state no-continental-growth Earth (in press)

Armstrong RL, Cooper JA (1971) Lead isotopes in island arcs. Bull Volcanol 35: 27–63

Arth J (1974) REE in the basalt-andesite-dacite-rhyolite suite of Talasea and Rabaul, New Britain. Geol Soc Am Abstr 6: 638

Arth JG (1976) Behavior of trace elements during magmatic processes – a summary of theoretical models and their applications. J Res US Geol Surv 4: 41–47

Aspinall WP, Sigurdsson H, Shepherd JB (1973) Eruption of Soufrière Volcano on St. Vincent Island, 1971–1972. Science 181: 117–124

Atwater T (1970) Implications of plate tectonics for the Cenozoic tectonic evolution of western North America. Geol Soc Am Bull 81: 3513–3535

Atwater T, Molnar P (1973) Relative motion of the Pacific and North American plates deduced from sea-floor spreading in the Atlantic, Indian, and South Pacific Oceans. In: Kovack RL, Nur A (eds) Proc Conf Tecton Probl San Andreas Fault Syst. Stanford Univ, Stanford, pp 136–148

Ayranci VB, Weibel M (1973) Zum Chemismus der Ignimbrite des Erciyes-Vulkans (Zentral-Anatolien). Schweiz Mineral Petrogr Mitt 53: 49–60

Ayuso RA, Bence AE, Taylor SR (1976) Upper Jurassic tholeiitic basalts from DSDP Leg 11. J Geophys Res 81: 4305–4326

Baker MCW (1974) Volcano spacing, fractures, and thickness of the lithosphere – discussion. Earth Planet Sci Lett 23: 161–163

Baker PE (1968) Petrology of Mt. Misery volcano, St. Kitts, West Indies. Lithos 1: 124–150

Baker PE (1978) The South Sandwich Islands: III. Petrology of the volcanic rocks. Br Antarct Surv Sci Rep No 93

Baker PE, Holland JG (1973) Geochemical variations in a pyroclastic succession on St. Kitts, West Indies. Bull Volcanol 37: 472–490

Baker PE, Harris PG, Reay A (1971) The geology of Tofua Island, Tonga. R Soc NZ Bull 8: 67–79

Baksi AK, Watkins ND (1973) Volcanic production rates: comparison of ocean ridges, islands, and the Columbia Plateau basalts. Science 180: 493–496

Baldridge WS, McGetchin TF, Frey FA (1973) Magmatic evolution of Hekla, Iceland. Contrib Mineral Petrol 42: 245–258

Balesta ST, Farberov AI, Smirnov VS, Tarakanovsky AA, Zubin MI (1977) Deep crustal structure of the Kamchatkan volcanic regions. Bull Volcanol 40: 260–266

Barazangi M, Isacks B (1972) Lateral variations of seismic wave attenuation in the upper mantle above the inclined earthquake zone of the Tonga island arc: deep anomaly in the upper mantle. J Geophys Res 76: 8493–8516

Barazangi M, Isacks B (1976) Spatial distribution of earthquakes and subduction of the Nazca Plate beneath South America. Geology 4: 686–692

Barazangi M, Isacks BL (1979a) Subduction of the Nazca plate beneath Peru: evidence from spatial distribution of earthquakes. Geophys J R Astron Soc 57: 537–555

Barazangi M, Isacks BL (1979b) A comparison of the spatial distribution of mantle earthquakes determined from data produced by local and by teleseismic network for the Japan and Aleutian arcs. Bull Seismol Soc Am 69: 1763–1770

Barazangi M, Pennington W, Isacks B (1975) Global study of seismic wave attenuation in the upper mantle behind island arcs using pP waves. J Geophys Res 80: 1079–1092

Barazangi M, Isacks B, Dubois J, Pascal G (1974) Seismic wave attenuation in the upper mantle beneath the southwest Pacific. Tectonophysics 24: 1–12

Barberi F, Gasparini P, Innocenti F, Villari L (1973) Volcanism of the southern Tyrrhenian Sea and its geodynamic implications. J Geophys Res 78: 5221–5232

Barberi F, Innocenti F, Ferrara G, Keller J, Villari L (1974) Evolution of Eolian arc volcanism (southern Tyrrhenian Sea). Earth Planet Sci Lett 21: 269–276

Barberi F, Santacroce R, Ferrara G, Treuil M, Varet J (1975) A transitional basalt-pantellerite sequence of fractional crystallization, the Boina Centre (Afar Rift, Ethiopia). J Petrol 16: 22–56

Barker F, Peterman ZE (1974) Bimodal tholeiitic-dacite magmatism and the early Precambrian crust. Precambrian Res 1: 1–12

Bartlett RW (1969) Magma convection, temperature distribution, and differentiation. Am J Sci 267: 1067–1082

Bartrum JA (1957) Interesting xenoliths from Whangarei Head, Auckland, New Zealand. Trans R Soc N Z 67: 251–280

Baskov Y, Vetshteyn V, Surikov S, Tolstikhin J, Malyuk G, Mishini T (1973) Isotope composition of H, O, C, Ar and He in hot springs and gases in the Kurile-Kamchatka volcanic region as indicators of formation conditions. Geochem Int 10: 130–138

Bass MM, Moberly RM, Rhodes M, Shih C, Church SE (1973) Volcanic rocks cored in the central Pacific, Leg 17, DSDP. Initial Rep DSDP 17: 429–504

Basu AR, Tatsumoto M (1980) Nd-isotopes in selected mantle-derived rocks and minerals and their implications for mantle evolution. Contrib Mineral Petrol 75: 43–54

Bauer GR (1970) The geology of Tofua Island, Tonga. Pac Sci 24: 333–350

Ben-Avraham Z, Emery KO (1973) Structural framework of Sunda Shelf. Bull Am Assoc Petrol Geol 57: 2323–2366

Bender JF, Hodges FN, Bence AE (1978) Petrogenesis of basalts from the Project Famous area: experimental study from 0 to 15 kbars. Earth Planet Sci Lett 41: 277–302

Benioff H (1954) Orogenesis and deep crustal structure: additional evidence from seismology. Geol Soc Am Bull 65: 385–400

Berg E, Sutton GH (1974) Dynamic interaction of seismic and volcanic activity of the Nazca plate edges. Phys Earth Planet Int 9: 175–182

Berger WH, Winterer EL (1974) Plate stratigraphy and the fluctuating carbonate line. In: Hsu KJ, Jenkyns HC (eds) Pelagic sediments: On land and under the sea. Int Assoc Sediment Spec Publ 1: 11–48

Best MG (1975) Migration of hydrous fluids in the upper mantle and potassium variation in calc-alkalic rocks. Geology 3: 429–432

Best MG, Mercy ELP (1967) Composition and crystallization of mafic mineral on the Guadalupe igneous complex, California. Am Mineral 52: 436–474

Beswick AE (1976) K and Rb relations in basalts and other mantle-derived materials. Is phlogopite the key? Geochim Cosmochim Acta 10: 1167–1183

Bice DC (1980) Eruption rate in Central American estimated from volumes of pumice deposits. EOS 61: 70

Birch WD, Gleadow AJW (1974) The genesis of garnet and cordierite in acid volcanic rocks: evidence from the Cerberean caldron, Central Victoria, Australia. Contrib Mineral Petrol 45: 1–13

Bird JM, Dewey JF (1970) Lithosphere plate: continental margin tectonics and the evolution of the Appalachian orogen. Geol Soc Am Bull 81: 1031–1060

Bird P (1977) Stress and temperature in the Tonga and Marianas subduction zones. EOS 58: 1233

Bischke RE (1974) A model of convergent plate margins based on the recent tectonics of Shikcku, Japan. J Geophys Res 79: 4845–4858

Black PM (1970) Observations on White Island Volcano, New Zealand. Bull Volcanol 24: 158–167

Black PM (1973) Mineralogy of New Caledonian metamorphic rocks I: Garnets from the Ouegoa District. Contrib Mineral Petrol 38: 221–235

Blake DH, Ewart A (1974) Petrography and geochemistry of the Cape Hoskins volcanoes, New Britain, Papua New Guinea. J Geol Soc Aust 21: 319–332

Bleahu M, Boccaletti M, Manetti P, Peltz S (1973) Neogene carpathian arc: a continental arc displaying the features of an island arc. J Geophys Res 78: 5025–5032

Bloomer S, Melchior J, Poreda R, Hawkins J (1979) Mariana arc-trench studies: petrology of boninites and evidence for a "boninite series". EOS 60: 968

Bloomfield K (1975) A late Quaternary monogenetic volcanic field in central Mexico. Geol Rundsch 64: 476–496

Blot C (1972) Volcanisme et séismes du manteau supérieur dans l'Archipel des Nouvelles-Hebrides. Bull Volcanol 36: 446–461

Blot C (1976) Volcanisme et sismicite dans les arcs insulaires. Collect Geophys no 13. Orstom, Paris, p 206

Boccaletti M, Manetti P, Peccerillo A, Peltz S (1973) Young volcanism in the Calimani-Harghita mountains (East Carpathians): evidence of a paleoseismic zone. Tectonophysics 19: 299–313

Boettcher AL (1973) Volcanism and orogenic belts – the origin of andesites. Tectonophysics 17: 223–240

Boettcher AL (1977) The role of amphibole and water in circum-Pacific volcanism. In: Manghnani MH, Akimoto S (eds) High-pressure research applications in geophysics. Academic Press, London New York, pp 107–126

Bogoyavlenskaya GE, Dukik YuM (1969) Andesite crystallization in the upper parts of volcanic canals. Bull Volcanol 33: 1269–1273

Bonatti E, Arrhenius G (1970) Acidic rocks on the Pacific Ocean floor. In: Maxwell AE (ed) The sea. Wiley-Interscience, New York, pp 445–464

Boronikhin VA, Baskina VA (1975) High pressure minerals in silicic volcanics of the maritime region as shown by X-ray spectral-microprobe analysis.Dokl ESS 223: 154–157

Born A (1933) Der geologische Aufbau der Erde. Handbuch der Geophysik, vol II. Berlin, pp 565–567

Bottari A, Federico B (1979) Estimate of the focal depths of deep earthquakes and some structural implications for the deep seismic activity in the Tyrrhenian Region. Bull Seismol Soc Am 69: 1193–1208

Bottinga Y, Weill DF (1970) Densities of liquid silicate systems calculated from partial molar volumes of oxide components. Am J Sci 269: 169–182

Bottinga Y, Kudo A, Weill D (1966) Some observations on oscillatory zoning and crystallization of magmatic plagioclase. Am Mineral 51: 792–806

Bowen NL (1928) The evolution of igneous rocks. University Press, Princeton

Bowin C, Lu RS, Lee C-S, Schouten H (1978) Plate convergence and accretion in Taiwan-Luzon region. AAPG Bull 62: 1645–1672

Boynton CH, Westbrook GK, Bott MHP, Long RE (1979) A seismic refraction investigation of crustal structure beneath the Lesser Antilles island arc. Geophys JR Astron Soc 58: 371–393

Brett CP (1977) Seismicity of the South Sandwich Islands region. Geophys J R Astron Soc 51: 453–464

Brew DA, Muffler LJP, Loney RA (1969) Reconnaissance geology of the Mt. Edgecumbe volcanic field, Kruzof Island, southeastern Alaska. US Geol Surv Prof Pap 650D: 1–18

Briqueu L, Lancelot JR (1979) Rb-Sr systematics and crustal contamination models for calc-alkaline igneous rocks. Earth Planet Sci Lett 43: 385–396

Brooks C, James DE, Hart SR (1976a) Ancient lithosphere: its role in young continental volcanism. Science 193: 1086–1094

Brooks C, Hart SR, Hofmann A, James DE (1976b) Rb-Sr mantle isochrons from oceanic regions. Earth Planet Sci Lett 32: 51–61

Brothers RN, Martin KR (1970) The geology of Macauley Island, Kermadec Group, southwest Pacific. Bull Volcanol 34: 330–346

Brothers RN, Searle EJ (1970) The geology of Raoul Island, Kermadec Group, southwest Pacific. Bull Volcanol 34: 7–37

Brousse R, Bizouard H, Salat J (1972) Grenats des andesites et des rhyolites de Slovaquie, origine des grenats dans les series andesitiques. Contrib Mineral Petrol 35: 201–213

Brouwer (1920) Geologische onderzoekingen op de Sangi-eilanden en op de eilander Ternate en Pisang. Jaarb Mijnwezen Ned Oost-Indie, Verh II, pp 1–68

Brown GM, Schairer JF (1971) Chemical and melting relations of some calc-alkaline volcanic rocks. Geol Soc Am Mem 130: 139–157

Brown GM, Holland JG, Sigurdsson H, Tomblin JF, Arculus PJ (1977) Geochemistry of the Lesser Antilles volcanic island arc. Geochim Cosmochim Acta 41: 785–801

Buddington AF, Lindsley DH (1964) Iron-titanium oxide minerals and synthetic equivalents. J Petrol 5: 310–357

Bultitude J, Johnson RW, Chappell BW (1978) Andesites of Bagana volcano, Papua New Guinea: chemical stratigraphy and a reference andesite composition. BMRJ Aust Geol Geophys 3: 281–289

Bunsen R (1851) Über die Prozesse der vulkanischen Gesteinsbildungen Islands. Ann Phys Chem 83: 197–272

Burnham CW (1967) Hydrothermal fluids at the magmatic stage. In: Barnes HL (ed) Geochemistry of hydrothermal ore deposits. Holt, Rinehart and Winston, pp 34–76

Burnham CW (1975) Water and magmas: a mixing model. Geochim Cosmochim Acta 40: 1077–1084

Burnham CW (1979) Magmas and hydrothermal fluids. In: Barnes HL (ed) Geochemistry of hydrothermal ore deposits, 2nd edn. Wiley-Interscience, New York, pp 71–137

Burns RG (1970) Mineralogical applications of crystal field theory. Cambridge University Press, Cambridge, 224 p

Burri C, Parga-Pondal I (1936) Neue Beiträge zur Kenntnis des granatführenden Cordieritandesites vom Hoyazo bei Nijar (Provinz Almeria, Spanien). Schweiz Mineral Petrogr Mitt 16: 226–263

Byerly GR, Nelson WG, Vogt RR (1976) Rhyodacites, andesites, and ferrobasalts and ocean tholeiites from the Galapagos spreading center. Earth Planet Sci Lett 30: 215–221

Byers FM Jr (1961) Petrology of three volcanic suits, Umnak and Bogoslof Islands, Aleutian Islands, Alaska. Geol Soc Am Bull 72: 93–128

Cameron KL, Cameron M, Bagby WC, Moll EJ (1980) Petrologic characteristics of mid-Tertiary igneous suites in Chihuahua, Mexico. Geology 8: 87–91

Capaldi G, Cortini M, Gasparini P, Pece R (1976) Short-lived radioactive disequilibria in freshly erupted volcanic rocks and their implications for the pre-eruption history of a magma. J Geophys Res 81: 350–358

Cardwell RK, Isacks BL (1978) Geometry of the subducted lithosphere beneath the Banda Sea in eastern Indonesia from seismicity and fault-plane solutions. J Geophys Res 83: 2825–2838

Cardwell RK, Isacks BI, Karig DE (1980) The spatial distribution of earthquakes, focal mechanism solutions, and plate boundaries in the Philippine and northeastern Indonesian islands. In: Hayes D (ed) Tectonics of Southeast Asia. AGU (in Press)

Carey SN, Sigurdsson H (1978) Deep sea evidence for distribution of tephra from the mixed magma eruption of the Soufrière on St. Vincent, 1902: ash turbidities and air fall. Geology 6: 271–274

Carmichael ISE (1964) The petrology of Thingmuli, a Tertiary volcano in eastern Iceland. J Petrol 5: 435–460

Carmichael ISE (1967a) The iron-titanium oxides of salic volcanic rocks and their associated ferromagnesian silicates. Contrib Mineral Petrol 14: 36–44

Carmichael ISE (1967b) The mineralogy of Thingmuli, a Tertiary volcano in eastern Iceland. Am Mineral 52: 1815–1841

Carmichael ISE, Nicholls J (1967) Iron-titanium oxides and oxygen fugacities in volcanic rocks. J Geophys Res 72: 4665–4687

Carmichael ISE, Turner FJ, Verhoogen J (1974) Igneous petrology. McGraw-Hill, New York, p 739

Carmichael ISE, Nicholls J, Spera FJ, Wood BJ, Nelson SA (1977) High temperature properties of silicate liquids: application to the equilibration and ascent of basic magma. Philos Trans R Soc London Ser A 286: 373–431

Carr MJ (1977) Volcanic activity and great earthquakes at convergent plate boundaries. Science 197: 655–657

Carr MJ, Stoiber RE (1973) Intermediate depth earthquakes and volcanic eruptions in Central America, 1961–1972. Bull Volcanol 37: 326–337

Carr MJ, Stoiber RE, Drake CL (1973) Discontinuities in the deep seismic zones under the Japanese arcs. Geol Soc Am Bull 84: 2917–2930

Carr MJ, Rose WI, Mayfield DG (1979) Potassium content of lavas and depth to the seismic zone in Central America. J Volcanol Geotherm Res 5: 387–401

Carter SR, Evensen NM, Hamilton PJ, O'Nions RK (1978) Neodymium and strontium isotope evidence for crustal contamination of continental volcanics. Science 202: 743–747

Cawthorn RG (1976) Some chemical controls on igneous amphibole compositions. Geochim Cosmochim Acta 40: 1319–1328

Cawthorn RG (1977) Petrological aspects of the correlation between potash content of orogenic magmas and earthquake depth. Mineral Mag 11: 173–182

Cawthorn RG, Brown PA (1976) A model for the formation and crystallization of corundum-normative calcalkaline magmas through amphibole fractionation. J Geol 84: 467–476

Cawthorn RG, O'Hara MJ (1976) Amphibole fractionation in calcalkaline magma genesis. Am J Sci 276: 309–329

Cawthorn RG, Curran EB, Arculus RJ (1973) A petrogenetic model for the origin of calc-alkaline suite of Grenada, Lesser Antilles. J Petrol 14: 327–337

Chai BHT (1972) Structure and tectonic evolution of Taiwan. Am J Sci 272: 389–422

Chase CG (1978) Plate kinematics: the Americas, East Africa, and the rest of the world. Earth Planet Sci Lett 37: 355–368

Chayes F (1960) On correlation between variables of constant sum. J Geophys Res 65: 4185–4193

Chayes F (1964a) A petrographic distinction between Cenozoic volcanics in and around the open oceans. J Geophhys Res 69: 1573–1588

Chayes F (1964b) Variance-covariance relations in some published Harker diagrams of volcanic suites. J Petrol 5: 219–237

Chayes F (1969) The chemical composition of Cenozoic andesite. In: McBirney AR (ed) Proc Andesite Conf. Dep Geol Min Res Oreg Bull 65: 1–11

Chayes F (1970) On the occurrence of corundum in the norms of the common volcanic rocks. Carnegie Inst Wash Yearb 68: 179–182

Chayes F (1971) A further look at the level of silica saturation in andesite. Carnegie Inst Wash Yearb 70: 204–205

Chayes F (1975) Statistical petrology. Carnegie Inst Wash Yearb 74: 542–550

Chayes F, Le Maitre RW (1972) Published analyses of igneous rocks. Nature (London) 236: 449–450

Cherdyntzev VV (1973) Modern volcanology. Acad Sci USSR Moscow, pp 76–106

Chow TJ, Patterson CC (1962) On the occurrence and significance of lead isotopes in pelagic sediments. Geochim Cosmochim Acta 26: 263–308

Christensen NI, Salisbury MH (1975) Structure and constitution of the lower oceanic crust. Rev Geophys Space Phys 13: 57–86

Christiansen RL, Lipman R (1972) Cenozoic volcanism and plate tectonic evolution of western United States, part 2, Late Cenozoic. Philos Trans R Soc London Ser A 271: 249–284

Christiansen RL, Kleinhampl FJ, Blakely RJ, Tuchek ET, Johnson FL, Conyoc MD (1977) Resource appraisal of the Mt. Shasta Wilderness Study area, Siskiyou Country, California. US Geol Surv Open-file Rep 77–250

Church SE (1973) Limits of sediment involvement in the genesis of orogenic volcanic rocks. Contrib Mineral Petrol 39: 17–32

Church SE (1976) The Cascade mountains revisited: a re-evaluation in light of new lead isotopic data. Earth Planet Sci Lett 29: 175–188

Church SE, Tilton GR (1973) Lead and strontium isotopic studies in the Cascade mountains: bearing on andesite genesis. Geol Soc Am Bull 84: 431–454

Clark RH, Cole JW (1976) Prediction studies on White Island Volcano, Bay of Plenty, New Zealand. In: Johnson RW (ed) Volcanism in Australasia. Elsevier, Amsterdam, pp 375–384

Coats RR (1952) Magmatic differentiation in Tertiary and Quaternary volcanic rocks from Adak and Kanaga Islands, Aleutian Islands, Alaska. Geol Soc Am Bull 63: 485–514

Coats RR (1956) Geology of northern Adak, Alaska. US Geol Surv Bull 1028-C: 47–67

Coats RR (1959) Geologic reconnaissance of the Semisopochnoi Island, western Aleutian Islands, Alaska. US Geol Surv Bull 1028-O

Coats RR (1962) Magma type and crustal structure in the Aleutian arc. In: Crust of the Pacific Basin. Geophys Monogr Am Geophys Union 6: 92–109

Cobbing EJ, Ozard JM, Snelling NJ (1977) Reconnaissance geochronology of the crystalline basement rocks of the coastal cordillera of southern Peru. Geol Soc Am Bull 88:241–246

Cohen RS, Evensen NM, Hamilton PJ, O'Nions RK (1980) U-Pb, Sm-Nd, and Rb-Sr systematics of mid-ocean ridge basalt glasses. Nature (London) 283: 149–153

Cole JW (1978) Andesites of the Tongariro volcanic centre, North Island, New Zealand. J Volcanol Geotherm Res 3: 121–154

Cole JW, Nairn JA (1975) Catalogue of the active volcanoes of the world, part 22, New Zealand, IAVCEI

Colley H, Ash RP (1971) The geology of Erromango. New Hebrides Cond Geol Surv Reg Rep, 112 pp

Colley H, Warden AJ (1974) The petrology of the New Hebrides. Geol Soc Am Bull 85: 1635–1646

Coney PJ, Reynolds SJ (1977) Cordilleran Benioff Zones. Nature (London) 270: 403–406

Condie KC (1976) Trace-element geochemistry of Archean greenstone belts. Earth Sci Rev 12: 393–417

Condie KC, Harrison NM (1976) Geochemistry of the Archean Bulawayan Group, Midlands Greenstone Belt, Rhodesia. Precambrian Res 3: 253–271

Condie KC, Hayslip DL (1975) Young bimodal volcanism at Medicine Lake volcanic center, northern California. Geochim Cosmochim Acta 39: 1165–1178

Condie KC, Moore JM (1977) Geochemistry of Proterozoic volcanic rocks from the Greenville Province, eastern Ontario. In: Volcanic regimes in Canada. Geol Assoc Can Spec Pap 16: 311–330

Condie KC, Potts MJ (1969) Calcalkaline volcanism and the thickness of the early Precambrian crust in North America. Can J Earth Sci 6: 1179–1184

Condie KC, Swenson DH (1973) Compositional variations in three Cascade stratovolcanoes: Jefferson, Ranier and Shasta. Bull Volcanol 37: 205–230

Condomines M, Bernat M, Allegre CJ (1976) Evidence for contamination of recent Hawaiian lavas from ^{230}Th-^{238}U data. Earth Planet Sci Lett 33: 122–125

Cooke RJS, McKee CO, Dent VF, Wallace DA (1976) Striking sequence of volcanic eruptions in the Bismarck volcanic area, Papua New Guinea. In: Johnson RW (ed) Volcanism in Australasia. Elsevier, Amsterdam, pp 149–172

Coombs DS (1963) Trends and affinities of basaltic magmas and pyroxenes as illustrated on the Diopside-Olivine-Silica diagram. Mineral Soc Am Spec Pap 1: 227–250

Coombs HA (1936) The geology of Mount Ranier National Park. Wash Univ Pub Geol 3: 131–212

Craig H, Lupton JE, Welhan JA, Poreda R (1978a) Helium isotope ratios in Yellowstone and Lassen Park volcanic gases. Geophys Res Lett 5: 897–900

Craig H, Lupton JE, Horibe Y (1978b) A mantle component in circum-Pacific volcanic gases: Hakone, the Marianas, and Mt. Lassen. Adv Earth Planet Sci 3: 3–16

Crosson RS (1976) Crustal structure modeling of earthquake data. 2. Velocity structure of the Puget Sound region, Washington. J Geophys Res 81: 3047–3054

Crowe BM, McLean H, Howell DG, Higgins RE (1976) Petrography and major-element chemistry of the Santa Cruz Island volcanics. In: Aspects of the geologic history of the California continental borderland. AAPG Misc Publ 24: 196–215

Cummings GL (1976) Lead isotope ratios of DSDP Leg 37 basalts. Earth Planet Sci Lett 31: 179–192

Curray JR, Shor GG Jr, Raitt RW, Henry M (1977) Seismic refraction and reflection studies of crustal structure of the eastern Sunda and western Banda arcs. J Geophys Res 82: 2479–2489

Dallwitz WB, Green DH, Thompson JE (1966) Clinoenstatite in a volcanic rock from the Cape Vogel area, Papua. J Petrol 7: 375–403

Daly RA (1933) Igneous rocks and the depths of the Earth. McGraw-Hill, New York, 598 pp

Dasch EJ, Hedge CE, Dymond J (1973) Effect of sea water interaction on strontium isotope composition of deep-sea basalts. Earth Planet Sci Lett 19: 177–183

Davies JN, House L (1979) Aleutian subduction zone seismicity, volcano-trench separation, and their relation to great thrust-type earthquakes. J Geophys Res 84: 4583–4591

Decker RW (1971) Icelandic versus Hawaiian volcanoes. Trans Am Geophys Union 52: 371

Deer WA, Howie RA, Zussman J (1963) An introduction to the rock-forming minerals. Longman, 1963

DeLaney JR, Muenow DW, Graham DG (1978) Abundance and distribution of water, carbon and sulfur in the glassy rims of submarine pillow basalts. Geochim Cosmochim Acta 42: 581–594

Delany JM, Helgeson HC (1978) Calculation of the thermodynamic consequences of dehydration in subducting oceanic crust to 100 kb and $>$800 C. Am J Sci 278: 638–686

De Long SE (1974) Distribution of Rb, Sr, and Ni in igneous rocks, Central and Western Aleutian Islands, Alaska. Geochim Cosmochim Acta 38: 245–266

De Long SE, Fox PJ (1977) Geological consequences of ridge subduction. In: Talwani M, Pitman WC III (eds) Island arcs, deep sea trenches, and back-arc basins. AGU, pp 221–228

De Long SE, Hodges FN, Arculus RJ (1975) Ultramafic and mafic inclusions, Kanaga Island, Alaska, and the occurrence of alkaline rocks in the island arcs. J Geol 83: 721–737

Denham D (1969) Distribution of earthquakes in the New Guinea-Solomon Islands region. J Geophys Res 74: 4290–4299

De Paolo DJ, Johnson RW (1979) Magma genesis in the New Britain island arc: constraints from Nd and Sr isotopes and trace-element patterns. Contrib Mineral Petrol 70: 367–380

De Paolo DJ, Wasserburg GJ (1977) The sources of island arcs as indicated by Nd and Sr isotopic studies. Geophys Res Lett 4: 465–468

Déruelle B (1978) Calc-alkaline and shoshonite lavas from five Andean volcanoes (between latitudes 21°45' and 24°30' S) and the distribution of the Plio-Quaternary volcanism of the south-central and southern Andes. J Volcanol Geotherm Res 3: 281–298

Déruelle B, Déruelle J (1974) Geologie des volcans quaternaires des Nevados de Chillan (Chile). Bull Volcanol 38: 425–444

Dewey JF, Pitman WC III, Ryan WBF, Bonnin J (1973) Plate tectonics and the Alpine system. Geol Soc Am Bull 84: 3137–3180

Dickinson WR (1968) Circum-Pacific andesite types. J Geophys Res 73: 2261–2269

Dickinson WR (1973) Widths of modern arc-trench gaps proportional to past duration of igneous activity in associated magmatic arcs. J Geophys Res 78: 3376–3389

Dickinson WR (1975) Potash-depth (K-h) relations in continental margin and intra-ocean magmatic arcs. Geology 3: 53–56

Dickinson WR, Hatherton T (1967) Andesite volcanism and seismicity around the Pacific. Science 157: 801–803

Dickinson WR, Snyder WS (1979a) Geometry of triple junctions related to San Andreas transform. J Geophys Res 84: 561–572

Dickinson WR, Snyder WS (1979b) Geometry of subducted slabs related to San Andreas transform. J Geol 87: 609–628

Dietrich V, Emmermann R, Oberhansli R, Puchelt H (1978) Geochemistry of basaltic and gabbroic rocks from the west Mariana basin and the Mariana trench. Earth Planet Sci Lett 39: 127–144

Di Paola GM (1974) Volcanology and petrology of Nisyros Island (Dodecanese, Greece). Bull Volcanol 38: 944–987

Di Scala CI (1969) Preliminary report of the 1968 eruption of the Cerro Negro volcano, Nicaragua. Smithsonian Inst Center for Short-Lived Phenomena

Dixon TH, Batiza R (1979) Petrology and chemistry of recent lavas in the northern Marianas: implications for the origin of island arc basalts. Contrib Mineral Petrol 70: 167–181

Doe BR (1970) Lead isotopes. Springer, Berlin Heidelberg New York, 137 pp

Doe BR, Zartman BE (1979) Plumbotectonics I, the Phanerozoic. In: Barnes H (ed) Geochemistry of hydrothermal ore deposits, 2nd edn. John Wiley and Sons, New York, pp 22–70

Donnelly TW, Rogers JJW (1980) Igneous series in island arcs: the northeastern Caribbean compared with worldwide island arc assemblages. Bull Volcanol 43: 347–382

Donnelly TW, Rogers JJW, Pushkar P, Armstrong RL (1971) Chemical evolution of the igneous rocks of the Eastern West Indies: an investigation of thorium, uranium, and potassium distributions, and lead and strontium isotope ratios. Geol Soc Am Mem 130: 181–224

Dorman JM, Ewing M, Oliver J (1960) Study of shear velocity distribution in the upper mantle by mantle Rayleigh waves. Bull Seismol Soc Am 50: 87–95

Dostal J, Zerbi M (1978) Geochemistry of the Savalan volcano (northwestern Iran). Chem Geol 22: 31–42

Dostal J, DuPuy C, Lefevre C (1977a) Rare earth distribution in Plio-quaternary volcanic rocks from southern Peru. Lithos 10: 173–183

Dostal J, Zentilli M, Caelles JC, Clark AH (1977b) Geochemistry and origin of volcanic rocks of the Andes (26–28 S). Contrib Mineral Petrol 63: 113–128

Dostal J, Capedri S, de Albuquerque CAR (1977c) Calc-alkaline volcanic rocks from NW Sardinia: evaluation of a fractional crystallization model. Bull Volcanol 40: 253–259

Drake CL, Nafe JE (1968) The transition from ocean to continent from seismic refraction data. Geophys Monogr Am Geophys Union 12: 174–188

Drake MJ (1975) The oxidation state of europium as an indicator of oxygen fugacity. Geochim Cosmochim Acta 39: 55–64

Drake MJ (1976) Plagioclase-melt equilibria. Geochim Cosmochim Acta 40: 457–465

Drake MJ, Weill DF (1975) Partition of Sr, Ba, Ca, Y, Eu^{2+}, $Eu3^+$, and other REE between plagioclase feldspar and magmatic liquid: an experimental study. Geochim Cosmochim Acta 39: 689–712

Drory A, Ulmer GC (1974) Oxygen fugacity determinations for Cascadian andesites. EOS 55: 487

Dubois J (1971) Propagation of P waves and Rayleigh waves in Melanesia: structural implications. J Geophys Res 76: 7217–7240

Duggan MB, Wilkinson JFG (1973) Tholeiitic andesite of high-pressure origin from the Tweed Shield volcano, northeastern New South Wales. Contrib Mineral Petrol 39: 267–276

Duncan AR, Taylor SR (1969) Trace element analyses of magnetites from andesitic and dacitic lavas from Bay of Plenty, New Zealand. Contrib Mineral Petrol 20: 30–33

Duncan RA, Compston W (1976) Sr-isotopic evidence for an old mantle source region for French Polynesian volcanism. Geology 4: 728–732

Dungan MA, Rhodes JM (1978) Residual glasses and melt inclusions in basalts from DSDP legs 45 and 46: evidence for magma mixing. Contrib Mineral Petrol 67: 417–431

Dungan MA, Long PE, Rhodes JM (1978) Magma mixing at mid-ocean ridges: evidence from Legs 45 and 46-DSDP. Geophys Res Lett 5: 423–425

DuPuy C, Dostal J (1978) Geochemistry of calc-alkaline volcanic rocks from southeastern Iran (Kouh-e-Shahsavaran). J Volcanol Geotherm Res 4: 363–373

DuPuy C, Lefèvre C (1974) Fractionnement des éléments en trace Li, Rb, Ba, Sr dans les séries andésitiques et shoshonitiques du Perou. Comparison avec d'autres zones orogéniques. Contrib Mineral Petrol 46: 147–157

Dupuy C, McNutt RH, Coulon C (1974) Détermination de $^{87}Sr/^{86}Sr$ dans les andésites cénozoiques et les lavas associées de Sardaigne Nord-Occidentale. Geochim Cosmochim Acta 38: 1287–1296

DuPuy C, Dostal J, Capredi S, Lefèvre C (1976) Petrogenetic implications of uranium abundances in volcanic rocks from southern Peru. Bull Volcanol 39: 363–370

Eade KE, Fahrig WF (1971) Chemical evolutionary trends of continental plates – preliminary study of the Canadian shield. Bull Geol Surv Can 179: 51 pp

Early TO, Silver LT (1973) Rb-Sr isotope systematics in the Peninsular Ranges batholith of southern and Baja California. EOS 54: 494

Eaton GP (1979) A plate-tectonic model for Late Cenozoic crustal spreading in the western United States. In: Riecker RE (ed) Rio Grande Rift: Tectonics and magmatism. AGU, pp 7–32

Eaton JP (1962) Crustal structure and volcanism in Hawaii. In: MacDonald GA, Kuno H (eds) The crust of the Pacific Basin. Geophys Monogr Am Geophys Union 6: 13–29

Eggler DH (1972a) Water-saturated and undersaturated melting relations in a Paricutin andesite and an estimate of water content in the natural magma. Contrib Mineral Petrol 34: 261–271

Eggler DH (1972b) Amphibole stability in H_2O-undersaturated calc-alkaline melts. Earth Planet Sci Lett 15: 28–34

Eggler DH (1974) Application of a portion of the system $CaAl_2Si_2O_8$-$NaAlSi_3O_8$-SiO_2-MgO-Fe-O_2-H_2O-CO_2 to genesis of the calcalkaline suite. Am J Sci 274: 297–315

Eggler DH (1975) CO_2 as a volatile component of the mantle: the system Mg_2SiO_4-SiO_2-H_2O-CO_2. Phys Chem Earth 9: 869–881

Eggler DH, Burnham CW (1973) Crystallization and fractionation trends in the system andesite-H_2O-CO_2-O_2 at pressures to 10 kb. Geol Soc Am Bull 84: 2517–2532

Eggler DH, Holloway JR (1977) Partial melting of peridotite in the presence of H_2O and CO_2: principles and review. In: Dick HJB (ed) Magma genesis. Oreg Dep Geol Mineral Ind Bull 96: 15–36

Eggler DH, McCallum ME (1974) Xenoliths in diatremes of the western United States. Carnegie Inst Wash Yearb 73: 294–300

Ehmann WD, Chyi LL, Garg AN, Ali MZ (1979) The distribution of Zr and Hf in terrestrial rocks, meteorites, and the moon. Phys Chem Earth 11: 247–260

Ehrlich EN (1968) Petrochimiya Kainozciskoi Kurilo-Kamchatskoi vulkanitsheskoi provincii. Nauka, Moskva, 277 pp

Eichelberger JC (1975) Origin of andesite and dacite: evidence of mixing at Glass Mountain in California and at other circum-Pacific volcanoes. Geol Soc Am Bull 86: 1381–1391

Eichelberger JC, Gooley R (1977) Evolution of silicic magma chambers and their relationship to basaltic volcanism. Geophys Monogr Am Geophys Union 20: 57–77

Elston W (1976) Tectonic significance of mid-Tertiary volcanism in the Basin and Range province: a critical review with special reference to New Mexico. In: Elston WE, Northrop SA (eds) Cenozoic volcanism in southwestern New Mexico. NM Geol Surv Spec Publ 5: 93–102

Engdahl ER (1973) Relocation of intermediate depth earthquakes in the Central Aleutians by seismic ray tracing. Nature Phys Sci 245: 23–25

Engdahl ER (1977) Seismicity and plate subduction in the central Aleutians. In: Talwani M, Pitman III WC (eds) Island arcs, deep sea trenches, and back arc basins. Am Geophys Union Monogr, pp 259–292

Engdahl ER, Scholz CH (1977) A double Benioff Zone beneath the central Aleutians: an unbending of the lithosphere. Geophys Res Lett 4: 473–476

Engdahl ER, Sleep NH, Lin MT (1977) Plate effects in North Pacific subduction zones. Tectonophysics 37: 95–116

England R, Wortel R (1980) Some consequences of the subduction of young slabs. Earth Planet Sci Lett 47: 403–415

Erlank AJ, Kable EJD (1976) The significance of incompatible elements in Mid-Atlantic Ridge basalts from 45 N with particular reference to Zr/Nb. Contrib Mineral Petrol 54: 281–291

Evans JR, Suyehiro K, Sacks IS (1978) Mantle structure beneath the Japan Sea – a re-examination. Geophys Res Lett 5: 487–490

Ewart A (1969) Petrochemistry and feldspar crystallization in the silicic volcanic rocks, central North Island, New Zealand. Lithos 2: 371–388

Ewart A (1971) Notes on the chemistry of ferromagnesian phenocrysts from selected volcanic rocks, Central Volcanic Region. N Z J Geol Geophys 14: 323–340

Ewart A (1976a) Mineralogy and chemistry of modern orogenic lavas – some statistics and implications. Earth Planet Sci Lett 31: 417–432

Ewart A (1976b) A petrological study of the younger Tongan andesites and dacites, and the olivine tholeiites of Niua Fo'ou Island, S.W. Pacific. Contrib Mineral Petrol 58: 1–21

Ewart A (1979) A review of the mineralogy and chemistry of Tertiary-Recent, dacitic, latitic, rhyolitic, and related salic volcanic rocks. In: Barker F (ed) Trondhjemites, dacites, and related rocks. Elsevier, Amsterdam, pp 13–121

Ewart A (1980) The mineralogy and petrology of Tertiary-Recent orogenic volcanic rocks: with special reference to the andesitic-basaltic compositional range. In: Thorpe RS (ed) Orogenic andesites. (In press)

Ewart A, Bryan WB (1972) The petrology and geochemistry of the igneous rocks from Eua, Tongan Island. Bull Geol Soc Am 83: 3281–3298

Ewart A, Bryan WB (1973) Petrology and geochemistry of the Tongan Islands. In: Coleman P (ed) Island arcs marginal seas, and geochemistry. W Aust Univ Press, pp 503–520

Ewart A, Stipp JJ (1968) Petrogenesis of the volcanic rocks of the Central North Island, New Zealand, as indicated by a study of $Sr^{87}/^{86}$ ratios, and Sr, Rb, K, U, and Th abundances. Geochim Cosmochim Acta 32: 699–736

Ewart A, Taylor SR (1969) Trace element geochemistry of the rhyolite volcanic rocks, Central North Island, New Zealand, Phenocryst data. Contrib Mineral Petrol 22: 127–146

Ewart A, Taylor SR, Capp AC (1968) Trace and minor element geochemistry of the rhyolitic volcanic rocks, Central North Island, New Zealand. Contrib Mineral Petrol 18: 76–104

Ewart A, Bryan W, Gill J (1973) Mineralogy and geochemistry of the younger volcanic islands of Tonga, S.W. Pacific. J Petrol 14: 429–465

Ewart A, Brothers RN, Mateen A (1977) Mineralogical and chemical evolution of the Tonga-Kermadec-New Zealand island arc. J Volcanol Geotherm Res 2: 205–250

Ewart A, Hildreth W, Carmichael ISE (1975) Quaternary acid magma in New Zealand. Contrib Mineral Petrol 51: 1–27

Faberov AI, Gorelchik VI, Zubkov SI (1973) On heterogeneities with reduced viscosity in the mantle under the Kamchatka volcanics according to seismological data. Bull Volcanol 37: 122–123

Fairbrothers GE, Carr MJ, Mayfield DG (1978) Temporal magmatic variation at Boqueron volcano, El Salvador. Contrib Mineral Petrol 67: 1–9

Fedotov SA (1968) On deep structure, properties of the upper mantle, and volcanism of the Kuril-Kamchatka island arc according to seismic data. In: Knopoff L, Drake CL, Hart PJ (eds) The crust and upper mantle of the Pacific Area. Am Geophys Union Mongr 12: 131–139

Fedotov SA (1976) Mechanism of magma ascent and deep feeding channels of island arc volcanoes. Bull Volcanol 39: 241–254

Fedotov SA, Tokarev PI (1973) Earthquakes, characteristics of the upper mantle under Kamchatka and their connection with volcanism (according to the data collected up to 1971). Bull Volcanol 37: 245–257

Fenner CN (1926) The Katmai magmatic province. J Geol 34: 673–772

Fenner CN (1931) The residual liquids of crystallizing magmas. Mineral Mag 22: 539–560

Fenner CN (1948) Immiscibility of igneous magmas. Am J Sci 246: 465–502

Fernandez CA, Hormann PK, Kussmaul S, Meave J, Pichler H, Subieta T (1973) First petrologic data on young volcanic rocks of S.W. Bolivia. Tschermaks Mineral Petrogr Mitt 19: 149–172

Finch RH, Anderson CA (1930) The quartz-basalt eruptions of Cinder Cone, Lassen Volcanic Park, California. Calif Univ Dep Geol Sci Bull 19: 245–273

Finlayson DM, Drummond BJ, Collins CDN, Connelly JB (1976) Crustal structure under the Mount Lamington region of Papua, New Guinea. In: Johnson RW (ed) Volcanism in Australasia. Elsevier, Amsterdam, pp 259–274

Finlayson DM, Cull JP, Wiebenga WA, Furumoto AS, Webb JP (1972) New Britain-New Ireland crustal seismic refraction investigations 1967 and 1969. Geophys JR Astron Soc 29: 245–253

Finnerty AA, Boyd FR (1978) Pressure-dependent solubility of calcium on forsterite coexisting with diopside and enstatite. Carnegie Inst Wash Yearb 77: 713–717

Fiske RS, Hopson CA, Waters AC (1963) Geology of Mount Rainier National Park, Washington. US Geol Surv Prof Pap 444: 93 pp

Fitch TJ (1972) Plate convergence, transcurrent faults, and internal deformation adjacent to southeast Asia and the western Pacific. J Geophys Res 77: 4432–4460

Fitton JG (1971) The generation of magmas in island arcs. Earth Planet Sci Lett 11: 63–67

Fitton JG (1972) The genetic significance of almandine-pyrope phenocrysts in the calc-alkaline Borrowdale Volcanic Group, northern England. Contrib Mineral Petrol 36: 231–248

Fitton JG, Hughes DJ (1970) Volcanism and plate tectonics in the British Ordovician. Earth Planet Sci Lett 8: 223–228

Fix JE (1975) The crust and upper mantle of Central Mexico. Geophys J R Astron Soc 43: 453–499

Flanagan F (1973) 1972 values for international geochemical reference standards. Geochim Cosmochim Acta 37: 1189–1200

Floran RJ, Grieve RAF, Phinney WC, Warner JL, Simonds CH, Blanchard DP, Dence MR (1978) Manicouagan Impact Melt, Quebec. 1. Stratigraphy, petrology, and chemistry. J Geophys Res 83: 2737–2759

Flynn RT, Burnham CW (1978) An experimental determination of rare earth partition coefficients between a chloride-containing vapor phase and silicate melts. Geochim Cosmochim Acta 42: 685–702

Fodor RU (1975) Petrology of basalt and andesite of the Black Range, New Mexico. Geol Soc Am Bull 86: 295–304

Forbes RB, Ray DK, Katsura T, Matsumoto H, Haramura H, Furst MJ (1969) The comparative chemical composition of continental vs. island arc andesites in Alaska. In: McBirney AR (ed) Proc Andesite Conf. Dep Geol Mineral Res Oreg Bull 65: 111–120

Forster H, Fesefeldt K, Kursten M (1972) Magmatic and orogenic evolution of the Central Iranian volcanic belt. 24th Int Geol Congr (Montreal) 2: 198–210

Forsyth DW (1975) Fault plane solutions and tectonics of the South Atlantic and Scotia Sea. J Geophys Res 80: 1429–1443

Foshag WF, Gonzalez JR (1954) Birth and development of Paricutin Volcano, Mexico. US Geol Surv Bull 965-D: 355–489

Fountain JC (1979) Geochemistry of Brokeoff volcano, California. Geol Soc Am Bull 90: 294–300

Francis DM (1976) The origin of amphibole in lherzolite xenoliths from Nunivak Island, Alaska. J Petrol 17: 357–378

Francis PW, Rundle CC (1976) Rates of production of the main magma types in the central Andes. Geol Soc Am Bull 87: 474–480

Francis PW, Moorbath S, Thorpe RS (1977) Strontium isotope data for recent andesites in Ecuador and North Chile. Earth Planet Sci Lett 37: 197–202

Francis PW, Roobol MJ, Walker GPL, Cobbold PR, Coward MP (1974) The San Pedro and San Pablo volcanoes of northern Chile and their hot avalanche deposits. Geol Rundsch 63: 357–388

Frangipane-Gysel M (1977) Analysis of multivariate data: an application to some Recent volcanics of the Central Andes. Schweiz Mineral Petrogr Mitt 57: 115–134

Frey FA, Prinz M (1978) Ultramafic inclusions from San Carlos, Arizona: petrologic and geochemical data bearing on their petrogenesis. Earth Planet Sci Lett 38: 129–176

Frey FA, Bryan WE, Thompson G (1974) Atlantic Ocean floor: geochemistry and petrology of basalts from Legs 2 and 3 of the Deep-Sea Drilling Project. J Geophys Res 79: 5507–5528

Frey FA, Green DH, Roy SD (1978) Integrated models of basalt petrogenesis: a study of quartz tholeiites to olivine melilites from southeastern Australia utilizing geochemical and experimental petrological data. J Petrol 19: 463–513

Friedricksen H, Hoernes S (1978) Stable isotope exchange between oceanic crust and ocean water. In: Short Pap 4th Int Conf, Geochronol, Cosmochronol, Isotope Geol. US Geol Surv Open-File Rep 78–701

Fudali RF (1965) Oxygen fugacities of basaltic and andesite magmas. Geochim Cosmochim Acta 29: 1063–1975

Fujimaki H (1975) Rare earth elements in volcanic rocks from Hakone volcano and northern Izu Peninsula, Japan. J Fac Sci Univ Tokyo Sect II 19: 81–93

Fujiioka K, Nasu N (1978) Chemical composition of the volcanic rocks from the Oyashio ancient landmass. EOS 59: 1183–1184

Furumoto AS, Hussong DM, Campbell JF, Sutton GH, Malahoff A, Rose JC, Woollard GP (1970) Crustal and upper mantle structure of the Solomon Islands as revealed by seismic refraction survey of Nov-Dec 1966. Pac Sci 24: 315–332

Fyfe WS (1978) The evolution of the earth's crust: modern plate tectonics to ancient hot spot tectonics. Chem Geol 23: 89–114

Fyfe WS, McBirney AR (1975) Subduction and the structure of andesitic volcanic belts. Am J Sci 275A: 285–297

Gansser A (1966) Catalog of the active volcanoes of the world. XVII, Appendix, Iran. IAVCEI

Gansser A (1971) The Taftan volcano (S.E. Iran). Eclogae Geol Helv 64: 319–334

Garcia MO, Liu NWK, Muenow DW (1979) Volatiles in submarine volcanic rocks from the Mariana island arc and trough. Geochim Cosmochim Acta 43: 305–312

Garrick RA (1968) A reinterpretation of the Wellington crustal refraction profile. N Z J Geol Geophys 11: 1280–1294

Garrison RE (1974) Radiolarian cherts, pelagic limestones, and igneous rocks in eugeosynclinal assemblages. In: Hsu HJ, Jenkyns HC (eds) Pelagic sediments: On land and under the sea. Int Assoc Sediment Spec Pap 1: 319–351

Gast PW (1968) Trace element fractionation and the origin of tholeiitic and alkaline magma types. Geochim Cosmochim Acta 32: 1057–1086

Georgalas GC (1962) Catalogue of the active volcanoes of the world, XII, Greece. IAVCEI

Ghiorso MS, Carmichael ISE (1980) A regular solution model for metaluminous silicate liquids: applications to geothermometry, liquid immiscibility, and the source regions of basic magmas. Contrib Mineral Petrol 71: 323–342

Gill JB (1970) Geochemistry of Viti Levu, Fiji, and its evolution as an island arc. Contrib Mineral Petrol 27: 179–203

Gill JB (1974) Role of underthrust oceanic crust in the genesis of a Fijian calc-alkaline suite. Contrib Mineral Petrol 43: 29–45

Gill JB (1976a) Composition and age of Lau Basin and ridge volcanic rocks: implications for evolution of an inter-arc basin and remnant arc. Geol Soc Am Bull 87: 1384–1395

Gill JB (1976b) From island arc to oceanic islands: Fiji, southwestern Pacific. Geology 4: 123–126

Gill JB (1976c) Evolution of the mantle: geochemical evidence from alkali basalt: comment. Geology 4: 625–626

Gill JB (1978) Role of trace element partition coefficients in models of andesite genesis. Geochim Cosmochim Acta 42: 709–724

Gill JB, Compston W (1973) Strontium isotopes in island arc volcanic rocks. In: Coleman P (ed) The Western Pacific – island arcs, marginal seas, geochemistry. W Aust Univ Press, pp 483–496

Gill JB, Gorton M (1973) A proposed geological and geochemical history of Eastern Melanesia. In: Coleman P (ed) The Western Pacific – island arcs, marginal seas, geochemistry. W Aust Univ Press, pp 543–566

Gill JB, Till AB (1978) Petrology and geochemistry of Namosi andesites and dacites, Fiji. In: Abstr Int Geodyn Conf, Tokyo, pp 240–241

Ginzburg IV, Maleyev YeF, Sidorenko GA, Teleshova RL (1964) Another find of pigeonite in the USSR. Akad Nauk SSSR Dokl 159: 99–102

Gledhill A, Baker PE (1973) Strontium isotope ratios in volcanic rocks from the South Sandwich Islands. Earth Planet Sci Lett 19: 369–372

Gnibidenko HS, Gorbachev SZ, Lebedev MM, Marakhanov VI (1972) Geology and deep structure of Kamchatka Peninsula. Pac Geol 7: 1–32

Goetze C (1977) A brief summary of our present-day understanding of the effect of volatiles and partial melt on the mechanical properties of the upper mantle. In: Manghnani MH, Akimoto S (eds) High-pressure research applications in geophysics. Academic Press, pp 3–24

Goodwin AM (1977) Archean volcanism in Superior Province, Canadian Shield. In: Volcanic regions in Canada. Geol Assoc Can Spec Pap 16: 205–242

Goodwin AM, Smith IEM (1980) Chemical discontinuities in Archean metavolcanic terrains and the development of Archean crust. Precambrian Res 10: 301–311

Goossens PJ, Rose WI Jr, Flores D (1977) Geochemistry of tholeiites of the Basic Igneous Complex of northwestern South America. Geol Soc Am Bull 88: 1711–1720

Gorshkov GS (1962) Petrochemical features of volcanism in relation to the types of the Earth's crust. In: MacDonald GA, Kuno H (eds) The crust of the Pacific Basin. Am Geophys Union Monogr 6: 110–115

Gorshkov GS (1970) Volcanism and the upper mantle. Plenum Press, New York, 385 pp

Gorshkov GS (1971) Prediction of volcanic eruptions and seismic methods of location of magma chambers, a review. Bull Volcanol 35: 198–210

Gorton M (1977) The geochemistry and origin of Quaternary volcanism in the New Hebrides. Geochim Cosmochim Acta 41: 1257–1270

Gough DI (1973) Dynamic uplift of Andean mountains and island arcs. Nature Phys Sci 242: 39–41

Gow AJ (1968) Petrographic and petrochemical studies of Mt.Egmont andesites. N Z J Geol Geophys 11: 166–190

Green DH (1970) A review of experimental evidence on the origin of basaltic and nephelinitic magmas. Phys Earth Planet Int 3: 221–235

Green DH (1973) Experimental melting studies on a model upper mantle composition at high pressure under water-saturated and water-undersaturated conditions. Earth Planet Sci Lett 19: 37–53

Green DH (1976) Experimental testing of "equilibrium" partial melting of peridotite under water-saturated, high-pressure conditions. Can Mineral 14: 255–268

Green DH, Ringwood AE (1967) The genesis of basaltic magmas. Contrib Mineral Petrol 15: 103–190

Green TH (1968) Experimental fractional crystallization of quartz diorite and its application to the problem of anorthosite genesis. In: Isachsen Y (ed) Origin of anorthosite and related rocks. NY Mus Sci Serv Mem 18: 23–30

Green TH (1970) High pressure experimental studies on the mineralogical composition of the lower crust. Phys Earth Planet Interiors 3: 441–450

Green TH (1972) Crystallization of calcalkaline andesite under controlled high-pressure hydrous conditions. Contrib Mineral Petrol 34: 150–166

Green TH (1977) Garnet in silicic liquids and its possible use as a P-T indicator. Contrib Mineral Petrol 65: 59–67

Green TH (1980) Island arc and continent-building magmatism – a review of key geochemical parameters and genetic processes. Tectonophysics 63: 367–385

Green TH, Ringwood AE (1968a) Genesis of the calc-alkaline igneous rock suite. Contrib Mineral Petrol 18: 105–162

Green TH, Ringwood AE (1968b) Origin of the garnet phenocrysts in calc-alkaline rocks. Contrib Mineral Petrol 18: 163–174

Green TH, Ringwood AE (1972) Crystallization of garnet-bearing rhyodacite under high pressure hydrous conditions. J Geol Soc Aust 19: 203–212

Green RC (1968) Petrography and petrology of volcanic rocks in the Mount Jefferson area, High Cascade Range, Oregon. US Geol Surv Bull 1251-G: 48 pp

Griggs DT (1972) The sinking lithosphere and the focal mechanism of deep earthquakes. In: Robertson EC (ed) The nature of the solid earth. McGraw-Hill Inc, New York, pp 361–384

Gromet LP, Silver LT (1977) Geographic variations of rare earth fractionations in plutonic rocks across the Peninsular Ranges, southern California. EOS 58: 532

Grow JA (1973) Crustal and upper mantle structure of the central Aleutian arc. Geol Soc Am Bull 84: 2169–2192

Grow JA, Qamar A (1973) Seismic wave attenuation beneath the central Aleutian arc. Bull Seismol Soc Am 63: 2155–2166

Gunn BM (1974) Systematic petrochemical difference in andesite suites. Bull Volcanol 38: 481–489

Gunn BM, Mooser F (1970) Geochemistry of the volcanics of central Mexico. Bull Volcanol 24: 577–616

Gunn BM, Roobal MJ, Smith AI (1974) Petrochemistry of the Peléean-type volcanoes of Martinique. Geol Soc Am Bull 85: 1023–1030

Gushchin AV, Kravchenko SM, Petrova MA (1976) Upper Pliocene Quaternary complex of Ishkhansar volcano, Lesser Caucasus, a representative of the trachyandesite association. Dokl ESS 231: 157–160

Gustafson LB (1978) Some major factors of porphyry copper genesis. Econ Geol 73: 600–607

Gustafson LB, Hunt JP (1975) The porphyry copper deposit at El Salvador, Chile. Econ Geol 70: 857–912

Haase CS, Chadam J, Feinn D, Ortoleva P (1980) Oscillatory zoning in plagioclase feldspars. Science 209: 272–274

Hall R (1976) Ophiolite emplacement and the evolution of the Taurus suture zone, southeastern Turkey. Geol Soc Am Bull 87: 1078–1088

Hallberg JA, Johnston C, Bye SM (1976) The Archean Marda Igneous Complex, Western Australia. Precambrian Res 3: 111–136

Hamelin B, Lambret B, Joron J-L, Treuil M, Allègre CJ (1979) Geochemistry of basalts from the Tyrrhenian Sea. Nature (London) 278: 832–834

Hamet J, Allègre CJ (1976) Rb-Sr systematics in granite from central Nepal (Manaslu): significance of the Oligocene age and high $^{87}Sr/^{86}Sr$ ratio in Himalayan orogeny. Geology 4: 470–489

Hamilton DL, Anderson GM (1968) Effects of water and oxygen pressures on the crystallization of basaltic magmas. In: Hess HH, Poldevaart A (eds) Basalts, vol I. Wiley Interscience Publ, New York, pp 445–482

Hamilton DL, Burnham CW, Osborn EF (1964) The solubility of water and effects of oxygen fugacity and water content on crystallization in mafic magmas. J Petrol 5: 21–39

Hamilton PJ, O'Nions RK, Evensen NM (1977) Sm-Nd dating of Archean basic and ultrabasic volcanics. Earth Planet Sci Lett 36: 263–268

Hamilton W (1970) The Uralides and the motion of the Russian and Siberian platforms. Geol Soc Am Bull 81: 2553–2576

Hamilton W (1979) Tectonics of the Indonesian region. US Geol Surv Prof Pap 1078

Handschumacher DW (1976) Post-Eocene plate tectonics of the Eastern Pacific. Am Geophys Union Monogr 19: 177–202

Hanson GN, Langmuir CH (1978) Modelling of major elements in mantle-melt systems using trace element approaches. Geochim Cosmochim Acta 42: 725–741

Hantke G, Parodi AI (1966) Catalogue of the active volcanoes of the world XIX: Columbia, Ecuador, and Peru. IAVCEI

Hanus V, Vanek J (1978) Morphology of the Andean Wadati-Benioff Zone, andesitic volcanism, and tectonic features of the Nazca plate. Tectonophysics 44: 65–77

Harris PG (1957) Zone refining and the origin of potassic basalts. Geochim Cosmochim Acta 12: 195–208

Hart RA (1973) A model for chemical exchange in the basalt-seawater system of oceanic layer II. Can J Earth Sci 10: 799–816

Hart RA (1976) Chemical variance in deep ocean basalts. Initial Rep DSDP 34: 301–336

Hart SR (1971) K, Rb, Cs, Sr, and Ba contents and Sr isotope ratios of ocean floor basalts. Philos Trans R Soc London Ser A 268: 573–587

Hart SR, Davis KE (1978) Nickel partitioning between olivine and silicate melt. Earth Planet Sci Lett 40: 203–219

Hart SR, Nalwalk AJ (1970) K, Rb, Cs and Sr relationships in submarine basalts from the Puerto Rico trench. Geochim Cosmochim Acta 34: 145–156

Hart SR, Staudigel H (1978) Oceanic crust: age of hydrothermal alteration. Geophys Res Lett 5: 1009–1012

Hart SR, Glassley WE, Karig DE (1972) Basalts and sea floor spreading behind the Mariana island arc. Earth Planet Sci Lett 15: 12–18

Hart SR, Brocks C, Krogh TE, Davis GL, Nava D (1970) Ancient and modern volcanic rocks: a trace element model. Earth Planet Sci Lett 10: 17–28

Hasewaga A, Sacks IS (1979) Subduction of the Nazca plate beneath Peru as determined from seismic observations. Carnegie Inst Wash Yearb 78: 276–284

Hasewaga A, Umino N, Takagi A (1978) Double-planed deep seismic zone and upper-mantle structure in the Northeastern Japan Arc. Geophys JR Astron Soc 54: 287–296

Haskin LA, Allen RO, Helmke PA, Paster TP, Anderson MR, Karotev RL, Zweifel KA (1970) Rare earth and other trace elements in Apollo 11 lunar samples. Proc Apollo 11 Lunar Sci Conf, pp 1213–1232

Hatherton T (1969) The geophysical significance of calc-alkaline andesites in New Zealand. N Z J Geol Geophys 12: 436–459

Hatherton T (1974) Active continental margins and island arcs. In: Burke CA, Drake CL (eds) The geology of continental margins. Springer, Berlin Heidelberg New York, pp 93–104

Hatherton T, Dickinson WR (1969) The relationship between andesite volcanism and seismicity in Indonesia, the Lesser Antilles, and other island arcs. J Geophys Res 74: 5301–5310

Haughton DR, Roeder PL, Skinner J (1974) Solubility of sulfur in mafic magmas. Econ Geol 69: 451–467

Hausel WD, Nash WP (1977) Petrology of Tertiary and Quaternary volcanic rocks, Washington County, southwestern Utah. Geol Soc Am Bull 88: 1831–1842

Hawkesworth CJ, O'Nions RK (1977) The petrogenesis of some Archean volcanic rocks from southern Africa. J Petrol 18: 487–520

Hawkesworth CJ, Powell M (1980) Magma series in the Lesser Antilles island arc. Earth Planet Sci Lett 51: 297–308

Hawkesworth CJ, Vollmer R (1979) Crustal contamination versus enriched mantle: ^{143}Nd/^{144}Nd and ^{87}Sr/^{86}Sr evidence from the Italian volcanics. Contrib Mineral Petrol 69: 151–165

Hawkesworth CJ, Norry MJ, Roddick JC, Baker PE, Francis PW, Thorpe RS (1979a) ^{143}Nd/^{144}Nd, ^{87}Sr/^{86}Sr, and incompatible trace element variations in calc-alkaline andesites and plateau lavas from South America. Earth Planet Sci Lett 42: 45–57

Hawkesworth CJ, Norry MJ, Roddick JC, Vollmer R (1979b) ^{143}Nd/^{144}Nd and ^{87}Sr/^{86}Sr ratios from the Azores and their significance in LIL-element enriched mantle. Nature (London) 280: 28–31

Hawkesworth CJ, O'Nions RK, Arculus RJ (1979c) Nd and Sr isotope geochemistry of island arc volcanics, Grenada, Lesser Antilles. Earth Planet Sci Lett 45: 237–248

Hawkesworth CJ, O'Nions RK, Pankhurst RJ, Hamilton PJ, Evensen NM (1977) A geochemical study of island arc and back arc tholeiites from the Scotia Sea. Earth Planet Sci Lett 36: 253–262

Hawkins JW, Divis AF (1975) Petrology and geochemistry of mid-Miocene volcanism on San Clemente and Santa Catalina islands and adjacent areas of the Southern California Borderland. Geol Soc Am Abstr 7: 323

Hawkins JW, Natland JH (1975) Nephelinites and basanites of the Samoan linear volcanic chains: their possible tectonic significance. Earth Planet Sci Lett 24: 427–439

Hawkins JW, Allison EC, Macdougall D (1971) Volcanic petrology and geologic history of Northeast Bank, southern California Borderland. Geol Soc Am Bull 82: 219–228

Hayatsu K (1976) Geological study of the Myoko volcanoes, Central Japan, part I. Stratigraphy. Mem Fac Sci Kyoto Univ Ser B 42: 131–170

Hayatsu K (1977) Geological study of the Myoko volcanoes, Central Japan, part II. Petrography. Mem Fac Sci Kyoto Univ Ser B 43: 1–48

Haynes SJ, McQuillan H (1974) Evolution of the Zagros suture zone, southern Iran. Geol Soc Am Bull 85: 739–744

Hearn BD Jr, Donnelly JM, Goff FE (1973) Geology and geochronology of the Clear Lake Volcanics, California. Proc 2nd UN Geotherm Symp

Hedervari P (1975) On the three-dimensional distribution of seismic domains beneath Volcano Kirisima, Kyushu, Japan. Pac Geol 9: 101–112

Hedge CE, Gorshkov GS (1978) Strontium-isotopic composition of volcanic rocks from Kamchatka. Dokl Akad Nauk SSSR 233: 163–166

Hedge CE, Knight RJ (1969) Lead and strontium isotopes in volcanic rocks from northern Honshu, Japan. Geochem J 3: 15–24

Hedge CE, Lewis JF (1971) Isotopic composition of strontium in three basalt-andesite centers along the Lesser Antilles Arc. Contrib Mineral Petrol 32: 39–47

Hedge CE, Hildreth RA, Henderson WT (1970) Strontium isotopes in some Cenozoic lavas from Oregon and Washington. Earth Planet Sci Lett 8: 434–438

Hellman PI, Green TH (1979) The role of sphene as an accessory phase in the high-pressure partial melting of hydrous mafic compositions. Earth Planet Sci Lett 42: 191–201

Helz RT (1973) Phase relations of basalts in their melting range at P_{H_2O} = 5 kb as a function of oxygen fugacity. J Petrol 14: 249–302

Helz RT (1976) Phase relations of basalts in their melting ranges at P_{H_2O} = 5 kb. Part II. Melt compositions. J Petrol 17: 139–193

Helz RT (1979) Alkali exchange between hornblende and melt: a temperature-sensitive reaction. Am Mineral 64: 953–965

Heming RF (1974) Geology and petrology of Rabaul Caldera, Papua New Guinea. Geol Soc Am Bull 85: 1253–1264

Heming RF (1977) Mineralogy and proposed P-T paths of basaltic lavas from Rabaul Caldera, Papua New Guinea. Contrib Mineral Petrol 61: 15–33

Heming RF, Carmichael ISE (1973) High-temperature pumice flows from the Rabaul caldera, Papua New Guinea. Contrib Mineral Petrol 38: 1–20

Heming RF, Rankin PC (1979) Ce-anomalous lavas from Rabaul caldera, Papua New Guinea. Geochim Cosmochim Acta 43: 1351–1355

Henriquez F, Martin RF (1978) Crystal-growth textures in magnetite flows and feeder dykes, El Laco, Chile. Can Mineral 16: 581–589

Hensen BJ, Green DH (1973) Experimental study of the stability of cordierite and garnet in pelite compositions at high pressures and temperatures. III. Synthesis of experimental data and geological applications. Contrib Mineral Petrol 38: 151–166

Herron EM (1972) Sea-floor spreading and the Cenozoic history of the east-central Pacific. Geol Soc Am Bull 83: 1671–1692

Hess HH (1941) Pyroxenes of common mafic magmas, Part II. Am Mineral 26: 573–592

Higgins MW (1973) Petrology of Newberry Volcano, central Oregon. Geol Soc Am Bull 84: 455–488

Higgins RE (1976) Major element chemistry of the Cenozoic volcanic rocks in the Los Angeles Basin and vicinity. In: Aspects of the geologic history of the California borderland. AAPG Misc Publ 24: 216–227

Hildreth W (1979) The Bishop Tuff: evidence for the origin of compositional zonation in silicic magma chambers. Geol Soc Am Spec Pap 180: 43–75

Hill DP (1979) Seismic evidence for the structure and Cenozoic tectonics of the Pacific Coast states. Geol Soc Am Mem 152: 145–174

Hill MD, Morris JD (1977) Near-trench plutonism in southwestern Alaska. Geol Soc Am Abstr 9: 436–437

Ho CS (1975) An introduction to the geology of Taiwan – explanatory text of the geologic map of Taiwan. Taipei, Taiwan, Republic of China Ministry of Economic Affairs, 153 pp

Hodges FN, Papike JJ (1976) DSDP Site 334: magmatic cumulates from oceanic layer 3. J Geophys Res 81: 4135–4151

Hofmann AW, Magaritz M (1976) Diffusion in silicate melts and glasses. Carnegie Inst Wash Yearb 75: 249–255

Holland HD (1972) Granites, solutions, and base metal deposits. Econ Geol 67: 281–301

Holland JG, Brown GM (1972) Hebridean tholeiitic magmas: a geochemical study of the Ardnamurchan cone sheets. Contrib Mineral Petrol 37: 139–160

Holloway JR (1971) Composition of fluid phase solutes in a basalt-H_2O-CO_2 system. Geol Soc Am Bull 82: 233–238

Holloway JR (1973) The system pargasite-H_2O-CO_2: a model for melting of a hydrous mineral with a mixed-volatile fluid. I. Experimental results to 8 kbar. Geochim Cosmochim Acta 37: 651–666

Holloway JR (1976) Fluids in the evolution of granitic magmas: consequences of finite CO_2 solubility. Geol Soc Am Bull 87: 1513–1518

Holloway JR, Burnham CW (1972) Melting relations of basalt with equilibrium water pressures less than total pressure. J Petrol 73: 1–29

Holloway JR, Ford CE (1975) Fluid-absent melting of the fluorohydroxyl amphibole pargasite to 35 kilobars. Earth Planet Sci Lett 25: 44–48

Holmes A (1932) The origin of igneous rocks. Geol Mag 69: 543–558

Homma F (1936) Classification of the zonal structure of plagioclase. Mem Fac Sci Kyoto Univ Ser B 11: 135–155

Honkura Y (1974) Electrical conductivity anomalies beneath the Japanese arc. J Geomagn Geoelectr 26: 147–171

Honkura Y, Koyama S (1979) Electrical conductivity structure beneath the central part of Japan as inferred from magnetotelluric fields at the Yatsugatake Magnetic Observatory. Bull Earthquake Res Inst 54: 491–501

Hoppenberger G von, Kiesl W (1975) Untersuchungen an süditalienischen Vulcaniten: Alicudi, Filicudi. Chem Erde 34: 185–195

Hopson CA (1971) Eruptive sequence at Mount St. Helens, Washington. Abstr Geol Soc Am 3: 138

Hopson CA, Crowder DF, Tabor RW, Cater FW, Wise WS (1965) Association of andesitic volcanoes in the Cascase Mountains with late Tertiary epizonal plutons. Abstr Geol Soc Am Annu Meet, p 79

Hormann PK, Pichler H, Zeil W (1973) New data on the young volcanism in the Puna of N.W. Argentina. Geol Rundsch 62: 397–418

Hotta H (1970) A crustal section across the Izu-Ogasawara arc and trench. J Phys Earth 18: 125–141

Hsu KJ et al (1977) History of the Mediterranean salinity crisis. Nature (London) 267: 399–403

Hsui AT, Toksoz MN (1979) The evolution of thermal structures beneath a subduction zone. Tectonophysics 60: 43–60

Hulme G (1974) The interpretation of lava flow morphology. Geophys J R Astron Soc 39: 361–383

Hutton CO (1964) Some igneous rocks from the New Plymouth area. Trans Proc R Soc N Z 74: 125–153

Ibrahim AK, Pontoise B, Latham G, Larue M, Chen T, Isacks B, Recy J, Louat R (1980) Structure of the New Hebrides arc-trench system. J Geophys Res 85: 253–266

Iida C (1961) Trace elements in minerals and rocks of the Izu-Hakone region, Japan, part II. Plagioclase. J Earth Sci Nagoya Univ 9: 14–28

Iida C, Juno H, Yamasaki K (1961) Trace elements in minerals and rocks of the Izu-Hakone region, Japan, part I. Olivine. J Earth Sci Nagoya Univ 9: 1–13

Imbo G (1965) Catalogue of the active volcanoes of the world. XVIII, Italy. IAVCEI

Innocenti F, Mazzuoli R, Pasquare G, Radicati di Brozolo F, Villari L (1975) The Neogene calcalkaline volcanism of Central Anatolia: geochronological data on Kayseri-Nidge area. Geol Mag 112: 349–360

Innocenti F, Mazuoli R, Pasquare G, Radicata di Brozolo F, Villari L (1976) Evolution of the volcanism in the area of interaction between the Arabian, Anatolian, and Iranian plates (Lake Van, eastern Turkey). J Volcanol Geotherm Res 1: 103–112

Irvine TN, Baragar WR (1971) A guide to the chemical classification of the common igneous rocks. Can J Earth Sci 8: 523–548

Irving AJ (1978) A review of experimental studies of crystal/liquid trace element partitioning. Geochim Cosmochim Acta 42: 743–770

Irving AJ, Frey FA (1978) Distribution of trace elements between garnet megacrysts and host volcanic liquids of kimberlitic to rhyolitic compositions. Geochim Cosmochim Acta 42: 771–788

Isacks BL, Barazangi M (1977) Geometry of Benioff Zones: lateral segmentation and downwards bending of subducted lithosphere. In: Talwani M, Pitman WC III (eds) Island arcs, deep sea trenches, and back-arc basins. AGU, pp 99–114

Isacks B, Molnar P (1971) Distribution of stresses in the descending lithosphere from a global survey of focal-mechanism solutions of mantle earthquakes. Rev Geophys Space Phys 9: 103–174

Isacks B, Oliver J, Sykes LR (1968) Seismology and the New Global Tectonics. J Geophys Res 73: 5855–5899

Ishii T (1975) The relations between temperature and composition of pigeonite in some lavas and their application to geothermometry. Mineral J 8: 48–57

Ishikawa K, Kanisawa S, Aoki K (1980) Content and behavior of fluorine in Japanese Quaternary volcanic rocks and petrogenetic application. J Volcanol Geotherm Res 8: 161–175

Ishikawa T (1951) Petrological significance of large anorthite crystals included in some pyroxene andesites and basalts in Japan. J Fac Sci Hokkaido Univ Ser IV 7: 339–354

Ishizaka K, Yanagi T, Hayatsu K (1977) A strontium isotopic study of the volcanic rocks of the Myoko volcano group, central Japan. Contrib Mineral Petrol 63: 295–307

Isshiki N (1963) Petrology of Hachijo-Jima volcanic group, seven Izu Islands, Japan. J Fac Sci Tokyo Univ 15: 91–134

Ivanov BV, Kadik AA, Maksimov AP (1978) Physicochemical conditions of crystallization of andesites of the Klyuchevskaya group of volcanoes (Kamchatka). Geochem Int 15: 100–115

Iwasaki B, Katsura T (1967) The solubility of hydrogen chloride in volcanic rock melts at a total pressure of one atmosphere and at temperatures of $1200^{\circ}C$ and $1290^{\circ}C$ under anhydrous conditions. Bull Chem Soc Jpn 40: 554–561

Iwasaki I, Katsura T, Yoshida M (1962) Geochemical investigations of the volcanoes in Japan. I. Titanomagnetite in volcanic rocks of Nekodake, Aso volcano. Bull Chem Soc Jpn 35: 448–452

Jackson J, Fitch TJ (1979) Seismotectonic implications of relocated aftershock sequences in Iran and Turkey. Geophys J R Astron Soc 57: 209–229

Jacob KH, Hamada K (1972) The upper mantle beneath the Aleutian island arc from purepath Rayleigh-wave dispersion data. Bull Seismol Soc Am 62: 1439–1454

Jacob KH, Quittmeyer RL (1979) The Makran region of Pakistan and Iran: trench-arc system with active plate subduction. In: Farah A, DeJong KA (eds) Geodynamics of Pakistan. Geol Surv Pakistan, pp 305–317

Jaegger JC (1964) Thermal effects of intrusions. Rev Geophys 2: 443–466

Jakeš P, Gill JB (1970) Rare earth elements and the island arc tholeiitic series. Earth Planet Sci Lett 9: 17–28

Jakeš P, Smith IEM (1970) High potassium calc-alkaline rocks from Cape Nelson, Eastern Papua. Contrib Mineral Petrol 28: 259–271

Jakeš P, White AJR (1969) Structure of the Melanesian Arcs and correlation with distribution of magma types. Tectonophysics 8: 223–236

Jakeš P, White AJR (1970) K/Rb ratios of rocks from island arcs. Geochim Cosmochim Acta 34: 849–856

Jakeš P, White AJR (1971) Composition of island arcs and continental growth. Earth Planet Sci Lett 12: 224–230

Jakeš P, White AJR (1972a) Major and trace element abundances in volcanic rocks of orogenic areas. Geol Soc Am Bull 83: 29–40

Jakeš P, White AJR (1972b) Hornblendes from calc-alkaline volcanic rocks of island arcs and continental margins. Am Mineral 57: 887–902

James DE (1971) Andean crustal and upper mantle structure. J Geophys Res 76: 3246–3271

James DE (1978) On the origin of the calc-alkaline volcanics of the central Andes: a revised interpretation. Carnegie Inst Wash Yearb 77: 562–590

James DE, Brooks C, Cuyubamba A (1976) Andean Cenozoic volcanism: magma genesis in the light of strontium isotopic composition and trace element geochemistry. Geol Soc Am Bull 87: 592–600

Jaques AL (1976) High-K_2O island arc volcanic rocks from the Finisterre and Adelbert Ranges, northern Papua New Guinea. Geol Soc Am Bull 87: 861–867

Jaques AL, Robinson GP (1977) The continent/island arc collision in northern Papua New Guinea. BMR J Aust Geol Geophys 2: 289–303

Jenkins DAL (1974) Detachment tectonics in western Papua New Guinea. Geol Soc Am Bull 85: 533–548

Jezek P, Hutchinson CS (1978) Banda Arc of Eastern Indonesia: petrology and geochemistry of the volcanic rocks. Bull Volcanol 41: 586–608

Jezek P, Gill J, Whitford D, Duncan R (1979) Reconnaissance geochemistry of the Sangihe-Sulawesi volcanic arc, Indonesia. EOS 60: 413

Johannes W (1978) Melting of plagioclase in the system $Ab-An-H_2O$ and $Qz-Ab-An-H_2O$ at $P_{H_2O} = 5$ kbars, an equilibrium problem. Contrib Mineral Petrol 66: 295–303

Johannsen A (1937) A descriptive petrography of the igneous rocks, vol III. University of Chicago Press, Chicago

Johnson RW (1976a) Late Cainozoic volcanism and plate tectonics at the southern margin of the Bismarck Sea, Papua New Guinea. In: Johnson RW (ed) Volcanism of Australasia. Elsevier, Amsterdam, pp 101–116

Johnson RW (1976b) Potassium variation across the New Britain volcanic arc. Earth Planet Sci Lett 31: 184–191

Johnson RW, Chappell BW (1979) Chemical analyses of rocks from the late Cainozoic volcanoes of north-central New Britain and the Witu Islands, Papua New Guinea. BMR Aust Rep 209: BMR Microform MF76

Johnson RW, Mackenzie DE, Smith IEM (1971) Seismicity and Late Cenozoic volcanism in parts of Papua New Guinea. Tectonophysics 12: 15–22

Johnson RW, Mackenzie DE, Smith IEM (1978a) Volcanic rock associations at convergent plate boundaries: reappraisal of the concept using histories from Papua New Guinea. Geol Soc Am Bull 89: 96–106

Johnson RW, Mackenzie DE, Smith IEM (1978b) Delayed partial melting of subduction-modified mantle in Papua New Guinea. Tectonophysics 46: 197–216

Johnson T, Molnar P (1972) Focal mechanisms and plate tectonics of the southwest Pacific. J Geophys Res 77: 5000–5032

Johnston DA (1978) Magma mixing prior to eruptions of Augustine volcano, Alaska. Abstr Geol Soc Am 10: 110–111

Johnston TH, Schilling J-G (1977) Lesser Antilles island arc: rare earth variations. EOS 58: 533

Jolly WT (1977) Relations between Archean lavas and intrusive bodies of the Abitibi Greenstone Belt, Ontario-Quebec. In: Volcanic regimes in Canada. Geol Assoc Can Spec Pap 16: 311–330

Jordan TH (1978) Composition and development of the continental tectosphere. Nature (London) 274: 544–548

Joron J-L, Treuil M (1977) Utilisation des propriétés des éléments fortement hygromagmatophiles pour l'étude de la composition chimique et de l'héterogénéité du manteau. Bull Soc Geol Fr 19: 1197–1205

Kaila KL, Krishna VG, Narain H (1971) Upper mantle P-wave velocity structure in the Japan region from travel-time studies of deep earthquakes using a new analytical method. Bull Seismol Soc Am 59: 1949–1967

Kaila KL, Krishna VG, Narain H (1974) Upper mantle shear-wave velocity structure in the Japan region. Bull Seismol Soc Am 64: 355–374

Kaminuma K (1975) P wave velocity anomaly beneath the Kirisima-Sakurazima volcanoes. Bull Volcanol Soc Jpn 19: 129–138

Kanamori H (1970) Mantle beneath the Japanese arc. Phys Earth Planet Int 3: 475–483

Kano H, Yahima R (1976) Almandine-garnets of acid magmatic origin from Yamanogawa, Fukushima Prefecture and Kamitazawa, Yamagata Prefecture. J Jpn Assoc Mineral Pet Econ Geol 71: 106–119

Karig DE (1971) Origin and development of marginal basins in the western Pacific. J Geophys Res 76: 2542–2561

Karig DE (1973) Plate convergence between the Philippines and the Ryuku islands. Mar Geol 14: 153–168

Karig DE (1975) Basin genesis in the Philippine Sea. Initial Rep DSDP 31: 857–880

Karig DE, Mammerickx J (1972) Tectonic framework of the New Hebrides island arc. Mar Geol 12: 187–205

Karig DE, Sharman GE III (1975) Subduction and accretion in trenches. Geol Soc Am Bull 86: 377–389

Karig DE, Caldwell JG, Parmentier EM (1976) Effects of accretion on the geomety of the descending lithosphere. J Geophys Res 81: 6281–6291

Katili JA (1975) Volcanism and plate tectonics in the Indonesian island arcs. Tectonophysics 26: 165–188

Kato Y (1964) Petrology of Nasu volcano. J Jpn Assoc Mineral Petrogr Econ Geol 51: 233–243

Katsui Y (1961) Petrochemistry of the Quaternary volcanic rocks of Hokkaido and surrounding areas. J Fac Sci Hokkaido Univ Ser IV 11: 1–58

Katsui Y (1963) Evolution and magmatic history of some Krakatoan calderas in Hokkaido, Japan. J Fac Sci Hokkaido Univ Ser IV 11: 631–650

Katsui Y (1971) List of the world active volcanoes. Volcanol Soc Japan and IAVCEI; Bull Volcanic Eruptions Spec Issue

Katsui Y, Gonzalez FO (1968) Geologia del area neovolcanica de las nevados de Payachata. Univ (Santiago) Chile Inst Geol Publ 29: 61 pp

Katsui Y, Ando S, Inaba K (1975) Formation and magmatic evolution of Mashu Volcano, East Hokkaido, Japan. J Fac Sci Hokkaido Univ Ser IV 16: 533–552

Katsui Y, Yamamoto M, Nemoto S, Niida K (1979) Genesis of calc-alkaline andesites from Oshima-oshima and Ichinomegata volcanoes, north Japan. J Fac Sci Hokkaido Univ Ser IV 19: 157–168

Katsui Y, Oba Y, Ando S, Nishimura S, Masuda Y, Kurasawa H, Fujimaki H (1978) Petrochemistry of the Quaternary volcanic rocks of Hokkaido, north Japan. J Fac Sci Hokkaido Univ Ser IV 18: 243–282

Katsumata M, Sykes LR (1969) Seismicity and tectonics of the western Pacific: Izu-Mariana-Caroline and Ryuku-Taiwan regions. J Geophys Res 74: 5923–5948

Katsura T, Kitayama K, Aoyagi R, Sasajima S (1976) High temperature experiments related to iron-titanium oxide minerals in volcanic rocks. Bull Volcanol Soc Jpn 21: 31–56

Kawano Y, Yagi K, Aoki K (1961) Petrography and petrochemistry of the volcanic rocks of Quaternary volcanoes of northeastern Japan. Tohoku Univ Sci Rep Ser III 7: 1–46

Kay RW (1977) Geochemical constraints on the origin of Aleutian magmas. In: Talwani M, Pitman WC III (eds) Island arcs, deep sea trenches and back-arc basins. AGU Ewing Ser 1: 229–242

Kay RW (1978) Aleutian magnesian andesites: melts from subducted Pacific Ocean crust. J Volcanol Geotherm Res 4: 117–132

Kay RW (1980) Volcanic arc magmas: implications of a melting-mixing model for element recycling in the crust-upper mantle system. J Geol 88: 497–522

Kay RW, Hubbard NJ (1978) Trace elements in ocean ridge basalts. Earth Planet Sci Lett 38: 95–116

Kay RW, Hubbard NJ, Gast PW (1970) Chemical characteristics and origin of oceanic ridge volcanic rocks. J Geophys Res 75: 1585–1614

Kay RW, Sun SS, Lee-Hu CN (1978) Pb and Sr isotopes in volcanic rocks from the Aleutian Islands and Pribilof Islands, Alaska. Geochim Cosmochim Acta 42: 263–273

Keith SB (1978) Paleosubduction geometries inferred from Cretaceous and Tertiary magmatic patterns in southwestern North America. Geology 6: 516–521

Kelleher J, McCann W (1976) Buoyant zones, great earthquakes, and unstable boundaries of subduction. J Geophys Res 81: 4885–4896

Keller J (1974) Petrology of some volcanic rock series of the Aeolian arc, southern Tyrrhenian Sea: calc-alkaline and shoshonite associations. Contrib Mineral Petrol 46: 29–47

Kennedy GC (1955) Some aspects of the role of water in rock melts. Geol Soc Am Spec Pap 62: 489–504

Kennett JP, McBirney AR, Thunell RC (1977) Episodes of Cenozoic volcanism in the circum-Pacific region. J Volcanol Geotherm Res 2: 145–163

Kepezhinskas VV (1970) Chemistry of Quaternary basalts of the Kurile-Kamchatka volcanic province. Dokl Akad Nauk SSSR 190: 110–113

Kesler SE, Jones LM, Walker RI (1975) Intrusive rocks associated with porphyry copper mineralization in island arc areas. Econ Geol 70: 515–526

Kiesl W von, Kluger F, Weinke HH, Scholl H, Klein P (1978) Untersuchungen an süditalienischen Vulkaniten: Lipari, Vulcano. Chem Erde 37: 40–49

Kimura M (1976) Major magmatic activity as a key to predicting earthquakes along the Sagami Trough, Japan. Nature (London) 260: 131–133

Kistler RW, Peterman ZE (1973) Variations in Sr, Rb, K, Na and initial $^{87}Sr/^{86}Sr$ in Mesozoic granitic rocks and intruded wall rocks in central California. Geol Soc Am Bull 84: 3489–3512

Klerkx J, Deutsch S, Pichler H, Zeil W (1977) Strontium isotopic composition and trace element data bearing on the origin of Cenozoic volcanic rocks of the central and southern Andes. J Volcanol Geotherm Res 2: 49–71

Klerkx J, Deutsch S, Hertogen J, DeWinter J, Gijbels R, Pichler H (1974) Comments on "Evolution of Eolian Arc Volcanism". Earth Planet Sci Lett 23: 297–303

Kluger VF, Weinke HH, Klein P, Kiesl W (1975) Bestimmung von Fluor in Vulkaniten von Filicudi und Alicudi (Aolische Inseln, Süditalien). Chem Erde 34: 168–174

Kobayashi T, Ohmori E, Ohmori T (1975) Petrochemistry of Ontake volcano. Bull Geol Surv Jpn 26: 7–22

Kogan MG (1975) Gravity field of the Kurile-Kamchatka arc and its relation to the thermal regime of the lithosphere. J Geophys Res 80: 1381–1390

Kohari H (1974) Volcanic rocks from Kayo-dake volcano. J Jpn Assoc Mineral Petrogr Econ Geol 69: 1–8

Komar PD (1972) Mechanical interactions of phenocrysts and flow differentiation of igneous dikes and sills. Geol Soc Am Bull 83: 973–985

Konečny V (1971) Evolutionary stages of the Banska Stiavnica caldera and its post-volcanic structures. Bull Volcanol 35: 95–116

Konečny V, Bagdasarjan GP, Vass D (1969) Evolution of Neogene volcanism in central Slovakia and its confrontation with absolute ages. Acta Geol Acad Sci Hung 13: 245–258

Kostuk VP (1958) Mineralogical character of the magmatic garnet in the volcanites of Transcarpathians. Mineral Sb Lvov Geol Soc 12: 280–296

Kramers JD (1977) Lead and strontium isotopes in Cretaceous kimberlites and mantle-derived xenoliths from southern Africa. Earth Planet Sci Lett 34: 419–427

Krause DC (1973) Crustal plates of the Bismarck and Solomon Seas. In: Fraser R (ed) Oceanography of the South Pacific 1972. N Z Ntl Comm UNSECO, Wellington, pp 271–280

Kroenke L (1972) Geology of the Ontong Java Plateau. Hawaii Inst Geophys Rep HIG 72-5

Krushensky RD (1972) Geology of the Istaru Quadrangle, Costa Rica. US Geol Surv Bull 1358: 46 pp

Kubota S, Berg E (1967) Evidence for magma in the Katmai volcanic range. Bull Volcanol 31: 175–214

Kubovics I (1973) Effect of different aggregates on the crystallization of microandesite and alveolar andesite melt. Acta Geol Acad Sci Hung 17: 287–306

Kudo AM, Weill DF (1970) An igneous plagioclase thermometer. Contrib Mineral Petrol 25: 52–65

Kuenen Ph H (1935) Contributions to the geology of the East-Indies from the Snellius expedition. I. Volcanoes. Leidsche Geol Med 7: 273

Kuno H (1950a) Petrology of Hakone volcano and the adjacent areas, Japan. Geol Soc Am Bull 61: 957–1020

Kuno H (1950b) Geology of Hakone volcano and the adjacent areas, part I. J Fac Sci Univ Tokyo Sect II 7: 257–279

Kuno H (1959) Origin of Cenozoic petrographic provinces of Japan and surrounding areas. Bull Volcanol 20: 37–76

Kuno H (1962) Catalogue of the active volcanoes of the world, part II. Japan, Taiwan, and Marianas. IAVCEI

Kuno H (1965) Fractionation trends of basalt magmas in lava flows. J Petrol 6: 302–321

Kuno H (1966) Review of pyroxene relations in terrestrial rocks in the light of recent experimental works. Mineral J 5: 21–43

Kuno H (1968a) Differentiation of basalt magmas. In: Hess HH, Poldevaart A (eds) Basalts, vol II. Wiley Intersciences Publ, New York, pp 624–688

Kuno H (1968b) Origin of andesite and its bearing on the island arc structure. Bull Volcanol 32: 141–176

Kuno H (1969) Pigeonite-bearing andesite and associated dacite from Asio, Japan. Am J Sci 267A: 257–268

Kuno H, Nagashima K (1952) Chemical composition of hypersthene and pigeonite in equilibrium in magma. Am Mineral 37: 1000–1006

Kuno H, Oki Y, Ogina K, Hirota S (1970) Structure of Hakone caldera as revealed by drilling. Bull Volcanol 34: 713–721

Kurasawa H (1968) Isotopic composition of lead and concentrations of uranium, thorium, and lead in volcanic rocks from Dogo of Oki Islands, Japan, Geochem J 2: 11–28

Kurasawa H, Michino K (1976) Petrology and chemistry of the volcanic rocks from the western and southern part of Izu Peninsula, central Japan. Bull Volcanol Soc Jpn 21: 11–29

Kuroda N, Shiraki K, Urano H (1978) Boninite as a possible calc-alkalic primary magma. Bull Volcanol 41: 563–575

Kushiro I (1960) Si-Al relation in clinopyroxenes from igneous rocks. Am J Sci 258: 548–554

Kushiro I (1972) Effect of water on the compositions of magmas formed at high pressures. J Petrol 13: 311–334

Kushiro I (1973) Crystallization of pyroxenes in Apollo 15 mare basalts. Carnegie Inst Wash Yearb 72: 647–650

Kushiro I (1974) Melting of hydrous upper mantle and possible generation of andesitic magma: an approach from synthetic systems. Earth Planet Sci Lett 22: 294–299

Kushiro I (1975) On the nature of silicate melt and its significance in magma genesis: regularities in the shift of the liquidus boundaries involving olivine, pyroxene, and silica minerals. Am J Sci 275: 411–431

Kushiro I (1978) Density and viscosity of hydrous calc-alkalic andesite magma at high pressures. Carnegie Inst Wash Yearb 77: 675–677

Kushiro I, Sato H (1978) Origin of calc-alkalic andesite in the Japanese islands. Bull Volcanol 41: 576–585

Kushiro I, Yoder HS Jr, Mysen BO (1976) Viscosities of basalt and andesite melts at high pressures. J Geophys Res 81: 6351–6356

Kussmaul S, Hormann PK, Ploskonka E, Subieta T (1977) Volcanism and structure of southwestern Bolivia. J Volcanol Geotherm Res 2: 73–111

Lachenbruch AH, Sass JH, Munroe RJ, Moses TH Jr (1976) Geothermal setting and simple heat conduction models for the Long Valley Caldera. J Geophys Res 81: 769–784

Lacroix A (1893) Les enclaves des roches volcaniques. Protat Frères, Macon

Lacroix A (1901) Sur deux nouveaux groupes d'enclaves des roches éruptives. Bull Soc Fr Mineral 24: 488–504

Lacroix A (1904) La montagne Pelée et ses éruptions. Masson et Cie, Paris

Lambert IB, Wyllie PJ (1972) Melting of gabbro (quartz eclogite) with excess water to 35 kilobars, with geological applications. J Geol 80: 693–720

Lambert IB, Wyllie PJ (1974) Melting of tonalite and crystallization of andesite liquid with excess water to 30 kilobars. J Geol 82: 88–97

Lambert RSt John, Holland JG (1974) Yttrium geochemistry applied to petrogenesis utilizing calcium-yttrium relationships in minerals and rocks. Geochim Cosmochim Acta 38: 1393–1414

Lambert RSt John, Holland JG, Owen PF (1974) Chemical petrology of a suite of calc-alkaline lavas from Mount Ararat, Turkey. J Geol 82: 419–438

Lancelot JR, Briqueu L, Girod M, Tatsumoto M (1978) Petrologie et géochimie (majeurs, traces, isotopes du Sr et du Pb) d'échantillons dragues aux Nouvelles Hébrides. Réunion Annuelle des Sciences de La Terre 6: 228

Langmuir CH, Vocke RD Jr, Hanson GN, Hart SR (1978) A general mixing equation with applications to Icelandic basalts. Earth Planet Sci Lett 37: 380–392

Larsen ES, Irving J, Gonyer FA, Larsen ES III (1937a) Petrologic results of a study of the minerals from the Tertiary volcanic rocks of the San Juan region, Colorado. 5. The amphiboles. Am Mineral 22: 889–898

Larsen ES, Gonyer FA, Irving J (1937b) Petrologic results of a study of the minerals from the Tertiary volcanic rocks of the San Juan region, Colorado, 6. Biotite. Am Mineral 22: 898-905

Larsen ES, Irving J, Gonyer FA, Larsen ES III (1938) Petrologic results of the minerals from the Tertiary volcanic rocks of the San Juan region, Colorado (Plagioclase feldspars). Am Mineral 23: 227–257

Larson EE, Reynolds RI, Merrill R, Levi S, Ozima M, Aoki Y, Kinoshita H, Zasshu S, Kawai N, Nakajima T, Hiraoka K (1975) Major-element pretrochemistry of some extrusive rocks from the volcanically active Mariana Islands. Bull Volcanol 39: 361–377

Larsson W (1941) Petrology of interglacial volcanoes from the Andes of northern Patagonia. Bull Geol Inst Univ Uppsala 28: 191–405

Lauer HV Jr, Morris RV (1977) Redox equilibria of multivalent ions in silicate glasses. J Am Ceramic Soc 60: 443–451

Launay J, Larue BM, Louat R, Maillet P, Monzier M (1979) The southern end of the New Hebrides arc: morphology, petrography, seismicity, gravity. In: Abstr 3rd SW Pac Symp, Sydney, Dec 1979

Leake BE (1968) A catalog of analyzed calciferous and subcalciferous amphiboles together with their nomenclature and associated minerals. Geol Soc Am Spec Pap 98: 210 pp

Leake BE (1978) Nomenclature of amphiboles. Can Mineral 16: 501–520

Le Bas MJ (1962) The role of aluminum in igneous clinopyroxenes with relation to their parentage. Am J Sci 260: 267–288

Leeman WP (1977) Comparison of Rb/Sr, U/Pb, and rare earth characteristics of sub-continental and sub-oceanic mantle regions. In: Dick HBJ (ed) Magma genesis. Oreg Dep Geol Mineral Ind Bull 96: 149–168

Leeman WP, Lindstrom DJ (1978) Partitioning of Ni^{2+} between basaltic and synthetic melts and olivines – an experimental study. Geochim Cosmochim Acta 42: 801–816

Leeman WP, Vitaliano CJ, Prinz M (1976) Evolved lavas from the Snake River Plain: Craters of the Moon National Monument, Idaho. Contrib Mineral Petrol 56: 35–60

Lefèvre C (1973) Les caractères magmatiques du volcanisme plio-quaternaire des Andes dans le Sud du Pérou. Contrib Mineral Petrol 41: 259–272

Le Maitre RW (1976a) The chemical variability of some common igneous rocks. J Petrol 17: 589–637

Le Maitre RW (1976b) A new approach to the classification of igneous rocks using the basalt-andesite-dacite-rhyolite suite as an example. Contrib Mineral Petrol 56: 191–203

Leo GW, Hedge CE, Marvin RF (1980) Geochemistry, Sr isotope data, and K-Ar ages of the andesite-rhyolite association in the Padang area, western Sumatra. J Volcanol Geotherm Res 7: 139–156

Leonova LL (1979) Geochemistry of the Quaternary and Recent volcanic rocks in the Kurile islands and Kamchatka. Geochem Int 16: 95–112

Leonova LL, Udal'tzova NI (1970) U, Th, Li, Rb, Cs in the volcanic rocks of the Kurile Isles and Kamchatka. Geokhimya 11: 1329–1334

Le Pichon X, Angelier J (1979) The Hellenic arc and trench system: a key to the neotectonic evolution of the Eastern Mediterranean area. Tectonophysics 60: 1–42

Lewis JF (1973) Petrology of the ejected plutonic blocks of the Soufriere volcano, St. Vincent, West Indies. J Petrol 14: 81–112

Lingenfelter RE, Schubert G (1976) Hot spot and trench volcano separations: evidence for whole mantle convection and depth of partial melting. In: Proc Symp Andean Antarct Volcanol Probl, IAVCEI, pp 762–764

Lipman PW, Christiansen RL, O'Connor JR (1966) A compositionally zoned ash-flow sheet in southern Nevada. US Geol Surv Prof Pap 524-F

Lipman P, Prostka HJ, Christiansen RL (1972) Cenozoic volcanism and plate tectonic evolution of western United States, part 1, Early and middle Cenozoic. Philos Trans R Soc London Ser A 271: 249–284

Lipman PW, Doe BR, Hedge CE, Steven TA (1978) Petrologic evolution of the San Juan volcanic field, southwestern Colorado: lead and strontium isotopic evidence. Geol Soc Am Bull 89: 59–82

Lisitzin AP (1972) Sedimentation in the World Ocean. Soc Econ Paleontol Mineral Spec Publ 17

Loeschke J (1979) Basalts of Oregon (USA) and the geotectonic environment. II. Petrochemistry of Tertiary and Quaternary basalts and andesites of the western and high Cascades. Neues Jahrb Mineral Abh 137: 135–161

Lofgren G (1974) Temperature-induced zoning in synthetic plagioclase feldspar. In: Mackenzie WS, Zussman J (eds) The feldspars, pp 362–376

Lopez-Escobar L, Frey FA, Vergara M (1976) Andesites from central-south Chile: trace-element abundances and petrogenesis. In: Proc Symp Andean Antarct Volcanol Probl, IAVCEI, pp 725–761

Lopez-Escobar L, Frey FA, Vergara M (1977) Andesites and high-alumina basalts from the Central-South Chile high Andes; geochemical evidence bearing on their petrogenesis. Contrib Mineral Petrol 63: 199–228

Lopez-Ruiz J, Badiola ER, Cacho LG (1977) Origine des grenats des roches calco-alcalines du Sud-Est de l'Espagne. Bull Volcanol 40: 141–152

Lowder GG (1970) The volcanoes and caldera of Talasea, New Britain: mineralogy. Contrib Mineral Petrol 26: 324–340

Lowder GG, Carmichael ISE (1970) The volcanoes and caldera of Talasea, New Britain. Geol Soc Am Bull 81: 17–38

Ludwig WJ, Nafe JE, Drake CL (1970) Seismic refraction. In: Maxwell AE (ed) The sea, vol IV, part 1. Wiley-Interscience, New York, pp 85–109

Luhr JF, Carmichael ISE (1980) The Colima volcanic complex, Mexico. Contrib Mineral Petrol 71: 343–372

Luyendyk B (1970) Dips of downgoing lithosphere beneath island arcs. Geol Soc Am Bull 81: 3411–3416

Luyendyk BP, MacDonald KC, Bryan WB (1973) Rifting history of the Woodlark Basin in the southwest Pacific. Geol Soc Am Bull 84: 1125–1134

Lyons JB, Faul H (1968) Isotope geochronology of the Northern Appalachians. In: Zen E, White WS, Hadley JB (eds) Studies of Appalachian geology: northern and maritime. Wiley-Interscience Publ, New York, pp 305–318

Maccarrone E (1963) Aspetti geochimico-petrografici di alcuni esemplari di andesite granatocordieritifera dell'Isola di Lipari. Period Mineral 32: 277–301

MacDonald GA (1960) Dissimilarity of continental and oceanic rock types. J Petrol 1: 172–177

MacDonald GA, Katsura T (1964) Chemical composition of Hawaiian lavas. J Petrol 5: 82–133

MacDonald GA, Katsura T (1965) Eruption of Lassen Peak, Cascade Range, California, in 1915: example of mixed magmas. Geol Soc Am Bull 76: 475–482

MacGregor AG (1938) The volcanic history and petrology of Montserrat with observations on Mt. Pelée in Martinique. Philos Trans R Soc London Ser B 557: 1–90

Machado F (1974) The search for magma reservoirs. In: Developments in solid earth geophysics, vol VI. Phys Volcanol, chap 11, pp 255–273

Mackenzie DE (1976) Nature and origin of Late Cenozoic volcanoes in western Papua New Guinea. In: Johnson RW (ed) Volcanism in Australasia. Elsevier, Amsterdam, pp 221–238

Mackenzie DE, Chappell BW (1972) Shoshonite and calc-alkaline lavas from the Highlands of Papua New Guinea. Contrib Mineral Petrol 35: 50–62

Mackevett EM Jr (1971) Stratigraphy and general geology of the McCarthy C-5 quadrangle, Alaska. US Geol Surv Bull 1323

Magaritz M, Taylor HP Jr (1976) Oxygen, hydrogen and carbon isotope studies of the Franciscan Formation, Coast Range, California. Geochim Cosmochim Acta 40: 215–237

Magaritz M, Whitford DJ, James DE (1978) Oxygen isotope and the origin of high-$^{87}Sr/^{86}Sr$ andesites. Earth Planet Sci Lett 40: 220–230

Makarov NN, Suprychev VA (1964) Xenogenic garnet (pyrope-almandine) from volcanic rock of the Crimea. Akad Dokl SSSR 157: 64–67

Makris J (1978) The crust and upper mantle of the Aegean region from deep seismic soundings. Tectonophysics 46: 269–284

Maksimov AP, Kadik AA, Korovushkina EYe, Ivanov BV (1978) Crystallization of an andesite melt with a fixed water content at pressures up to 12 kbar. Geochem Int 15: 20–29

Malin MC (1977) Comparison of volcanic features of Elysium (Mars) and Tibesti (Earth). Geol Soc Am Bull 88: 908–919

Mallik DIJ (1973) Some petrological and structural variations in the New Hebrides. In: Coleman PJ (ed) The Western Pacific: Island arcs, marginal seas, geochemistry. W Aust Univ Press, pp 193–212

Manghnani MH, Ramanantoandro R, Clark SP Jr (1974) Compressional and shear wave velocities in granulite facies rocks and eclogites to 10 kbars. J Geophys Res 79: 5427–5446

Markhinin EK (1968) Volcanism as an agent of formation of the Earth's crust. In: The crust and upper mantle of the Pacific area. Am Geophys Union Monogr 12: 413–422

Markhinin EK, Stratula DS (1973) Relationship between chemical composition of volcanic rocks and depth of the seismofocal layer as shown by the Kliucheskaya Volcanic Group (Kamchatka) and the Kurile-Kamchatka Island Arc. Bull Volcanol 37: 175–182

Marlow MS, Scholl DW, Buffington EC, Alpha TR (1973) Tectonic history of the Central Aleutian Arc. Geol Soc Am Bull 84: 1555–1574

Marsh BD (1976a) Some Aleutian andesites: their nature and source. J Geol 84: 27–45

Marsh BD (1976b) Mechanism of Benioff zone magmatism. Am Geophys Union Monogr 19: 337–350

Marsh BD (1978) On the cooling of ascending andesitic magma. Philos Trans R Soc London Ser A 288: 611–625

Marsh BD (1979) Island arc development: some observations, experiments, and speculations. J Geol 87: 687–714

Marsh BD, Carmichael ISE (1974) Benioff zone magmatism. J Geophys Res 79: 1196–1206

Marsh BD, Leitz RE (1979) Geology of Amak island, Aleutian islands, Alaska. J Geol 87: 715–723

Marshak RS, Karig DE (1977) Triple junctions as a cause for anomalously near-trench igneous activity between the trench and volcanic arc. Geology 5: 233–236

Marshall P (1912) Address by President. Rep Australas Assoc Adv Sci 13: 90–99

Mason DR (1978) Compositional variations in ferromagnesian minerals from porphyry copper-generating and barren intrusions of the western Highlands, Papua New Guinea. Econ Geol 73: 878–890

Masuda A, Nagasawa S (1975) Rocks with negative cerium anomalies dredged from Shatsky Rise. Geochem J 9: 227–233

Masuda Y (1979) Lateral variation of trace element contents in Quaternary volcanic rocks across northeast Japan. Bull Univ Osaka Ser A 28: 105–125

Masuda Y, Aoki K (1978) Two types of island arc tholeiite in Japan. Earth Planet Sci Lett 39: 298–302

Masuda Y, Aoki K (1979) Trace element variations in the volcanic rocks from the Nasu zone, northeast Japan. Earth Planet Sci Lett 44: 139–149

Masuda Y, Nishimura S, Ikeda T, Katsui Y (1975) Rare-earth and trace elements in the Quaternary volcanic rocks of Hokkaido, Japan. Chem Geol 15: 251–271

Mathews WH (1957) Petrology of Quaternary volcanics of the Mount Garibaldi map area, southwestern British Columbia. Am J Sci 255: 400–415

Mathez EA (1973) Refinement of the Kudo-Weill plagioclase thermometer and its application to basaltic rocks. Contrib Mineral Petrol 41: 61–72

Mathez EA (1976) Sulfur solubility and magmatic sulfides in submarine basalts. J Geophys Res 81: 4269–4276

Matsubaya O, Ueda A, Kusakabe M, Matsuhisa Y, Sakai H, Sasaki A (1975) An isotopic study of the volcanoes and the hot springs in Satsuma Iwo-jima and some areas in Kyushu. Bull Geol Surv Jpn 26: 1–18

Matsuda J-I, Zashu S, Ozima M (1976) Sr isotopic studies of volcanic rocks from island arcs in the western Pacific. Tectonophysics 37: 141–151

Matsuhisa Y (1979) Orxygen isotopic compositions of volcanic rocks from the East Japan island arcs and their bearing on petrogenesis. J Volcanol Geotherm Res 5: 271–296

Matsuhisa Y, Matsubaya O, Sakai H (1973) Oxygen isotope variations in magmatic differentiation processes of the volcanic rocks in Japan. Contrib Mineral Petrol 39: 277–288

Matsushima S, Akeni K (1977) Elastic wave velocities in the Ichinomegata ultramafic nodules: composition of the uppermost mantle. In: Manghnani MH, Akimoto S-I (eds) High-pressure research applications in geophysics. Academic Press, London New York, pp 65–76

Matumoto T (1971) Seismic body waves observed in the vicinity of Mount Katmai, Alaska, and evidence for the existence of molten chambers. Geol Soc Am Bull 82: 2905–2920

Matumoto T, Ohtake M, Latham G, Umana J (1977) Crustal structure in southern Central America. Bull Seismol Soc Am 67: 121–134

Mavko G, Nur A (1975) Melt squirt in the asthenosphere. J Geophys Res 80: 1444–1448

Mayfield DG, Walker JA, Carr MJ (1980) Patterns of chemical variation at recent volcanoes in El Salvador. (Preprint)

Mazzulo LJ, Bence AE (1976) Abyssal tholeiites from DSDP Leg 34: The Nazca Plate. J Geophys Res 81: 4327–4352

McBirney AR (1968) Petrochemistry of the Cascade andesite volcanoes. In: Dole HM (ed) Andesite Conference Guidebook. Oreg Dep Geol Mineral Ind Bull 62: 101–107

McBirney AR (1969) Andesitic and rhyolitic volcanism of orogenic belts. Am Geophys Union Monogr 13: 501–506

McBirney AR (1973) Factors governing the intensity of explosive andesitic eruptions. Bull Volcanol 37: 443–453

McBirney AR (1976) Some geologic constraints on models for magma generation in orogenic environments. Can Mineral 14: 245–254

McBirney AR (1980) Mixing and unmixing of magmas. J Volcanol Geotherm Res 7: 357–371

McBirney AR, Williams H (1965) Volcanic history of Nicaragua. Univ Calif Publ Geol Sci 55

McBirney AR, Williams H (1969) Geology and petrology of Galapagos island. Geol Soc Am Mem 118

McDougall I (1976) Geochemistry and origin of basalt of the Columbia River Group, Oregon and Washington. Geol Soc Am Bull 87: 777–792

McKee CO, Cooke RJS, Wallace DA (1976) 1974–75 eruptions of Karkar Volcano, Papua New Guinea. In: Johnson RW (ed) Volcanism in Australasia. Elsevier, Amsterdam, pp 173–190

McKenzie DP (1972) Active tectonics of the Mediterranean region. Geophys J R Astron Soc 30: 109–185

McKenzie DP, Morgan WJ (1969) Evolution of triple junctions. Nature (London) 224: 125–133

McKenzie DP, Sclater JG (1968) Heat flow inside the island arcs of the northwestern Pacific. J Geophys Res 73: 3173–3179

McNutt RH, Clark AH, Zentilli M (1979) Lead isotopic compositions of Andean igneous rocks, latitudes 26° to 29°S: petrologic and metallogenic implications. Econ Geol 74: 827–837

McNutt RH, Crocket JH, Clark AH, Caelles JS, Farrar E, Haynes SJ, Zentilli M (1975) Initial $^{87}Sr/^{86}Sr$ ratios of plutonic and volcanic rocks of the central Andes between latitudes 26° and 29° south. Earth Planet Sci Lett 27: 305–313

Meijer A (1976) Pb and Sr isotopic data bearing on the origin of volcanic rocks from the Mariana island-arc system. Geol Soc Am Bull 87: 1358–1369

Meijer A (1978) Petrology of volcanic rocks from the Mariana arc trench gap and gabbros from the Mariana inter-arc basin, IPOD Leg 60. EOS 59: 1182

Meijer A (1980) Primitive arc volcanism and a boninite series: examples from western Pacific island arcs. In: Hayes D (ed) Tectonics of Southeast Asia. AGU, in press

Melson WG, Saenz R (1973) Volume, energy and cyclicity of eruptions of Arenal volcano, Costa Rica. Bull Volcanol 37: 416–437

Melson WG, Jarosewich E, Lundquist CA (1970) Volcanic eruption at Metis Shoal, Tonga, 1967–1958: description and petrology. Smithsonian Contrib Earth Sci No 4

Melson WG, Jarosewich E, Switzer G, Thompson G (1972) Basaltic nuées ardentes of the 1970 eruption of Ulawan volcano, New Britain. Smithsonian Contrib Earth Sci 9: 15–32

Melson WG, Vallier TL, Wright TL, Byerly G, Nelson J (1976) Chemical diversity of abyssal volcanic glass erupted along Pacific, Atlantic, and Indian ocean sea-floor spreading centers. Am Geophys Union Monogr 19: 351–368

Menard HW (1967) Sea floor spreading, topography, and the second layer. Science 157: 923–924

Menzies M, Murthy VR (1980) Nd and Sr isotope geochemistry of hydrous mantle nodules and their host alkali basalts: implications for local heterogeneities in metasomatically veined mantle. Earth Planet Sci Lett 46: 323–334

Merrill RB, Wyllie PJ (1975) Kaersutite and kaersutite eclogite from Kakanui, New Zealand: water-excess and water-deficient melting to 30 kilobars. Geol Soc Am Bull 86: 555–570

Mertzman SA Jr (1977) The petrology and geochemistry of the Medicine Lake Volcano, California. Contrib Mineral Petrol 62: 221–247

Mertzman SA Jr (1978) A tschermakite-bearing high-alumina olivine tholeiite from the southern Cascades, California. Contrib Mineral Petrol 67: 261–266

Middlemost EAK (1972) A simple classification of volcanic rocks. Bull Volcanol 36: 382–397

Mihaliková A, Šimová M (1965) Final basalt vulcanism in West Carpathians. Geologicke Pr Spravy 36: 257–264

Milhollen GL, Irving AJ, Wyllie PJ (1974) Melting interval of peridotite with 5.7 percent water to 30 kilobars. J Geol 82: 575–587

Miller TP, Smith RL, Richter DH, Lanphere MH, Dalrymple GB (1980) Volcanic history and geothermal characteristics of Mount Drum volcano, Alaska. J Volcanol Geotherm Res (in press)

Milsom J, Smith IEM (1975) Southeastern Papua: generation of thick crust in a tensional environment? Geology 3: 117–120

Minakami T, Utibori S, Hiraga S, Miyazaki T, Gyoda N, Utsunomiya T (1970a) Seismometrical studies of volcano Asama, Part 1. Seismic and volcanic activities of Asama during 1934–1969. Bull Earthquake Res Inst 48: 235

Minakami T, Utibori S, Miyazaki T, Hiraga S, Terao H, Hirai K (1970b) Seismometrical studies of volcano Asama, part 2. Anomalous distribution of the P arrival times and some information of the velocity of the P wave propagating through the volcano. Bull Earthquake Res Inst 48: 431

Minakami T, Utibori S, Yamaguchi M, Gyoda N, Utsunomiya T, Hagiwara M, Hirai K (1969) The Ebino earthquake swarm and the seismic activity in the Kirisima volcanoes, in 1968–1969, part 1. Bull Earthquake Res Inst 47: 721

Minster JF, Minster JB, Treuil M, Allègre CJ (1977) Systematic use of trace elements in igneous processes, part II. Inverse problem of the fractional crystallization process in volcanic suites. Contrib Mineral Petrol 61: 49–77

Mitronovas W, Isacks B, Seeber L (1969) Earthquake locations and seismic wave propagation in the upper 250 km of the Tonga island arc. Bull Seismol Soc Am 59: 1115–1135

Miyashiro A (1957) Cordierite-indialite relations. Am J Sci 255: 43–62

Miyashiro A (1974) Volcanic rock series in island arcs and active continental margins. Am J Sci 274: 321–355

Miyashiro A (1975) Island arc volcanic rock series: a critical review. Petrologie 1: 177–187

Miyashiro A, Shido F (1975) Tholeiitic and calc-alkalic series in relation to the behaviors of titanium, vanadium, chromium, and nickel. Am J Sci 275: 265–277

Miyashiro A, Shido F (1976) Behavior of nickel in volcanic rocks. In: Aoki H (ed) Volcanoes and tectosphere. Tokyo Univ Press, Tokyo, pp 115–121

Mogi K (1969) Relationship between the occurrence of great earthquakes and tectonic structures. Bull Earthquake Res Inst 47: 429–451

Mogi K (1970) Recent horizontal deformation of the earth's crust and tectonic activity in Japan. Bull Earthquake Res Inst 48: 413–430

Mohr PA, Wood CA (1976) Volcano spacings and lithospheric attenuations in the eastern rift of Africa. Earth Planet Sci Lett 33: 126–144

Molnar P, Sykes LR (1969) Tectonics of the Caribbean and Middle America regions from focal mechanisms and seismicity. Geol Soc Am Bull 80: 1639–1684

Molnar P, Wyss M (1972) Moments, source dimensions and stress drops of shallow focus earthquakes in the Tonga-Kermadec arc. Phys Earth Planet Int 6: 263–278

Montigny R, Javoy M, Allègre CJ (1969) Le problème des andésites. Etude du volcanisme quaternaire du Costa Rica (Amérique centrale) a l'aide des traceurs couples $^{87}Sr/^{86}Sr$ et $^{18}O/^{16}O$. Bull Soc Geol Fr 11: 794–799

Moorbath S, Thorpe RS, Gibson II (1978) Strontium isotope evidence for petrogenesis of Mexican andesites. Nature (London) 271: 437–438

Moore HJ, Arthur DWG, Schaber GG (1978) Yield strengths of flows on the Earth, Mars and Moon. Proc 9th Lunar Planet Sci Conf 3351-78

Moore JC (1975) Selective subduction. Geology 3: 530–532

Moore JG, Fabbi BP (1971) An estimate of the juvenile sulfur content of basalt. Contrib Mineral Petrol 33: 118–127

Moore JG, Melson WG (1969) Nuées ardentes of the 1968 eruption of Mayon volcano, Philippines. Bull Volcanol 33: 600–620

Moore JM (1977) Orogenic volcanism in the Proterozoic of Canada. In: Volcanic regimes in Canada. Geol Assoc Can Spec Pap 16: 127–148

Mooser F, Meyer-Abich H, McBirney A (1958) Catalogue of the active volcanoes of the world, VI. Central America. IAVCEI

Morgan WR (1966) A note on the petrology of some lava types from East New Guinea. J Geol Soc Aust 13: 583–591

Morimoto R (1953) Note on the inclusions of some andesites from Setouchi region, southwestern Japan. Proc 7th Pac Sci Congr 2: 301–307

Morris J, Hart SR (1980) Transverse geochemical variations across the Aleutian arc: Cold Bay to Amak. EOS 61: 400

Muehlenbachs K, Hoering TC (1972) Oxygen isotope geochemistry of some rocks from island arcs. Carnegie Inst Wash Yearb 71: 545–548

Muenow DW, Graham DG, Liu NWK, Delaney JR (1979) The abundance of volatiles in Hawaiian tholeiite submarine basalts. Earth Planet Sci Lett 42: 71–76

Muir TD, Tilley CE (1964) Iron-enrichment and pyroxene fractionation in tholeiites. Geol J 4: 143–156

Murase T, McBirney AR (1973) Properties of some common igneous rocks and their melts at high temperatures. Geol Soc Am Bull 84: 3563–3592

Murauchi S, Den N, Asano S, Hotta H, Yoshii T, Asanuma T, Hagiwara K, Ichikawa K, Sato I, Ludwig WJ, Ewing JI, Edgar NT, Houtz RE (1968) Crustal structure of the Philippine Sea. J Geophys Res 73: 3143–3171

Mysen BO (1976) The role of volatiles in silicate melts: solubility of carbon dioxide and water in feldspar, pyroxene and feldspathoid melts to 30 kb and 1625°C. Am J Sci 276: 969–996

Mysen BO (1977) Solubility of volatiles in silicate melts under the pressure and temperature conditions of partial melting in the upper mantle. In: Dick HJB (ed) Magma genesis. State Oreg Dep Geol Mineral Ind Bull 96: 1–14

Mysen BO (1978) Experimental determination of rare earth element partitioning between hydrous silica melt, amphibole, and garnet peridotite minerals at upper mantle pressures and temperatures. Geochim Cosmochim Acta 42: 1253–1263

Mysen BO (1979) Trace element partitioning between garnet peridotite minerals and water-rich vapor: experimental data from 5 to 30 kbar. Am Mineral 64: 274–287

Mysen BO, Boettcher AL (1975) Melting of a hydrous mantle: II. Geochemistry of crystals and liquids formed by anatexis of mantle peridotite at high pressures and high temperature as a function of controlled activities of water, hydrogen, and carbon dioxide. J Petrol 16: 549–593

Mysen BO, Arculus RJ, Eggler DH (1975) Solubility of carbon dioxide in melts of andesite, tholeiite, and olivine nephelinite composition at 30 kbar pressure. Contrib Mineral Petrol 53: 227–239

Mysen BO, Kushiro I, Fujii T (1978a) Preliminary experimental data bearing on the mobility of H_2O in crystalline upper mantle. Carnegie Inst Wash Yearb 77: 793–797

Mysen BO, Kushiro T, Nicholls IA, Ringwood AE (1974) A possible mantle region for andesite magmas: discussion and replies. Earth Planet Sci Lett 21: 221–229

Mysen BO, Virgo D, Hoover J, Sharma SK (1978b) Experimental data bearing on Eu^{2+}/Eu^{3+} in silicate melts and crystals. Carnegie Inst Wash Yearb 77: 677–682

Nagasawa H (1970) Rare earth concentration in zircon and apatite and their host dacites and granites. Earth Planet Sci Lett 5: 47–51

Nagasawa H, Wakita H (1968) Partition of uranium and thorium between augite and host lavas. Geochim Cosmochim Acta 32: 917–921

Nakada S, Takahashi M (1979) Regional variation in chemistry of the Miocene intermediate to felsic magmas in the Outer Zone of the Setouchi Province of southwest Japan. J Geol Soc Jpn 85: 571–582

Nakamura K (1974) Preliminary estimate of global volcanic production rate. In: Golp JL, Furumoto AS (eds) Utilization of volcanic energy. Sandia Laboratories, Albuquerque, NM, pp 273–285

Nakamura K (1977) Volcanoes as possible indicators of tectonic stress orientations: principle and proposal. J Volcanol Geotherm Res 2: 1–16

Nakamura K, Jacob KH, Davies JN (1977) Volcanoes as possible indicators of tectonic stress orientation – Aleutians and Alaska. P A Geophys 115: 87–112

Nakamura N (1974) Determination of REE, Ba, Fe, Mg, Na, and K in carbonaceous and ordinary chondrites. Geochim Cosmochim Acta 38: 757–775

Nakamura Y (1978) Geology and petrology of Bandai and Nekoma volcanoes. Sci Rep Tohoku Univ Ser III 14: 67–119

Nakamura Y, Kushiro I (1970a) Compositional relations of coexisting orthopyroxene, pigeonite and augite in a tholeiitic andesite from Hakone volcano. Contrib Mineral Petrol 26: 265–275

Nakamura Y, Kushiro I (1970b) Equilibrium relations of hypersthene, pigeonite and augite in crystallizing magmas: microprobe study of a pigeonite andesite from Weiselberg, Germany. Am Mineral 55: 1999–2015

Nakamura Y, Kushiro I (1974) Composition of the gas phase in Mg_2SiO_4-SiO_2-H_2O at 15 kbar. Carnegie Inst Wash Yearb 73: 255–259

Naldrett AJ, Goodwin AM, Fisher TL, Ridler RG (1978) The sulfur content of Archean volcanic rocks and a comparison with ocean floor basalts. Can J Earth Sci 15: 715–728

Negendank JFW (1972) Volcanics of the Valley of Mexico. Part II: The opaque mineralogy. Neues Jahrb Mineral Abh 117: 183–195

Nelson SA, Carmichael ISE (1979) Partial molar volumes of oxide components in silicate liquids. Contrib Mineral Petrol 71: 117–124

Neumann van Padang M (1951) Catalogue of the active volcanoes of the world, part 1, Indonesia. IAVCEI

Neuman van Padang M (1953) Catalogue of the active volcanoes of the world, part II, Philippine Islands and Chochin China. IAVCEI

Newhall CG (1979) Temporal variations in the lavas of Mayan volcano, Philippines. J Volcanol Geotherm Res 6: 61–83

Niazi M, Asudeh I, Ballard G, Jackson J, King G, McKenzie D (1978) The depth of seismicity in the Kermanshah region of the Zagros Mountains (Iran). Earth Planet Sci Lett 40: 270–274

Nicholls IA (1971a) Petrology of Santorini volcano, Cyclades, Greece. J Petrol 12: 67–119

Nicholls IA (1971b) Calcareous inclusions in lavas and agglomerates of Santorini Volcano. Contrib Mineral Petrol 30: 261–276

Nicholls IA (1974) Liquids in equilibrium with peridotitic mineral assemblages at high water pressures. Contrib Mineral Petrol 45: 289–316

Nicholls IA, Harris KL (1980) Experimental rare earth element partition coefficients for garnet, clinopyroxene, and amphibole co-existing with andesitic and basaltic liquids. Geochim Cosmochim Acta 44: 287–308

Nicholls IA, Lorenz V (1973) Origin and crystallization history of Permian tholeiites from the Saar-Nahe Trough, SW Germany. Contrib Mineral Petrol 40: 327–344

Nicholls IA, Ringwood AE (1973) Effect of water on olivine stability in tholeiites and the production of silica-saturated magmas in the island arc environment. J Geol 81: 285–300

Nicholls IA, Whitford DJ (1976) Primary magmas associated with Quaternary volcanism in the western Sunda arc, Indonesia. In: Johnson RW (ed) Volcanism in Australasia. Elsevier, Amsterdam, pp 77–90

Nicholls J, Carmichael ISE (1972) The equilibration temperature and pressure of various lava types with spinel- and garnet-peridotite. Am Mineral 57: 941–959

Nicholls J, Carmichael ISE, Stormer JC Jr (1971) Silica activity and Ptotal in igneous rocks. Contrib Mineral Petrol 33: 1–20

Nielson DR, Stoiber RE (1973) Relationship of potassium content in andesitic lavas and depth to the seismic zone. J Geophys Res 78: 6887–6892

Ninkovich D, Hays JD (1972) Mediterranean island arcs and origin of high potash volcanoes. Earth Planet Sci Lett 16: 331–345

Nishimura S (1970) Disequlibrium of the [238]U series in recent volcanic rocks. Earth Planet Sci Lett 5: 199–206

Noble DC, Bowman HR, Hebert AJ, Silberman ML, Heropoulos CE, Fabbi BP, Hedge CE (1975) Chemical and isotopic constraints on the origin of low-silica latite and andesite from the Andes of central Peru. Geology 3: 501–504

Nockolds SR, Allen R (1956) The geochemistry of some igneous rock series, part 3. Geochim Cosmochim Acta 9: 34–77

Oba (1966) Geology and petrology of Usu volcano, Hokkaido, Japan. J Fac Sci Hokkaido Univ IV 18: 185–236

Ocala LC, Meyer RP (1972) Crustal low velocity zone under the Peru Bolivia Altiplano. Geophys JR Astron Soc 30: 199–209

Ocala LC, Aldrich IT, Gettrust JF, Meyer RP, Ramirez JE (1975) Project Narino I. Crustal structure under southern Columbia-northern Ecuador Andes from seismic refraction data. Bull Seismol Soc Am 65: 1681–1696

Okamoto K (1979) Geochemical study on magmatic differentiation of Asama volcano, central Japan. J Geol Soc Jpn 85: 525–535

Oliver RL (1956) The origin of garnets in the Borrowdale volcanic series and associated rocks, English Lake District. Geol Mag 93: 121–139

O'Nions RK, Carter SR, Evensen NM, Hamilton PJ (1979) Geochemical and cosmochemical applications of Nd isotope analysis. Annu Rev Earth Planet Sci 7: 11–38

O'Nions RK, Carter SR, Cohen RS, Evensen NM, Hamilton PJ (1978) Pb, Nd, and Sr isotopes in oceanic ferromanganese deposits and ocean floor basalts. Nature (London) 273: 435–438

Ono K (1962) Chemical composition of volcanic rocks in Japan. Geol Surv Jpn, 441 pp

Ooshima O (1970) Compositional variation of magnetite during the eruption and its bearing on the stage of crystallization of magma of Futatsu-dake, Haruna volcano. Mineral J 6: 249–263

Osborn EF (1959) Role of oxygen pressure in the crystallization and differentiation of basaltic magma. Am J Sci 257: 609–647

Osborn EF (1962) Reaction series for subalkaline igneous rocks based on different oxygen pressure conditions. Am Mineral 47: 211–226

Osborn EF (1969) The complementariness of orogenic andesite and alpine peridotite. Geochim Cosmochim Acta 33: 307–324

Osborn EF (1976) Origin of calc-alkali magma series of Santorini volcano type in the light of recent experimental phase-equilibrium studies. Proc Int Congr Therm Waters, Geotherm Energy Vulcanism Med Area, Athens, vol III, pp 154–167

Osborn EF (1979) The reaction principle. In: Yoder HS Jr (ed) The evolution of the igneous rocks – fiftieth anniversary perspectives. Univ Press, Princeton, pp 133–170

Osborn EF, Watson EB (1977) Studies of phase relations in subalkaline volcanic rock series. Carnegie Inst Wash Yearb 76: 472–478

Osborn EF, Watson EB, Rawson SA (1979) Composition of magnetite in subalkaline volcanic rocks. Carnegie Inst Wash Yearb 78: 475–481

Ota R, Dincel A (1975) Volcanic rocks of Turkey. Bull Geol Surv Jpn 26: 19–46

Oversby VM (1972) Genetic relationships among the volcanic rocks of Reunion: chemical and lead isotopic evidence. Geochim Cosmochim Acta 36: 1167–1179

Oversby VM, Ewart A (1972) Lead isotopic composition of Tonga-Kermadec volcanics and their petrogentic significance. Contrib Mineral Petrol 37: 181–210

Oversby VM, Gast PW (1970) Isotopic composition of lead from oceanic islands. J Geophys Res 75: 2097–2114

Oxburgh ER, Turcotte DL (1970) Thermal structure of island arcs. Geol Soc Am Bull 81: 1665–1688

Oxburgh ER, Turcotte DL (1971) Origin of paired metamorphic belts and crustal dilation in island arc regions. J Geophys Res 76: 1315–1327

Oyarzun J, Villalobos J (1969) Recopilacion de analisis quimicos de rocas Chilenas. Univ (Santiago) Chile Inst Geol Publ 33: 47

Page RW (1976) Geochronology of igneous and metamorphic rocks in the New Guinea Highlands. Aust Bur Mineral Res Bull 162

Page RW, Johnson RW (1974) Strontium isotope ratios of Quaternary volcanic rocks from Papua New Guinea. Lithos 7: 91–100

Pankhurst RJ (1969) Strontium isotope studies related to petrogenesis in the Caledonian basic igneous province of NE Scotland. J Petrol 10: 115–143

Papazachos BC (1973) Distribution of seismic foci in the Mediterranean and surrounding area and its tectonic implication. Geophys J R Astron Soc 33: 421–430

Papazachos BC, Comninakis PE, Drakopoulos JC (1966) Preliminary results of an investigation of crustal structure in southeastern Europe. Bull Seismol Soc Am 56: 1241–1268

Papike JJ, Cameron KL, Baldwin K (1974) Amphiboles and pyroxenes: characterization of other than quadrilateral components and estimates of ferric iron from microprobe data. Abstr Geol Soc Am 6: 1053–1054

Papike JJ, Hodges FN, Bence AE, Cameron M, Rhodes JM (1976) Mare basalts: crystal chemistry, mineralogy, and petrology. Rev Geophys Space Phys 14: 475–540

Pascal G, Isacks BL, Barazangi M, Dubois J (1978) Precise relocations of earthquakes and seismotectonics of the New Hebrides island arc. J Geophys Res 83: 4957–4973

Paul A, Douglas RW (1965) Ferrous-ferric equilibrium in binary alkali silicate glasses. Phys Chem Glasses 6: 207–211

Paulo A, Narebski W, Bakun-Czubarow N, Prochazka K, Wichrowski Z (1979) Geology, geochemistry and petrogenesis of volcanics of Cotopaxi (Ecuador). Pol Acad Sci (Cracow) Mineral Trans 61: 1–62

Pe GG (1973) Petrology and geochemistry of volcanic rocks of Aegina, Greece. Bull Vulcanol 37: 491–514

Pe GG (1974) Volcanic rocks of Methana, South Aegean Arc, Greece. Bull Volcanol 38: 270–290

Pe GG (1975) Strontium isotope ratios in volcanic rocks from the northwestern part of the Hellenic arc. Chem Geol 15: 53–60

Pe GG, Gledhill A (1975) Strontium isotope ratios in volcanic rocks from the southeastern part of the Hellenic arc. Lithos 8: 209–214

Peacock MA (1931) Classification of igneous rock series. J Geol 39: 54–67

Pearce JA (1980) Geochemical evidence for the genesis and eruptive setting of lavas from Tethyan ophiolite. In: Proc Int Ophiolite Symp (in press)

Pearce JA, Cann JR (1973) Tectonic setting of basic volcanic rocks determined using trace element analyses. Earth Planet Sci Lett 19: 290–300

Pearce JA, Norry MJ (1979) Petrogenetic implications of Ti, Zr, Y, and Nb variations in volcanic rocks. Contrib Mineral Petrol 69: 33–47

Peccerillo A, Taylor SR (1976) Geochemistry of Eocene calc-alkaline volcanic rocks from the Kastamonu area, northern Turkey. Contrib Mineral Petrol 58: 63–81

Peterman ZE, Heming RF (1974) Sr^{87}/Sr^{86} ratios of calcalkalic lavas from the Rabaul caldera, Papua New Guinea. Geol Soc Am Bull 85: 1265–1268

Peterman ZE, Lowder GG, Carmichael ISE (1970a) Sr^{87}/Sr^{86} ratios of the Talasea series, New Britain, Territory of New Guinea. Geol Soc Am Bull 81: 39–40

Peterman ZE, Carmichael ISE, Smith AL (1970b) Sr^{87}/Sr^{86} ratios of Quaternary lavas of the Cascade Range, Northern California. Geol Soc Am Bull 81: 311–318

Peterman ZE, Doe BR, Prostka HJ (1970c) Lead and strontium isotopes in rocks of the Absaroka volcanic field, Wyoming. Contrib Mineral Petrol 27: 121–130

Philpotts AR (1967) Origin of certain iron-titanium oxide and apatite rocks. Econ Geol 62: 303–315

Philpotts AR (1978) Textural evidence for liquid immiscibility in tholeiites. Mineral Mag 42: 417–426

Philpotts JA, Schnetzler CC (1970) Phenocryst-matrix partition coefficients for K, Rb, Sr, and Ba, with applications to anorthosite and basalt genesis. Geochim Cosmochim Acta 36: 1131–1166

Philpotts JA, Martin W, Schnetzler CC (1971) Geochemical aspects of some Japanese lavas. Earth Planet Sci Lett 12: 89–96

Pichler H (1967) Neue Erkenntnisse über Art und Genese des Vulkanismus der Aolischen Inseln (Sizilien). Geol Rundsch 57: 102–126

Pichler H, Kussmaul S (1972) The calc-alkaline volcanic rocks of the Santorini Group (Aegean Sea, Greece). Neues Jahrb Mineral Abh 116: 268–307

Pichler H, Zeil W (1969) Die quartäre „Andesite"-Formation in der Hochkordillere Nord-Chiles. Geol Rundsch 58: 866–903

Pichler H, Zeil W (1972) The Cenozoic rhyolite-andesite association of the Chilean Andes. Bull Volcanol 35: 424–452

Pike RJ (1978) Volcanoes on the inner planets: some preliminary comparisons of gross topography. Proc 9th Lunar Planet Sci Conf, pp 3239–3273

Pilger RH Jr, Henyey TL (1979) Pacific-North American plate interaction and Neogene volcanism in coastal California. Tectonophysics 57: 189–209

Pinkerton H, Sparks RSJ (1978) Field measurements of the rheology of lava. Nature (London) 276: 383–384

Piskunov BN (1975) Vulcanism Bolshoy Kurilskoy Gryady i Petrologiya porod Vysokoglinozemistoy Serii. Novosibirsk, 185 pp

Piwinskii AJ (1968) Experimental studies of igneous rock series: Central Sierra Nevada batholith, California. J Geol 76: 548–570

Plafker G (1976) Tectonic aspects of the Guatemala earthquake of 4 February 1976. Science 193: 1201–1208

Poldervaart A (1955) Chemistry of the earth's crust. Geol Soc Am Mem 62: 119–144

Ponce RF (1976) The Hudson Volcano. In: Proc Symp Andean Antarct Volcanol Probl. IAVCEI, pp 78–87

Powell M (1978) Crystallization conditions of low-pressure cumulate nodules from the Lesser Antilles island arc. Earth Planet Sci Lett 39: 162–172

Powell R, Powell M (1977) Geothermometry and oxygen barometry using co-existing iron-titanium oxides: a reappraisal. Mineral Mag 41: 257–263

Presnall DC (1966) The join forsterite-diopside-iron oxide and its bearing on the crystallization of basaltic and ultramafic magmas. Am J Sci 264: 753–809

Presnall DC, Dixon SA, Dixon JR, O'Donnell TH, Brenner NL, Schrock RL, Dycus DW (1978) Liquidus phase relations on the join diopside-forsterite-anorthite from 1 atm to 20 kbar: their bearing on the generation and crystallization of basaltic magma. Contrib Mineral Petrol 66: 203–220

Prevot M, Mergoil J (1973) Crystallization trend of titanomagnetites in an alkali basalt from Saint-Clement (Massif Central, France). Mineral Mag 39: 474–481

Pringle GJ, Trembath LT, Pajari GJ Jr (1974) Crystallization history of a zoned plagioclase. Mineral Mag 39: 867–877

Pushkar P (1968) Strontium isotope ratios in volcanic rocks of three island arc areas. J Geophys Res 73: 2701–2713

Pushkar P, Stoeser DB (1975) $^{87}Sr/^{86}Sr$ ratios in some volcanic rocks and some semifused inclusions of the San Francisco volcanic field. Geology 3: 669–671

Pushkar P, Steuber AN, Tomblin JF, Julian GM (1973) Strontium isotopic ratios in volcanic rocks from St. Vincent and St. Lucia, Lesser Antilles. J Geophys Res 78: 1279–1287

Rafferty WJ, Heming RF (1979) Quaternary alkalic and sub-alkalic volcanism in South Auckland, New Zealand. Contrib Mineral Petrol 71: 139–150

Rajamani V, Naldrett AJ (1978) Partitioning of Fe, Co, Ni, and Cu between sulfide liquid and basaltic melts and the composition of Ni-Cu sulfide deposits. Econ Geol 73: 82–93

Rea DK, Scheidegger KF (1979) Eastern Pacific spreading rate fluctuation and its relation to area volcanic episodes. J Volcanol Geotherm Res 5: 135–148

Rea WJ (1974) The volcanic geology and petrology of Monteserrat, West Indies. J Geol Soc 130: 341–366

Research Group for Explosion Seismology (1977) Regionality of the upper mantle around northeastern Japan as derived from explosion seismology observations and its seismological implications. Tectonophysics 37: 131–139

Rhodes JM, Dungan MA, Blanchard DP, Long PE (1979) Magma mixing at mid-ocean ridges: evidence from basalts drilled near 22°N on the Mid-Atlantic Ridge. Tectonophysics 55: 35–61

Rice A (1978) Soret convection and isotope ratios. EOS 60: 95

Rikitake T (1969) The undulation of an electrically conductive layer beneath the islands of Japan. Tectonophysics 7: 257–264

Ringwood AE (1975) Composition and petrology of the earth's mantle. McGraw-Hill, New York, 618 pp

Ritchey JL (1980) Divergent magmas at Crater Lake, Oregon: products of fractional crystallization and vertical zoning in a shallow, water-undersaturated chamber. J Volcanol Geotherm Res 7: 373–386

Ritchey JL, Eggler DH (1978) Amphibole stability in a differentiated calc-alkaline magma chamber. an experimental investigation. Carnegie Inst Wash Yearb 77: 790–793

Rittman A (1953) Magmatic character and tectonic position of the Indonesian volcanoes. Bull Volcanol 14: 45

Rittman A (1962) Volcanoes and their activity. Interscience Publ, New York, 305 pp

Roa HM (1976) The Upper Cenozoic volcanism in the Andes of southern Chile (40° 00' to 41° 30' S). In: Proc Symp Andean Antarct Volcanol Probl, IAVCEI, pp 143–171

Robson GR, Tomblin JF (1966) Catalogue of the active volcanoes of the world, XX. West-Indies. IAVCEI

Rodgers KA, Brothers RN, Searle EJ (1975) Ultramafic nodules and their host rocks from Auckland, New Zealand. Geol Mag 112: 163–174

Roedder E (1978) Silicate liquid immiscibility in magmas and the system K_2O-FeO-Al_2O_3-SiO_2: an example of serendipity. Geochim Cosmochim Acta 42: 1597–1618

Roedder RL, Emslie RF (1970) Olivine-liquid equilibrium. Contrib Mineral Petrol 29: 275–289

Roedder RL, Osborn EF (1966) Experimental data for the system MgO-FeO-Fe_2O_3-CaA_{12}-Si_2O_8-SiO_2 and their petrologic implications. Am J Sci 264: 428–480

Roman C (1970) Seismicity in Romania: evidence for the sinking lithosphere. Nature (London) 228: 1176–1178

Rona PA (1978) Criteria for recognition of hydrothermal mineral deposits in oceanic crust. Econ Geol 73: 135–160

Roobol MJ, Smith AL (1976) A comparison of the recent eruptions of Mt. Pelée, Martinique and Soufrière, St. Vincent. Bull Volcanol 39: 214–254

Roobol MJ, Ridley WI, Rhodes MJ, Walker GPL, Francis RW, Cobbley T (1976) Physicochemical characters of the Andean volcanic chain between latitudes 21° and 22° S. In: Proc Symp Andean Antarct Volcanol Probl, IAVCEI, pp 450–464

Rose WI Jr (1972a) Santiaguito volcanic dome, Guatemala. Geol Soc Am Bull 83: 1413–1434

Rose WI Jr (1972b) Notes on the 1902 eruption of Santa Maria volcano, Guatemala. Bull Volcanol 36: 29–45

Rose WI Jr (1973) Pattern and mechanism of volcanic activity at the Santiaquito volcanic dome, Guatemala. Bull Volcanol 37: 73–94

Rose WI Jr, Anderson AT Jr, Woodruff LG, Bonis SB (1978) The October 1974 basaltic tephra from Fuego volcano: description and history of the magma body. J Volcanol Geotherm Res 4: 3–54

Rose WI Jr, Grant NK, Hahn GA, Lange IM, Powell JL, Easter J, De Graff JM (1977) The evolution of Santa Maria Volcano, Guatemala. J Geol 85: 63–88

Ross M, Huebner JS (1979). Temperature-composition relationships between naturally occurring augite, pigeonite, and orthopyroxene at one bar pressure. Am Mineral 64: 1133–1155

Rowley K (1978) Late Pleistocene pyroclastic deposits of Soufrière volcano, St. Vincent, West Indies. Geol Soc Am Bull 89: 825–835

Sacks IS, Okada H (1974) A comparison of the anelasticity structure beneath western South America and Japan. Phys Earth Planet Int 9: 211–220

Sakhno VG, Lagovskaya YeA (1970) Garnets in Late Mesozoic extrusives of the Badzhal'skiy Range (Khabarovsk Kray). Dokl Akad Nauk SSSR 193: 122–123

Sakuyama M (1977) Lateral variation of H_2O contents in Quaternary magma of northeastern Japan. Bull Volcanol Soc Jpn 22: 263–271

Sakuyama M (1978) Evidence of magma mixing: petrological study of Shiroumaoike calc-alkaline andesite volcano, Japan. J Volcanol Geotherm Res 5: 179–208

Sakuyama M, Kushiro I (1979) Vesiculation of hydrous andesite melt and transport of alkalies by separated vapor phase. Contrib Mineral Petrol 71: 61–66

Sass JH, Munroe RJ (1970) Heat flow from deep boreholes on two island arcs. J Geophys Res 75: 4387–4395

Sato H (1975) Diffusion coronas around quartz xenocrysts in andesites and basalt from Tertiary volcanic regions in northeastern Shikoku, Japan. Contrib Mineral Petrol 50: 49–64

Sato K (1975) Unilateral isotopic variation of Miocene oreleads from Japan. Econ Geol 70: 800–805

Sato K, Sato J (1977) Estimation of gas-releasing efficiency of erupting magmas from $^{226}Ra/^{222}Rn$ disequilibrium. Nature (London) 266: 439–440

Saunders AD, Tarney J (1979) The geochemistry of basalts from a back-arc spreading center in the East Scotia Sea. Geochim Cosmochim Acta 43: 555–572

Saunders AD, Tarney J, Weaver D (1980) Transverse geochemical variations across the Antarctic Peninsula: implications for the genesis of calcalkaline magmas. Earth Planet Sci Lett 46: 344–360

Scarfe CM (1973) Viscosity of basic magmas at a varying pressure. Nature Phys Sci 241:101–102

Scarfe CM, Mysen BO, Rai CS (1979) Invariant melting behavior of mantle material: partial melting of two lherzolite nodules. Carnegie Inst Wash Yearb 78: 498–502

Scheidegger KF, Kulm LD (1975) Late Cenozoic volcanism in the Aleutian arc: information from ash layers in the northeastern Gulf of Alaska. Geol Soc Am Bull 86: 1407–1412

Schilling J-G, Unni CK, Bender ML (1978) Origin of chlorine and bromine in the oceans. Nature (London) 273: 631–636

Schmucker U (1969) Conducting anomalies, with special reference to the Andes. In: Runcorn SD (ed) The application of modern physics to the earth and planetary interiors. Wiley-Interscience, New York, pp 125–138

Schnetzler CC, Philpotts JA (1970) Partition coefficients of rare-earth elements between igneous matrix material and rock-forming mineral phenocrysts – II. Geochim Cosmochim Acta 34: 331–340

Schock HH (1979) Distribution of rare-earth and other trace elements in magnetites. Chem Geol 26: 119–133

Scholl DW, Marlow MS (1974) Sedimentary sequence in modern Pacific trenches and the deformed circum-Pacific eugeosyncline. Soc Econ Paleontol Mineral Spec Publ 19:193–211

Scholl DW, Marlow MS, Cooper AK (1977) Sediment subduction and offscraping at Pacific margins. In: Talwani M, Pitman WC III (eds) Island arc, deep sea trenches, and back-arc basins. AGU, pp 199–210

Schweickert RA, Cowan DS (1975) Early Mesozoic tectonic evolution of the western Sierra Nevada, California. Geol Soc Am Bull 86: 1329–1336

Segawa J, Tomada Y (1976) Gravity measurements near Japan and study of the upper mantle beneath the oceanic trench – marginal sea transition zones. Am Geophys Union Monogr 19: 35–52

Sekine T, Katsura T, Aramaki S (1979) Water-saturated phase relations of some andesites with application to the estimation of the initial temperature and water pressure at the time of eruption. Geochim Cosmochim Acta 43: 1367–1376

Seno T, Kurita K (1979) Focal mechanisms and tectonics in the Taiwan-Philippine region. In: Uyeda S, Murphy RW, Kobayashi K (eds) Geodynamics of the Western Pacific. Jpn Sci Soc Press, pp 249–262

Shaw DM (1968) A review of K-Rb fractionation trends by covariance analysis. Geochim Cosmochim Acta 32: 573–601

Shaw HR (1963) Obsidian-H_2O viscosities at 1000 and 2000 bars in the temperature range $700°$ to $900°C$. J Geophys Res 68: 6337–6343

Shaw HR (1969) Rheology of basalt in the melting range. J Petrol 10: 510–535

Shaw HR (1974) Diffusion of H_2O in granite liquids. Part I. Experimental data, part II. Mass transfer in magma chambers. In: Hoffman AW et al (eds) Geochemical transport and kinetics. Carnegie Inst, Washington, pp 139–170

Shaw HR, Wright TL, Peck DI, Okamura R (1968) The viscosity of basaltic magma: an analysis of field measurements in Makaopuhi lava lake, Hawaii. Am J Sci 266: 225–264

Sheppard RA (1967) Petrology of a late Quaternary potassium-rich andesite flow from Mt. Adams, Washington. US Geol Surv Prof Pap 575C: 55–63

Shimizu N (1969) Lead isotopic studies on granitic rocks of the Abukuma and Sidara areas in the Ryoke-Abukuma metamorphic belt, central Japan. Geochem J 3: 25–34

Shiono K (1974) Travel time analysis of relatively deep earthquakes in southwest Japan with special reference to the underthrusting of the Philippine Sea plate. J Geosci Osaka Univ 18: 37–59

Shiraki K, Kuroda N, Maruyama S, Urano H (1978) Evolution of the Tertiary volcanic rocks in the Izu-Mariana arc. Bull Volcanol 41: 548–562

Shiraki N, Kuroda N, Urano H, Maruyama S (1980) Clinoenstatite in boninites from the Bonin Islands, Japan. Nature (London) 285: 31–32

Shor GG Jr, Kirk HK, Menard HW (1971) Crustal structure of the Melanesian area. J Geophys Res 76: 2562–2586

Shuto K (1974) The strontium isotopic study of the Tertiary acid rocks from the southern part of northeast Japan. Sci Rep Tokyo Kyoiku Daigaku Ser C 12: 75–140

Sibley DF, Vogel TA, Walker BM, Byerly G (1976) The origin of oscillatory zoning in plagioclase: a diffusion and growth-controlled model. Am J Sci 276: 275–284

Siegers A, Pichler H, Zeil W (1969) Trace element abundances in the "Andesite" formation of northern Chile. Geochim Cosmochim Acta 33: 882–887

Sigvaldason GE (1974) The petrology of Hekla and origin of silicic rocks in Iceland. In: The eruption of Hekla, 1947–1948, vol I. Visindafelag Islendinga, Reykjavik, pp 1–44

Sigvaldason GE, Oskarsson N (1976) Chlorine in basalts from Iceland. Geochim Cosmochim Acta 40: 777–789

Sillitoe RH (1972) Relation of metal provinces in western America and the subduction of oceanic lithosphere. Geol Soc Am Bull 83: 813–818

Sillitoe RH (1976) Andean mineralization: a model for the metallogeny of convergent plate margins. In: Strong DF (ed) Metallogeny and plate tectonics. Geol Assoc Can Spec Pap 14: 59–100

Sillitoe RH (1977) Metallic mineralization affiliated to subaerial volcanism: a review. In: Volcanic processes in ore genesis. Geol Soc London Spec Publ 7: 99–116

Silver EA (1971) Small plate tectonics in the northeastern Pacific ocean. Geol Soc Am Bull 82: 3491–3496

Silver EA, Moore JC (1978) The Molucca Sea collision zone, Indonesia. J Geophys Res 83: 1681–1691

Silver EA, Case JE, Macgillavry HJ (1975) Geophysical study of the Venezuelan borderland. Geol Soc Am Bull 86: 213–226

Sinha AK, Hart SR (1972) A geochemical test of the subduction hypothesis for generation of island arc magmas. Carnegie Inst Wash Yearb 71: 309–312

Smith AL, Carmichael ISE (1968) Quaternary lavas from the southern Cascades, western U.S.A. Contrib Mineral Petrol 19: 212–238

Smith IEM (1976) Peralkaline rhyolites from the D'Entrecasteaux Islands, Papua New Guinea. In: Johnson RW (ed) Volcanism in Australasia. Elsevier, Amsterdam, pp 275–286

Smith IEM, Davies HI (1976) Geology of the Southeast Papuan Mainland. Aust Bur Mineral Res Bull 165

Smith JV (1974) Feldspar minerals, vol II. Springer, Berlin Heidelberg New York, 690 pp

Snoke JA, Sacks IS, Okada H (1977) Determination of the subducting lithosphere boundary by use of converted phases. Seismol Soc Am Bull 67: 1051–1060

Snoke JA, Sacks IS, James D (1979) Subduction beneath western South America: evidence from converted phases. Geophys JR Astron Soc 59: 219–225

Snyder GL (1959) Geology of Little Sitkin Island, Alaska. US Geol Surv Bull 1028-H: 169–210

Snyder WS, Dickinson WR, Silberman ML (1976) Tectonic implications of space-time patterns of Cenozoic magmatism in the western United States. Earth Planet Sci Lett 32: 91–106

Souther JG (1970) Volcanism and its relationship to recent crustal movements in the Canadian Cordillera. Can J Earth Sci 7: 553–568

Souther JG (1977) Volcanism and tectonic environments in the Canadian Cordillera – a second look. In: Baragar WRA, Coleman LC, Hall JM (eds) Volcanic regimes in Canada. Geol Assoc Can Spec Pap 16: 3–24

Sparks SRJ, Sigurdsson H, Wilson L (1977) Magma mixing: a mechanism for triggering acid explosive eruptions. Nature (London) 267: 315–318

Spence W (1977) The Aleutian arc: tectonic blocks, episodic subduction, strain diffusion, and magma generation. J Geophys Res 82: 213–230

Spera F (1980) Thermal evolution of plutons: a parameterized approach. Science 207: 299–301

Spooner ETC (1976) The strontium isotopic composition of sea water and sea water-oceanic crust interaction. Earth Planet Sci Lett 31: 167–174

Stanton RL (1967) A numerical approach to the andesite problem. Koninkl Nederl Akad Wet Proc Ser B 70: 176–216

Stanton RL, Bell JD (1969) Volcanic and associated rocks of the New Georgia Group, British Solomon Islands Protectorate. Overseas Geol Mineral Resources 10: 113–145

Stauder W (1975) Subduction of the Nazca plate under Peru as evidenced by focal mechanisms and by seismicity. J Geophys Res 80: 1053–1064

Steinberg GS, Zubin MI (1965) Geological structure of the Avachinsky Group of volcanoes according to geophysical data. Bull Volcanol 28: 1–8

Steiner A (1958) Petrogenetic implications of the 1954 Ngauruhoe lava and its xenoliths. N Z J Geol Geophys 1: 325–363

Stern CR, Wyllie PJ (1978) Phase compositions through crystallization intervals in basalt-andesite-H_2O at 30 kb with implications for subduction zone magmas. Am Mineral 63: 641–663

Stern CR, Huang W, Wyllie PJ (1975) Basalt-andesite-rhyolite-H_2O: crystallization intervals with excess H_2O and H_2O-undersaturated liquidus surfaces to 35 kilobars, with implications for magma genesis. Earth Planet Sci Lett 28: 189–196

Stern CR, Skewes MA, Duran MA (1976) Volcanismo orogenico en Chile. Austral Prim Congr Geol Chileno

Stern RJ (1979) On the origin of andesite in the Northern Mariana Island Arc: implications from Agrigan. Contrib Mineral Petrol 68: 207–219

Stewart DC (1975) Crystal clots in calc-alkaline andesites as breakdown products of high-Al amphiboles. Contrib Mineral Petrol 53: 195–204

Stewart DC, Thornton CP (1975) Andesites in oceanic regions. Geology 3: 565–584

Stipp JJ (1968) The geochronology and petrogenesis of the Cenozoic volcanics of the North Island, New Zealand. Unpubl Ph D Thesis, ANU

St John VP (1970) The gravity field and structure of Papua and New Guinea. J Aust Petrol Explor Assoc 10: 41–55

Stoiber RE, Carr MJ (1973) Quaternary volcanic and tectonic segmentation of Central America. Bull Volcanol 37: 304–325

Stoiber RE, Jepsen A (1973) Sulfur dioxide contributions to the atmosphere by volcanoes. Science 182: 577–578

Stoiber RE, Rose WI Jr (1973) Cl, F, and SO_2 in Central American volcanic gases. Bull Volcanol 37: 454–460

Streckeisen A (1979) Classification and nomenclature of volcanic rocks, lamprophyres, carbonatites, and melilitic rocks: Recommendations and suggestions of the IUGS Subcommission of the Systematics of Igneous Rocks. Geology 7: 331–335

Studt FE, Thompson GEK (1969) Geothermal heat flow in the North Island of New Zealand. N Z J Geol Geophys 12: 673–683

Sugimura A (1960) Zonal arrangement of some geophysical and petrological features in Japan and its environs. J Fac Sci Tokyo Univ Ser 2 12: 133–153

Sugimura A (1968) Spatial relations of basaltic magmas in island arcs. In: Hess HH, Poldevaart A (eds) Basalts, vol II. Wiley Interscience Publ, New York, pp 537–571

Sugimura A (1973) Multiple correlation between composition of volcanic rocks and depth of earthquake foci. In: Coleman P (ed) The Western Pacific: Island arcs, marginal seas and geochemistry. W Aust Univ Press, pp 471–482

Sugimura A, Uyeda S (1973) Island arcs: Japan and its environs. Elsevier, Amsterdam, 247 pp

Sugisaki R (1972) Tectonic aspects of Andesite Line. Nature Phys Sci 240: 109–111

Sugisaki R (1976) Chemical characteristics of volcanic rocks: relation to plate movements. Lithos 9: 17–30

Sugiura T (1968) Bromine to chlorine ratios in igneous rocks. Bull Chem Soc Jpn 41:1133–1139

Sun S (1980) Lead isotopic study of young volcanic rocks from mid-ocean ridges, oceanic islands, and island arcs. Philos Trans R Soc London A 297: 409–445

Sun S, Nesbitt RW (1978) Geochemical regularities and genetic significance of ophiolitic basalts. Geology 6: 689–693

Suyehiro K, Sacks IS (1978) An anomalous low-velocity region above the deep earthquakes in the Japan subduction zone. Carnegie Inst Wash Yearb 77: 505–511

Sykes LR (1966) The seismicity and deep structure of island arcs. J Geophys Res 71:2981–3006

Sykes LR, Ewing M (1965) The seismicity of the Caribbean region. J Geophys Res 70: 5065–5074

Tabor RW, Crowder DF (1969) On batholiths and volcanoes-intrusion and eruption of late Cenozoic magmas in the Glacier Peak area, North Cascade, Washington. US Geol Surv Prof Pap 604: 67 pp

Tagiri M, Onuki H, Yamazaki T (1975) Mineral paragenesis of argillaceous xenoliths in andesite rocks from Nijo-san and Amataki-yama districts, southwest Japan. J Jpn Assoc Mineral Pet Econ Geol 70: 305–314

Takahashi E (1978) Petrological model of the upper mantle and lower crust of the island arc: petrology of mafic and ultramafic xenoliths in Cenozoic alkali basalts of the Oki-Dogo Island in the Japan Sea. Bull Volcanol 41: 529–547

Takeshita H, Oji Y (1968) Hornblende gabbroic inclusions in the calcalkaline andesites from the northern district of Nagano Prefecture, Japan, I., II. J Jpn Assoc Mineral Pet Econ Geol 60: 1–26, 57–74

Takeuchi A (1978) The Pliocene stress field and tectonism in the Shin-Etsu region, central Japan. J Geosci Osaka Univ 21: 37–52

Taneda S (1977) Significance of the change of magma composition and the upward moving of the magma chamber of Sakura-jima volcano. Bull Volcanol Soc Jpn 22: 61–64

Taniguchi H (1972) Petrological study on Towada volcano. J Jpn Assoc Mineral Pet Econ Geol 67: 128–138

Tarney J, Saunders AD, Weaver SD (1977) Geochemistry of volcanic rocks from the island arcs and marginal basins of the Scotia Arc region. In: Talwani M, Pitman WC III (eds) Island arcs, deep sea trenches, and back-arc basins. AGU, pp 367–378

Tatsumoto M (1969) Lead isotopes in volcanic rocks and possible ocean floor thrusting beneath island arcs. Earth Planet Sci Lett 6: 369–376

Tatsumoto M (1978) Isotopic composition of lead in oceanic basalt and its implication to mantle evolution. Earth Planet Sci Lett 38: 63–87

Tatsumoto M, Knight RJ (1969) Isotopic composition of lead in volcanic rocks from central Honshu – with regard to basalt genesis. Geochem J 3: 53–86

Taylor GA (1958) The 1951 eruption of Mount Lamington, Papua. Aust B M R Bull 38

Taylor HP Jr (1968) The oxygen isotope geochemistry of igneous rocks. Contrib Mineral Petrol 19: 1–71

Taylor HP Jr (1980) The effects of assimilation of country rocks by magmas on $^{18}O/^{16}O$ and $^{87}Sr/^{86}Sr$ systematics in igneous rocks. Earth Planet Sci Lett 47: 243–254

Taylor RW (1963) Liquidus temperatures in the system $FeO-Fe_2O_3-TiO_2$. J Am Ceramic Soc 46: 276–279

Taylor RW (1964) Phase equilibria in the system $FeO-Fe_2O_3-TiO_2$ at $1300°C$. Am Mineral 49: 1016–1030

Taylor SR (1967) The origin and growth of continents. Tectonophysics 4: 17–34

Taylor SR (1969) Trace element chemistry of andesites and associated calc-alkaline rocks. In: McBirney AR (ed) Proc Andesite Conf. Dep Geol Mineral Res Oreg Bull 65: 43–64

Taylor SR, Hallberg JA (1978) Rare-earth elements in the Marda calc-alkaline suite: an Archean geochemical analogue of Andean-type volcanism. Geochim Cosmochim Acta 41: 1125–1129

Taylor SR, White AJR (1966) Trace element abundances in andesites. Bull Volcanol 29: 177–194

Taylor SR, Capp AC, Graham AL, Blake DH (1969a) Trace element abundances in andesites, II. Saipan, Bougainville and Fiji. Contrib Mineral Petrol 23: 1–26

Taylor SR, Kaye M, White AJR, Duncan AR, Ewart A (1969b) Genetic significance of Co, Cr, Ni, Sc, and V in andesites. Geochim Cosmochim Acta 33: 275–286

Thiele R, Katsui Y (1969) Contribucion al conocimiento del volcanism post-Miocenico de los Andes en la provincia de Santiago, Chile. Univ (Santiago) Chile Inst Geol Publ 35: 3–32

Thompson G, Bryan WB, Frey FA, Sung CM (1974) Petrology and geochemistry of basalts and related rocks from Sites 214, 215, 216, DSDP Leg 22, Indian Ocean. Initial Rep DSDP 22: 459–468

Thompson RN (1972) Evidence for a chemical discontinuity near the basalt-andesite transition in many anorogenic volcanic suites. Nature (London) 236: 106–110

Thompson RN (1973) One-atmosphere melting behavior and nomenclature of terrestrial lavas. Contrib Mineral Petrol 41: 197–204

Thompson RN (1975) The 1-atmosphere liquidus oxygen fugacities of some tholeiitic intermediate, alkalic, and ultra-alkalic lavas. Am J Sci 275: 1049–1072

Thorarinsson S (1967) The eruptions of Hekla in historical times. The eruption of Hekla 1947–1948, vol I. Soc Sci Islandica, Reykjavik, pp 1–170

Thorarinsson S, Sigvaldason GE (1972) The Hekla eruption of 1970. Bull Volcanol 36: 269–288

Thorpe RS (1977) Tectonic significance of alkaline volcanism in eastern Mexico. Tectonophysics 40: 719–726

Thorpe RS, Francis PW (1976) Volcan Ceboruco: a major composite volcano of the Mexican volcanic belt. Bull Volcanol 39: 201–213

Thorpe RS, Francis PE (1979) Petrogenetic relationships of volcanic and intrusive rocks of the Andes. In: Atherton MP, Tarney J (eds) Origin of granite batholiths geochemical evidence. Shiva Publ, Kent, pp 65–75

Thorpe RS, Potts PJ, Francis PW (1976) Rare earth data and petrogenesis of andesite from the North Chilean Andes. Contrib Mineral Petrol 54: 65–78

Thorpe RS, Francis PW, Moorbath S (1979) Strontium isotope evidence for petrogensis of Central American andesites. Nature (London) 277: 44–45

Tilley CE (1950) Some aspects of magmatic evolution. J Geol Soc 106: 37–61

Tilley CE, Yoder HS Jr, Schairer JF (1967) Melting relations of volcanic rock series. Carnegie Inst Wash Yearb 65: 260–269

Tilley CE, Yoder HS Jr, Schairer JF (1968) Melting relations of igneous rocks. Carnegie Inst Wash Yearb 66: 450–457

Tilton GR (1979) Isotopic studies of Cenozoic Andean calc-alkaline rocks. Carnegie Inst Wash Yearb 78: 298–304

Tokarev PI, Zobin VM (1970) Peculiarities of seismic wave distribution of near earthquakes within the earth's crust and upper mantle in the region of the Klyuchevskaya group of volcanoes, Kamchatka. Bull Vulkanol Stantsiy SSSR 46: 17–23

Toksöz MN, Minear JW, Julian BR (1971) Temperature field and geophysical effects of a downgoing slab. J Geophys Res 76: 1113–1138

Toksöz MN, Sleep NH, Smith AT (1973) Evolution of the downgoing lithosphere and the mechanism of deep focus earthquakes. Geophys J R Astron Soc 35: 285–310

Tomita K (1964) The existence of oxyhornblende built up from two different lattices. Mem Coll Sci Kyoto Univ Ser B 30: 1–5

Tomita T (1935) On the chemical compositions of the Cenozoic alkaline suite of the circum-Japan Sea region. J Shanghai Sci Inst, Sect II, 1: 227–306

Tomkeiff SI (1949) The volcanoes of Kamchatka. Bull Volcanol 8: 87

Tovish A, Schubert G (1978) Island arc curvature, velocity of convergence, and angle of subduction. Geophys Res Lett 5: 329–332

Tsuya H (1955) Geological and petrological studies of volcano Fuji. V. On the 1707 eruption. Bull Earthquake Res Inst 33: 341–384

Turcotte DL, Ahren JL (1978) A porous flow model for magma migration in the asthenosphere. J Geophys Res 83: 767–772

Turcotte DL, Schubert G (1973) Frictional heating of the descending lithosphere. J Geophys Res 78: 5876–5886

Tuttle OF, Bowen NL (1958) Origin of granite in light of experimental studies in the system $NaAlSi_3O_8$-$KAlSi_3O_8$-SiO_2-H_2O. Geol Soc Am Mem 74

Ui T (1972) Recent volcanism in Masaya-Granada area, Nicaragua. Bull Volcanol 36: 176–190

Ui T, Aramaki S (1978) Relationship between chemical composition of Japanese island-arc volcanic rocks and gravimetric data. Tectonophysics 45: 249–259

Ujike O (1974) Post-eruption oxidation of hornblende phenocryst from Kaitaku, Shodo-shima, Japan. J Jpn Assoc Mineral Pet Econ Geol 69: 426–433

Ujike O (1975) Petrogenetic significance of normative corundum in calc-alkaline volcanic rock series. J Jpn Assoc Mineral Pet Econ Geol 70: 85–92

Ujike O (1977) Chemical compositions of amphibole phenocrysts in calcalkaline volcanic rocks: a compilation of 95 analyses. J Jpn Assoc Mineral Pet Econ Geol 72: 85–93

Ujike O, Onuki H (1976) Phenocrystic hornblende from Tertiary andesites and dacites, Kagawa Prefecture, Japan. J Jpn Assoc Mineral Pet Econ Geol 71: 389–399

Untung M, Sato M (1978) Gravity and geological studies in Java, Indonesia. Geol Surv Indonesia Spec Publ 6

Utnasin VK, Abdurakhimov AI, Anasov GI, Balesta ST, Budyanskiy YuA, Markhinin YeK, Fedorenko VI (1975) Deep structure of Klyuchevskoy group of volcanoes and problem of magmatic hearths. Int Geol Rev 17: 791–806

Utsu T (1971) Seismological evidence for anomalous structure of island arcs with special reference to the Japanese region. Rev Geophys Space Phys 9: 839–890

Uyeda S, Horai K (1964) Terrestrial heat flow in Japan. J Geophys Res 69: 2121–2141

Uyeda S, Kanamori H (1979) Back-arc opening and the mode of subduction. J Geophys Res 84: 1049–1061

Uyeda S, Miyashiro A (1974) Plate tectonics and the Japanese Islands: a synthesis. Geol Soc Am Bull 85: 1159–1170

Vance JA (1965) Zoning in igneous plagioclase: patchy zoning. J Geol 73: 636–651

Vergara M (1969) Rocas volcanicas y sedimentario-volcanicos, Mesozoicas y Cenozoicas, en la latitud 34° 30' S, Chile. Univ (Santiago) Chile Inst Geol Publ 32: 36 pp

Vergara M, Katsui Y (1969) Contribution a la geologia y petrologia del volcan Antuco, Cordillera de los Andes, Chile central. Univ (Santiago) Chile Inst Geol Publ 35: 25–47

Vergara M, Munizaga F (1974) Age and evolution of the Upper Cenozoic andesite volcanism in central-south Chile. Geol Soc Am Bull 85: 603–606

Verhoogen J (1937) Mount Saint Helens, a recent Cascade volcano. Univ Calif Publ Geol Sci Bull 24: 263–302

Verhoogen J (1951) Mechanics of ash formation. Am J Sci 249: 729–739

Villari L (1972) L'isola di Filicudi ed il suo significato magmatologico. Rend Soc Ital Mineral Pet 28; 475–506

Villari L, Nathan S (1978) Petrology of Filicudi, Aeolian Archipelago. Bull Volcanol 41: 81–96

Virgo D, Ross M (1973) Pyroxenes from Mull andesites. Carnegie Inst Wash Yearb 72: 535–540

Vlodavetz VI, Piip BI (1959) Catalogue of the active volcanoes of the world, part 8, Kamchatka and continental areas of Asia. IAVCEI

Vogt PR (1972) Evidence for global synchronism in mantle plume convection, and possible significance for geology. Nature (London) 240: 338–342

Vogt PR (1974) Volcano spacing, fractures, and thickness of the lithosphere. Earth Planet Sci Lett 21: 235–251

Vogt PR, Lowrie A, Bracey DR, Hey RW (1976) Subduction of aseismic oceanic ridges: effects on shape, seismicity, and other characteristics of consuming plate boundaries. Geol Soc Am Spec Pap 172

Von Buch L (1836) Über Erhebungskrater und Vulkane. Ann Phys 2nd Ser 37: 169–190

Von der Borch CC (1979) Continent-island arc collision in the Banda arc. Tectonophysics 54: 169–193

Von Huene R, Moore GW, Moore JC (1979) Cross section, Alaska Peninsula-Kodiak Island, Aleutian Trench. Geol Soc Am Map Chart Ser MC-28 A: Sheet 2

Wadati K (1955) On the activity of deep-focus earthquakes in the Japan Island and neighbourhoods. Geophys Mag 8: 305–325

Waff HS (1974) Theoretical considerations of electrical conductivity in a partially molten mantle and implications for geothermometry. J Geophys Res 79: 4003–4010

Wager LR (1962) Igneous cumulates from the 1902 eruption of Soufrière, St. Vincent. Bull Volcanol 24: 93–99

Wager LR, Deer WA (1939) The petrology of the Skaergaard intrusion, Kangerdlugssuag, East Greenland. Medd Groenl 105: no 4

Wakita H, Fujii N, Matsuo S, Notsu K, Nagao K, Takaoka N (1978) "Helium spots": caused by a diapiric magma from the Upper Mantle. Science 200: 430–432

Walcott RI (1978) Geodetic strains and large earthquakes in the axial tectonic belt of North Island, New Zealand. J Geophys Res 83: 4419–4430

Walker D, Stolper EM, Hays HF (1978) A numerical treatment of melt/solid segregation: size of the eucrite parent body and stability of the terrestrial low-velocity zone. J Geophys Res 83: 6005–6013

Walker D, Kirkpatrick RJ, Longhi J, Hays JF (1976a) Crystallization history of lunar picritic basalt sample 12002: phase-equilibria and cooling-rate studies. Geol Soc Am Bull 87: 646–656

Walker D, Longhi J, Kirckpatrick RJ, Hays JF (1976b) Differentiation of an Apollo 12 picrite magma. Proc 7th Lunar Sci Conf, pp 1365–1389

Walker KR (1969) The Palisades Sill, New Jersey: a reinvestigation. Geol Soc Am Spec Pap 111

Wang Y (1970) Variation of potash content in the Pleistocene andesites from Taiwan. Proc Geol Soc China 13: 41–50

Warden AJ (1967) The geology of the Central Islands. New Hebrides Ann Rep Geol Surv 5: 106 pp

Warren DH, Healy JH (1973) Structure of the crust in the coterminous United States. Tectonophysics 20: 203–213

Watanabe T, Longseth MG, Anderson RN (1977) Heat flow in back-arc basins of the Western Pacific. In: Talwani M, Pitman WC III (eds) Island arcs, deep sea trenches, and back-arc basins. AGU, pp 137–162

Watson EB (1979a) Calcium content of forsterite coexisting with silicate liquid in the system Na_2O-CaO-MgO-Al_2O_3-SiO_2. Am Mineral 64: 824–829

Watson EB (1979b) Zircon saturation in felsic liquids: experimental results and application to trace element geochemistry. Contrib Mineral Petrol 70: 407–419

Weaver SD, Sceal JSC, Gibson IL (1972) Trace element data relevant to the origin of trachytic and pantelleritic lavas in the East African Rift System. Contrib Mineral Petrol 36: 181–194

Weaver SD, Saunders AD, Pankhurst RJ, Tarney J (1979) A geochemical study of magmatism association with the initial stages of back-arc spreading. Contrib Mineral Petrol 68: 151–169

Weertman J (1971) Theory of water-filled crevices in glaciers applied to vertical magma transport beneath oceanic ridges. J Geophys Res 76: 1171–1183

Weill DF, Drake MJ (1973) Europium anomaly in plagioclase feldspar: experimental results and semiquantitative model. Science 180: 1059–1060

Wells PRA (1977) Pyroxene thermometry in simple and complex systems. Contrib Mineral Petrol 62: 129–139

Wendlandt RF, Eggler DH (1980) The origin of potassic magmas, 2. Am J Sci 280: 421–458

Westercamp D (1976) Petrology of the volcanic rocks of Martinique, West Indies. Bull Volcanol 39: 175–200

Westerweld J (1952) Quaternary volcanism on Sumatra. Geol Soc Am Bull 63: 561–594

Weyl R (1961) Die Geologie Mittelamerikas. Borntraeger, Berlin, 226 pp

White AJR, Chappell BW (1977) Ultrametamorphism and granitoid genesis. Tectonophysics 43: 7–22

White CM, McBirney AR (1979) Some quantitative aspects of orogenic volcanism in the Oregon Cascades. Geol Soc Am Mem 152: 369–388

White D, Waring GA (1963) Volcanic emanations. Data of geochemistry. US Geol Surv Prof Pap 440-K: 1–29

Whitford DJ (1975) Strontium isotopic studies of the volcanic rocks of the Sunda Arc, Indonesia, and their petrogenetic implications. Geochim Cosmochim Acta 39: 1287–1302

Whitford DJ, Bloomfield K (1976) Geochemistry of the late Cenozoic volcanic rocks from the Nevado de Toluca area, Mexico. Carnegie Inst Wash Yearb 75: 207–213

Whitford DJ, Jezek P (1979) Origin of late-Cenozoic lavas from the Banda arc, Indonesia: trace element and Sr isotope evidence. Contrib Mineral Petrol 68: 141–150

Whitford DJ, Nicholls IA (1976) Potassium variations in lavas across the Sunda arc in Java and Bali. In: Johnson RW (ed) Volcanism in Australasia. Elsevier, Amsterdam, pp 63–76

Whitford DJ, Foden J, Varne R (1978) Sr isotope geochemistry and alkaline lavas from the Sunda arc in Lombok and Sumbawa, Indonesia. Carnegie Inst Wash Yearb 77: 613–620

Whitford DJ, Nicholls IA, Taylor SR (1979a) Spatial variations in the geochemistry of Quaternary lavas across the Sunda arc in Java and Bali. Contrib Mineral Petrol 70: 341–356

Whitford DJ, Compston W, Nicholls JA, Abbott MJ (1977) Geochemistry of Late-Cenozoic lavas from eastern Indonesia; the role of subducted sediments in petrogenesis. Geology 5: 571–575

Whitford DJ, White WM, Jezek PA, Nicholls IA (1979b) Nd isotope composition of recent andesites from Indonesia. Carnegie Inst Wash Yearb 78: 304–308

Wichmann A (1925) Geologische Ergebnisse der Skiboga Expedition. Skiboga Exped 66

Wiebenga WA (1973) Crustal structure of the New Britain-New Zealand region. In: Coleman PJ (ed) The Western Pacific. Western Australia Univ Press, pp 163–177

Wilcox RE (1954) Petrology of Paricutin Volcano, Mexico. US Geol Surv Bull 965-C

Wilcox WR, Kuo VHS (1973) Gas bubble nucleation during crystallization. J Cryst Growth 19: 221–228

Wilkinson JFG (1971) The petrology of some vitrophyric calc-alkaline volcanics from the Carboniferous of New South Wales. J Petrol 12: 587–619

Wilkinson JFG, Binns RA (1977) Relatively iron-rich lherzolite xenoliths of the Cr-diopside suite: a guide to the primary nature of anorogenic tholeiitic andesite magmas. Contrib Mineral Petrol 65: 199–212

Williams CEF, Warden AJ (1964) Table of chemical analyses. New Hebrides Geol Surv Prog Rept 1959–62

Williams H (1942) Geology of Crater Lake National Park, Oregon. Carnegie Inst Wash Publ No 540: 162 pp

Wilshire HG, Pike JEN (1975) Upper-mantle diapirism: evidence from analogous features in alpine peridotite and ultramafic inclusions in basalt. Geology 3: 467–470

Wise WS (1969) Geology and petrology of the Mt Hood area: a study of High Cascase volcanism. Geol Soc Am Bull 80: 969–1006

Wood BJ, Banno S (1973) Garnet-orthopyroxene and orthopyroxene-clinopyroxene relationships in simple and complex systems. Contrib Mineral Petrol 42: 109–124

Wood CA (1978) Morphometric evolution of composite volcanoes. Geophys Res Lett 5: 437–439

Wood CP (1974) Petrogenesis of garnet-bearing rhyolites from Canterbury, New Zealand. N Z J Geol Geophys 17: 759–788

Wood DA, Jron J-L, Treuil M, Norry M, Tarney J (1979a) Elemental and Sr isotope variations in basic lavas from Iceland and the surrounding ocean floor. Contrib Mineral Petrol 70: 319–339

Wood DA, Joron J-L, Treuil M (1979b) A re-appraisal of the use of trace elements to classify and discriminate between magma series erupted in different tectonic settings. Earth Planet Sci Lett 45: 326–336

Wood DA, Joron J-L, Marsh NG, Tarney J, Treuil M (1980) Major and trace element variations in basalts from the North Philippine Sea DSDP Leg 58: a comparative study of back-arc-basin basalts with lava series from Japan and mid-ocean ridges. Initial Rep DSDP Leg 58, pp 873–894

Woodruff LG, Rose WI Jr, Rigot W (1979) Contrasting fractionation patterns for sequential magmas from two calc-alkaline volcanoes in Central America. J Volcanol Geotherm Res 6: 217–240

Wright JB (1971) Volcanism and the Earth's crust. In: Gass IG et al (eds) Understanding the Earth. MIT Press, Cambridge, pp 301–313

Wright TL, Fiske RS (1971) Origin of the differentiated and hybrid lavas of Kilauea volcano, Hawaii. J Petrol 12: 1–65

Wright TL, Peck DL, Shaw HR (1976) Kilauea lava lake: natural laboratories for study of cooling, crystallization and differentiation of basaltic magmas. Am Geophys Union Monogr 19: 375–390

Wright TL, Swanson DA, Helz RT, Byerly GR (1979) Major oxide, trace element, and glass chemistry of Columbia River basalt samples collected between 1971 and 1977. US Geol Surv Open File Rep 79-711

Wu FT (1979) Recent tectonics of Taiwan. In: Uyeda S, Murphy RW, Kobayashi K (eds) Geodynamics of the Western Pacific. Jpn Sci Soc Press, pp 265–299

Wyllie PJ (1971) The dynamic earth. Wiley and Sons, New York, 416 pp

Wyllie PJ (1977) From crucibles through subduction to batholiths. In: Saxena SA, Bhattacharji S (eds) Energetics of geological processes. Springer, Berlin Heidelberg New York, pp 389–433

Wyllie PJ (1979) Magmas and volatile components. Am Mineral 64: 469–500

Wyss M (1973) The thickness of deep seismic zones. Nature (London) 242: 255–256

Yagi K, Takeshita H, Oba Y (1972) Petrological study of the 1970 eruption of Akita-Komagatake Volcano, Japan. J Fac Sci Hokkaido Univ Ser IV 15: 109–138

Yajima T, Higuchi H, Nagasawa H (1972) Variation of rare earth concentrations in pigeonitic and hypersthenic rock series from Izu-Hakone region, Japan. Contrib Mineral Petrol 35: 235–244

Yamamoto K, Akimoto S (1977) The system $MgO-SiO_2-H_2O$ at high pressures and temperatures-stability field for hydroxyl-chondrodite, hydroxyl-clinohumite and 10 A-phase. Am J Sci 277: 288–312

Yamashina K, Nakamura K (1978) Correlations between tectonic earthquakes and volcanic activity of Izu-oshima volcano, Japan. J Volcanol Geotherm Res 4: 233–250

Yamazaki T, Onuki H, Tiba T (1966) Significance of hornblende gabbroic inclusions in calc-alkali rocks. J Jpn Assoc Mineral Petrol Econ Geol 55: 87–103

Yanagi T, Ishizaka K (1978) Batch fractionation model for the evolution of volcanic rocks in an island arc: an example from central Japan. Earth Planet Sci Lett 40: 252–262

Yasui Y (1963) Summary of forecasting the explosions of volcano Sakurajima. Geophys Mag 31: 491–504

Yoder HS Jr (1969) Calcalkalic andesites: experimental data bearing on the origin of their assumed characteristics. In: McBirney AR (ed) Proc Andesite Conf. Dep Geol Mineral Res Oreg Bull 65: 43–64

Yoder HS Jr (1973) Contemporaneous basaltic and rhyolitic magmas. Am Mineral 58: 153–172

Yoder HS Jr (1976) Generation of basaltic magma. US Natl Acad Sci, Washington, 265 pp

Yoder HS Jr, Kushiro I (1972) Origin of calc-alkalic peraluminous andesite and dacites. Carnegie Inst Wash Yearb 71: 411–413

Yoder HS Jr, Tilley CE (1962) Origin of basalt magmas: an experimental study of natural and synthetic rock systems. J Petrol 3: 342–532

Yoder HS Jr, Stewart DB, Smith JR (1957) Feldspars. Carnegie Inst Wash Yearb 66: 477–478

Yoshida M, Takahashi K, Yonehara N, Ozawa T, Iwasaki I (1971) The fluorine, chlorine, bromine, and iodine contents of volcanic rocks in Japan. Bull Chem Soc Jpn 44: 1844–1850

Yoshii T (1979) A detailed cross section of the deep seismic zone beneath northeast Honshu, Japan. Tectonophysics 55: 349–360

Zentilli M, Dostal J (1977) Uranium in volcanic rocks from the Central Andes. J Volcanol Geotherm Res 2: 251—258

Zielinski RA, Lipman PW (1976) Trace element variations at Summer Coon volcano, San Juan Mountains, Colorado, and the origin of continental-interior andesite. Geol Soc Am Bull 87: 1477—1485

Zies EG (1946) Temperature measurements at Paricutin volcano. Am Geophys Union Trans 27: 178—180

Zimmerman C, Kudo AM (1979) Geochemistry of andesites and related rocks, New Mexico. In: Riecker RE (ed) Rio Grande Rift: Tectonics and magmatism. AGU, pp 355—381

Zolotarev BP, Sobolev SF (1971) Longitudinal geochemical zonation of the basalt-andesite series in island arcs and its relation to the crust and mantle structure. Dokl Akad Nauk SSSR 197: 222—225

Subject Index

Numbers in *italics* refer to tables or figures
Numbers in parentheses identify volcanoes in the Appendix

Minerals and Rocks

Editor in Chief: P. J. Wyllie
Editors: A. El Goresy, W. von Engelhard,
T. Hahn

A Selection

Volume 9
J. Hoefs
Stable Isotope Geochemistry

2nd completely revised and updated edition.
1980. 52 figures, 23 tables. XII, 208 pages
ISBN 3-540-09917-4

"The book is well written and superbly organized... Hoefs has done well in the pages
allotted to him, and one can only hope that
the editors of this series will continue to update
it..." *American Scientist*

Volume 10
J. T. Wasson
Meteorites

Classification and Properties

1974. 70 figures. X, 136 pages
ISBN 3-540-06744-2

"The book is a very successful attempt at including all aspects of meteorite studies in one
thin volume. It is almost unbelievable that
scientific material, originating from so many
sciences and of such large extent, can be compressed into one relatively small book. The
author achieved this without detriment to
clearness, and the book is written definitely
and distinctly... (The author) has been one of
the leading specialists in meteorite research
during the past decade. He is responsible for a
good deal of progress in meteorite science. His
survey book on meteorites will be used by
experienced specialists as a good reference
book and by starting specialists and students
as an excellent introduction into meteorite
research. Moreover, it will be a handbook for
everybody, who wants to classify new or unknown meteorite samples."
J. of the British Interplanetary Society

Volume 11
W. Smykatz-Kloss
Differential Thermal Analysis

Application and Results in Mineralogy

1974. 82 figures, 36 tables. XIV, 185 pages
ISBN 3-540-06906-2

"This is a compact, critical, and authoritative
treatment of DTA, ...References are conveniently listed alphabetically and the subjects are
well indexed...
The book will be useful to materials scientists
and mineralogists and invaluable to specialists
in raw materials and thermochemistry."
Ceramic Abstracts

Springer-Verlag
Berlin
Heidelberg
New York

Volume 12
R. G. Coleman

Ophiolites

Ancient Oceanic Lithosphere?

1977. 72 figures, 18 tables. IX, 229 pages
ISBN 3-540-08276-X

"...The general style and presentation of the subject is attractive and very readable with considerable balance in subject matter. There are chapters on igneous and metamorphic petrology, ore deposits associated with ophiolites, and the geologic character, plate tectonics, and emplacement tectonics of ophiolites. The book concludes with a lucid presentation of the complex geologic, tectonic, and petrologic nature of four examples: the Bay of Islands, Troodos, Semail, and Papua ophiolites...
Students of ophiolites will find the book especially valuable for its focus on the principal areas of needed research...
The authoritative manner in which the author has outlined the present status of research on ophiolites should help to influence and mould the course of investigations over the next several years... As a guide for future studies of these rocks, *Ophiolites* is an invaluable book for careful reading and consideration. It also serves as an excellent introduction to the subject for the student as well as for the general reader." *The American Mineralogist*

Volume 14
A. K. Gupta, K. Yagi

Petrology and Genesis of Leucite-Bearing Rocks

1980. 99 figures, 43 tables. XV, 252 pages
ISBN 3-540-09864-X

This volume is a much needed review-synthesis of the extensive geochemical, petrological and experimental studies on leucite-bearing mafic and ultramafic rocks.
The first five chapters of the book summarize the mineralogy, major and minor element geochemistry, strontium and oxygen isotopic studies, distribution, and conditions surrounding the formation of leucitic rocks. The next twelve chapters present a detailed account of the phase equilibria studies of synthetic and natural leucite-bearing rock systems in air and under variable pressures (in presence or absence of water), analcitization of leucite, formation of pseudoleucite, genetic relationship between kimberlites and leucitic rocks, and structure and tectonic control of volcanism associated with leucite-bearing rocks. In the last chapter, trace element geochemistry as well as field and laboratory data are used to elucidate the origin of this interesting suite of rocks.
This book will be of great value to a wide range of earth scientists and advanced students with particular interest in mineralogy, geochemistry and experimental petrology.

Volume 13
M. S. Paterson

Experimental Rock Deformation – The Brittle Field

1978. 56 figures, 3 tables. XII, 254 pages
ISBN 3-540-08835-0

"...Professor Paterson, who has built up a flourishing school of rock mechanics in Canberra, has successfully set out to assemble a concise source-book. The bibliography is very extensive indeed and a scan through dates shows how modern a subject this is ... This bibliography will be of lasting value ... The book is exceptionally good value for money..." *Nature*

Springer-Verlag
Berlin
Heidelberg
NewYork